Biology of the Insect Midgut

Biology of the Insect Midgut

Edited by

M.J. Lehane

School of Biological Sciences
University of Wales
Bangor
UK

and

P.F. Billingsley

Department of Zoology
University of Aberdeen
Aberdeen
UK

CHAPMAN & HALL

London · Weinheim · New York · Tokyo · Melbourne · Madras

Published by Chapman & Hall, 2–6 Boundary Row, London SE1 8HN, UK

Chapman & Hall, 2–6 Boundary Row, London SE1 8HN, UK

Chapman & Hall GmbH, Pappelallee 3, 69469 Weinheim, Germany

Chapman & Hall USA, Fourth Floor, 115 Fifth Avenue, New York NY 10003, USA

Chapman & Hall Japan, ITP-Japan, Kyowa Building, 3F, 2-2-1 Hirakawacho, Chiyoda-ku, Tokyo 102, Japan

DA Book (Aust.) Pty Ltd, 648 Whitehorse Road, Mitcham 3132, Victoria, Australia

Chapman & Hall India, R. Seshadri, 32 Second Main Road, CIT East, Madras 600 035, India

---

First edition 1996

Typeset in 10/12pt Palatino by Cambrian Typesetters, Frimley, Surrey

Printed in Great Britain by The University Press, Cambridge

ISBN 0 412 61670 X

A Catalogue record for this book is available from the British Library

Library of Congress Catalog Card Number: 96–86091

∞ Printed on permanent acid-free text paper, manufactured in accordance with ANSI/NISO Z39.48-1992 (Permanence of Paper)

Contents

Contributors

D.E. Ashhurst
Department of Anatomy
St George's Hospital Medical School
University of London
Tooting
London
SW17 3RE
UK

J.E. Baker
US Grain Marketing Research Laboratory
USDA ARS
1515 College Avenue
Manhattan
Kansas 66502
USA

K.M. Baldwin
Department of Anatomy
College of Medicine
Howard University
Washington DC 20059
USA

C.B. Beard
Division of Parasitic Diseases
Center for Disease Control and Prevention
Atlanta
Georgia 30333
USA

P.F. Billingsley
Department of Zoology
University of Aberdeen
Aberdeen
AB24 2TZ
UK

K. Crailsheim
Department of Zoology
Karl-Franzens-University
Universitätsplatz 2
A 8010 Graz
Austria

A. Crisanti
Department of Biology
Imperial College of Science
Technology and Medicine
London
SW7 2BB
UK

R. Dallai
Dipartimento di Biologia Evolutiva
Via Mattioli
4 Universita di Siena
Siena 53100
Italy

R.J. Dillon
Department of Entomology
Natural History Museum
Cromwell Road
London
SW7 5BD
UK

A.E. Douglas
Department of Biology
University of York
Heslington
York
Y01 5YW
UK

G.W. Felton
Department of Entomology
University of Arkansas
Fayetteville
AR 72701
USA

C. Ferreira
Departamento de Bioquimica

Instituto de Quimica
Universidade de São Paulo
CP 26077
05599-970 São Paulo
Brazil

J.A. Gatehouse
Department of Biological Sciences
University of Durham
South Road
Durham
DH1 3LE
UK

S.S. Gill
Department of Entomology
University of California-Riverside
Riverside
CA 92521
USA

R. Hakim
Department of Anatomy
College of Medicine
Howard University
Washington DC 20059
USA

B.P. Jordão
Departamento di Bioquimica
Instituto de Quimica
Universidade de São Paulo
CP 26077
05599-970 São Paulo
Brazil

D.C. Kaslow
Malaria Vaccines Section
National Institutes of Health
Bethesda
MD 20892–0425
USA

U. Klein
Institute of Zoology
University of Munich
D-80022 München
Germany

A. Koch
Molecular Physiology Laboratory
Washington State University
Pullman
WA 99164
USA

N.J. Lane
Department of Zoology
Downing Street
University of Cambridge
Cambridge
CB2 3EJ
UK

M.J. Lehane
School of Biological Sciences
University of Wales
Bangor
Gwynedd
LL57 2UW
UK

M. Loeb
Insect Neurobiology and Hormone Laboratory
USDA
BARC East
Beltsville
MD 20705
USA

D.F. Moffett
Molecular Physiology Laboratory
Washington State University
Pullman
WA 99164
USA

H.M. Müller
Institute of Parasitology
Universita di Roma La Sapienza
P Aldo Moro
5/I-00185
Rome
Italy

P.V. Pietrantonio
Department of Entomology
Texas A&M University
College Station
TX 77843-2475
USA

V.F. Sacchi
Laboratoria di fisiologia delle membrane
Via Trentacosta 2
I-20134
Milan
Italy

S. Sadrud-Din
Department of Anatomy
College of Medicine
Howard University
Washington DC 20059
USA

F. Sehnal
Institute of Entomology
AVČR
Branišovská 31
37005 České Budějovice
Czech Republic

R. Tellam
CSIRO-Division of Tropical Animal Production
Private Mail Bag 3
Indooroopilly
Brisbane
Queensland 4151
Australia

W.R. Terra
Departamento de Bioquimica
Instituto de Quimica
Universidade de São Paulo
CP 26077
05599-970 São Paulo
Brazil

S. Turunen
Department of Bioscience
University of Helsinki
POB 17
SF 00014
Helsinki
Finland

S. Welburn
Institute of Biomedical and Life Sciences
Division of Molecular Genetics
University of Glasgow
Laboratory of Genetics
Church Street
Glasgow
G11 5SJ
UK

M.G. Wolfersberger
Department of Biology
Temple University
Philadelphia
PA 19122
USA

P. Willadsen
CSIRO-Division of Tropical Animal Production
Private Mail Bag 3
Indooroopilly
Brisbane
Queensland 4068
Australia

D. Žitňan
Institute of Zoology
Slovak Academy of Sciences
Dúbravská cesta 9
84206 Bratislava
Slovak Republic

Preface

We are sure that this book arose, like so many others, from the typical statement 'its about time somebody . . .'. In this case, it was about time that we tried to get a like-minded group of people together within a conference, and from there it was 'about time that somebody' drew together all the various avenues of research into the insect midgut. *Biology of the Insect Midgut* is the end result. We have taken the unusual step of inviting co-authors to contribute when the individuals may not normally collaborate let alone write together. Consequently, we feel the various chapters represent balanced views of the current state of knowledge in each subject area.

The book is divisible into four conceptual areas – the structural biology of the midgut (Chapters 1–5), digestion and transport (Chapters 6–11), the midgut as a target for control strategies (Chapters 12–14) and the midgut as an environment for other organisms (Chapters 15 and 16). Within these sections we have tried to encompass the broad and the specific, the classical and the modern, the observational and the experimental. It is impressive that what many regard as a single tissue can invite such diverse contributions both in subject matter and interdisciplinary approach.

Editing a book like this is educational and it is rewarding to see that so much work about which one is uninformed has progressed so far. At the same time, there is clearly much more to do. We hope therefore that you will use this book as a reference for getting acquainted with the field as it stands now, and use it as a springboard into the literature and an exciting research field. We hope you agree with us that the authors have done a fine job. Our thanks to them for all that these chapters have entailed, not least the considerable time and effort, and to others who have contributed information and figures. Finally, our thanks to the editorial widows, widowers and orphans who put up with much in the interests of our pursuit of science.

Mike Lehane
Peter Billingsley
December 1995

Part One

Structural Biology of the Midgut

1

Structure and ultrastructure of the insect midgut

P.F. Billingsley and M.J. Lehane

This chapter provides a brief overview of insect midgut structure and ultrastructure, and thereby sets the book into context; an exhaustive review of the subject is neither appropriate nor necessary here. Rather, we aim to describe the structural components of the midgut, introduce some terminology, and provide a framework for the subsequent chapters. Studies on midgut structure are probably the least topical of all those covered in this text and so knowledge is moving forward only slowly. We therefore refer the readers to more detailed review articles (Bignell, 1981; Andries, 1982; Mori, 1983; Cioffi, 1984; Martoja and Ballan-Dufrançais, 1984; Dow, 1986; Billingsley, 1990). The limited advances in this area of research in the past decade or so makes them still useful for reference purposes, but given the progress in the related areas of cell and molecular biology, many of the interpretations in the reviews are worthy of reconsideration. Where appropriate, we will attempt to do this, but such advances are usually more fully considered in subsequent chapters.

1.1 MORPHOLOGY AND ORGANIZATION OF THE INSECT MIDGUT

The midgut is composed of a single layered epithelium resting on a continuous basal lamina or basement membrane (Figure 1.1). The

Biology of the Insect Midgut.
Edited by M.J. Lehane and P.F. Billingsley.
Published in 1996 by Chapman & Hall, London.
ISBN 0 412 61670 X.

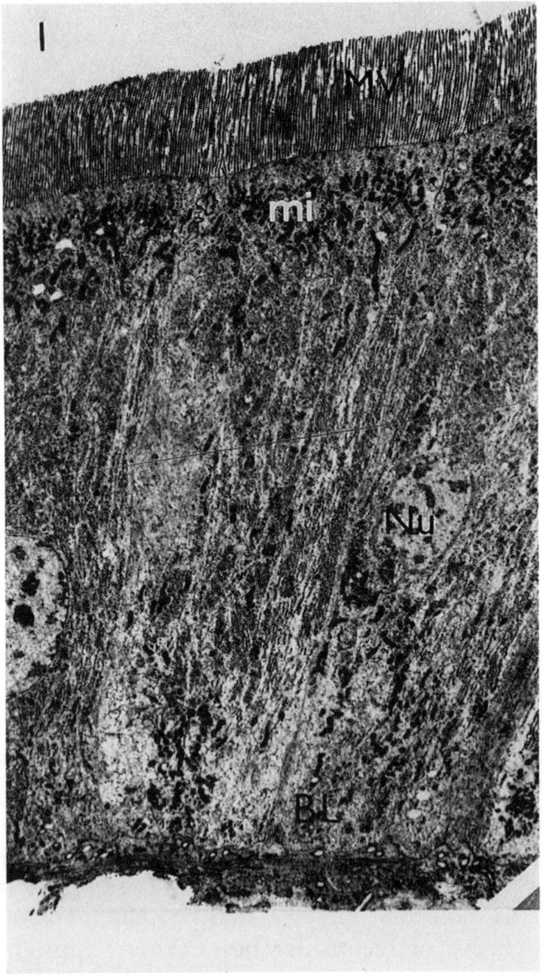

Figure 1.1 Typical insect midgut epithelial cells from *Locusta migratoria*. A brush border of microvilli (MV) lines the lumen, and mitochondria (mi) are concentrated in the apical and basal parts of the cells. Nuclei (Nu) are located centrally, and the cytoplasm is rich in endoplasmic reticulum. The basolateral plasma membrane is thrown into convoluted folds to form the basal labyrinth (BL) and the whole epithelium rests on a continuous basement membrane (× 3200). (Modified from Nasiruddin and Mordue (Luntz), 1993.)

ultrastructure of the various cell types found within the epithelium is described below, and the ultrastructure and molecular composition of the basement membrane and its associated connective tissue are reviewed in Chapter 5. On the haemolymph side of the epithelium is a discontinuous latticework of inner circular and outer longitudinal muscles, the layering and organization of which vary considerably according to species and diet. Typically in fluid-feeding insects the muscle layers are poorly formed, whereas in insects with solid diets, more extensive muscle layers and connective tissues are present (Chapter 5). These muscles are relatively poor in mitochondria, possess a sparse tracheation, have a reduced sarcoplasmic reticulum and T system, and have 12 actin filaments around each myosin filament. Thus they are designed for slow contraction and perhaps supercontraction. The midgut is poorly innervated, nerves often limited to a small number stretching back from the stomatogastric system over the proventriculus and extreme anterior end of the midgut. In some insects nerves may traverse the length of the midgut (see Chapter 3), but true innervation is usually sparse. In locusts only the longitudinal muscles are innervated (Anderson and Cochrane, 1978), and current evidence suggests that such limited polyneuronal motor innervation of the musculature is typical.

Unlike many other animal groups, the insects do not produce mucin, instead the meal in the midgut lumen is separated from the epithelium by a peritrophic matrix or membrane (see Chapter 4). This may be produced either by all cells along the midgut (type I peritrophic matrix) or by specialized cells just posterior to the proventriculus (type II peritrophic matrix).

Not surprisingly in a phylum as large and diverse as the Insecta there is considerable variation in midgut arrangement and structure, but this occurs within an underlying pattern. The midgut forms the conduit between foregut and hindgut. In plant feeders, the midgut is usually short and broad allowing for a high throughput rate of food which is often abundant but of a relatively low nutrient content. In protein-feeding insects, the midgut tends to be longer allowing longer retention of the more nutritious meals which are processed more fully. Interstadial differences in diet result in interstadial differences in midgut organization and structure.

The midgut is commonly a simple tube varying in diameter along its length (Figure 1.2). In some groups the midgut bears diverticula, most commonly with two to six blunt-ending caeca at the extreme anterior end (see Chapter 8). In the Acrididae, these are divided at their base and there is a short posterior and a longer anterior branch. Diverticula may occur at other midgut positions. The lamellicorn beetle larvae possess two circles of caeca at the anterior end and another circle of caeca at the posterior end. Caeca are limited to the posterior end in many families of the Heteroptera where, in different species, they vary from typical blunt-ending caeca to asymmetrically positioned, long tubes of variable length.

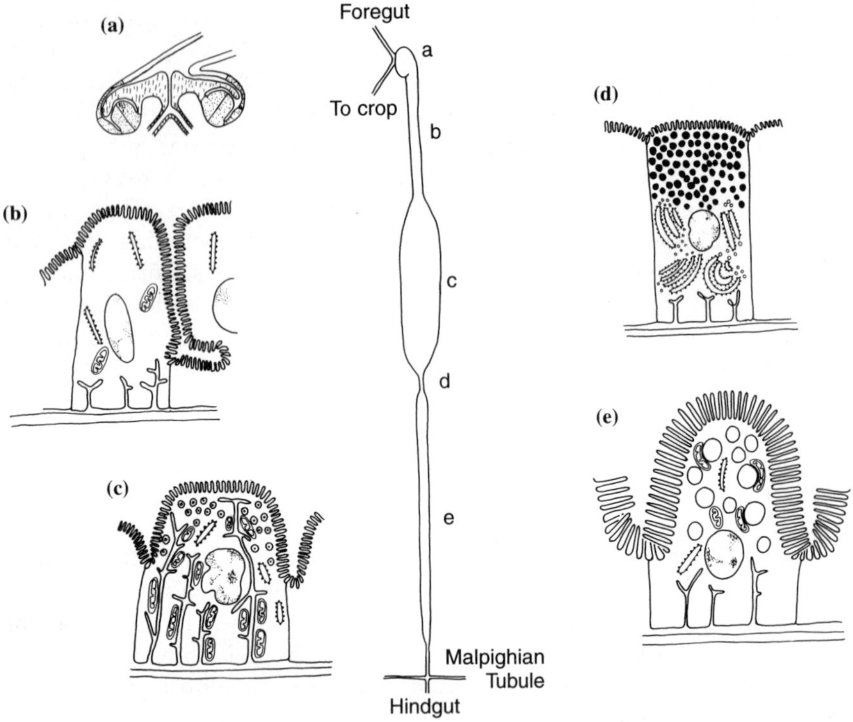

Figure 1.2 Diagram of cells associated with different regions of the midgut of adult *Stomoxys calcitrans*. The foregut and midgut meet at the proventriculus (a) which has cells specialized for production of peritrophic matrix (see also Figure 1.9). Blood from the crop passes through the proventriculus and thoracic midgut (b), where cells are unspecialized, into the reservoir zone (c), where cells possess many mitochondria, an extensive basal lamina and are specialized for water absorption. In the short opaque zone (d), where digestive proteinases are synthesized and secreted, the cells are rich in secretory vesicles and rough endoplasmic reticulum. Finally, the lipoid zone (e) is responsible mainly for nutrient absorption, and is characterized by a large surface area of microvilli and the accumulation of lipid vesicles in the cytoplasm.

Some insects, in particular the haematophagous species, ingest extremely large meals at intermittent intervals, and the midgut must be designed to cope with this. Part of the midgut, normally the anterior section, often displays ultrastructural specializations consistent with the abilities to distend and to rapidly transfer water from the blood meal to the haemolymph. The cells become greatly distorted by the blood meal, but epithelial integrity is retained by cell–cell junctions (Chapter 5;

Hecker, 1977) (Figure 1.6) and by the unfolding of the basement membrane (Billingsley and Downe, 1983, 1989a).

1.2 CELL TYPES AND ULTRASTRUCTURE

In most insects the midgut is structurally and functionally differentiated along its length (Figure 1.2). In others although there is little apparent structural regional differentiation, careful morphometric analyses can identify temporal differentiation of cell function (Hecker, 1977; Billingsley, 1988). Thus, although it is possible to ascribe function to various midgut regions, it is not always possible to determine which cellular characteristics confer functionality. As a result, we can only describe the 'typical' midgut cell types and provide examples.

1.2.1 Digestive epithelial cells

The predominant cells in the insect midgut are those responsible for processing of the diet, the digestive and absorptive epithelial cells (Figure 1.1). The apical plasma membrane of midgut digestive cells is highly folded, usually regularly, into microvilli (Thomas and Bockeler, 1994). The microvilli have an internal cytoskeleton (Bonfanti *et al.*, 1992) composed of actin-like microfibrils (Chapter 5) and may be formed early in the maturation of the midgut cells (Hecker *et al.*, 1971; Andries and Torpier, 1982; Raes *et al.*, 1994). On the luminal surface of the microvilli is a glycocalyx, composed typically of a regular, carbohydrate-rich electron-dense layer (Houk *et al.*, 1986a; Rudin and Hecker, 1989) but elaborate modifications to this arrangement have been observed (Chapter 5). Microvilli may possess blebs or portasome-like structures associated with secretory activity (Figure 1.3) (see below and Chapters 7 and 8).

The basolateral plasma membrane is usually highly and irregularly convoluted, the infoldings sometimes stretching into the apical third of the cells (Lehane, unpublished observations; Figure 1.2). The status of these folded membranes is important to cell function (Berridge, 1970). When tightly folded so that they remain juxtaposed they appear to play no role in transport mechanisms (Figure 1.4). However, when the membranes become separated (as in the blood-feeding insects after a blood meal; for review see Billingsley, 1990) they appear to function in ion and water transport out of the lumen (Figure 1.5). This exposes to the haemolymph a large surface area of membrane containing integral sodium pumps (MacVicker, 1993). The concentration of mitochondria in the apical and basal thirds of the cells supports this active transport role. The apical third of the basolateral plasma membrane is usually involved integrally with adhesion to neighbouring cells, and contains junctional

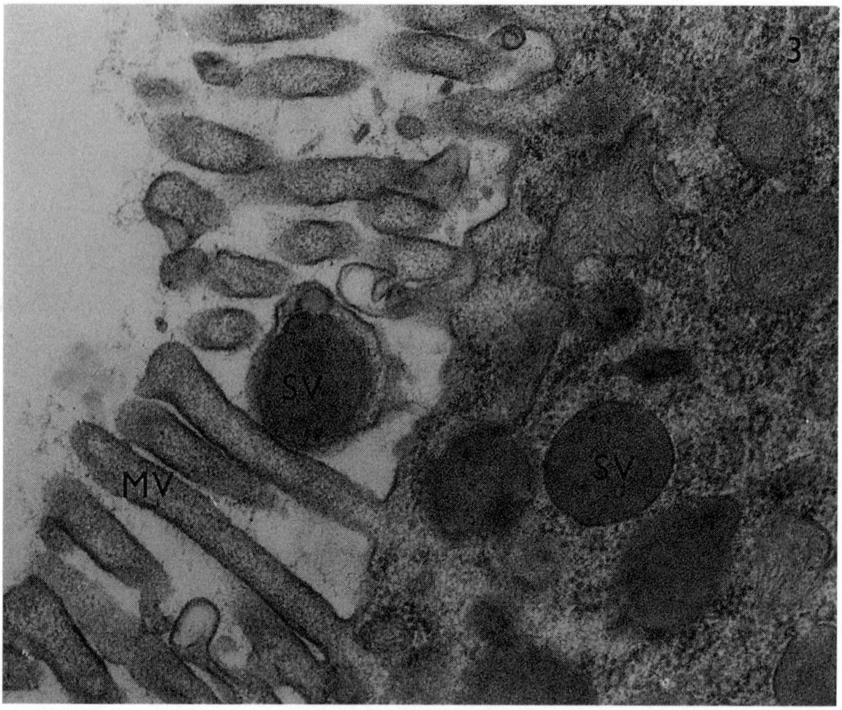

Figure 1.3 Secretory vesicles (SV) in the opaque zone of *Stomoxys calcitrans* release their contents in a variety of ways. Here a complete secretory vesicle is budding from the surface of the cell where blebbing is apparent on some microvilli (MV) (× 35 550).

complexes (Chapter 5). These junctions are quite leaky, allowing the passage of such large molecules as IgG (Hatfield, 1988; Vaughan *et al.*, 1990; Allingham *et al.*, 1992), horseradish peroxidase (Fishman and Zlotkin, 1984; Modespacher *et al.*, 1986) and cobra venom toxin (Fishman *et al.*, 1984). This permeability is dependent on midgut region, indicating differences in the nature of cell–cell contact along the midgut.

A large part of the central portion of the cell is occupied by a prominent, often lobed, nucleus. In both light and electron microscopy, heterochromatin can be observed (Figure 1.1), but condensed chromosomes, indicative of cell division, are rarely observed. Unusual nuclear structures have been described, in particular an electron lucent halo around the midgut nuclei of *Glossina* (Jenni and Bohringer, 1976; Ellis *et al.*, 1981) and the intranuclear crystals of the flea midgut (Richards and Richards, 1969). The nuclei in the midguts of haematophagous insects

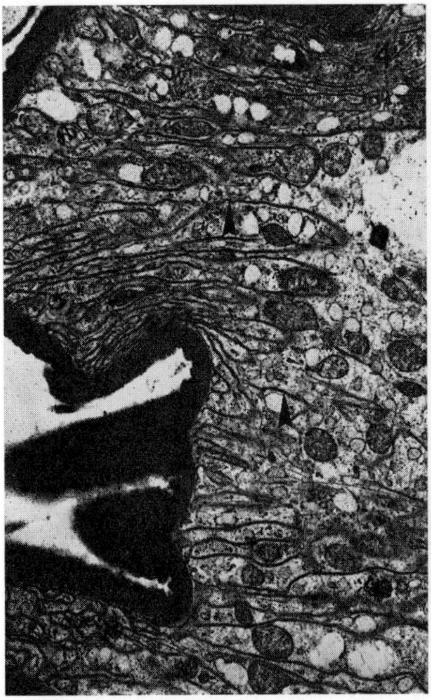

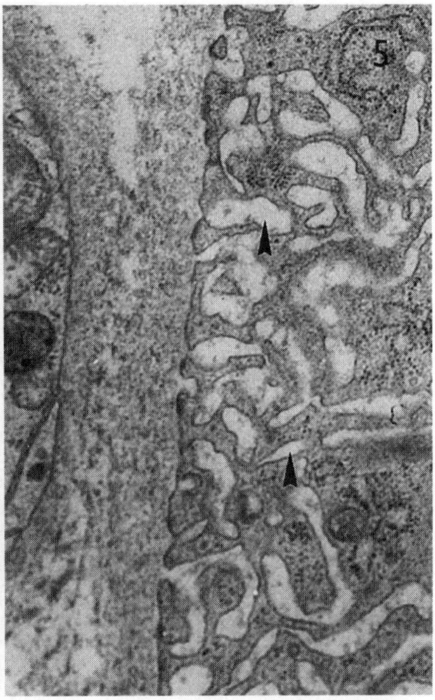

Figure 1.4 The basal labyrinth of *Rhodnius prolixus* anterior midgut before feeding. The extensive folds of the basolateral plasma membrane are tightly apposed to one another (arrowheads) such that no extracellular spaces are present and the surface area of plasma membrane exposed to the haemolymph is very small (× 7740). (Modified from Billingsley and Downe, 1989a).

Figure 1.5 The basal labyrinth of *Stomoxys calcitrans* lipoid zone after feeding. The folds of the basolateral plasma membrane separate from one another to form extensive extracellular spaces (arrowheads) that are continuous with the haemolymph (× 18 450).

increase in size after feeding (Böhringer-Schweizer, 1977; Rudin and Hecker, 1979) and in *Rhodnius prolixus* this apparent nuclear activity correlates with a decrease in ploidy (Billingsley, 1988, 1989) and *de novo* enzyme synthesis.

The other major structures in digestive epithelial cells are usually those associated with the endoplasmic reticulum. The rough endoplasmic reticulum (RER) is often highly organized into elaborate stacks or concentric whorls (Figure 1.6). These whorls are common in, but not restricted to, the midguts of blood-feeding species (for review see Billingsley, 1990), and may be associated with storage, lysosomal or secretory vesicles (Andries, 1977; Billingsley and Downe, 1986). Under

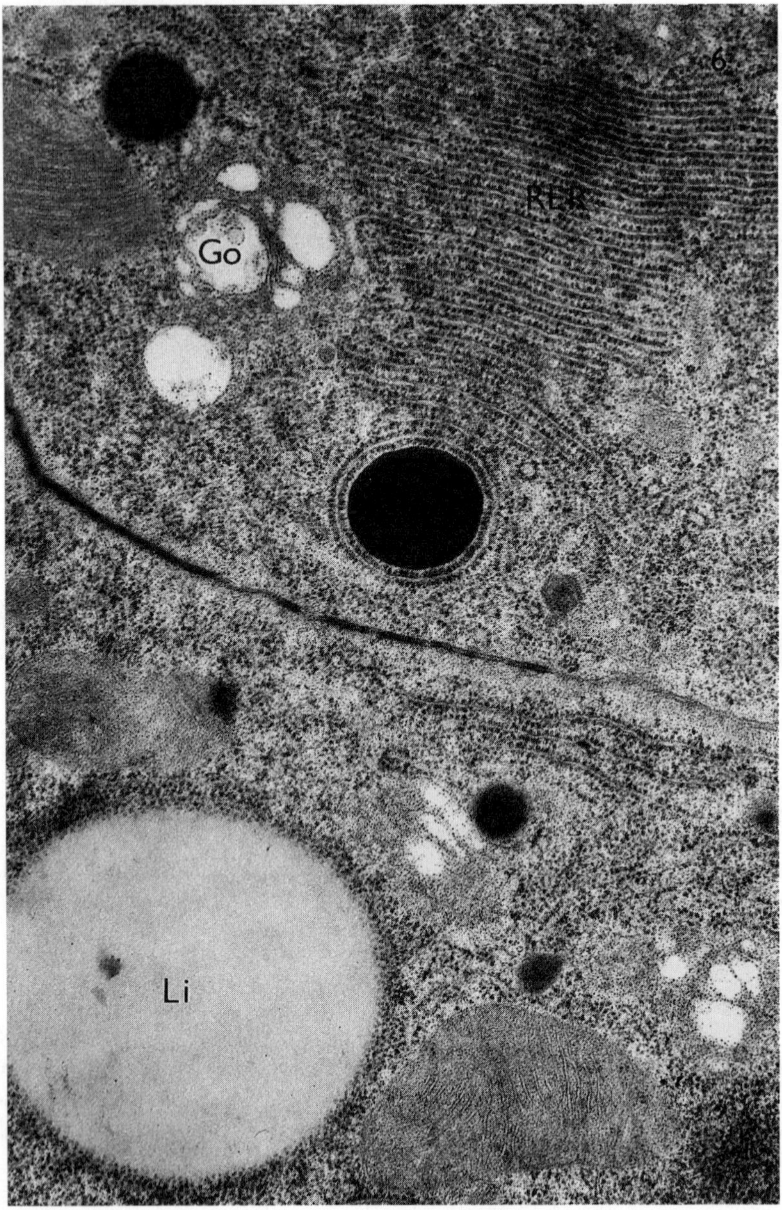

Figure 1.6 Cells at the junction of the lipoid and opaque zones of adult *Stomoxys calcitrans* share features of each zone. Large stacks of rough endoplasmic reticulum (RER) are found usually in the central part of the cell and often in close proximity to Golgi apparatus (Go). The stacks of RER disperse throughout the cells after feeding. Lipid vesicles (Li) are conspicuous and electron-dense junctions occur where the basolateral plasma membranes are apposed (× 25 075).

certain physiological conditions, these whorls become modified, as in the period following blood feeding by the mosquito, *Aedes aegypti*, and the stable fly, *Stomoxys calcitrans*; unfolding of a whorl occurs at times of increased synthetic activity and involves 'both its opening out rather like a multi-petal flower, together with an unrolling of its coiled and possibly multi-thonged, strap-like petals in manner of reversibly protusible paper whistles of carnival novelties' (Bertram and Bird, 1961). Whorl formation may be under juvenile hormone control, and the whorls themselves may act as a store of intracellular, membrane-associated ribosomes, in which associated mRNA may be protected from degradation, allowing the cells to respond rapidly to a change in lumen content by *de novo* synthesis (Gooding, 1973; Hecker and Rudin, 1979; Felix *et al.*, 1991) (Chapter 7). When folded, the inner layers of the RER whorl are essentially separated from the cytoplasm, and their unfolding results in a greater effective surface area of RER in the cell (Hecker and Rudin, 1976; Gander *et al.*, 1980; Billingsley, 1988). In common with many other eukaryotic cells, close association has often been observed between the RER and Golgi apparatus (Figure 1.6). Unusual smooth endoplasmic reticulum (SER) arrangements have also been observed (Andries, 1977), but analysing SER structure and function has been compounded by its similarity in electron microscopy to other complex membrane arrangements, in particular the basal labyrinth.

1.2.2 Ultrastructure of secretion in midgut cells

Insects must secrete material from the midgut cells to the lumen in order to manage and digest the meal. The perceived primary role of the digestive epithelial cells is, of course, synthesis and secretion of digestive enzymes (see Chapters 6–8), usually achieved via secretory vesicles released near the base of the microvilli (Figure 1.12) or via other small secretory structures observed within or on the edge of the microvilli (Figure 1.3). The other major secretion derived from midgut cells is the peritrophic matrix (PM) and like digestive enzymes, different modes of secretion have been observed. Such materials as lectins, haemolysins and anticoagulants are also secreted, but despite this, there are surprisingly few reports of midgut digestive cells accumulating large quantities of material for secretion (Figure 1.2d). Therefore the regulated route of secretion, in which stored material is released in response to an appropriate signal, is not predominant in the insect midgut. In continuously feeding insects there are probably few benefits to be gained from regulated rather than constitutive secretion.

Cells undertaking regulated secretion accumulate large quantities of secretory material, usually in the form of secretory vessels, and these are released into the midgut lumen upon induction. One of the best studied examples of this is the release of proteinases by digestive epithelial cells in the opaque zone of the midgut in *Stomoxys calcitrans* (Figures 1.2d and 1.3). Secretory proteins follow the classical route of synthesis in the rough endoplasmic reticulum, processing in the Golgi apparatus, holding in intermediate vesicles and release from secretory vesicles (Lehane, 1976, 1989; Jordão *et al.*, 1996a,b). Transfer of newly synthesized material from the rough endoplasmic reticulum to the Golgi begins about 13 min after radiolabelling of the fly, and from the Golgi to the secretory granules after about 25 min. Secretion of newly synthesized material may begin as early as 30 min after labelling of the fly but certainly no later than 60 min. These times are similar to those in vertebrates, except for the Golgi phase where processing takes 40–60 min in cells (pancreatic acinus) producing a similar product. Storage of such large amounts of material is unusual, even for intermittent feeders. Electron-dense granules, stored in midgut cells of starving *Aeshna cyanea* larvae, are secreted from the base of the microvilli after feeding (Andries, 1976) and regenerated during the following 24 h.

The release of the secretory granules occurs largely by classical exocytosis (Wood and Lehane, 1991) (Figure 1.3). In *S. calcitrans* there is a burst of secretory activity immediately after feeding during which contents of hundreds of vesicles are released into the ectoperitrophic space of the lumen. This burst of activity, although very intense, does not affect the area of apical plasma membrane as would be expected (Wood and Lehane, 1991). In general, excess membrane added during exocytosis is thought to be recycled after endocytosis, but the rapid putative recycling rate here is suggestive of other mechanisms operating to balance the membrane budget. Conversely, in anopheline mosquitoes, merocrine secretion of the PM also occurs within minutes of feeding (Berner *et al.*, 1983; Billingsley and Rudin, 1992), and this is accompanied by a doubling of microvillar surface area within one day of feeding (Hecker, 1977).

Apocrine secretion, in various forms, may also be involved in digestive enzyme release from midgut cells. The commonly observed apical cell extrusions in insect midgut cells (Nopanitaya and Misch, 1974; Baker *et al.*, 1984; Santos *et al.*, 1986) may be responsible for release of secretory material, although they often do not contain secretory granules. These extrusions can be formed in response to starvation, during degenerative changes in the gut, or may sometimes be fixation artefacts (Brunings and DePriester, 1971). Thus great care needs to be taken in any claims that apocrine secretion occurs in insects, and is never likely to be proven by ultrastructural evidence alone. Lehane (1988) adapted morphometric techniques to show results consistent with the occurrence of apocrine secretion in *S. calcitrans*. It is possible that the cellular extrusions containing secretory granules seen in *S. calcitrans* represent apocrine secretion although this is likely to play a secondary role to the classical exocrine route (Wood and Lehane, 1991). Similar mechanisms may be occurring in the heteropteran, *Nepa cinerea*, where secretion of large numbers of lysosome-like bodies contributes to formation of the elaborate extracellular membrane system (Andries and Torpier, 1982; see also Chapters 3 and 12). In these cases more work is required before we can conclude that the extrusions are natural events and to determine their function. More often, microvilli also display extrusions or blebs consistent with a 'microapocrine' secretory process. Columnar cells of the lepidopteran larva, *Erinnyis ello*, release microvillar extrusions containing vesicles into the midgut lumen, and also bud off small, membrane-bounded spherical bodies (Santos *et al.*, 1984). However, terminal swelling of the microvilli are more likely associated, in some way, with ion regulation by specialized cells (Terra *et al.*, 1988; Dimitriadis, 1991).

Less is known of secretion by the constitutive route in insects (Jordão *et al.*, 1996b) although this is probably the major route for secretion used by the majority of insects. Cells undertaking constitutive secretion are characterized by the presence of suitable amounts of endoplasmic reticulum and Golgi, but few secretory granules as they are not stored after synthesis. Theoretically, the insects most likely to benefit from adopting the regulated pathway are those feeding intermittently, such as adult mosquitoes, but these provide the best ultrastructural evidence for constitutive secretion. In *Aedes aegypti*, trypsin is localized to secretory vesicles originating in the Golgi and later fusing with the apical plasma membrane at the base of the microvilli (Graf *et al.*, 1986). Electron-dense secretory vesicles are also present in midgut cells of the mosquito, *Culex tarsalis*, where they originate from the Golgi and fuse with the apical plasma membrane (Houk and Hardy, 1982), though their appearance is not coincident with digestion of a blood meal. Similar vesicles are present in posterior midgut cells of *Rhodnius prolixus* and their number correlates well with activity of digestive proteinases (Billingsley, 1988; Billingsley and Downe, 1989c). More commonly in continuous feeders, secretory vesicles are found at all stages of processing (for example see Baker *et al.*, 1984) and the relatively small number of vesicles present at any given time reflects their constant turnover. However, in all these examples, definitive radiolabelling studies are conspicuous by their absence.

1.2.3 Absorptive and storage epithelial cells

The apparent structure of absorptive epithelial cells is often difficult to distinguish from that of digestive cells even where midgut function seems to be clearly regionalized (Chapter 8). In some insects such as mosquitoes, one cell type can undertake all these functions; a detailed model of how this is achieved was developed by Hecker and colleagues and is reviewed by Billingsley (1990). In general, however, absorptive and storage cells will display greater elaborations of the basolateral membrane, contain such storage products as lipids and glycogens, and have little or no apparent secretory activity (Figure 1.2) (Billingsley, 1990).

Absorptive cells of the midgut sometimes contain multilayered, mineral concretions or spherites composed of such cations as Ca^{2+}, Mg^{2+}, K^{+}, Zn^{2+}, Cu^{2+}, Pb^{3+} and Sr^{2+} in the form of phosphates, carbonates and chlorides (Martoja and Ballan-Dufrançais, 1984). These may be produced by the rough endoplasmic reticulum or Golgi body. The lysosomes of the cuprophilic cells of some dipteran larvae accumulate copper. The purpose of these concretions in midgut cells is unknown, but their size and number increase with age and are correlated with dietary cations, suggesting their role as sinks for storage excretion. This role is more transient in *Rhodnius prolixus* where the concretions are resorbed via autophagic lysosomes around 20 days after blood feeding (Billingsley and Downe, 1989a).

The absorption of lipids is complicated by the need first to solubilize the lipids in the gut lumen. Theoretically this might be achieved by the presence of fat-binding proteins in the intestine, by saponification under alkaline conditions, emulsifications using phospholipids (Turunen, 1988) or the incorporation of dietary lipids into polar glycolipids in the midgut lumen (see Chapter 11). The onset of meal digestion is often associated with the appearance of vacuoles containing lipid (Figures 1.2 and 1.6) (Lehane, 1977). The origin of these lipid vacuoles remains unclear. They are often associated with Golgi apparatus or RER; in both cases this suggests that they may be the result of synthetic processes within the cell and that lipid accumulation may not represent direct lipid absorption (Lehane, 1977; Billingsley and Downe, 1989a).

Glycogen deposits are often found associated with absorptive cells or with the absorptive phases of multifunctional midgut epithelial cells (Billingsley, 1990b). These tend to form large, electron-dense structures of no particular shape near the central part of the midgut cells. The carbohydrate nature of these deposits in mosquito midgut has been confirmed by lectin staining (Rudin and Hecker, 1989) and by incorporation of radiolabelled glucose (Schneider *et al.*, 1987).

Other absorptive characteristics of cells are probably less specific and can only be identified by more detailed, comparative analyses. Although the microvilli, for example, appear qualitatively similar throughout the midgut of *Rhodnius* (Billingsley and Downe, 1983, 1989a), morphometric analyses identify some differences. Those in the anterior midgut and posterior intestine, both considered major absorptive regions, show increases in their surface area and volume ratios associated with identifiable absorptive events (Billingsley, 1990). The same is true of the timing of separation of the basolateral plasma membrane folds in the basal labyrinth.

1.2.4 Endocrine cells

Distributed throughout the epithelium are cells which bear the hallmarks of endocrine function. These cells are of variable shape being pyramidal, bowl-shaped, oval or fusiform (Andries and Beauvillain, 1988) and are often of different electron density from neighbouring cells (Figure 1.7). The basolateral plasma membrane has few or no basal infoldings, lying tightly apposed to the basement membrane (see Chapter 3). Some endocrine cells, the so-called open-type (Fujita and Kobayashi, 1977), are in direct contact with the midgut lumen via a narrow extension from the broader basal part of the cell (Brown, 1980; Brown *et al.*, 1985; Sivasubramanian, 1992), suggestive of a recepto-secretory role (Endo and Nishiitsutsuji-Uwo, 1981). Other endocrine cells, the 'closed' type, do not extend through the epithelial layer as far as the lumenal surface (Hecker, 1977; Billingsley and Downe, 1986; Glättli *et al.*, 1987) and even the contact with the basement membrane may be minimal (Andries and Tramu, 1985). However, without full three-dimensional reconstruction of these cells, their shape and relationships with basal and apical cell surfaces are often unclear. Toward their basal end these cells contain electron-dense granules varying in size from cell to cell between about 0.1 μm and 0.5 μm. In the basal region these cells contain electron-dense granules varying in diameter from 0.1 μm to 0.5 μm. Midgut endocrine cells are usually more common in the posterior midgut (Brown, 1980; Brown *et al.*, 1985; Billingsley and Downe, 1986), and even within a single midgut, up to ten cell types have been identified morphologically based on cell shape, cytoplasmic density, and secretory vesicle characteristics (Andries and Tramu, 1985). In addition, many of these cells stain positively with antibodies raised against vertebrate hormones (Andries and Tramu, 1985; Brown *et al.*, 1986; Glättli *et al.*, 1987; Montuenga *et al.*, 1989; Sivasubramanian, 1992; Chapter 3).

1.2.5 Regenerative cells

Also scattered throughout the epithelium are relatively undifferentiated cells found singly, paired or in groups (also called nidi). These cells have a regenerative role, and are characterized by a dense cytoplasm with few differentiated organelles, and none of the elaborations of the apical and basal plasma membranes described above (Figure 1.8). In some insects,

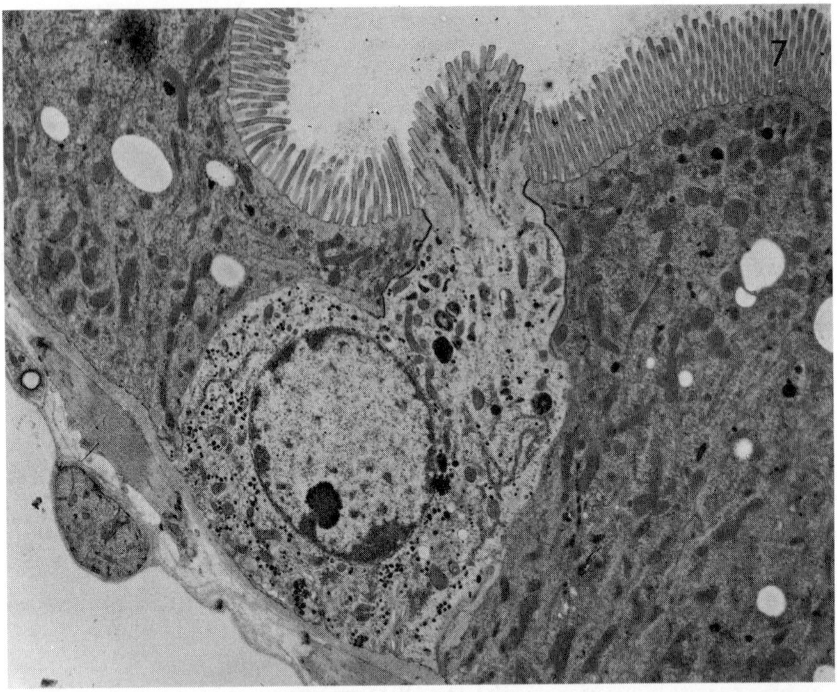
7

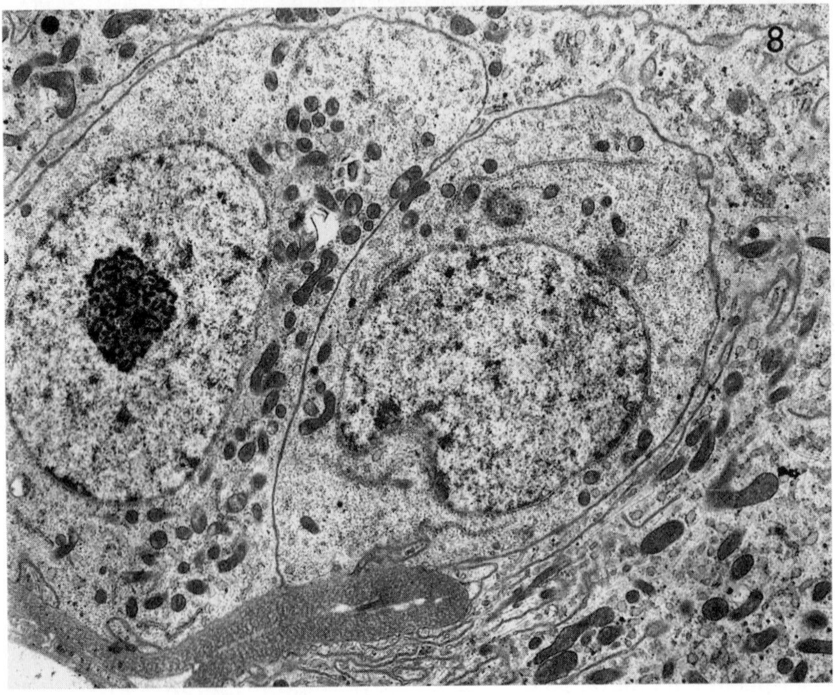
8

the cockroaches for instance, the fully formed midgut is renewed continuously from regenerative cells, and the whole midgut epithelium is replaced on a four-day cycle in *Tenebrio molitor*. In other insects intrastadial cell replacement is rare or even absent (Hecker, 1977; Billingsley and Downe, 1986) even though regenerative cells are common. The cells in the regenerative nidi may also be used in replacement following injury (Spies and Spence, 1985), or can be triggered into mitosis and epithelial replacement by specific physiological events, such as blood feeding in fleas (Reinhardt *et al.*, 1972a). There may even be more sophisticated co-ordination of cell replacement where a tetrad of regenerative cells are joined to each other by cytoplasmic bridges and initial formation of the microvilli occurs at the juncture of the four cells (Raes *et al.*, 1994). Conversely, in the tsetse fly, *Glossina*, the apparent absence of regenerative cells in the adult (Böhringer-Schweizer, 1977), means that the cells live as long as the adult fly, up to 200 days. Regenerative cells also appear to be absent from the midgut of some dipteran larvae (Terra *et al.*, 1988), leaving intriguing questions concerning replacement of the midgut epithelium during interstadial development.

1.2.6 Goblet cells

The goblet cells are typical of ephemeropteran and lepidopteran midguts, and are so-called because of their invaginated shape. Each goblet cell encloses a large extracellular lumen that is continuous with the midgut lumen, via a labyrinthine apical valve formed by the interdigitating microvilli. The enclosed lumen of the goblet cell is filled with flocculent material formed of about 20% solids, probably sulphated glycosaminoglycans. The ultrastructure and function of these cells are fully discussed in Chapters 2 and 9. In the anterior and middle portions of the midgut the irregular microvilli of the goblet cells contain mitochondria, a feature unique to arthropod epithelia (Cioffi, 1984), and characteristic of active ion-transporting tissues (Klein *et al.*, 1991; Wieczorek *et al.*, 1991). In keeping with their absorptive function, goblet

---

Figure 1.7 Endocrine cells of the 'open' type in the lipoid zone of *Stomoxys calcitrans*. The cell typically has a less electron-dense cytoplasm than neighbouring epithelial cells and no basal labyrinth. Secretory vesicles accumulated in the basal third of the cytoplasm release their contents into the haemolymph (× 18 750).

Figure 1.8 A pair of regenerative cells from the anterior intestine of *Rhodnius prolixus* characterized by the paucity of organelles and an undifferentiated state (× 9200). (Modified from Billingsley and Downe, 1986.)

cells may also have a basolateral plasma membrane that folds right into the body of the cell. As a consequence, the goblet cells often have very little cytoplasm which forms the thinnest of cellular layers between the midgut lumen and the haemocoel. Portasomes are commonly observed in the microvilli, and their function is discussed elsewhere (Chapters 8–10).

1.2.7 Cuprophilic cells

The cuprophilic cells of some dipteran larvae have interesting similarities to the lepidopteran goblet cell, and are associated anatomically with very acidic (pH 3 ± 1) midgut zones in some species (Terra *et al.*, 1988). The apical invaginations of cuprophilic cells are lined with microvilli, but the invaginations are poorly structured compared to those in goblet cells and there may be more than one invagination per cell. The invaginations may be almost completely occluded by the irregular microvilli, just below which there is a zone of apical cytoplasm which is less dense and has no mitochondria. In contrast, the rest of the cell is extremely rich in mitochondria. The cells possess carbonic anhydrase activity and the microvilli are studded on the cytoplasmic side with portasome-like particles. There is some infolding of the basolateral plasma membranes, but not to the degree seen in absorptive or goblet cells.

1.3 SPECIALIZED STRUCTURES IN THE INSECT MIDGUT

Much of the literature concerning insect midgut structure and ultrastructure describes cell types, their responses to feeding or diet, or some of the more unusual or specialized structures. These may be an integral part of the midgut or associated with the midgut in some way. We have chosen just a few examples of the more specialized structures. The peritrophic matrix represents perhaps the most discussed, non-cellular structure in the midgut; its structure will be introduced here and its molecular composition and role is discussed in detail in Chapter 4. The homopteran filter chamber represents one of the most elaborate modifications of structure to diet, and the tsetse mycetome perhaps the closest cellular association between insect midgut and the symbionts that it harbours. The midgut caeca are now known to play important roles in the spatial separation of digestive events and in midgut fluid fluxes (Chapter 8); what little is known about their ultrastructure has been reviewed in some detail by Peters (1992) and will not be covered here.

1.3.1 The proventriculus and peritrophic matrix*

Peritrophic matrices (PMs) are found in most insects and have been reviewed in detail by Peters (1992). Two types of PM formation occur; type I PMs are most common and secreted by cells along the entire length of the midgut, whereas type II PMs are produced by a group of specialized cells at the anterior end of the midgut and collectively known as the cardia or proventriculus. A type II PM passes down the length of the midgut as an unbroken cylinder surrounding the food (see Chapter 4, and for an alternative type structure see Chapters 5 and 8).

In many insect groups, especially the higher Diptera, the proventriculus is formed in a press-like arrangement with the secretory midgut cells curved around a mushroom-shaped extension of foregut cells termed the valvula cardiaca (Figure 1.9). This arrangement is also present in larval lower Diptera and Lepidoptera, even though the respective adults produce type I PMs (Volkmann and Peters, 1989). Type II PMs are formed where the apical surfaces of the cells from the two gut regions meet; the foregut cells are characterized by a thin cuticular layer which separates the cell apices and may act as a template for the forming PM (Peters *et al.*, 1978). The PMs are synthesized constantly in the proventriculus, and more matrices or layers are added to the midgut cell side of the PM as it passes through the proventriculus. Each layer of the PM or separate matrix appears to be produced by a single ring of cells around the midgut; thus the more elaborate or complex the PM, the more cell layers are associated with its manufacture. Up to 400 cells (10 rings of approximately 40 cells each) may form the PM in some species (Richards and Richards, 1971). These cells have rarely been examined in any fine structural detail, but may vary in their size, cytoplasmic density, microvillar arrangement, mitochondrial content, and organization of the secretory apparatus (Platzer-Schultz and Welsch, 1970; Richards and Richards, 1971; Peters, 1979). The diverse and elaborate structural arrangements of the proventriculus have been reviewed recently (Peters, 1992).

Three separate matrices are produced in some Calliphoridae whereas only two are produced in many Sarcophagidae, Muscidae and Tachinidae (Zhuzhikov, 1963). The blood-sucking dipterans, *Stomoxys calcitrans* (Lehane, 1976b) and *Glossina* spp., are unusual in having only a single PM (although this still has the multilayered structure typical of other dipteran peritrophic matrices) and this reduction may be related to

* In keeping with the growing evidence concerning its structural molecules and composition, here and throughout the rest of the book we have adopted the term peritrophic matrix rather than the traditional 'peritrophic membrane'. Like many others, we are very aware that the PM is not a 'membrane' in any biological sense of the word, and see most evidence for it functioning as an extracellular matrix.

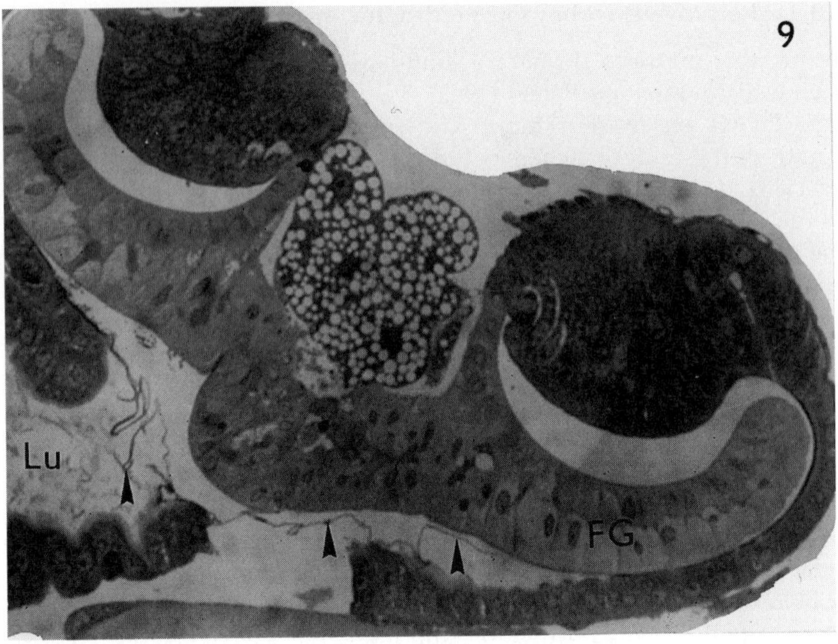
9
Lu
FG

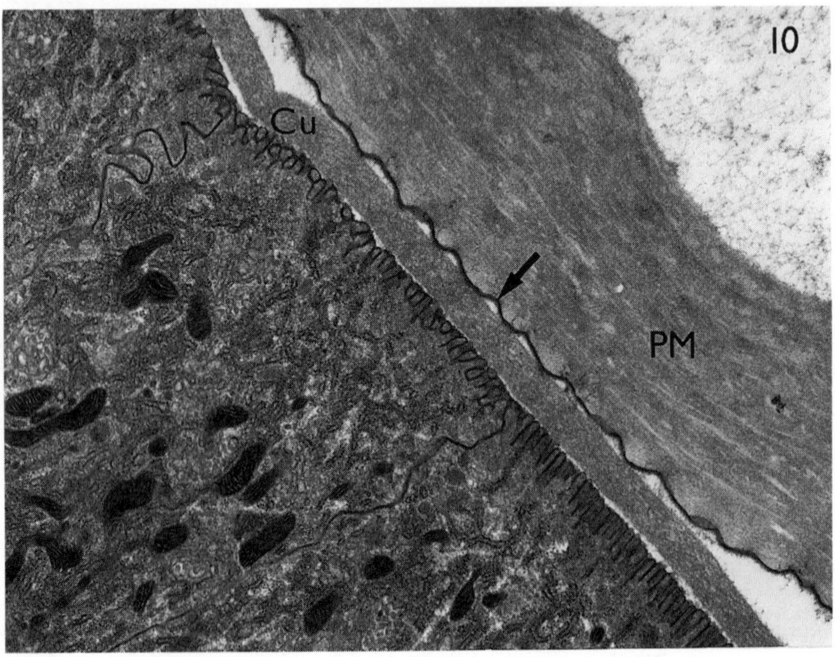
10
Cu
PM

feeding habit. The type II PM of *S. calcitrans* is between 150 and 400 nm thick. All layers arise from the midgut cells of the proventriculus. The food-bordering layers of the type II PMs often resemble epicuticle (Figure 1.10). In *S. calcitrans* the electron-lucent layer immediately abutting the food is ca. 3–4 nm thick and resembles outer epicuticle. Inside this is a 3 nm electron-dense layer resembling cuticulin, and inside again is a 20–35 nm layer which may be equivalent to inner epicuticle. Mucopolysaccharides (or proteoglycans) are present within some of these layers (Steiger, 1973; Zimmermann *et al.*, 1975; Lehane, 1976b) and biochemical analyses of *Calliphora erythrocephala* PM showed them to contain 36% protein and 20% galactosamine and glucosamine by weight (Zimmermann *et al.*, 1975). The innermost of the three layers contains chitin, but the layers immediately bordering the lumen probably do not (Nagel and Peters, 1991).

There is less detail available concerning the structural organization and composition of type I PMs, partly because their structures are more variable, and partly because type I PMs are more difficult to dissect for analysis. There may be interstadial differences in the type of PM produced. Larvae of both *Aedes aegypti* and *Anopheles stephensi* produce similar type II PMs (Peters, 1979), but the adult type I PMs are quite different. *Ae. aegypti* PM is produced *de novo*, and has an obvious concentric multilaminar arrangement, whereas that of *An. stephensi* is produced mainly from pre-formed secretory vesicles stored within digestive epithelial cells and has no obvious laminar structure (Berner *et al.*, 1983; Perrone and Spielman, 1988; Billingsley and Rudin, 1992).

1.3.2 The filter chamber

Hemipteran insects feeding on xylem, a fluid diet of low osmolarity compared to haemolymph, must concentrate and conserve nutrients and essential ions while efficiently eliminating unwanted water. These insects have developed a specialized arrangement of the midgut, Malpighian tubules and hindgut known collectively as a filter chamber. The filter chamber is formed from a distended, reception region of

---

Figure 1.9 The proventriculus of *Stomoxys calcitrans*. Cells of the foregut (FG) epithelium fold into the lumen of the midgut. The midgut cells (MG) secrete material that forms peritrophic matrix (arrowheads) as it passes through the press-like arrangement and into the lumen of the anterior midgut (Lu) (× 585).
Figure 1.10 The peritrophic matrix (PM) of *Culicoides nubeculosis* larva is relatively simple. The single-layered PM is lined on the midgut lumen side by a thin epicuticular-like layer (arrow) that appears to condense against the cuticle (Cu) lining the foregut (× 13 800). Micrograph courtesy of K. Darling and A.J. Mordue (Luntz).

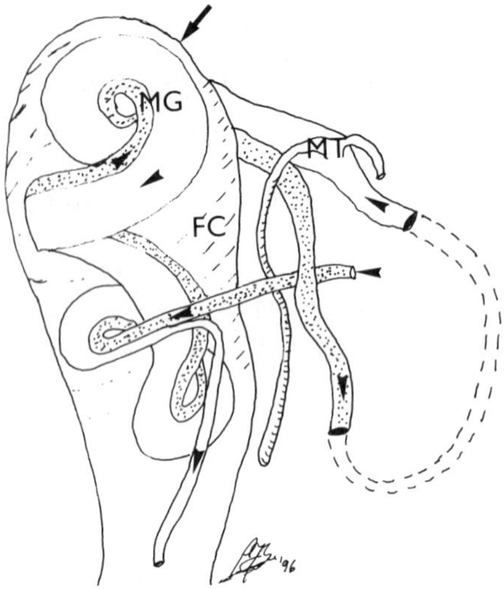

Figure 1.11 The homopteran filter chamber (FC) is a complex midgut arrangement contained within a connective tissue sheath (arrow). Anterior (dotted region) and posterior (clear region) portions of the midgut are tightly juxtaposed within the filter chamber and countercurrent flow occurs (the arrowheads show the direction of the fluid meal movement along the midgut). Malpighian tubules (MT) lie outside the filter chamber. This structure allows an isolated portion of the midgut to establish steep osmotic gradients across epithelia without the need to modify the osmotic characteristics of the whole haemolymph. (Redrawn from Gullan and Cranston, 1994.)

anterior midgut to which Malpighian tubules and regions of posterior midgut or anterior hindgut are tightly apposed, the whole being enclosed in a thin, cellular, connective sheath (Figure 1.11). Water is moved from the midgut lumen across the epithelia to the lumens of the Malpighian tubules and posterior intestine for subsequent excretion. The epithelia of all three tissues are extremely thin presenting minimum barriers to water movements, and the cells possess the extensive basal infoldings so typical of ion and water transporting tissues (Marshall, 1983). The filter chamber of *Cicadella viridis* possesses intramembranous proteins, possibly water channels (Guillam *et al.*, 1992), which may be associated with the regularly arranged intramembranous particles revealed by electron microscopy.

1.3.3 The mycetome

Although symbiotic bacteria are associated commonly with the insect midgut (see Chapter 15), in some insect species midgut epithelial cells

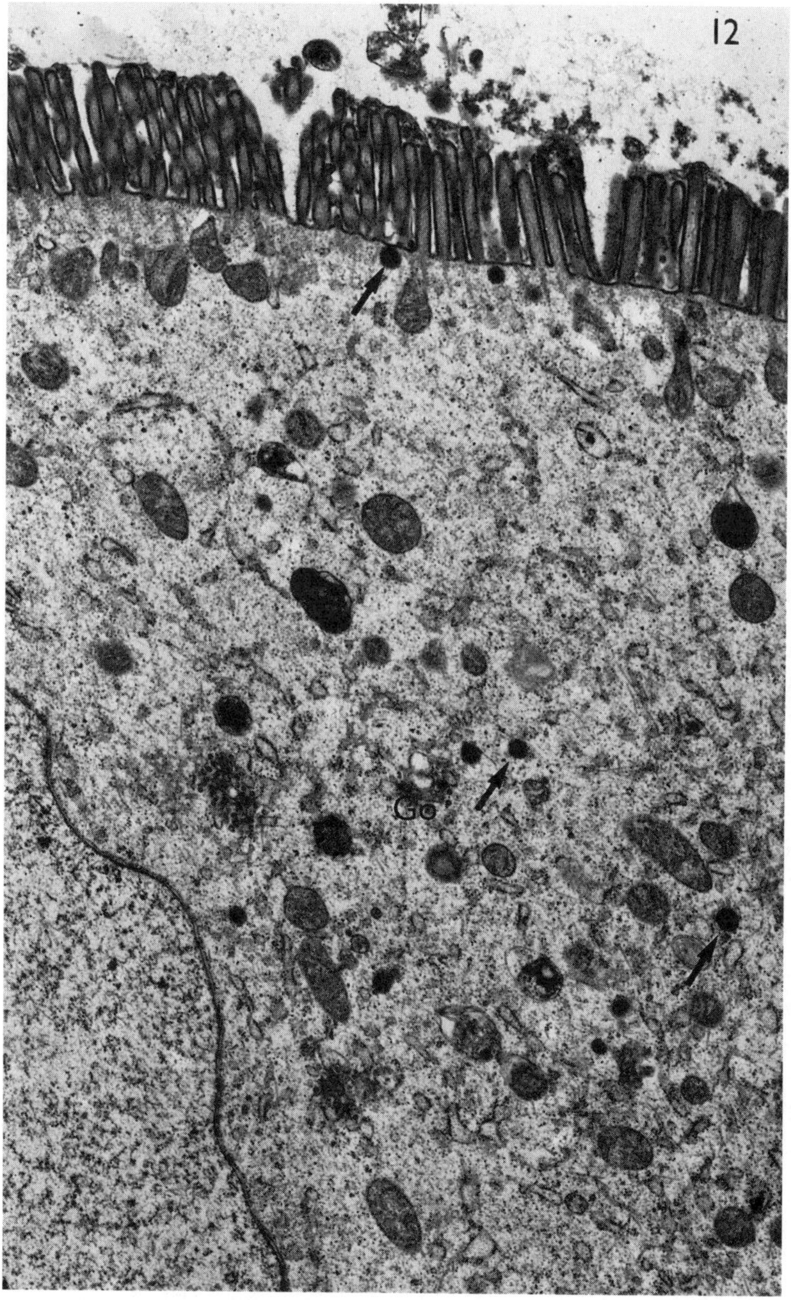

Figure 1.12 The passage of secretory vesicles (arrows) can be seen in this anterior midgut cell of *Rhodnius prolixus*. Vesicles produced from the Golgi apparatus (Go), pass to the apex of the cell where they fuse with the apical plasma membrane near the base of the microvilli (× 12 750). (Modified from Billingsley and Downe, 1988.)

become highly modified and intimately associated with the symbionts. In the tsetse fly, *Glossina morsitans*, two groups of hypertrophied cells or mycetocytes comprise the mycetome found as an annular structure in the anterior midgut. The mycetocytes almost occlude the midgut lumen from one side of the midgut, whereas the opposing side of the epithelium comprises apparently normal cells (Reinhardt *et al.*, 1972b). These mycetocytes rest on the same continuous basal lamina, and are characterized by the presence of intracellular symbiotic bacteria with a dense, amorphous cytoplasm. Intracellular *Rickettsia*-like organisms are also observed, usually surrounded by a halo of electron-lucent material. More commonly, bacteria are extracellular, but may still be restricted to an annulus in the anterior midgut or proventriculus by such structures as the PM or foregut cuticle (Peters, 1992).

1.4 FUTURE PROSPECTS

The diversity of midgut ultrastructure in insects is very apparent and has led to an equivalent diversity of studies, most of which are descriptive. The time is ripe for studies on common mechanisms, structure–function relationships and trends to be sought and described. It seems unlikely that this will be achieved by ultrastructural studies alone. The benefits of systematic approaches to studying insect digestion are apparent throughout this book, and techniques for a similar approach are available to researchers looking at midgut structure; these include histochemical localization of enzyme activities (Billingsley and Downe, 1985), detection of surface charge (Houk *et al.*, 1986b) and epithelial permeability (Fishman and Zlotkin, 1984), and sugar-based lectin histochemistry (Rudin and Hecker, 1989). As more immunological and molecular probes are produced for studies in digestion, these can be applied by appropriate techniques to midgut cells in known physiological states (Graf *et al.*, 1986). Confocal electron microscopy should benefit such studies significantly. These probes could and arguably should be applied systematically rather than necessarily within the confines of a single researcher's interests. Along with complementary data, e.g. molecular sequences, western blotting or activity assays, detailed structure-based models can be developed that have clear relation to phylogeny and mode of feeding.

ACKNOWLEDGEMENTS

PFB was supported by the Royal Society during the writing of this chapter. We thank Dr Jenny Mordue and Mr Kevin Darling for the use of micrographs, and Mr Dan Gare for his help in preparing the manuscript.

REFERENCES

Allingham, P.G., Kerlin, R.L., Tellam, R.L. *et al.* (1992) Passage of host immunoglobulin across the midgut epithelium into the haemolymph of blood-fed buffalo flies *Haematobia irritans exigua*. *J. Insect Physiol.*, **38**, 9–17.

Andries, J.C. (1976) Variations ultrastructurales au sein des cellules épithéliales mésentérique d'*Aeshna cyanea* (Insecte, Odonate) en fonction de la prise de nourriture. *Cytobiology*, **13**, 451–68.

Andries, J.C. (1977) Specialization of the endoplasmic reticulum in the apex of the midgut cells of *Aeshna cyanea*. *Cell Tissue Res.*, **178**, 97–101.

Andries, J.C. (1982) L'Intestin moyen des insectes. *Ann. Biol.*, **21**, 143–86.

Andries, J.C. and Beauvillain, J.C. (1988) Ultrastructural study of cholecystokinin-like immunoreactivity in endocrine cells of the insect midgut. *Cell Tissue Res.*, **254**, 75–81.

Andries, J.C. and Torpier, G. (1982) An extracellular brush border coat of lipid membranes in the midgut of *Nepa cinerea* (Insecta, Heteroptera): ultrastructure and genesis. *Biol. Cell.*, **46**, 195–202.

Andries, J.C. and Tramu, G. (1985) Ultrastructural and immunohistochemical study of endocrine cells in the midgut of the cockroach *Blaberus craniifer* (Insecta, Diptera). *Cell Tissue Res.*, **240**, 323–32.

Baker, J.E., Woo, S.M. and Byrd, R.V. (1984) Ultrastructural features of the gut of *Sitophilus granarius* (L.) (Coleoptera: Curculionidae) with notes on distribution of proteinases and amylases in crop and midgut. *Can. J. Zool.*, **62**, 1251–9.

Berner, R., Rudin, W. and Hecker, H. (1983) Peritrophic membranes and protease activity in the midgut of the malaria mosquito, *Anopheles stephensi* (Liston) (Insecta: Diptera) under normal and experimental conditions. *J. Ultrastruct. Res.*, **83**, 195–204.

Berridge, M.J. (1970) A structural analysis of intestinal absorption, in *Symposia of the Royal Entomological Society of London. 5. Insect Ultrastructure* (ed. A.C. Neville), pp. 135–51.

Bertram, D.S. and Bird, R.G. (1961) Studies on mosquito-borne viruses in their vectors. 1. The normal fine structure of the midgut epithelium of the adult female *Aedes aegypti* (L.) and the functional role of its modifications following a blood meal. *Trans. R. Soc. Trop. Med. Hyg.*, **55**, 404–23.

Bignell, D.E. (1981) Nutrition and digestion, in *The American Cockroach* (eds W.J. Bell and K.G. Adiyodi), Chapman & Hall, London, pp. 57–86.

Billingsley, P.F. (1988) Morphometric analysis of *Rhodnius prolixus* Stål (Hemiptera: Reduviidae) midgut cells during blood digestion. *Tissue Cell*, **20**, 291–301.

Billingsley, P.F. (1989) Endopolyploidy and digestion in the midgut of *Rhodnius prolixus* Stål (Hemiptera: Reduviidae). *Ann. Trop. Med. Parasitol.*, **83**, 93–9.

Billingsley, P.F. (1990) The midgut ultrastructure of hematophagous insects. *Ann. Rev. Entomol.*, **35**, 219–48.

Billingsley, P.F. and Downe, A.E.R. (1983) Ultrastructural changes in posterior midgut cells associated with blood-feeding in adult female *Rhodnius prolixus* Stål (Hemiptera: Reduviidae). *Can. J. Zool.*, **61**, 2574–86.

Billingsley, P.F. and Downe, A.E.R. (1985) Cellular localisation of aminopeptidase in the midgut of *Rhodnius prolixus* Stål (Hemiptera: Reduviidae) during blood digestion. *Cell Tissue Res.*, **241**, 421–8.

Billingsley, P.F. and Downe, A.E.R. (1986) Nondigestive cell types in the midgut epithelium of *Rhodnius prolixus* (Hemiptera: Reduviidae). *J. Med. Entomol.*, **23**, 212–16.

Billingsley, P.F. and Downe, A.E.R. (1989a) Changes in the anterior midgut cells of adult female *Rhodnius prolixus* (Hemiptera: Reduviidae) after feeding. *J. Med. Entomol.*, **26**, 104–8.

Billingsley, P.F. and Downe, A.E.R. (1989b) The effects of artificial diets on the anterior intestinal cell ultrastructure of *Rhodnius prolixus* (Hemiptera: Reduviidae). *Int. J. Parasitol.*, **19**, 291–9.

Billingsley, P.F. and Downe, A.E.R. (1989c) Ultrastructural localisation of cathepsin B in the midgut of *Rhodnius prolixus* Stål (Hemiptera: Reduviidae) during blood digestion. *Int. J. Insect Morphol. Embryol.*, **17**, 295–302.

Billingsley, P.F. and Rudin, W. (1992) The role of the mosquito peritrophic membrane in blood meal digestion and infectivity of *Plasmodium* species. *J. Parasitol.*, **78**, 430–40.

Böhringer-Schweizer, S. (1977) Digestion in the tsetse fly: an ultrastructural analysis of structure and function of the midgut epithelium of *Glossina morsitans morsitans* (Machado) (Diptera: Glossinidae). PhD Thesis, University of Basel, Switzerland.

Bonfanti, P., Colombo, A., Heintzelman, M.B. *et al.* (1992) The molecular architecture of an insect midgut brush border cytoskeleton. *Eur. J. Cell. Biol.*, **57**, 298–307.

Brown, M.R., Crim, J.W. and Lea, A.O. (1986) FMRFamide- and pancreatic polypeptidase-like immunoreactivity of endocrine cells in the midgut of a mosquito. *Tissue Cell*, **18**, 419–28.

Brown, M.R., Raikhel, A.S. and Lea, A.O. (1985) Ultrastructure of midgut endocrine cells in the adult mosquito, *Aedes aegypti*. *Tissue Cell*, **17**, 709–21.

Brown, R.P. (1980) Ultrastructure and function of midgut epithelium in the tsetse *Glossina morsitans* Westw. (Diptera: Glossinidae). *J. Ent. Soc. S. Afr.*, **43**, 195–214.

Brunings, E.A. and DePriester, W. (1971) Effects of mode of fixation on formation of extrusions in the midgut epithelium of *Calliphora erythrocephala* (Diptera: Calliphoridae). *Cytobiologie*, **4**, 487–92.

Cioffi, M. (1984) Comparative ultrastructure of arthropod transporting epithelia. *Am. Zool.*, **24**, 139–56.

Dimitriadis, V.K. (1991) Fine structure of the midgut of adult *Drosophila auraria* and its relationship to the sites of acidophilic secretion. *J. Insect Physiol.*, **37**, 167–77.

Dow, J.A.T. (1986) Insect midgut function. *Adv. Insect Physiol.*, **19**, 187–328.

Ellis, D.S., Young, C., Stamford, S. and Lehane, M.J. (1981) Notes on midgut cell nuclear coats in various Tsetse species. *J. Trop. Med. Hyg.*, **84**, 209–14.

Endo, Y. and Nishiitsutsuji-Uwo, J. (1981) Gut endocrine cells in insects: the ultrastructure of the gut endocrine cells of the lepidopterous species. *Biomed. Res.*, **2**, 270–80.

Felix, C.R., Betschart, B., Billingsley, P.F. and Freyvogel, T.A. (1991) Post-feeding induction of trypsin in the midgut of *Aedes aegypti* L. (Diptera: Culicidae) is separable into two cellular phases. *Insect Biochem.*, **21**, 197–203.

Fishman, L. and Zlotkin, E. (1984) A diffusional route of transport of horseradish peroxidase through the midgut of a fleshfly. *J. Exp. Zool.*, **229**, 189–95.

Fujita, T. and Kobayashi, S. (1977) Structure and function of gut endocrine cells. *Int. Rev. Cytol.*, **6**, 187–233.

Gander, E.S., Schoenenberger, M.C. and Freyvogel, T.A. (1980) Ribosomes and ribosome-function in the midgut cells of *Aedes aegypti*. *Insect Biochem.*, **10**, 441–7.

Glättli, E., Rudin, W. and Hecker, H. (1987) Immunoelectron microscopic demonstration of pancreatic polypeptide in midgut epithelium of haematophagous dipterans. *J. Histochem. Cytochem.*, **35**, 891–6.

Gooding, R.H. (1973) The digestive processes of haematophagous insects IV. Secretion of trypsin by *Aedes aegypti* (Diptera: Culicidae). *Can. Entomol.*, **105**, 599–603.

Graf, R., Raikhel, A.S., Brown, M.R. *et al.* (1986) Mosquito trypsin: immunocytochemical localization in the midgut of blood-fed *Aedes aegypti* (L.). *Cell Tissue Res.*, **245**, 19–27.

Guillam, M.T., Beuron, F., Grandin, N. *et al.* (1992) Expression of RNA isolated from the water-shunting complex of a sap-sucking insect increases the membrane permeability for water in *Xenopus* oocytes. *Exp. Cell Res.*, **200**, 301–5.

Gullan, P.J. and Cranston, P.S. (1994) *The Insects. An Outline of Entomology*, Chapman & Hall, London, 491 pp.

Hatfield, P.R. (1988) Detection and localization of antibody ingested with a mosquito blood meal. *Med. Vet. Entomol.*, **2**, 339–45.

Hecker, H. (1977) Structure and function of midgut epithelial cells in Culicidae mosquitoes (Insecta, Diptera). *Cell Tissue Res.*, **184**, 321–41.

Hecker, H., Freyvogel, T.A., Briegel, H. and Steiger, R. (1971) Ultrastructural differentiation of the midgut epithelium in female. *Aedes aegypti* (L.) (Insecta, Diptera) imagines. *Acta Trop.*, **28**, 80–104.

Hecker, H. and Rudin, W. (1976) Dynamics of rough endoplasmic reticulum (RER) in midgut cells of the mosquito *Aedes aegypti*. *Sixth European Congress on Electron Microscopy*, pp. 289–91.

Hecker, H. and Rudin, W. (1979) Normal versus α-amanitin induced cellular dynamics of the midgut epithelium in female *Aedes aegypti* L. (Insecta, Diptera) in response to blood-feeding. *Eur. J. Cell Biol.*, **19**, 160–7.

Houk, E.J. and Hardy, J.L. (1982) Midgut cellular responses to bloodmeal digestion in the mosquito, *Culex tarsalis* Coquillett (Diptera: Culicidae). *Int. J. Insect Morphol. Embryol.*, **11**, 109–19.

Houk, E.J., Hardy, J.L. and Chiles, R.E. (1986a) Histochemical staining of the complex carbohydrates of the midgut of the mosquito, *Culex tarsalis* coquillett. *Insect Biochem.*, **16**(4), 667–75.

Houk, E.J., Hardy, J.L. and Chiles, R.E. (1986b) Mesenteronal epithelial cell surface charge of the mosquito, *Culex tarsalis* Coquillett. Binding of colloidal iron hydroxide, native ferritin and cationised ferritin. *J. Submicrosc. Cytol.*, **18**, 385–96.

Jenni, L. and Bohringer, S. (1976) Nuclear coat and viruslike particles in the midgut epithelium of *Glossina morsitans* sspp. *Acta Trop.*, **38**, 380–9.

Jordão, B.P., Lehane, M.J., Terra, W.R. *et al.* (1996a) An immunocytochemical investigation of trypsin secretion in the midgut of *Stomoxys calcitrans*. *Insect Biochem. Mol. Biol.* (in press).

Jordão, B.P., Terra, W.R., Ribeiro, A.F. (1996b) Trypsin secretion in *Musca domestica* larval midguts. A biochemical and immunological study. *Insect Biochem. Mol. Biol.* (in press).

Klein, U., Loffelmann, G. and Wieczorek, H. (1991) The midgut as a model for insect K^+ transporting epithelia – immunocytochemical localisation of a vacuolar type H^+ pump. *J. Exp. Biol.*, **161**, 61–75.

Lehane, M.J. (1976a) Digestive enzyme secretion in *Stomoxys calcitrans* (Diptera: Muscidae). *Cell Tissue Res.*, **170**, 275–87.

Lehane, M.J. (1976b) Formation and histochemical structure of the peritrophic membrane in the stablefly, *Stomoxys calcitrans*. *J. Insect Physiol.*, **22**, 1551–7.
Lehane, M.J. (1977) Transcellular absorption of lipids in the midgut of the stablefly. *J. Insect Physiol.*, **23**, 945–54.
Lehane, M.J. (1988) Evidence for secretion by the release of cytoplasmic extrusions from midgut cells of *Stomoxys calcitrans*. *J. Insect Physiol.*, **34**, 949–53.
Lehane, M.J. (1989) The intracellular pathway and kinetics of digestive enzyme secretion in an insect midgut cell. *Tissue Cell*, **21**, 101–11.
MacVicker, J.A.K., Billingsley, P.F. and Djamgaz, M.B.A. (1993) Na^+/K^+-ATPases in the midguts of haematophagus insects: biochemical and immunochemical studies. PhD Thesis, University of London.
Marshall, A.T. (1983) X-ray microanalysis of the filter chamber of the cicada, *Cyclochila australasia* Don. A water shunting epithelial complex. *Cell Tissue Res.*, **231**, 215–17.
Martoja, R. and Ballan-Dufrançais, C. (1984) The ultrastructure of the digestive and excretory organs, in *Insect Ultrastructure* Vol. 2 (eds R.C. King and H. Akai), Plenum Press, New York.
Modespacher, U.-P., Rudin, W., Jenni, L. and Hecker, H. (1986) Transport of peroxidase through the midgut epithelium of *Glossina m. morsitans* (Diptera, Glossinidae). *Tissue Cell*, **18**, 429–36.
Montuenga, L.M., Barrenechea, M.A., Sesma, P. *et al.* (1989) Ultrastructure and immunohistochemistry of endocrine cells in the midgut of the desert locust, *Schistocerca gregaria* (Forskal). *Cell Tissue Res.*, **258**, 577–83.
Mori, H. (1983) Origin, development, morphology, functions and phylogeny of the embryonic midgut epithelium in insects. *Entomol. Genet.*, **8**, 135–54.
Nagel, G. and Peters, W. (1991) Formation, properties and degradation of the peritrophic membranes of larval and adult fleshflies, *Sarcophaga barbata* (Insecta, Diptera). *Zoomorphology*, **111**, 103–11.
Nasiruddin, N. and Mordue (Luntz), A.J. (1993) The effect of azadirachtin on the midgut histology of the locusts, *Schistocerca gragaria* and *Locusta migratoria*. *Tissue Cell*, **25**, 875–84.
Nopanitaya, W. and Misch, D.W. (1974) Developmental cytology of the midgut in the fleshfly *Sarcophaga bullata* (Parker). *Tissue Cell*, **6**, 487–502.
Perrone, J.B. and Spielman, A. (1988) Time and site of assembly of the peritrophic membrane of the mosquito *Aedes aegypti*. *Cell Tissue Res.*, **252**, 473–8.
Peters, W. (1979) The fine structure of peritrophic membranes of mosquito and blackfly larvae of the genera *Aedes*, *Anopheles*, *Culex* and *Odagmia* (Diptera: Culicidae/Simuliidae). *Entomol. Genet.*, **5**, 289–99.
Peters, W. (1992) *Peritrophic Membranes*, Springer-Verlag, Berlin.
Peters, W., Herlet, N. and Thienemann, H. (1978) Bildung und Feinstruktur peritrophischer Membranen bei Muckenlarven aus dem Bereich der Tipulo-, Psychodo- und Bibiomorpha (Diptera: Nematocerca). *Entomol. Genet.*, **4**, 33–54.
Platzer-Schultz, I. and Welsch, U. (1970) Aprokrine Sekretion der peritrophischen Membran von *Chironomus thummi piger* Str. (Diptera). *Z. Zellforsch. Mikrosk. Anat.*, **104**, 530–40.
Raes, H., Verbeke, M., Meulemans, W. and De-Coster, W. (1994) Organisation and ultrastructure of the regenerative crypts in the midgut of the adult worker honeybee (L. *Apis mellifera*). *Tissue Cell*, **26**, 231–8.
Reinhardt, C., Schultz, U., Hecker, H. and Freyvogel, T.A. (1972a) Zur

ultrastruktur des Mitteldarmepithels bei Flöhen (Insecta, Siphonaptera) *Rev. Suisse Zool.*, **79**, 1130–7.

Reinhardt, C., Steiger, R. and Hecker, H. (1972b) Ultrastructural study of the midgut mycetome-bacteroids of the tsetse flies *Glossina morsitans*, *G. fuscipes* and *G. brevipalpis* (Diptera, Brachycera). *Acta Trop.*, **29**, 280–8.

Richards, A.G. and Richards, P.A. (1971) Origin and composition of the peritrophic membrane of the mosquito, *Aedes aegypti*. *J. Insect Physiol.*, **17**, 2253–75.

Richards, P.A. and Richards, A.G. (1969) Intranuclear crystals in the midgut epithelium of a flea. *Ann. Entomol. Soc. Am.*, **62**, 249–50.

Rudin, W. and Hecker, H. (1979) Functional morphology of the midgut of *Aedes aegypti* L. (Insecta, Diptera) during blood digestion. *Cell Tissue Res.*, **200**, 193–203.

Rudin, W. and Hecker, H. (1989) Lectin-binding sites in the midgut of the mosquitoes *Anopheles stephensi* Liston and *Aedes aegypti* L. (Diptera: Culicidae). *Parasitol. Res.*, **75**, 268–79.

Santos, C.D., Ribeiro, A.F., Ferreira, C. and Terra, W.R. (1984) The larval midgut of the cassava hornworm (*Erinnyis ello*): ultrastructure, fluid fluxes and the secretory activity in relation to the organization of digestion. *Cell Tissue Res.*, **237**, 565–74.

Santos, C.D., Ribeiro, A.F. and Terra, W.R. (1986) Differential centrifugation, calcium precipitation, and ultrasonic disruption of midgut cells of *Erinnyis ello* caterpillars. Purification of cell microvilli and inferences concerning secretory mechanisms. *Can. J. Zool.*, **64**, 490–500.

Schneider, M., Rudin, W. and Hecker, H. (1987) Absorption and transport of radioactive tracers in the midgut of the malaria mosquito, *Anopheles stephensi*. *J. Ultrastruct. Mol. Struct. Res.*, **97**, 50–63.

Sivasubramanian, P. (1992) Localization of FMRFamide-like immunoreactivity in the larval midgut of the fly, *Sarcophaga bullata*. *Comp. Biochem. Physiol.* [C], **102**, 555–60.

Spies, A.G. and Spence, K.D. (1985) Effect of sublethal *Bacillus thuringiensis* crystal endotoxin treatment on the larval midgut of a moth, *Manduca*: SEM study. *Tissue Cell*, **17**, 379–94.

Steiger, R.F. (1973) On the ultrastructure of *Trypanosoma* (*Trypanozoon*) *brucei* in the course of its life cycle and some related aspects. *Acta Trop.*, **30**, 64–168.

Terra, W.R., Espinoza-Fuentes, F.P., Ribeiro, A.F. and Ferreira, C. (1988) The larval midgut of the housefly (*Musca domestica*): ultrastructure, fluid fluxes and ion secretion in relation to the organisation of digestion. *J. Insect Physiol.*, **34**, 463–72.

Thomas, G. and Bockeler, W. (1994) Investigation of the intestinal spherocrystals of different Cephalobaenida (Pentastomida). *Parasitol. Res.*, **80**, 420–5.

Turunen, S. (1988) Uptake of dietary lipids: a novel pathway in *Pieris brassicae*. *Insect Biochem.*, **18**, 499–505.

Vaughan, J.A., Wirtz, R.A., do Rosario, V.E. and Azad, A.F. (1990) Quantitation of antisporozoite immunoglobulins in the hemolymph of *Anopheles stephensi* after bloodfeeding. *Am. J. Trop. Med. Hyg.*, **42**, 10–16.

Volkmann, A. and Peters, W. (1989) Investigations on the midgut caeca of mosquito larvae. I. Fine structure. *Tissue Cell*, **21**, 243–51.

Wieczorek, H., Putzenlechener, M., Zeiske, W. and Klein, U. (1991) A vacuolar type proton pump energises K^+/H^+ antiport in an animal plasma membrane. *J. Biol. Chem.*, **266**, 15340–7.

Wood, A.R. and Lehane, M.J. (1991) Relative contributions of apocrine and eccrine secretion to digestive enzyme release from midgut cells of *Stomoxys calcitrans* (Insects: Diptera). *J. Insect Physiol.*, **37**, 161–6.

Zhuzhikov, D.P. (1963) The structure of peritrophic membranes of Diptera. *Vestn. Mosk. Univ.*, **1963**, 24–35.

Zimmermann, U., Mehlan, D. and Peters, W. (1975) Investigations on the transport function and structure of peritrophic membranes. V. Amino acid analysis and electron microscopic investigations of the peritrophic membranes of the blowfly *Calliphora erythrocephala* Mg. *Comp. Biochem. Physiol.*, **51**, 181–6.

2

Midgut development

K.M. Baldwin, R.S. Hakim, M.J. Loeb and S.Y. Sadrud-Din

2.1 INTRODUCTION

New information about insect midgut development mostly falls into three areas: genetic and molecular biology of *Drosophila* embryogenesis, cellular differentiation in embryonic and molting larva of *Manduca sexta*, and cell culture of larval *Manduca* midgut epithelial cells. It is the intention of the authors to briefly discuss information which has been covered in a number of earlier reviews (Mori, 1983; Dow, 1986; Schwalm, 1988) and to concentrate the discussion on material which has been published in the last 10 years.

2.2 EMBRYONIC DEVELOPMENT

The cellular origin of the epithelium of the midgut varies depending upon the species of insect. In the more 'primitive' species, such as apterygotes, the midgut epithelium is formed from yolk cells (vitellophages), whereas in the more 'advanced' species (most pterygotes), the epithelial cells are derived from bipolar rudiments which invaginate from the germ band at gastrulation (Figure 2.1).

2.2.1 Apterygotes

As the germ band spreads to enclose the yolk, cells present within the yolk move peripherally to form a complete layer of cells at the yolk

Biology of the Insect Midgut.
Edited by M.J. Lehane and P.F. Billingsley.
Published in 1996 by Chapman & Hall, London.
ISBN 0 412 61670 X.

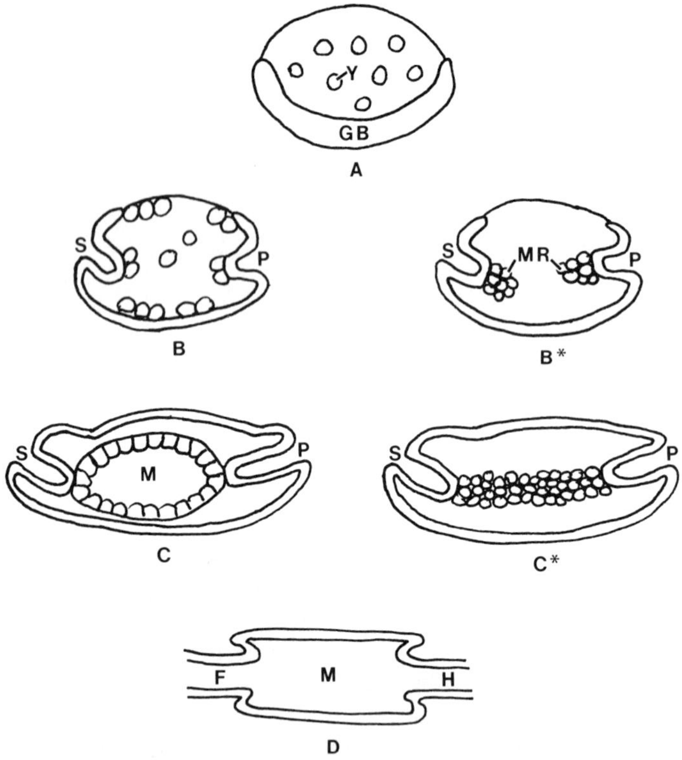

Figure 2.1 Formation of the embryonic midgut. In apterygotes (A, B, C and D), as the germ band (GB) spreads to enclose the embryo, yolk cells (Y) migrate to a peripheral position in the yolk. Eventually the yolk cells form a complete layer which surrounds the yolk and lies between the developing stomodeum (S) and proctodeum (P). This layer of cells forms the embryonic midgut (M) epithelium. Later the foregut (F) and hindgut (H) fuse with the midgut to form a complete digestive tube. In pterygotes (A, B*, C* and D), the two midgut rudiments (MR) are derived from the germ band (see discussion in text). They form masses of cells adjacent to the developing stomodeum and proctodeum, and then migrate towards each other. Later, they spread to enclose the yolk, first ventrally and then dorsally. As with apterygotes, the midgut fuses with the foregut and hindgut to form the digestive tube. (Redrawn from Miya, 1976; Machida and Ando, 1981; Dow, 1986.)

surface (Figure 2.1B). As the yolk is reduced by digestion, it and its surrounding layer of cells become incorporated within the developing embryo and form an epithelial sac which lies between the stomodeal invagination (which will form the foregut) and the proctodeal invagination (which will form the hindgut) (Figure 2.1C). The layer of cells at the surface of the yolk thus forms the midgut epithelium. Later, the foregut and hindgut fuse with the midgut, leaving a continuous tube through the interior of the animal (Figure 2.1D).

2.2.2 Pterygotes

In the germ band, there are two groups of cells which are destined to form the midgut epithelium, one at the anterior pole of the embryo and one at the posterior pole. At gastrulation, these cells invaginate into the interior of the developing embryo and come to lie adjacent to the invaginating stomodeum and proctodeum, respectively (Figure 2.1B*). After invaginating, the cells of the two midgut primordia lose their epithelial organization and appear as two solid masses of cells (Figures 2.1B*, 2.2A,B). These proliferate and migrate towards each other on

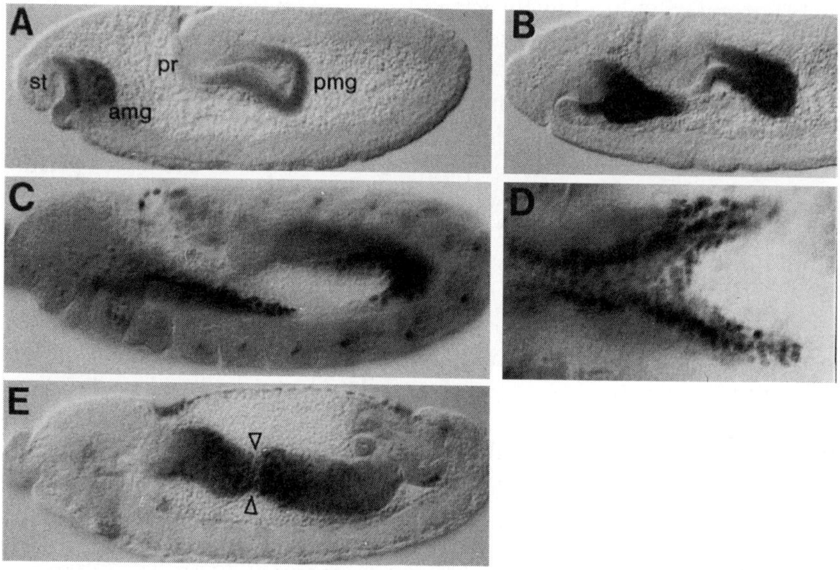

Figure 2.2 *Drosophila* midgut development as demonstrated with enhancer trap line A490.2M3 and immunostained for β-galactosidase. All figures show lateral views except D, which is a horizontal view. A, In early stage 10, the anterior midgut rudiment (amg) is a mass of cells adjacent to the stomodeum (st) while the posterior midgut rudiment (pmg) is still part of the proctodeum (pr). B, At late stage 10, both midgut rudiments form masses of cells. C, At stage 11, the two rudiments have begun migrating towards each other. D, Horizontal view of the anterior midgut rudiment showing the cells migrating on either side of the yolk. E, In early stage 13, the two midgut rudiments meet (open arrowheads). (Reprinted from Reuter *et al.*, 1993, with permission of the Company of Biologists Ltd.)

either side of the yolk, between the yolk and the visceral mesoderm (Figure 2.2C,D). After meeting, at a variable distance from the midline, the anterior and posterior primordia form two lateral bands which extend from the developing stomodeum to the developing proctodeum (Figures 2.1C*, 2.2E). The midgut primordial cells now regain their epithelial organization and spread to enclose the yolk, first ventrally and then dorsally. As with apterygotes, the midgut now consists of an epithelial sac which lies between the stomodeum and proctodeum. The midgut later fuses with the foregut and hindgut to form a continuous digestive tube (Figure 2.1D).

A number of species have been described in which the midgut development differs from the apterygote and pterygote schemes (discussed in Mori, 1983; Dow, 1986; Schwalm, 1988). There are some species in which the endodermal anlagen form much of the midgut epithelium, but cells from the yolk are also incorporated, sometimes only temporarily. In a few insects, the midgut epithelium forms from the germ band, but from only one rudiment rather than following the bipolar scheme.

Although it is clear that in most pterygotes the midgut forms from two different anlagen, the source of these cells is not firmly established. In a report on the embryonic development of the midgut of the lepidopteran, *Neomicropteryx nipponensis*, Kobayashi and Ando (1983) stated that the anterior and posterior midgut epithelium precursors arose from the blind ends of the stomodeum and proctodeum, respectively. However, since they did not label the midgut precursors, it is not possible to be certain whether the midgut cells actually arose directly from the stomodeum and proctodeum or migrated from other locations. In fact this same caveat must be made about all early studies on insect midgut development, including the ones on apterygotes. To be really certain of the origin of midgut epithelial cells in any insect, the precursors must be labelled specifically and then the migration of the labelled cells into the midgut followed.

Technau and Campos-Ortega (1986) injected *Drosophila* midgut precursor cells with horseradish peroxidase (HRP) and then traced their locations in the developing embryo. They found that the cells in the anterior and posterior anlagen became committed to forming midgut quite early, at the beginning of gastrulation. The anterior midgut precursors invaginated in two groups: one as a separate invagination of 'endodermal' cells, which secondarily associated with the stomodeum, and the other 'ectodermal' group, which invaginated later as a part of the stomodeum itself. The posterior midgut anlage invaginated as the basal part of the cup-shaped proctodeal invagination. Even though initially part of the same invagination, the posterior midgut anlage was considered to be endodermal while the hindgut anlage was considered to be ectodermal (Technau and Campos-Ortega, 1986; Skaer, 1993). Interestingly, cells from the 'ectodermal' anterior midgut anlage could be transplanted to the posterior 'endodermal' anlage and vice versa. The transplanted cells developed normally with the host location (Technau and Campos-Ortega, 1986). This, plus the apparent artificiality of dividing the proctodeum into ectodermal and endodermal portions, has led to some discussion in the literature as to whether the distinction between ectoderm and endoderm is appropriate in insects (Dow, 1986; Skaer, 1993). This issue is further complicated by inconsistencies in the older literature as to whether the midgut epithelium in pterygotes is described as ectodermal or endodermal.

In light of this debate, it is interesting that, when the completed midgut tube is first formed in *Manduca sexta* embryos, the midgut epithelial cells are held together by pleated septate junctions typical of ectodermally derived cells (Baldwin and Hakim, 1987). Later, when the epithelium begins to differentiate into its larval form, the pleated septate junctions are replaced by smooth septate junctions, thought characteristic of endodermally derived tissue. Thus the same cell population shows both ectodermal and endodermal features at different times.

Once the midgut precursor cells have invaginated, their further development has been elegantly demonstrated using *Drosophila* enhancer trap lines (Figure 2.2). Enhancer trap lines are *Drosophila* in which the gene for *Escherichia coli* β-galactosidase has been inserted randomly into the genome (Bellen *et al.*, 1989). Depending on exactly where in the genome this 'reporter gene' is located, it becomes activated at different times of development and in different tissues. The presence of the β-galactosidase is then detected by antibody staining. Bellen *et al.* (1989) examined hundreds of such lines and found that in many of them, the β-galactosidase was expressed in organ-, tissue- or even cell-specific manners. Several of these lines label midgut epithelial cells specifically (Bellen *et al.*, 1989; Hartenstein and Jan, 1992; Reuter *et al.*, 1993). When the midgut precursors in these lines first label, the anterior anlage is already a mesenchymal mass adjacent to the stomodeum, but the posterior anlage is still forming the base of the proctodeal invagination (Figure 2.2A). The transformation of the posterior midgut rudiment to a mesenchymal arrangement and the migration of the anterior and posterior rudiments can be seen quite clearly in these embryos (Figure 2.2).

Genetic studies of *Drosophila* mutants have shown a number of genes which are required for normal midgut development (reviewed in Skaer, 1993). In most cases, the mechanism of the defect is not known. However, interesting information has been obtained using two mutant lines, *snail* and *twist* (Reuter *et al.*, 1993). In these embryos, the mesoderm does not form, but the two midgut anlagen invaginate normally and take on a mesenchymal arrangement. However, the cell masses do not migrate towards each other nor do they regain their epithelial character. From these data, Reuter *et al.* (1993) suggest that the endodermal cells require mesoderm to direct their migration and to regain their epithelial arrangement. They further suggest that the endodermal cells migrate by directly contacting the visceral mesoderm. To strengthen their argument they show light microscopic pictures of *Drosophila* embryos in which the endoderm and visceral mesoderm are closely opposed. A fine structural analysis of midgut cellular differentiation in *Manduca sexta* lends support to this idea (Hakim *et al.*, 1988). At 56 hours of development (hatching is at 104 hours), the midgut consists of a newly complete epithelial tube surrounded by visceral mesoderm; the epithelial cells have not yet begun to differentiate. They make direct contact with the mesodermal cells as there is no intervening basal lamina (Figure 2.3). There are occasional cell junctions between the epithelial cells and the underlying mesoderm. Hence in *Manduca*, the early midgut epithelium is well placed to receive direct signals from the mesoderm. By 66 hours of development, the two tissue layers are separated by a basal lamina, and cellular differentiation of the midgut epithelium has begun.

2.2.3 Cellular differentiation in the embryo

There have been relatively few studies of embryonic cellular differentiation in the midgut, but those few would suggest species variations occur. The earlier literature on this subject has been reviewed by Mori (1983), who concludes that the variations in embryonic cytodifferentiation seen in different insect species correlate with the feeding behaviour at hatching. Those forms which begin feeding immediately have a well-differentiated midgut epithelium at hatching whereas those which continue to live off the yolk for some time may continue cytodifferentiation well into the larval period. Kadiri and Louvet (1982) describe the cellular differentiation in *Clitumnus extradentatus* as a late embryonic development, occurring during the last half of the embryonic period. They observed cellular changes as beginning with the muscle layers and then in the epithelial cell base (with development of basal infoldings) and moving apically so that microvilli and septate junctions appeared later. In *Manduca sexta* embryos, cell differentiation in the midgut also occurred in the later half of the embryonic period, but the basal to apical pattern did not hold (Hakim *et al.*, 1988). In lepidoptera, the larval midgut epithelium is formed primarily by two cell types: an absorptive 'columnar cell' and an ion-transporting 'goblet cell' which contains a large apical cavity or 'goblet'. Early signs of cellular differentiation in *Manduca* embryos were the appearance of goblet vesicles in the apices of

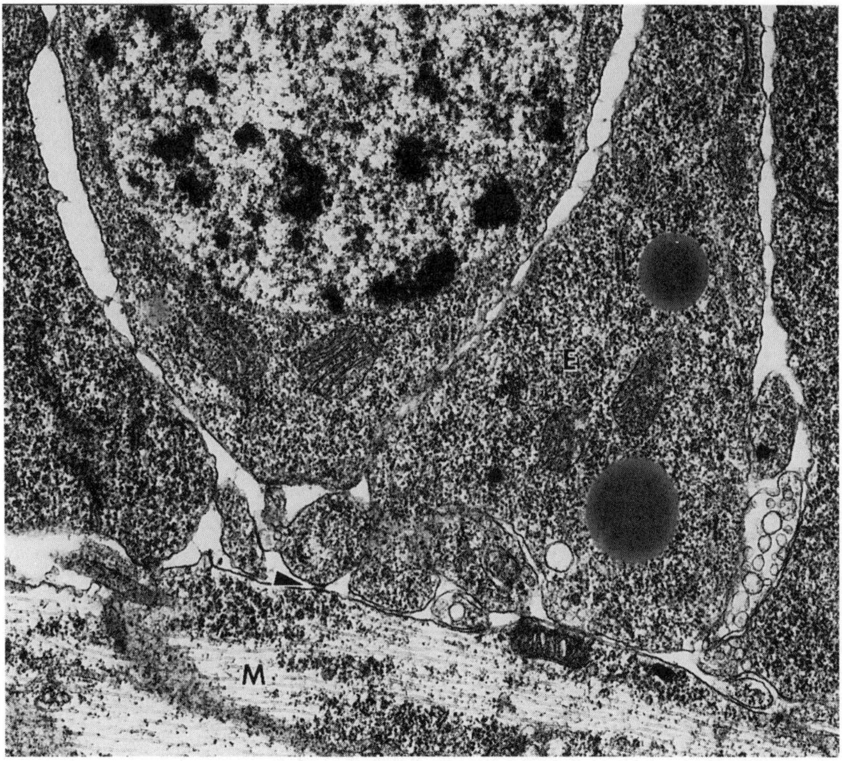

Figure 2.3 *Manduca sexta* embryo at 56 hours of development showing the close contact between the midgut epithelial cells (E) and the visceral mesoderm (M) with no intervening basal lamina. Occasional small desmosomal attachments are seen (arrowhead). The presumptive visceral muscles contain large numbers of microtubules at this stage (× 25 200).

the forming goblet cells and the appearance of vacuoles (presumably digestive in nature) in the apical cytoplasm of the columnar cells; basal infoldings appeared much later, just before hatching. Furthermore, myofilaments did not develop in the muscle cells until the end of the embryonic period (Hakim *et al.*, 1988). An important finding of the *Manduca* studies was that the goblet cavities first appeared as closed cytoplasmic vesicles, with no connection to the exterior of the cell (Figure 2.4A). About 20 hours later, the membrane limiting the goblet became contiguous with the luminal plasma membrane through a 'valve' (Figure 2.4B) (Hakim *et al.*, 1988).

Using freeze fracture analysis of *Manduca* embryos, we found that gap junctions were present from the time the midgut epithelium was first

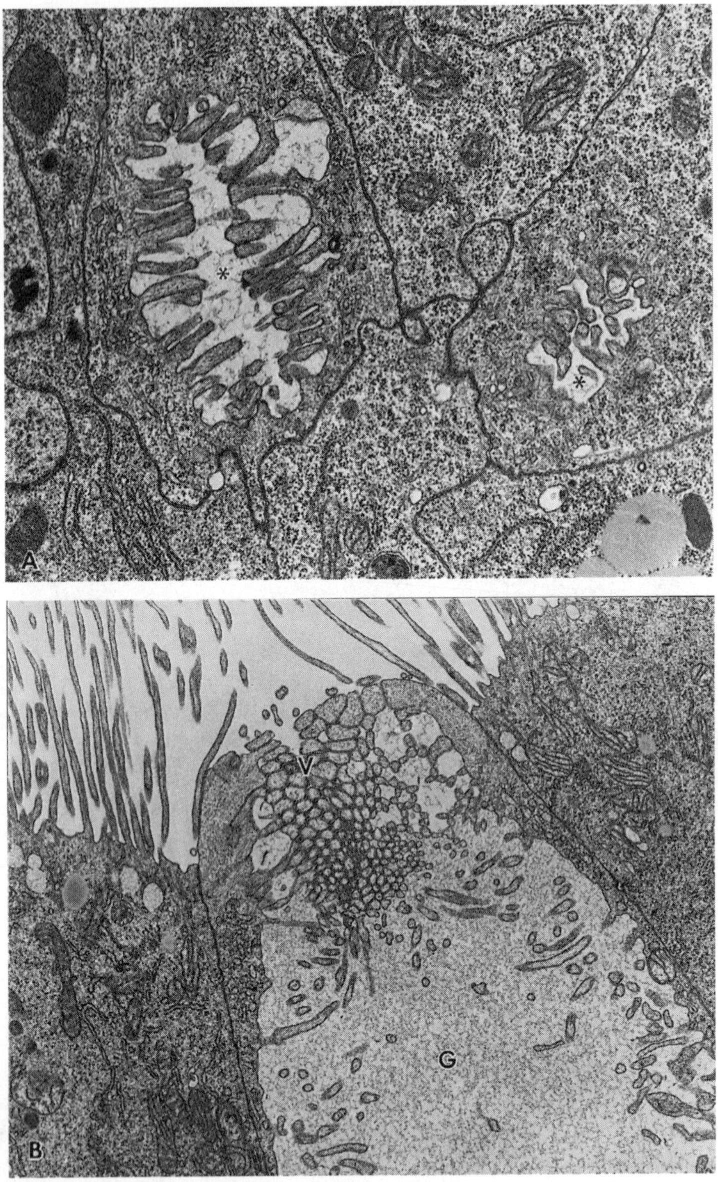

Figure 2.4 A, Horizontal section through the apices of the developing goblet cells at mid moult (see Figure 2.6B). Small goblets first appear as closed cytoplasmic vesicles (*). B, At the end of the moult period (see Figure 2.6D), the newly differentiated goblet cell (G) reaches the luminal surface of the epithelium; the membrane lining the goblet is contiguous with the luminal membrane through a 'valve' (V). Columnar cells are on either side of the goblet cell (A, × 13 430; B, × 8925).

formed. Smooth septate junctions typical of larval midgut appeared at approximately 70 hours of development, about the time the cellular differentiation was first visible (Baldwin and Hakim, 1987). Surprisingly, the smooth septate junctions replaced pleated septate junctions which were present when the epithelium was first recognizable as a complete tube. Kadiri and Louvet (1982) reported that no (pleated) septate junctions preceded the appearance of the continuous (smooth septate) junctions in the *Clitumnus* embryos. They did not, however, do a freeze fracture analysis. We found in our studies with *Manduca* that the pleated septate junctions were not obvious in standard sectioned material.

The patterning of the two cell types present in larval *Manduca* midgut epithelium is quite striking. Goblet cells occur singly and are always surrounded by a one-cell thick reticulum of columnar cells. This pattern is present in the embryo as soon as the two cell types can be distinguished (see further discussion on pattern formation below).

2.2.4 Stem or imaginal cells

Little is known about the origin of the population of cells which proliferate to enlarge or replace the existing population of midgut epithelial cells as insects progress through their life cycles. In embryos of *Calliphora erythrocephala*, van der Starr-van der Molen *et al.* (1973) recorded the presence of two types of infrequently occurring cells. One of these they suggested was a stem (imaginal) cell and the other a presumptive 'granular' (enteroendocrine) cell. In lepidoptera, the stem cells are often so inconspicuous that they have been incorrectly declared absent (Anderson and Harvey, 1966; Turbeck, 1974; Hakim *et al.*, 1988).

In their study of HRP-injected midgut precursor cells in *Drosophila* embryos, Technau and Campos-Ortega (1986) discovered a population of cells which were spindle shaped and scattered infrequently along the apical surface of the other midgut epithelial cells in the early embryo (Figure 2.5); a similar population of cells is labelled in one of the *Drosophila* enhancer trap lines (Hartenstein and Jan, 1992). These cells have been shown to migrate from their earlier apical location to a basal location within the midgut epithelium. Because of their location and spatial distribution, they have been identified as adult midgut precursor (imaginal or stem) cells (Hartenstein and Jan, 1992). Interestingly, this same population of cells can be stained with an antibody to the product of the *asense* (*ase*) gene, which is believed to be activated in neural precursor cells (Brand *et al.*, 1993). Furthermore, this population of cells is affected in certain mutants with defective neurogenic genes (Hartenstein *et al.*, 1992). In *Notch* gene mutants, the development of the midgut is defective, and in particular, the population of putative

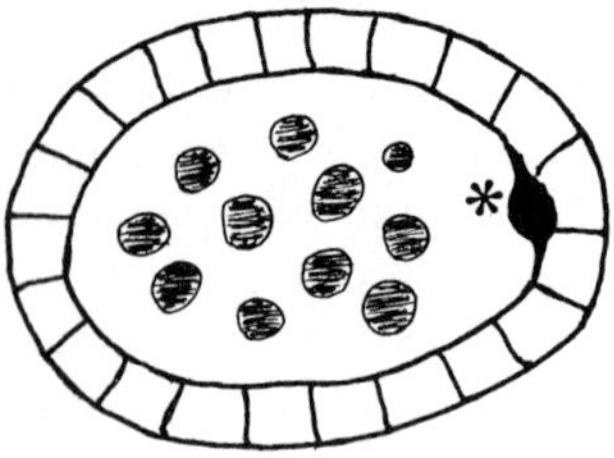

Figure 2.5 Drawing of a section of a stage 16 *Drosophila* midgut showing the epithelium and luminal yolk droplets. The labelled cell (*) is one of a scattered population which first occupies an apical position in the epithelium and later migrates to a basal location. These cells are thought to be adult precursor cells (see text). (Drawn from Technau and Campos-Ortega, 1986.)

imaginal cells is abnormally increased. These mutants also show increases in a variety of neural precursor cells.

2.3 DEVELOPMENTAL CHANGES DURING MOULTING

Tomocerus minor Lubbock (Collembola) moults throughout life and in this species the entire midgut epithelium is shed at the moult. It is replaced by stem cells which proliferate and differentiate prior to the moult (Humbert, 1979). This pattern is characteristic for apterygotes and allows the midgut to function in excretion as well as in digestion.

Werner *et al.* (1991) described the midgut of the waterstrider *Gerris najas* in which they found nests of differentiating cells in different states of development among the mature cells. They interpreted their findings to support the idea that stem cells are continuously added to the midgut epithelium to replace degenerated cells, similar to the situation in vertebrates.

Lepidoptera have been studied more intensively than many other insects, in part due to their usefulness in ion transport studies and in part due to their unusual gap junctions (see below). At hatching, *Manduca* are equipped with a fully differentiated larval midgut epithelium, and the two main cell types (columnar cells and goblet cells) appear unchanged throughout the feeding larval life (Baldwin and Hakim, 1991a). What does change is the size of the midgut. From the first instar to the fifth instar, the surface area of the midgut increases by a factor of approximately 200. This means that the number of cells in the epithelium must also increase by a factor of approximately 200, all within a period of 2–3 weeks. This is accomplished in *Manduca* by stem cells which form an inconspicuous part of the larval epithelium until near the beginning of the next moult. At this point, the stem cells proliferate until they outnumber the differentiated cells by a ratio of roughly 3:1 (Figure 2.6). Most of these cells then differentiate and intercalate among the already differentiated cells, whereas a very few remain as stem cells for the next round of growth (Baldwin and Hakim, 1991a). Cellular differentiation generally follows the pattern seen in the embryo, with the appearance of tiny goblets in the apices of the goblet cells as an early sign of differentiation (Figure 2.4A), and development of microvilli and basal infoldings occurring later. As in the embryo, the goblet is formed initially as a closed cytoplasmic vesicle and much later forms the 'valve' through which the goblet membrane becomes contiguous with the luminal membrane of the cell (Figure 2.4) (Baldwin and Hakim, 1991a). This is consistent with the observation that the goblet membrane contains a vacuolar-type H^+-ATPase (Schweikl *et al.*, 1989) characteristic of intracellular vacuoles.

While the stem cells of moulting *Manduca* third and fourth instars are differentiating, they grow in height from about 10 μm to 70 μm (Figure 2.6). Meanwhile, some dramatic changes are occurring in the surrounding population of mature (already differentiated) epithelial cells. At mid moult, both mature goblet and mature columnar cells develop large numbers of autophagic vacuoles (Figure 2.7), and the columnar cells develop large apical blebs (Figure 2.6C). Late in the moult period, the number of autophagic vacuoles decreases, and there is a marked reduction in the amount of stored metabolites, glycogen and lipid droplets, in the cytoplasm of the absorptive columnar cells. The apical blebs disappear (Figure 2.6D); presumably, they are shed into the lumen by apocrine secretion (Baldwin and Hakim, 1991a). The end result is that the mature cells (particularly the absorptive columnar cells) have reduced dramatically their cytoplasmic mass at the same time that the newly differentiating cells are increasing theirs (cf. Figures 2.6A and 2.6D). This scheme has two advantages. First, the mature cells can supply nutrients digested from their own cytoplasm at a time of great need for the differentiating cells. This is particularly important since the caterpillars do not eat during the moulting period. Second, the caterpillars grow continuously, but increase their midgut epithelial cell number periodically at the beginning of each moult. By decreasing the size of the mature cells, the caterpillar completes the mould with a modestly increased gut surface area, but a greatly increased cell number. The midgut surface area can then continue to enlarge by expanding the size of the existing cells until the next moult, when the cycle repeats.

The appearance of large apical blebs and autophagic vacuoles has been observed in insect midgut epithelia under a variety of conditions, and often has been taken as sign of cell degeneration (Humbert, 1979; Werner *et al.*, 1991). Although that is undoubtedly true in some cases, it is not true of moulting *Manduca* larvae, and with good reason. Clearly,

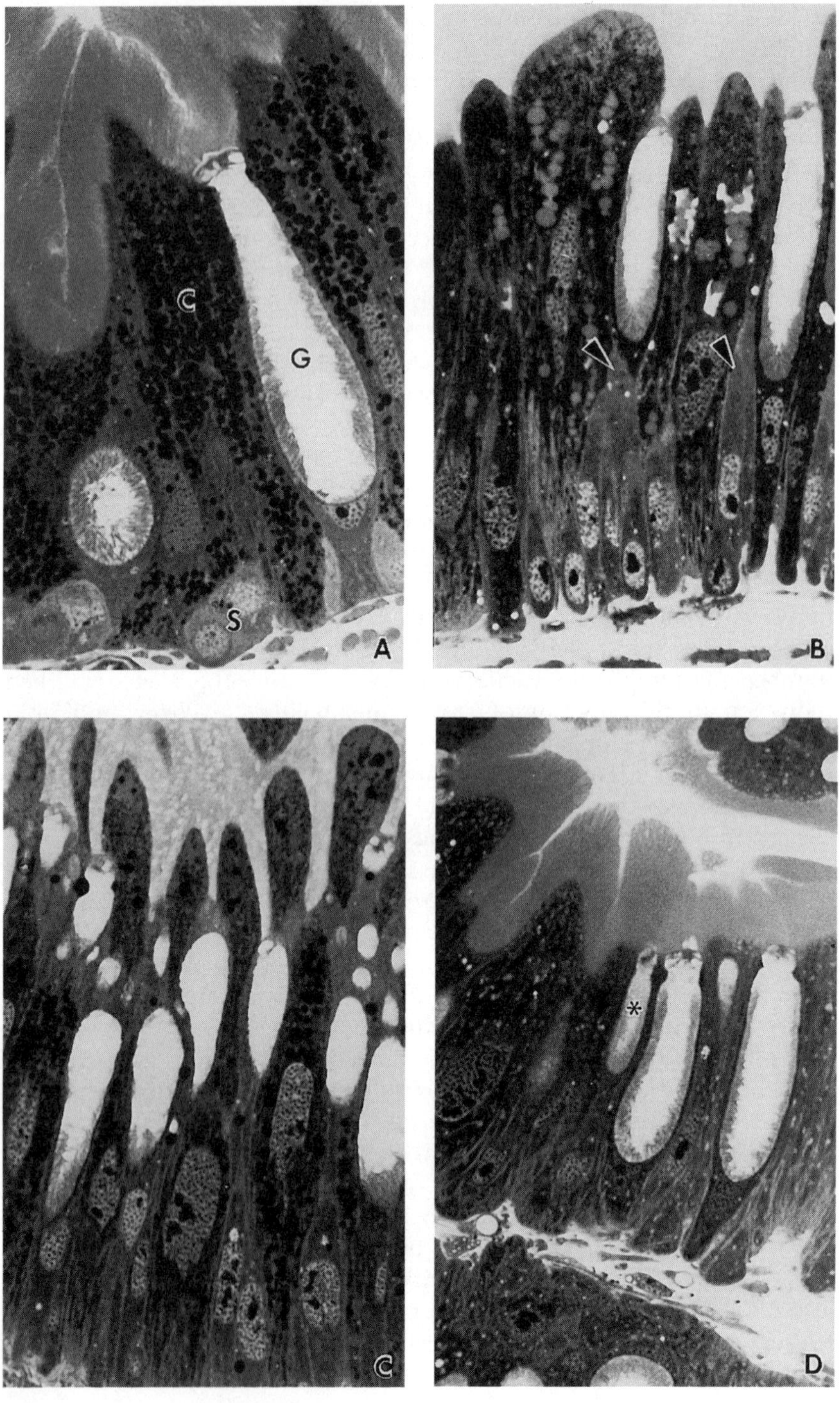
C
G
S
A
B
*
C
D

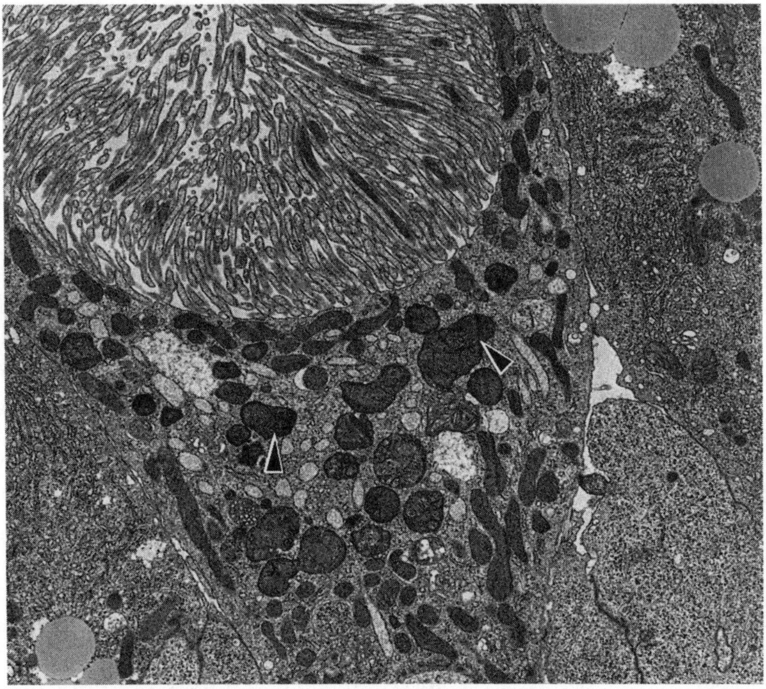

Figure 2.7 At mid moult, large numbers of autophagic vacuoles are present in a mature goblet cell (arrowheads). Similar structures are present in the apices of the mature columnar cells also (not shown) (× 6880).

any cell degeneration at a time when the cell number must increase rapidly would not be to the advantage of the larvae.

For reasons that are not clear, it appears that the mature cells increase in ploidy at the time the stem cells are proliferating. We observed considerable uptake of [^{3}H]thymidine by mature cells in the moult

---

Figure 2.6 Changes in the midgut epithelium of *Manduca* during moulting from the fourth to the fifth larval instar. A, Early in the moult, the midgut epithelium consists mostly of columnar (C) and goblet (G) cells, but stem cells (S) are beginning to proliferate. B, At mid moult, the stem cells (arrowheads) are elongating and differentiating. C, This slightly oblique section of a late moult larva exaggerates the height of the epithelium, but shows the apical blebs formed by the mature columnar cells. Differentiating cells are now tall enough to reach the lumen and are intercalated among the mature cells. Note that the diameter of the mature cells is reduced when compared to the early moult stages. D, In a newly moulted fifth instar larva, the new goblets (*) are still smaller than the mature goblets. The whole epithelium has more, thinner cells when compared to the premoult appearance (× 790).

period, but never saw them divide (Baldwin and Hakim, 1991b). It is possible that the higher larval instars have a midgut epithelium which consists of cells with a wide range in ploidy. If true, this would explain the wide range of nuclear sizes observed in the differentiated epithelial cells.

2.3.1 Pattern formation and cell communication in larval *Manduca* midgut

As soon as the two differentiated cell types can be distinguished in *Manduca* embryos, they exhibit a distinctive pattern: each goblet cell is surrounded by four to six columnar cells (Hakim *et al.*, 1988). When the epithelium is expanded at each larval moult, this pattern is maintained, but with an apparently random mix of newly differentiated cells and mature cells (Figure 2.8). This type of pattern, in which single cells of one type (goblet cells in this case) are surrounded by a continuous reticulum of cells of another type (columnar cells) is also seen in the development of bristle cells in *Drosophila* (Heitzler and Simpson, 1991). It is generally believed to occur when all cells are programmed for the single cell fate (e.g. goblet cell); once some cells are determined to be goblet cells, they then signal their immediate neighbours to follow an alternative fate pathway (as columnar cells) (reviewed by Greenwald and Rubin, 1992). If this is the correct mechanism for pattern formation in the *Manduca* larval midgut, the signals must come from mature goblet cells as well as newly differentiating ones, since the expanded larval pattern holds despite the mix of old and new cells.

We are interested in determining whether gap junctional communication could play a role in such signalling between cells. Non-moulting larval midgut epithelium is unusual in that the cells are not dye-coupled, even though gap junctions are present (Baldwin *et al.*, 1993). Lucifer yellow CH (LY) remains in the injected cell and does not pass to any of its neighbours (Figure 2.9A). This remains true through the early moult period, except that small groups of small cells which were dye coupled were occasionally seen among the non-dye-coupled cells (Figure 2.9B). These small cells appeared to be proliferating stem cells. At mid moult, at the time the new cells were differentiating, the whole epithelium became dye coupled; not only were the stem cells communicating, but the mature cells were also (Figure 2.9C). Late in the moult, the whole epithelium reverted to its non-communicating state, typical of the non-moulting stages. This means that the epithelium has a short period of gap junctional communication at the time the new cells are differentiating and the pattern is being formed in the enlarging epithelium. We concluded that even though gap junctional communication is restricted much of the time in this epithelium, it could play a role in pattern formation during moulting (Baldwin *et al.*, 1993).

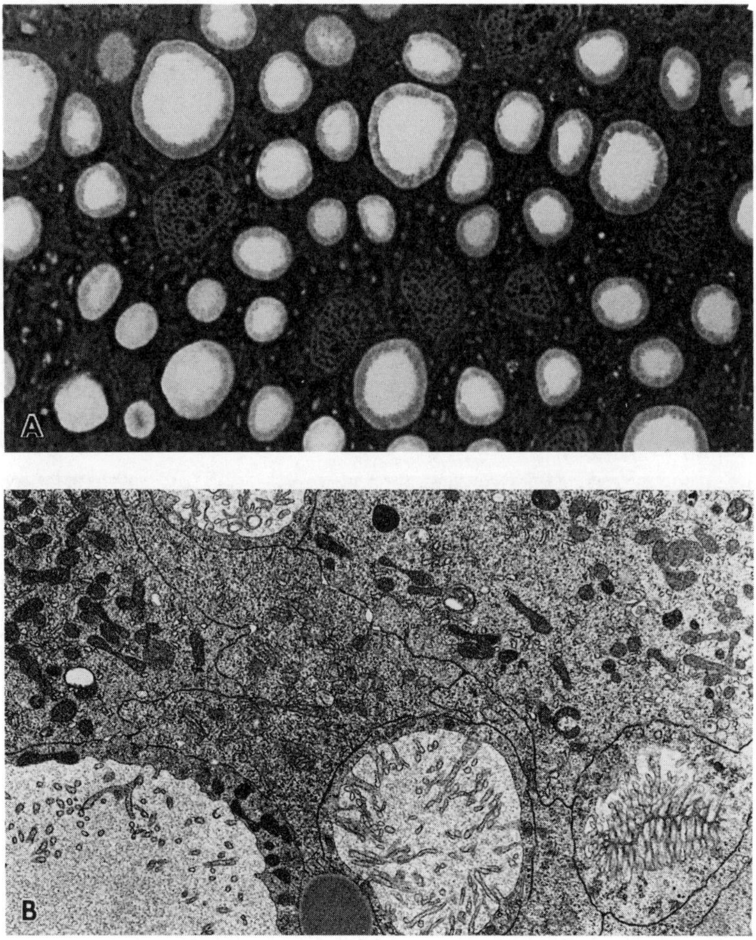

Figure 2.8 A, Horizontal section of a newly moulted fifth instar larva. At low magnification the random mix of new (small) and mature (large) goblet cells is apparent. B, Horizontal section of a late moult larva at higher magnification. New and mature goblet cells are surrounded by a reticulum of columnar cells so that no two goblet cells touch (A, × 890; B, × 6550).

Other types of cell communication also appear to be involved in cell patterning. In *Drosophila* there is evidence that cell signalling during pattern formation is dependent on the presence of membrane proteins which are expressed on the cell surface. Mutations in the *Notch* or *Delta* genes lead to incorrect patterning in proneural clusters (Heitzler and Simpson, 1991). The products of these genes are transmembrane proteins which are believed to take part in the signalling process.

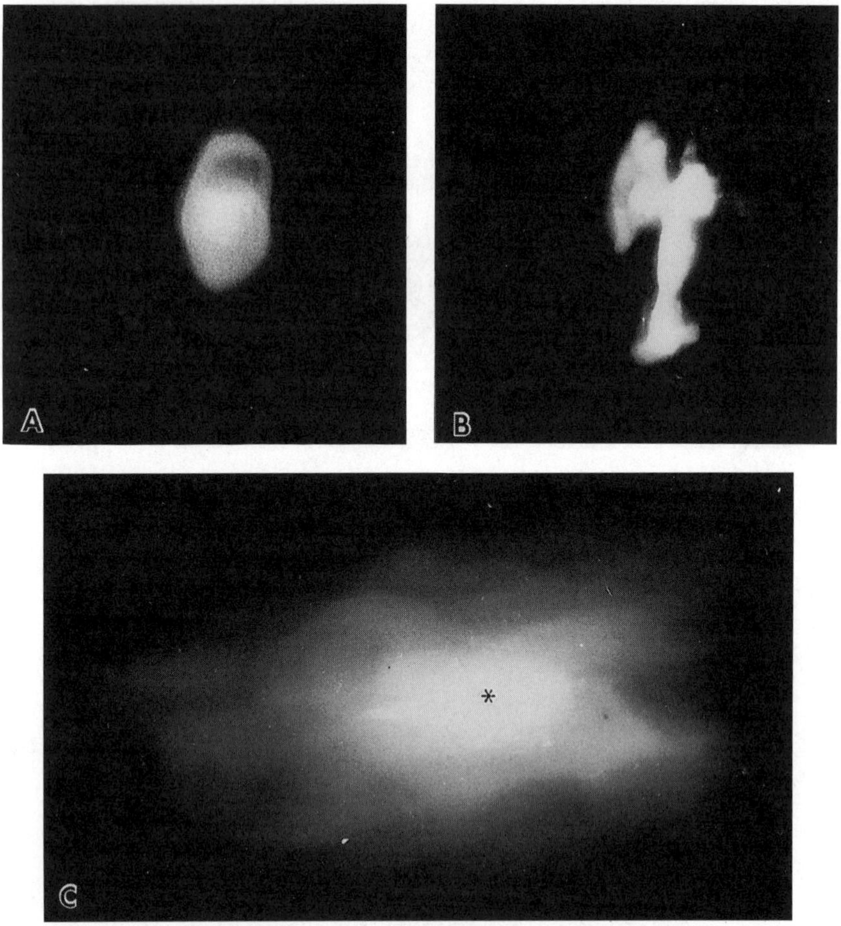

Figure 2.9 Whole mounts of *Manduca* midgut injected with LY and observed by fluorescence microscopy. A, Non-moulting fourth instar larva. Dye remains in the injected cell and does not spread to adjacent cells. B, In early moult larvae, small groups of small, dye-coupled cells were sometimes observed. However, most cells were not dye-coupled (as in A). C, At mid moult, most cells were extensively dye coupled, as indicated by spread of LY from the injected cell (*) to many adjacent cells. In the late moult period, cells returned to the non-dye-coupled state (as in A) (× 567).

2.4 METAMORPHOSIS

There have been few published studies of changes in the midgut at metamorphosis, and most of those have been of lepidoptera. All reports show a common pattern (Figure 2.10); the epithelium present in the last

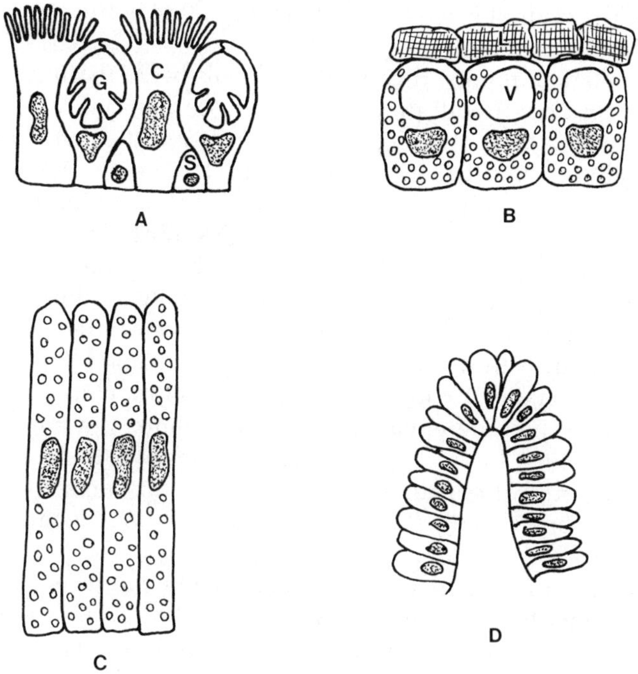

Figure 2.10 Diagram illustrating metamorphosis in the midgut epithelium of a lepidopteran. A, The larval epithelium consists mainly of columnar (C) and goblet (G) cells, with a few stem (S) cells located basally. B, In the prepupal stage, the stem cells have proliferated and differentiated to form a complete layer of large cells basally, while the remaining larval cells (L) are degenerating in an apical location. The prepupal cells have a large apical vacuole (V) which contains lysozyme and the cytoplasm is filled with mineral-containing spherites. C, In the pupa, the degenerating larval cells have been shed into the lumen and the contents of the apical vacuole secreted. The pupal epithelial cells have a tall, thin columnar shape. D, The adult epithelium consists of small cells in a highly folded arrangement. These are apparently the same cells as the pupal cells, but have shed cytoplasm into the lumen to reduce their size. The spherites are no longer present. (Redrawn after Judy and Gilbert, 1970; Waku and Sumimoto, 1971; Russell and Dunn, 1991.)

larval instar degenerates and is sloughed off into the lumen along with the peritrophic membrane. A completely new pupal epithelium is formed from the stem cells prior to the larval/pupal moult. In the adult, the epithelial cells are reduced in size, but are apparently the same cells as in the pupal epithelium (Waku and Sumimoto, 1971). Drastic changes in the structure of the midgut epithelium at metamorphosis are to be expected; not only does the insect's feeding behaviour change, but also the size of the midgut is reduced dramatically, first in the pupa and then even more in the adult.

As in the larval moults, the pupal stem cells begin to proliferate prior to the larval/pupal moult; they then form a continuous layer of rather large cells at the base of the epithelium (Judy and Gilbert, 1970; Waku and Sumimoto, 1971; Russell and Dunn, 1991). The degenerating larval cells are now completely separated from the basal lamina and accumulate large numbers of autophagic vacuoles (Radford and Misch, 1971; Kömüves *et al.*, 1985). They also form apical blebs which are shed into the lumen (Judy and Gilbert, 1970). It is generally assumed that these apical blebs and autophagic vacuoles are part of the degeneration process, but since they are also seen in non-degenerating larval cells (Baldwin and Hakim, 1991a), it seems likely that they provide a supply of nutrients to the differentiating pupal epithelium at a time when the insect is not eating. Interestingly, the appearance of autophagic vacuoles can be triggered prematurely by injecting the animals with 20-hydroxyecdysone (20HE) (Radford and Misch, 1971; Kömüves *et al.*, 1985).

The developing pupal cells accumulate large numbers of spherical granules (spherites) (Waku and Sumimoto, 1971, 1974; Russell and Dunn, 1991) which are composed predominantly of magnesium phosphate (Miya, 1976; Dow, 1986). These spherites begin to accumulate in the prepupal period and occupy much of the cytoplasm of the pupal epithelial cells. Their contents are secreted late in the pupal stage and are sometimes seen in adult epithelial cells (Waku and Sumimoto, 1971, 1974). Turbeck (1974) reported sometimes seeing spherites in stem cells prior to larval moults in various lepidoptera raised on natural diets. We did not see them in *Manduca* embryos or larvae raised on an artificial diet. Waku and Sumimoto (1971) reported that spherites did not appear until the prepupal period in *Bombyx mori* raised on a natural diet, but Miya (1976) saw them in differentiating midgut epithelial cells of *Bombyx* embryos.

At the time the prepupal cells are accumulating the mineral-containing spherites, they also are developing a very large secretion vacuole (Judy and Gilbert, 1970; Russell and Dunn, 1991). This vacuole is first seen in the basal part of the cell; it then moves apically and is finally secreted into the lumen at the larval/pupal ecdysis. This vacuole contains lysozyme and plays a protective antibacterial role, necessary after the loss of the peritrophic membrane (Russell and Dunn, 1991).

The mechanism of shortening the midgut in the pupa and adult has not been studied, but the change in shape of the epithelial cells during these periods certainly would accommodate this shortening. When first formed, the prepupal epithelium consists of large cuboidal cells (Figure 2.10B); these later become tall, thin columnar cells (Judy and Gilbert,

1970; Waku and Sumimoto, 1971) which reduces the surface area of the epithelium (Figure 2.10C). In the adult, the epithelial cells become even smaller, apparently by shedding cytoplasm into the lumen by an apocrine secretory mechanism (Figure 2.10D). As some of the shed cytoplasmic vesicles were nucleated, there must be loss of whole cells during this period as well (Judy and Gilbert, 1970). The adult midgut epithelium finally becomes more highly folded than in earlier stages.

2.5 CELL CULTURE STUDIES

Developmental studies involving chemical or physical manipulations of the midgut are difficult to carry out *in vivo*. Therefore, methods have been developed for growing primary midgut cultures, and highly enriched stem cell cultures *in vitro* (Baines *et al.*, 1994). *Manduca* midgut epithelial cells from pharate fourth instar larvae (Sadrud Din *et al.*, 1994, 1996), and from wandering (mid) and prepupal (late) fifth instar larvae have been successfully cultured (Figure 2.11) (Loeb and Hakim, unpublished). To grow these cultures it was necessary to add fat-body derived growth factors and the insect metamorphosis hormone, 20HE. Cells were maintained as mixed cultures containing a distribution of stem, developing and differentiated cells, or as highly enriched stem cell cultures. The cells were identified by their resemblance to their *in vivo* counterparts (Figures 2.10 and 2.11). Cultures have been maintained and subcultured for periods of three to six months.

The cultures derived from pharate fourth instar larvae confirmed that midgut growth occurs through the proliferation and differentiation of stem cells. The stem cells in mixed cultures divided, and yet the total number of stem cells in each culture did not increase exponentially, as in typical cell cultures. Instead, the number of stem cells increased and decreased in a cyclic fashion; the number of goblet and columnar cells increased concomitant with a decrease in stem cell number. At least three cycles were observed over a period of 60 days. Thus, while stem cells proliferated, some of them also differentiated to columnar and goblet cells (Sadrud-Din *et al.*, 1994). Proliferating stem cells take up the thymidine analogue, bromodeoxyuridine (BrdU) (Loeb and Hakim, unpublished). Interestingly, when stem cells divided and the two daughter cells remained attached, usually one daughter took up BrdU and the other did not, suggesting that one daughter was continuing to proliferate while the other was committed to differentiation (Loeb and Hakim, unpublished). This behaviour is typical of stem cell populations (Burgess and Nicola, 1983).

Cultures which are highly enriched in stem cells (>99%) have been developed using pharate fourth instar and wandering fifth instar larval

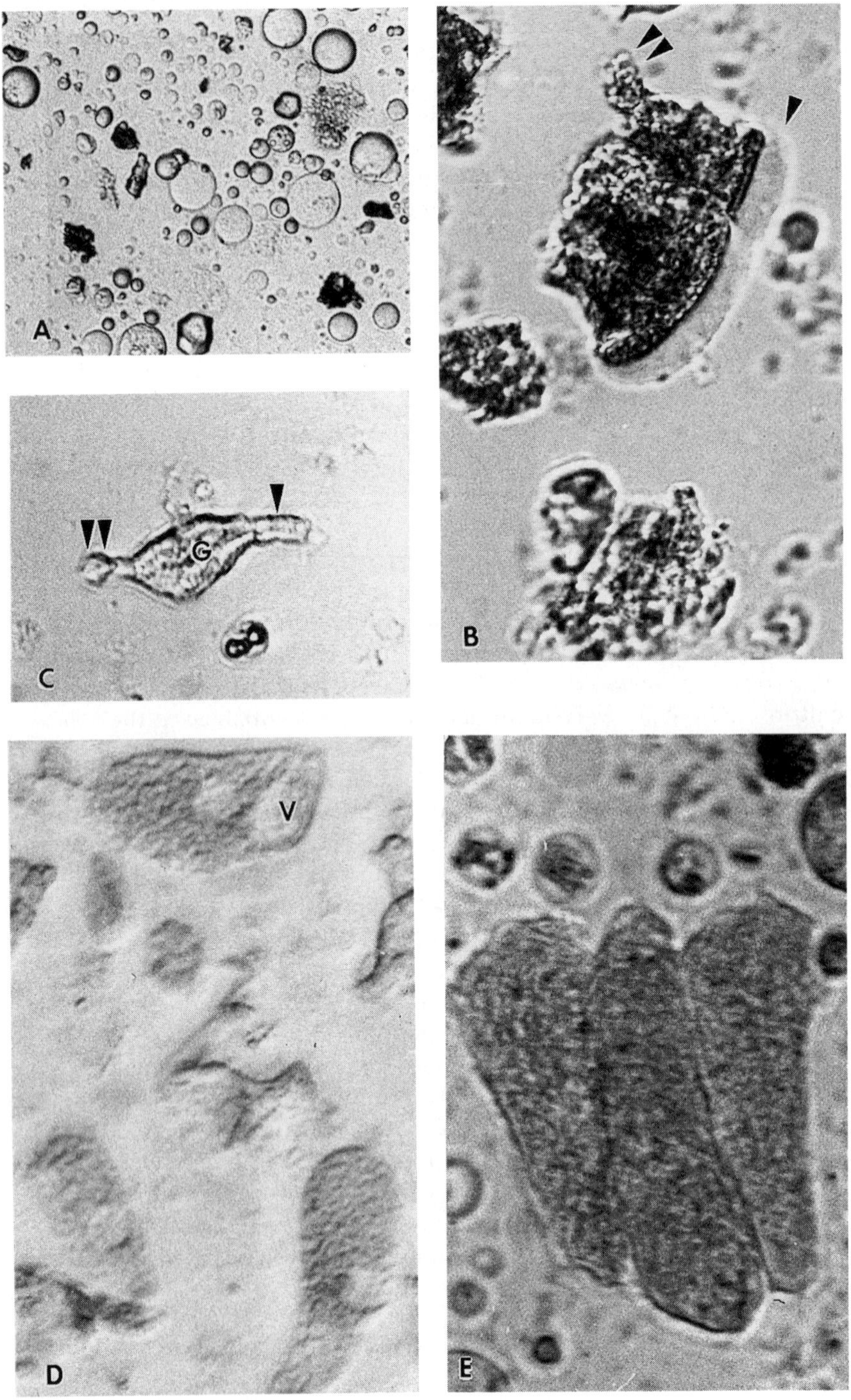
A
B
C
G
D
V
E

midgut epithelium (Sadrud-Din *et al.*, 1996; Loeb and Hakim, unpublished). As in mixed cultures, stem cells in the highly enriched stem cell cultures proliferated in the presence of 20HE and either fat body or cell-free fat body extract, but almost no differentiation was observed. However, when cell-free medium from established mixed pharate fourth instar cultures was added to the stem cell cultures, differentiation of stem cells into columnar and goblet cells did occur. Thus, the 'conditioned medium' contained differentiation factor(s) produced by mature goblet and/or columnar cells. The factor(s) is a small, heat-stable peptide-like material with a molecular mass of 10 kDa or less (Sadrud-Din *et al.*, 1996). Recent experiments have shown that 'conditioned medium' obtained from wandering fifth instar midgut cell cultures caused differentiation of some of the pharate fourth instar stem cells into cells typical of wandering fifth instar midgut cells (Figure 2.11D) (Hakim and Loeb, unpublished). Thus, differentiation factors derived from different developmental stages can direct the development of stem cells to form the corresponding stage-specific, mature cell types (Loeb and Hakim, unpublished). This suggests that *in vivo* differentiation of stem cells may be both initiated and directed by mature cells in the surrounding epithelium.

Midgut stem cells are somewhat different from stem cells of vertebrate epithelia, in that the cells from *Manduca sexta* did not attach to laminin, collagen, collagen plus poly-L-lysine or fibronectin substrates (Loeb, unpublished). In fact, cell proliferation on laminin-coated dishes was lower than that in control cultures (Sadrud-Din *et al.*, 1994). In addition, polarized goblet and columnar cells differentiated in culture in the absence of fixed attachment to the culture dish, without the presence of

---

Figure 2.11 *Manduca* midgut epithelial cells in culture. A–C, Cells cultured from moulting pharate fourth instar larvae. Stem cells (A) have a large nucleus (difficult to see in this phase-contrast image) and a clear cytoplasmic rim. They come in a range of sizes; when collected from caterpillars just entering the moult, the largest cells predominate. These cells will transform to goblet and columnar cells under the proper conditions (see text). B, A pair of columnar cells; note the prominent brush border (arrowhead) on these cells and the expanded apical surface. The basal knob (at double arrowhead) is common *in vitro*, but is not seen *in vivo*. C, A goblet cell (G) with a narrow apical neck (arrowhead). *In vitro*, it is common for large goblet cells to have a knoblike constriction (double arrowhead) at its basal surface as well; this does not appear *in vivo*. D, Cells cultured from wandering (mid) fifth instar larvae are low columnar to pyramidal in shape. These cells have a granular cytoplasm and a prominent apical vacuole (V). E, Cells taken for culture one day prior to the larval/pupal moult have a tall columnar shape, granular cytoplasm and no apical vacuole (× 392).

a distinct basal lamina and often without attachment to other cells. This ability to form fully mature polarized cells in the absence of attachment to a substrate or other cells is unusual among epithelial cells (Klein *et al.*, 1988; Rodriguez-Boulan and Nelson, 1989) and suggests that development of polarity is an intrinsic property of these cells.

REFERENCES

Anderson, E. and Harvey, W.R. (1966) Active transport by the *Cecropia* midgut. II. Fine structure of the midgut epithelium. *J. Cell Biol.*, **31**, 107–34.

Baines, D., Brownwright, A. and Schwartz, J.L. (1994) Establishment of primary and continuous cultures of epithelial cells from larval lepidopteran midguts. *J. Insect Physiol.*, **40**, 347–57.

Baldwin, K.M. and Hakim, R.S. (1987) Change of form of septate and gap junctions during development of the insect midgut. *Tissue Cell*, **19**, 549–58.

Baldwin, K.M. and Hakim, R.S. (1991a) Growth and differentiation of the larval midgut epithelium during molting in the moth, *Manduca sexta*. *Tissue Cell*, **23**, 411–22.

Baldwin, K.M. and Hakim, R.S. (1991b) Cell proliferation and maintenance of pattern during molting in midgut of Tobacco Hornworm. *Anat. Rec.*, **229**, 7A–8A.

Baldwin, K.M., Hakim, R.S. and Stanton, G.S. (1993) Cell–cell communication correlates with pattern formation in molting *Manduca* midgut epithelium. *Dev. Dynam.*, **197**, 239–43.

Bellen, H.J., O'Kane, C.J., Wilson, C. *et al.* (1989) P-element-mediated enhancer detection: a versatile method to study development in *Drosophila*. *Genes Dev.*, **3**, 1288–300.

Brand, M., Jarman, A.P., Jan, L.Y. and Jan, Y.N. (1993) *asense* is a *Drosophila* neural precursor gene and is capable of initiating sense organ formation. *Development*, **119**, 1–17.

Burgess, A. and Nicola, N. (1983) *Growth Factors and Stem Cells*, Academic Press, New York, pp. 1–22.

Dow, J.A.T. (1986) Insect midgut function. *Adv. Insect Physiol.*, **19**, 187–328.

Greenwald, I. and Rubin, G.M. (1992) Making a difference: the role of cell–cell interactions in establishing separate identities for equivalent cells. *Cell*, **68**, 271–81.

Hakim, R.S., Baldwin, K.M. and Bayer, P.E. (1988) Cell differentiation in the embryonic midgut of the tobacco hornworm, *Manduca sexta*. *Tissue Cell*, **20**, 51–62.

Hartenstein, A.Y., Rugendorff, A., Tepass, U. and Hartenstein, V. (1992) The function of the neurogenic genes during epithelial development in the *Drosophila* embryo. *Development*, **116**, 1203–20.

Hartenstein, V. and Jan, Y.N. (1992) Studying *Drosophila* embryogenesis with P-*lacZ* enhancer trap lines. *Roux's Arch. Dev. Biol.*, **201**, 194–220.

Heitzler, P. and Simpson, P. (1991) The choice of cell fate in the epidermis of *Drosophila*. *Cell*, **64**, 1083–92.

Humbert, W. (1979) The midgut of *Tomocerus minor* Lubbock (Insecta, Collembola): ultrastructure, cytochemistry, ageing and renewal during a moulting cycle. *Cell Tissue Res.*, **196**, 39–57.

Judy, K.J. and Gilbert, L.I. (1970) Histology of the alimentary canal during the metamorphosis of *Hyalophora cecropia* (L.). *J. Morphol.*, **131**, 277–300.

Kadiri, Z. and Louvet, J.-P. (1982) Ultrastructural study of midgut embryonic cytodifferentiation in the phasmid *Clitumnus extradentatus*. *Br. J. Morphol.*, **172**, 323–34.

Klein, G., Langegger, M., Timpl, R. and Ekblom, P. (1988) Role of laminin A chain in the development of epithelial cell polarity. *Cell*, **55**, 331–41.

Kobayashi, Y. and Ando, H. (1983) Embryonic development of the alimentary canal and ectodermal derivatives in the primitive moth, *Neomicropteryx nipponensis* Issiki (Lepidoptera, Micropterygidae). *J. Morphol.*, **176**, 289–314.

Kömüves, L.G., Sass, M. and Kovács, J. (1985) Autophagocytosis in the larval midgut cells of *Pieris brassicae* during metamorphosis. *Cell Tissue Res.*, **240**, 215–21.

Machida, R. and Ando, H. (1981) Formation of midgut epithelium in the jumping bristletail *Pedetontus unimaculatus* Machida (Archaeognatha: Machidae). *Int. J. Insect Morphol. Embryol.*, **10**, 297–308.

Miya, K. (1976) Ultrastructural changes of embryonic cells during organogenesis in the silkworm, *Bombyx mori*. II. The alimentary canal and the Malpighian tubules. *J. Iwate Daigaku. Nogaku-bu*, **13**, 95–122.

Mori, H. (1983) Origin, development, morphology, functions and phylogeny of the embryonic midgut epithelium in insects. *Entomol. General.*, **8**, 135–54.

Radford, S.V. and Misch, D.W. (1971) The cytological effect of ecdysterone on the midgut cells of the flesh-fly *Sarcophaga bullata*. *J. Cell Biol.*, **49**, 702–11.

Reuter, R., Grunewald, B. and Leptin, M. (1993) A role for the mesoderm in endodermal migration and morphogenesis in *Drosophila*. *Development*, **119**, 1135–45.

Rodriguez-Boulan, E. and Nelson, W.J. (1989) Morphogenesis of the polarized epithelial cell phenotype. *Science*, **245**, 718–25.

Russell, V.W. and Dunn, P.E. (1991) Lysozyme in the midgut of *Manduca sexta* during insect metamorphosis. *Arch. Insect Biochem. Physiol.*, **17**, 67–80.

Sadrud-Din, S.Y., Hakim, R.S. and Loeb, M.J. (1994) Proliferation and differentiation of midgut epithelial cells from *Manduca sexta, in vitro*. *Invert. Reprod. Dev.*, **26**, 197–204.

Sadrud-Din, S., Loeb, M.J. and Hakim, R.S. (1996) *In vitro* differentiation of isolated stem cells from the midgut of *Manduca sexta* larvae. *J. Exp. Biol.*, **199**, 319–25.

Schwalm, F. (1988) Insect morphogenesis. *Monogr. Dev. Biol.*, **20**, 196–201.

Schweikl, H., Klein, U., Schindlbeck, M. and Wieczorek, H. (1989) A vacuolar-type ATPase, partially purified from potassium transporting plasma membranes of tobacco hornworm midgut. *J. Biol. Chem.*, **264**, 11136–42.

Skaer, H. (1993) The alimentary canal, in *The Development of* Drosophila melanogaster Vol. II (eds M. Bate and A. Martinez Arias), Cold Spring Harbor Laboratory Press, Plainview, NY, pp. 941–1012.

Technau, G.M. and Campos-Ortega, J.A. (1986) Lineage analysis of transplanted individual cells in embryos of *Drosophila melanogaster*. III. Commitment and proliferative capabilities of pole cells and midgut progenitors. *Roux's Arch. Dev. Biol.*, **195**, 489–98.

Turbeck, B.O. (1974) A study of the concentrically laminated concretions, 'spherites', in the regenerative cells of the midgut of lepidopterous larvae. *Tissue Cell*, **6**, 627–40.

van der Starr-van der Molen, L.G., Planque-Huidekiper, B. and de Priester, W. (1973) Embryogenesis of *Calliphora erythrocephala*. III. Ultrastructure of the midgut epithelial cells during late embryonic development. *Z. Zellforsch.*, **144**, 117–38.

Waku, Y. and Sumimoto, K.I. (1971) Metamorphosis of midgut epithelial cells in the silkworm (*Bombyx mori* L.) with special regard to the calcium salt deposits in the cytoplasm. I. Light microscopy. *Tissue Cell*, **3**, 127–36.

Waku, Y. and Sumimoto, K.I. (1974) Metamorphosis of midgut epithelial cells in the silkworm (*Bombyx mori* L.) with special regard to the calcium salt deposits in the cytoplasm. II. Electron microscopy. *Tissue Cell*, **6**, 127–36.

Werner, K., Moutairou, I. and Werner, G. (1991) Formation and structure of the surface coat in the midgut of a waterstrider, *Gerris najas* Deg. (Heteroptera: gerridae). *Int. J. Insect Morphol. Embryol.*, **20**, 69–77.

3

Midgut endocrine cells

F. Sehnal and D. Žitňan

3.1 EVOLUTIONARY ASPECTS

3.1.1 Gastrointestinal hormones and their sources: a lesson from the research in mammals

At the beginning of this century, Bayliss and Starling (1902) described a product of intestinal mucosa that stimulated secretion of pancreatic juice. The product was named secretin and its discovery led to the concept of hormones as circulating regulators derived from specific endocrine cells or glands. Several other mammalian gastrointestinal hormones were subsequently identified as extracts eliciting certain physiological effects. The peptidic nature of these hormones was soon recognized, but their amino acid sequence could be determined only in the last 30 years. Today we know structures of nearly three dozen hormones produced in the digestive tract of mammals. Most of them are straight-chain peptides consisting of less than 40 amino acids, few are biogenic amines. Similar compounds were detected in the gut wall of all Chordata (Rawdon and Andrew, 1990). Immunohistochemical techniques revealed that they are produced in the neurons of neural plexuses located chiefly in the gut musculature and/or in the endocrine cells that are scattered in gut mucosa (McGuigan, 1968).

Electron microscopy techniques showed that the gastrointestinal endocrine cells (GEC) are distinguished from other cells of the digestive tract by the presence of peptidergic secretory granules (150–450 nm in

Biology of the Insect Midgut.
Edited by M.J. Lehane and P.F. Billingsley.
Published in 1996 by Chapman & Hall, London.
ISBN 0 412 61670 X.

diameter). There are two morphologically distinct types of GEC. The apical end of the 'open type' GEC reaches the gut lumen and is folded into a brush border, whereas the 'closed type' GEC are embeded in the epithelium and lack the luminar contact. Both cell types sometimes contain one or more basal cytoplasmic processes, which are indicative of paracrine function (secretion acting on the neighbouring cells). The digestive tract of humans contains at least 18 kinds of GEC differing in their hormonal secretions (Desbuquois, 1990). Coexistence of several peptides, biogenic amines, and/or peptides and amines in a single GEC or an enteric neuron is common. Most GEC do not contain amines but all appear to possess enzymes for the decarboxylation of neuroamine precursors. They also contain a neuron-specific isoform of the glycolytic enzyme enolase.

The histochemical properties of GEC encouraged speculation that they are of neuroectodermal origin. According to the APUD (amine precursor uptake and decarboxylation) theory, GEC are immigrants from the neural plate (Pearse, 1969). The term paraneuron was coined to embrace all cells producing regulators similar to those in neurons, and to accentuate the neuroectodermal origin of these cells. 'Paraneurons' are found in various organs, occurring in largest numbers and variety in the digestive tract as GEC (Fujita *et al.*, 1988). For GEC it was proven, however, that they originate from the same precursor cells as the digestive epithelium and are thus of endodermal origin (Fontaine and LeDouarin, 1977; Andrew, 1981).

The enteric nervous system (ENS) and GEC are interlinked and jointly control gut movements, production of digestive fluids, rate of replacement of the gut epithelium and blood flow to the gut (Dockray, 1988). According to Fujita *et al.* (1988), GEC function as primary sensors, those of the open type registering the nutrient contents of the gut, and those of the closed type complementing nervous perception of the gut wall tension. Hormones are liberated upon direct sensory stimulation, or in response to circulating regulators or nervous stimuli. Most hormones probably act in the immediate vicinity of the releasing sites. These paracrine secretions act on the adjacent muscle cells, digestive cells, and apparently also on the nerve termini in the subepithelial nervous reticulum, by which the humoral signal is transduced into nervous stimuli and eventually causes changes in the nervously controlled functions, including behaviour. There is evidence that gastrin, somatostatin and secretin are also released into the gut lumen and possibly act on the apical site of the digestive cells. Finally, some GEC products enter the body circulation and exert hormonal effects on distal targets. Owing to the various routes of action, a single hormone can exert a variety of effects. For example, cholecystokinin controls gall bladder contractions, pancreatic enzyme secretion and gastric emptying, and inhibits food intake. A variety of the effects of gastrointestinal peptides is due to their action on the central nervous system (CNS); this concerns the control of appetite, certain metabolic functions, locomotor activity, nociception, and others (Dockray, 1988).

Some gastrointestinal hormones, such as gastrin, cholecystokinin, and vasoactive intestinal peptide, also occur in considerable amounts in the CNS, whereas compounds known originally from the CNS (somatostatin, neurotensin, substance P, enkephalins) have since been identified in GEC (Desbuquois, 1990). Peptidic hormones that are produced within the digestive tract and in the CNS received the name brain/gut peptides (Fujita *et al.*, 1981).

3.1.2 Phylogenetic conservation of brain/gut peptides

Ultrastructural studies on the digestive tract of various invertebrates invariably led to the discoveries of cells containing peptidergic granules. In the coelenterates, all such cells, including those embeded in the endoderm and opened into the digestive cavity, are classified as neurons (Westfall 1973). They are devoid of acetylcholine, catecholamines and serotonin but contain a variety of peptides related to FMRFamide (Grimmelikhuijzen *et al.*, 1989). In the plathelminthes (Reuter and Gustafsson, 1989) and more advanced triblastic animals, it is possible to distinguish enteric neurons from GEC. Both these cell types can be revealed with antisera to mammalian brain/gut peptides (Dockray, 1979; Fujita *et al.*, 1981). A particularly large array of antigens related to mammalian hormones has been recognized in annelids (Al-Yousuf, 1990). Research on insects confirmed that such antigens occur both in the CNS (El-Salhy *et al.*, 1980) and in the ENS and GEC (Iwanaga *et al.*, 1981). Most families of the peptidic hormones of mammals seem to be represented in the insects (De Loof, 1987; Sehnal and Žitňan, 1990).

Two of the known types of brain/gut peptides, the FMRFamide and proctolin, were first identified in the invertebrates. Homologues of the molluscan cardioactive peptide FMRFamide (Price and Greenberg, 1977) were later found in various animals from anemones to mammals (Greenberg *et al.*, 1988). Demonstration of the hormonal function of this class of peptides in mammals (Raffa, 1988) brought further evidence for the conservation of functional brain/gut peptides in animal evolution. Proctolin was isolated from cockroaches (Brown and Starratt, 1975) and its occurrence in other arthropods was proven (Orchard *et al.*, 1989), but only one preliminary study reported its immunohistochemical detection in mammals, notably in rat brain (Holets *et al.*, 1984).

3.2 INNERVATION AND ENDOCRINE CELLS OF INSECT MIDGUT

3.2.1 General morphology

Axonal processes of the ENS terminate at numerous sites in the foregut, midgut and hindgut musculature; most of these sites are probably neuromuscular junctions, others function as neurohaemal organs. ENS may be subdivided into the anterior stomatogastric and the posterior caudal nervous systems that are each connected to CNS (Kirby *et al.*, 1984; Penzlin, 1985). A pair of frontal connectives extends from the brain to the frontal ganglion, from which a single recurrent nerve runs to the distal stomatogastric ganglia. These usually include the unpaired hypocerebral, paired or unpaired ingluvial, and paired proventricular ganglia, plus a various number of loose neurons. Arrangement of the stomatogastric ganglia is considerably modified in some insect orders or families. The frontal ganglion is well developed in nearly all insects, whereas the hypocerebral ganglion is in some groups rudimentary. The ingluvial and proventricular ganglia are in some species replaced by loose neurons. Gastric nerves stretching from the proventricular ganglia or a neural plexus at the foregut/midgut boundary innervate musculature of the entire midgut and often contain scattered neurons. The stomatogastric nervous system serves as a neurohaemal organ for some neurohormones produced in the CNS, e.g. the diuretic peptide corazonin (Cantera *et al.*, 1994).

Innervation of the hindgut is provided by the paired proctodeal nerves that begin in the last ganglion of the nerve cord, include intrinsic neurons, and branch in a variety of ways (Reinecke *et al.*, 1973; Kirby *et al.*, 1984). They run along the hindgut and merge with gastric nerves in a meshwork at the midgut/hindgut boundary (Žitňan *et al.*, 1995). Terminals of the proctodeal nerves contain neurohaemal areas for the liberation of several hormones produced in the CNS (Truman and Copenhaver, 1989; Chen *et al.*, 1994).

Ultrastructural studies identified in insect midgut small cells characterized by dense core granules (Hecker *et al.*, 1971; Kobayashi, 1971; de Priester, 1971; Cassier *et al.*, 1972). They were recognized as GEC only when immunohistochemical investigations demonstrated that such cells of cockroach midgut contain antigens related to mammalian gut peptides (Iwanaga *et al.*, 1981). Endo and Nishiitsutsuji-Uwo (1981) pointed out that insect GEC are also either of the 'open type', extending with their apical end into the gut lumen, or of the 'closed type', without a direct contact with the lumen, and that some of them also form basal cytoplasmic extensions. Ultrastructural (Cassier and Fain-Maurel, 1977; Nishiitsutsui-Uwo and Endo, 1981; Duve and Thorpe, 1982; Brown *et al.*, 1986) and immunohistochemical (Iwanaga *et al.*, 1981; Schoofs *et al.*,

1988) investigations revealed the existence of several types of GEC whose abundance and distribution are species- and stage-specific. Insect GEC are usually single, but in some taxa one finds them in pairs or small groups.

Axons supplying midgut muscles were also shown to contain a variety of neurosecretory granules (Nishiitsutsui-Uwo and Endo, 1981). Reaction with antibodies to the mammalian gut peptides was further detected in the neurons of ENS (Duve and Thorpe, 1982; Endo *et al.*, 1982). Subsequent investigations revealed that ENS and GEC share some, but not all antigens (Žitňan *et al.*, 1993). Most ENS neurons and GEC seem to produce more than one regulatory peptide.

3.2.2 Ontogeny of ENS and GEC

The embryonic origin of *Manduca sexta* ENS was described in detail by Copenhaver and Taghert (1989a,b, 1991). The stomatogastric nervous system differentiates from three ectodermal placodes on the dorsal foregut surface. Some neurons migrate posteriorly when the axons of gastric nerves grow along the major longitudinal muscle bands of the midgut. The ENS of other insects originates from similar placodes of the ectodermal gut sections (Penzlin, 1985). The midgut epithelium of arthropods is derived from the yolk cells that represent embryonic endoderm (Chapter 2). In Zygentoma and Odonata, terminal parts of ectodermal invaginations for the foregut and hindgut contribute a small number of cells to the midgut (Ivanova-Kazas, 1981), but since this occurs in just two insect orders it is unlikely that it is of general significance for the generation of GEC. Endo *et al.* (1983) showed beyond doubt that cockroach GEC descend from the dividing cells of regenerative midgut nidi. GEC are thus of endodermal origin, and probably differentiate from the same stem cells as the digestive epithelium. GEC seem to persist and function for varying lengths of time, new ones being continuously generated from the nidi (Endo *et al.*, 1990). GEC typically assume their endocrine function only outside the nidi and become scattered among the digestive cells. With a few exceptions described below, only very small and obviously still differentiating GEC can be detected within the nidi.

At metamorphosis of the waxmoth, *Galleria mellonella*, all GEC are destroyed when the larval midgut degenerates. They are sloughed off with larval midgut epithelium into the lumen of proliferating pupal midgut, but there they preserve their hormonal antigens until a new population of pupal GEC is differentiated (Žitňan *et al.*, 1993). Similarly, columnar and goblet cells degenerate in the starving caterpillars of *Manduca sexta*, but the GEC retain their integrity and antigenic properties even after one week of starvation (Žitňan *et al.*, 1995).

3.2.3 Comparison of ENS and GEC morphology in major insect taxa

(a) Archaeognatha (brittletails)

Examination of the brittletails confirmed their close relationship to the insects in *sensu stricto*. All examined species of the family Machilidae possess a short simple foregut and a baggy midgut with six caeca. The ingluvial ganglion is located at the foregut/midgut junction and contains 10–30 FRMFamide immunopositive cells. The ganglion sends off one dorsal and one ventral gastric nerve which arborize into numerous longitudinal and transverse branches over the caecal and other midgut surfaces. GEC were detected exclusively in the centre of regenerative nidi. Most of them are single but some are paired; if so, at least one of them is of the closed type. Paired endocrine cells were observed for the first time by Cassier *et al.* (1972) in the machilid *Petrobius maritimus*. Distribution of nidi containing GEC and the types of GEC are species specific. For example, both the open and closed GEC without paracrine processes were found in the first and last quarters of the midgut in *Machilis helleri*, whereas *Lepismachilis notata* contains two rings of FMRFamide-positive GEC only in the last quarter of the midgut; the proximal ring consists of closed-type amoeboid GEC and the distal ring accommodates open-type plus a few closed-type GEC.

(b) Zygentoma (silverfish)

The voluminous foregut of both examined species of Lepismatidae is as long as the midgut and consists of a large crop and a small proventriculus. Midgut caeca are well developed in *Thermobia domestica* but rudimentary in *Lepisma saccharina*. The oesophageal nerve with 2–5 FRMFamide-immunoreactive perikarya and numerous side branches runs into the ingluvial ganglion, in which 12–15 neurons were detected in *Thermobia* (Figure 3.1A) but not in *Lepisma*. Gastric nerves (11 pairs in *Thermobia* and 12 pairs in *Lepisma*) beginning in the ingluvial ganglion evenly circumvent the midgut and in its last three quarters anastomose into a network of fine nerves. GEC occur in the regenerative nidi throughout the midgut (Figure 3.1B). Up to four small closed-type GEC adhere to one central GEC, which may be open or closed. GEC in the caeca of *Thermobia* are usually single (exceptionally paired) and smaller than those in the remaining midgut. A double-staining technique distinguishing gastrin-like antigen(s) from those detected with antisera to FMRFamide, Arg-vasopressin and vasoactive intestinal peptide (VIP) was applied to *Lepisma*. Several kinds of GEC were identified: (a) single VIP-positive cells; (b) single gastrin-positive cells with or without an adjacent VIP-positive cell; (c) single or grouped cells reacting with antisera to FMRFamide and Arg-vasopressin. It is worth noting that

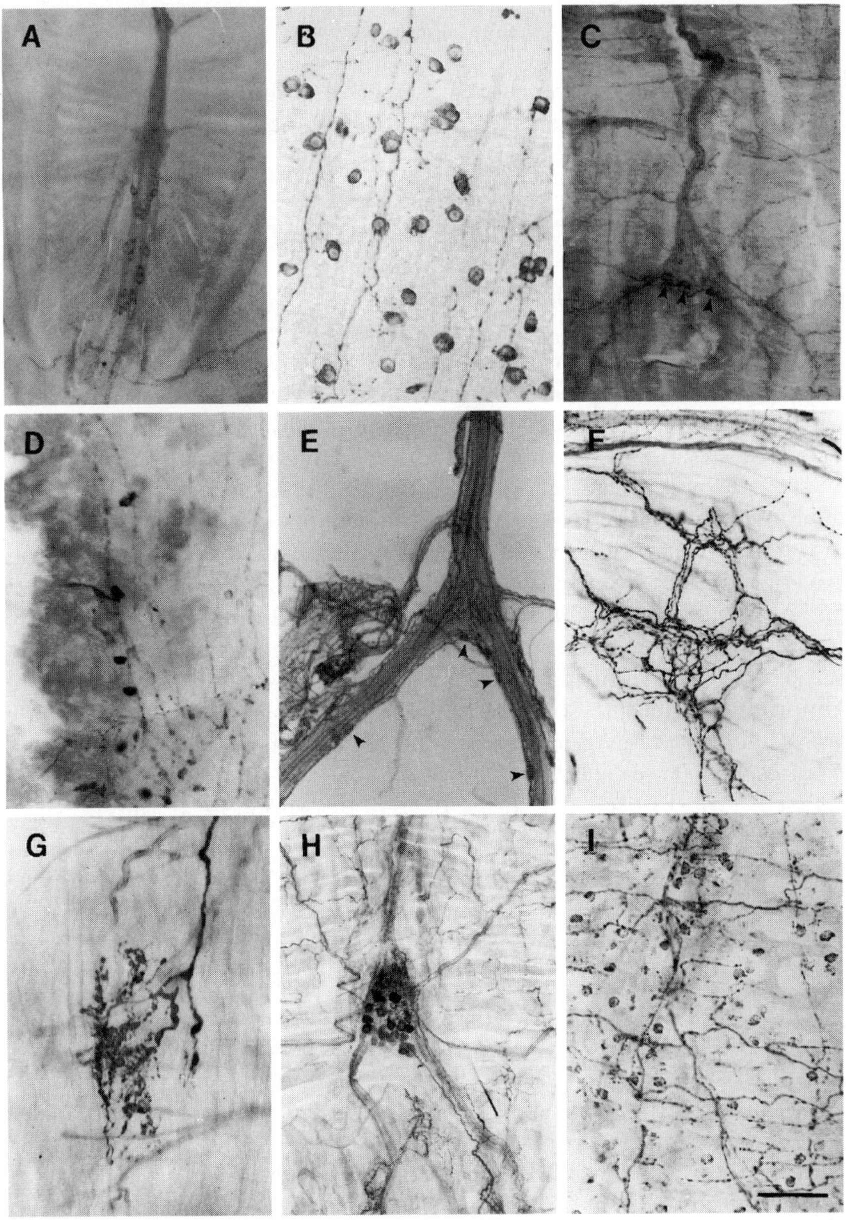

Figure 3.1 FMRFamide-immunoreactivity in the ENS and GEC of a firebrat, dragonfly and polyneopteran insects. A, ingluvial ganglion of *Thermobia* with 15 immunoreactive neurons; B, single, paired, and tripled GEC in *Thermobia;* C, neurons in the ingluvial ganglion and D, open and closed GEC in *Sympetrum*; E, branching point of the oesophageal and two ingluvial nerves with scattered neurons (arrowheads) and F, meshwork of axon terminals on the foregut of *Nauphoeta*; G, ENS terminals in *Nauphoeta*; H, the ingluvial ganglion and I, the gastric nerves and GEC of *Mantis*. Bar = 100 μm in A, C, E, H; 200 μm in B, D, F, G. I.

gastrin-like antigen was never detected simultaneously with the FMRFamide-, Arg-vasopressin- and VIP-like antigens.

(c) Palaeoptera

In our investigations of the ENS of mayfly larvae (*Cleon* sp., Ephemeroptera) we found only a weak immunoreaction in the gastric nerves. Strongly reacting GEC, mostly of the closed type, were at the very anterior of the midgut whereas mostly open-type, long GEC were detected in about one third of the midgut. By contrast, ENS of the examined Odonata was well stained. In adult damselfly, *Agrion* sp. (Zygoptera), the oesophageal nerve was traced to the ingluvial ganglion that contains 4–5 FRMFamide immunopositive neurons (Figure 3.1C), and two proventricular ganglia, each with 2–5 neurons were found. In the dragonfly, *Sympetrum* sp. (Anisoptera), the oesophageal nerve ends in the ingluvial ganglion with 8–12 immunopositive neurons. Diffused poventricular ganglia supply numerous nerves to the meshwork of midgut innervation which resembles that of the apterygotes. Singly scattered GEC are either long and open or round and closed (Figure 3.1D); many of the latter possess paracrine processes. In *Agrion* they are uniformly distributed along the entire midgut length. In *Sympetrum* they reach highest density in the central and posterior midgut regions. Andriès and Tramu (1985) noted that GEC of the dragonfly *Aeschna cyanea* are most numerous in the posterior portion of the midgut.

(d) Polyneoptera

All examined representatives of Polyneoptera exhibit strong immunoreactivity with antiserum to FMRFamide. The arrangement of the frontal and hypocerebral ganglia is similar, but ENS organization in the posterior foregut and in the midgut regions may differ at the level of families. In cockroaches, *Nauphoeta cinerea* and *Blabera craniifer* (Blattodea), the single oesophageal nerve splits at the oesophagus/crop junction into dorsal and ventral ingluvial nerves. The ingluvial and proventricular ganglia are wanting but the wide ingluvial nerves contain numerous neurons (Figure 3.1E) whose fine axons anastomose over the crop and proventriculus (Figure 3.1F). The ingluvial nerves eventually ramify into a number of gastric nerves that form a meshwork innervation of the midgut. Some axons terminate in varicosities indicative of a neurohaemal function (Figure 3.1G). In *Periplaneta americana*, a single ingluvial ganglion was identified at the branching point of ingluvial nerves and two proventricular ganglia were found at the end of the nerves at the foregut/midgut boundary (Penzlin, 1985). Single, open or closed GEC without cytoplasmic processes are scattered throughout the midgut (Žitňan *et al.*, 1993).

In the praying mantis, *Mantis religiosa* (Mantodea), and the embians, *Adelembia* sp. and *Heoembia* sp. (Embioptera), a single dorsal oesophageal nerve, which sends numerous side branches along its entire course, terminates in the ingluvial ganglion. The ingluvial ganglion of *Mantis* contains 8–10 immunoreactive perikarya and sends off three pairs of ingluvial nerves (Figure 3.1H). Two lateral pairs innervate muscles of the posterior foregut, whereas the posterior pair forms the network of midgut gastric nerves (Figure 3.1I). Small GEC of both open and closed types are distributed throughout the midgut and many of them contain basal paracrine processes (Figure 3.1I). Embians show immunoreactivity in 2–4 neurons of the ingluvial ganglion, both ingluvial nerves, and 4–5 neurons in each of the two proventricular ganglia. A network of gastric nerves emanates from the latter ganglia (Žitňan *et al.*, 1993). Amoeboid GEC of both open and closed types occur in the females, but only closed GEC are found in the males (Figure 3.2A).

In Orthoptera, one oesophageal nerve connects the hypocerebral ganglion with the dorsal, and the other nerve with the ventral ingluvial ganglia. In the primitive Jerusalem cricket, *Ceuthophilus* sp., the true cricket, *Gryllus bimaculatus*, and the long-horn grasshopper, *Tettigonia viridissima* (suborder Ensifera), each ingluvial ganglion contains 5–10 immunopositive cells and sends posteriorly two nerves: an ingluvial nerve that innervates the crop, and a caecal nerve that innervates the midgut. Both the dorsal and ventral caecal nerves arborize on the surface of two large caeca (Figure 3.2C) and run posteriorly along the whole midgut. In the short-horned grasshoppers, *Melanopus bivitatus* and *Dissosteira carolina* (suborder Caelifera), each ingluvial ganglion contains 0–2 immunopositive cells but a number of stained neurons lie along the four ingluvial and the two caecal nerves that extend from the ganglion (Figure 3.2B). A pair of dorsal and a pair of ventral caecal nerves proceed posteriorly in grooves separating the six caeca. At the caecal base, the nerves send off several anterior and posterior side branches. Anterior branches innervate the caeca, whereas posterior branches ramify into numerous parallel gastric nerves that are distributed evenly over the midgut surface. All mentioned orthopterans contain GEC of both closed and open types. The GEC of *Ceuthophilus* are scattered throughout the midgut and most are characterized by central location of the nucleus (Figure 3.2D). In *Gryllus*, the highest density of GEC occurs in a midgut zone just behind the caeca, where one finds clusters of very small GEC in the regenerative nidi, and single long and open-type GEC dispersed among the columnar cells (Figure 3.2E). Large numbers of bottle-shaped or rounded GEC or *Tettigonia* are located in the posterior midgut region. On the other hand, GEC are evenly distributed in the midgut of both examined short-horned grasshoppers,

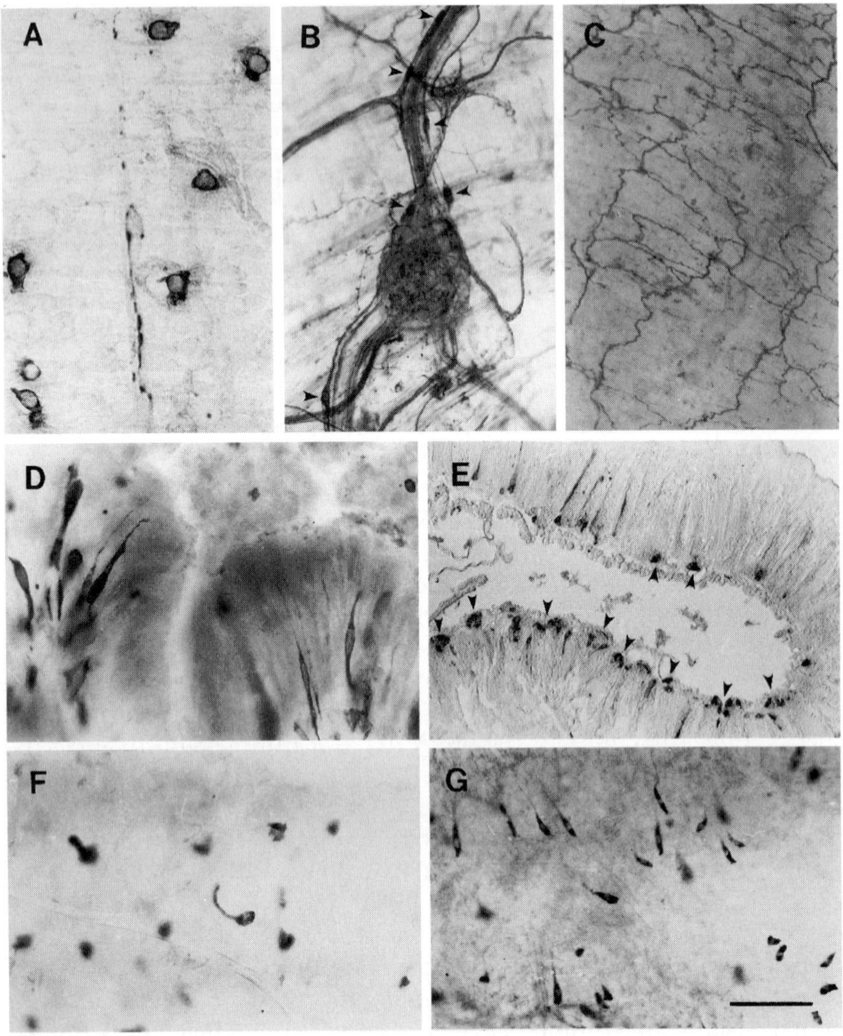

Figure 3.2 FMRFamide-immunoreactivity in the ENS and GEC of the polyneopteran and paraneopteran insects. A, closed type GEC and a gastric axon in male *Heoembia*; B, ingluvial ganglion and loose neurons (arrowheads) in *Dissosteira*; C, axonal network on the caecal surface in *Oecanthus*; D, GEC with centrally located nuclei in *Ceuthophilus*; E, section through the midgut of *Gryllus* with most of the small rounded GEC in the nidi; F, GEC of the *Tibicen* cicada; G, GEC with long and narrow apical cytoplasmic processes in *Dysdercus*. Bar = 200 μm.

except that those with basal cytoplasmic projections are most common in the caecal region.

A close relative of Orthoptera, the walking stick, *Carausius morosus* (Phasmatodea), possesses only one oesophageal nerve, which ends in the single (dorsally located) ingluvial ganglion. Immunoreactive neurons are scattered along the oesophageal nerve, around the ingluvial ganglion (Figure 3.3A,B) and on the surface of the anterior third of the midgut. Similarly as in *Mantis*, the ingluvial ganglion sends off two pairs of anterior nerves, which innervate crop, and two posterior nerves, which ramify into the gastric nerves. The latter are revealed with the FMRFamide antibody only in the first third of the midgut, where they run along prominent longitudinal muscles. No immunopositive axons were found in the central and posterior midgut regions. However, strong staining occurs in the axons that innervate six longitudinal

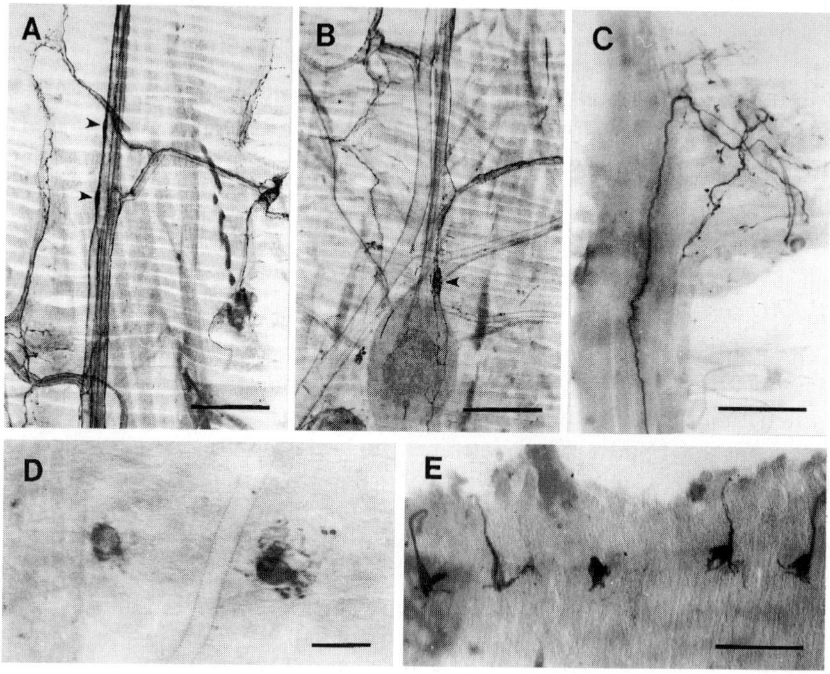

Figure 3.3 FMRFamide-immunoreactivity in the ENS and GEC of the stick insect, *Caraussius*. A, the oesophageal nerve and B, the ingluvial ganglion (arrows point to neurons); C, the proctodeal nerve and its anastomosing terminal axons; D, closed GEC in the anterior and E, open GEC in the central midgut regions (note the long and branching cytoplasmic processes in all GEC). Bar = 200 μm in A, B; 100 μm in C, D, E.

muscles of the hindgut (Figure 3.3C). Immunoreactive GEC of *Carausius morosus* have very distinct shape and distribution. A few big and mostly closed GEC are stained in the anterior midgut (Figure 3.3D). Most midguts contain numerous GEC; they are large and open in the central region (Figure 3.3E), and smaller and mostly closed in the posterior regions. All GEC have a characteristic asteroid shape with long basal cytoplasmic projections (Figure 3.3D, E).

(e) Paraneoptera

No gut innervation is identified with FMRFamide antiserum in Homoptera and Heteroptera, whereas GEC show very strong staining in several midgut regions of all examined species except for the aphids (Žitňan *et al.*, 1993). In the cicada, *Tibicen canicularis* (Homoptera), rounded and exceptionally bottle-shaped open-type GEC are singly scattered in a long and narrow band of midgut epithelium (Figure 3.2F). Few GEC are paired and some have paracrine processes. Most immunoreactive GEC of the bugs, *Dysdercus cingulatus* and *Oncopeltus fasciatus* (Heteroptera), are restricted to the rear half of the sack-like anterior midgut. Only very few GEC are stained in the posterior narrow part of the midgut. Most bug GEC are drop-like in shape, with long and narrow apical projection to the gut lumen (Figure 3.2F). This is in contrast to the blood-feeding *Rhodnius prolixus*, where GECs were concentrated in the posterior midgut (Billingsley and Downe, 1986).

(f) Oligoneoptera (Holometabola)

Oligoneoptera are diversified in respect to the ENS organization, whereas the bottle-shaped open-type, or rounded and closed GEC are quite similar in all examined species. In *Chrysopa* sp., *Myrmeleon* sp. and *Asalaphus* sp. (Neuroptera), a pair of ingluvial nerves, which innervate the crop, dichotomize into several dorsal and ventral gastric nerves that run along the longitudinal muscles of the blind bulbous midgut. No ganglia seem to be present at the foregut/midgut boundary but individual neurons are embedded in the ingluvial and gastric nerves. About 10 perikarya are localized in the rear foregut region and up to 40 are distributed along the gastric nerves of the midgut in *Chrysopa*. In *Myrmeleon* and *Ascalaphus*, about 10–20 immunoreactive neurons and their axons are scattered over the entire midgut surface, including the most posterior blind region (Figure 3.4A). Some axons derived from gastric nerves encircle and terminate around the basal ampullae of the Malpighian tubules (Figure 3.4B). Most GEC occur in groups (Figure 3.4C) but those in the most posterior narrow midgut region are single (Figure 3.4D). ENS of the carnivorous beetle, *Carabus* sp. (Coleoptera,

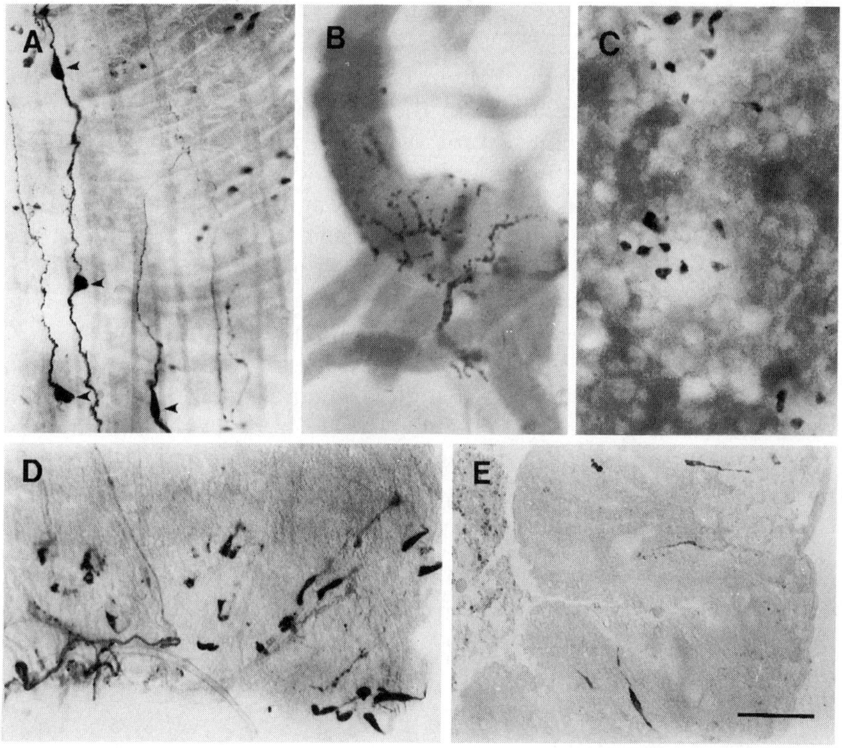

Figure 3.4 FMRFamide-immunoreactivity in the ENS and GEC of neuropteran insects and a beetle. A, neurons (arrowheads) and axons at the posterior part of the blind midgut in *Myrmeleon*; B, basal ampulla of a Malpighian tubule encircled by axon terminals of the gastric nerves in *Myrmeleon*; C, grouped GEC in the central part of *Ascalaphus* midgut; D, single GEC in the most posterior blind end of the *Myrmeleon* midgut; E, long GEC with centrally located nucleus in *Carabus*. Bar = 200 μm.

Adephaga) reacts poorly, whereas ENS of the mealworm beetle, *Tenebrio molitor* (Cleoptera, Polyphaga), shows a very strong reaction with the FMRFamide antiserum. Foregut is innervated by a single oesophageal nerve, which ramifies into an enteric plexus with 6–8 immunoreactive neurons at the foregut/midgut boundary. The plexus sends off numerous branching gastric nerves that innervate the whole midgut and eventually fuse with a pair of lateral proctodeal nerves into a neural meshwork at the midgut/hindgut boundary. Single GEC (Figure 3.4E) are most abundant in the central and posterior sections of the midgut in both species.

The wasp, *Vespa* sp., and bumblebee, *Bombus* sp. (Hymenoptera), show only very weak reaction in few varicosities in the nerves of anterior midgut. Strong staining is detected in the GEC that are singly dispersed through the midgut.

The posterior part of the ENS in Lepidoptera forms a ring encircling the foregut/midgut boundary. In the waxmoth, *Galleria mellonella*, the ring includes a single ingluvial and paired proventricular ganglia (Žitňan *et al.*, 1989), whereas only an assembly of loose neurons and nerves, called the enteric plexus (Copenhaver and Taghert, 1989a) is found there in the moths, *Manduca sexta*, *Bombyx mori* and *Lymantria dispar* (Žitňan *et al.*, 1989), and the butterfly, *Danaus plexipus*. Singly distributed GEC are most numerous in the posterior half of the midgut (Žitňan *et al.*, 1995) and are of similar appearance in all the aforementioned species.

Paired oesophageal nerves of the cranefly, *Tipula* sp. (Diptera, Nematocera), terminate in two ingluvial ganglia with four immunoreactive cells (Figure 3.5A). Eight gastric nerves beginning in these ganglia pass through the grooves between the four caeca and continue

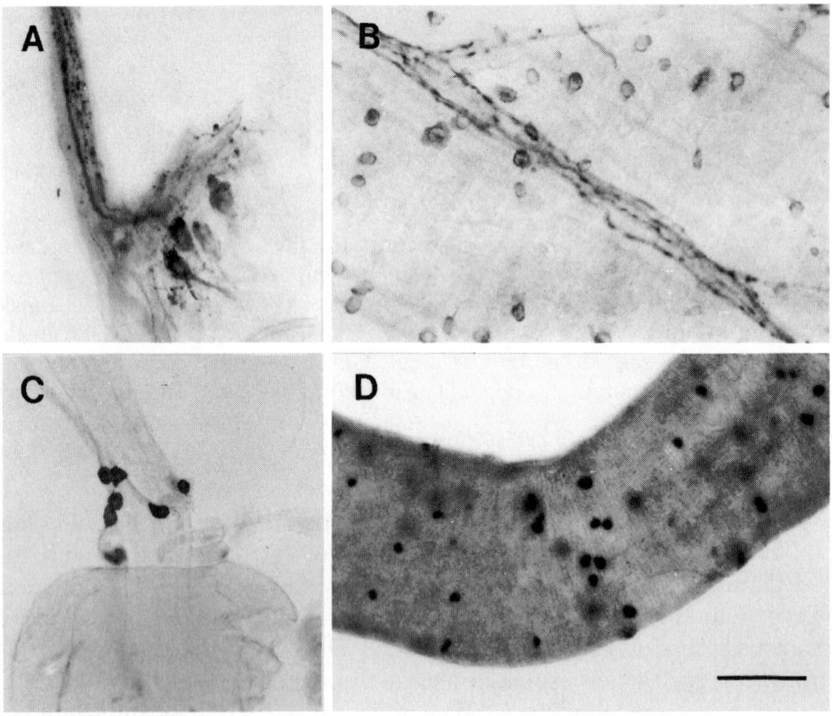

Figure 3.5 FMRFamide-immunoreactivity in the ENS and GEC of Diptera. A, four stained neurons in the ingluvial ganglion, and B, the gastric nerve and GEC in *Tipula*; C, neurons in the hypocerebral ganglion of *Drosophila* larva; D, GEC in the central portion of midgut in adult *Drosophila*. Bar = 200 μm.

over the midgut (Figure 3.5B). Immunoreactive GEC are present in all parts of the midgut (Figure 3.5B), and are most abundant in the caeca. *Drosophila melanogaster* (Diptera, Orthorrhapha) has no frontal ganglion and the stomatogastric nervous system is connected with CNS by the recurrent nerve that runs from the brain to the hypocerebral ganglion. In the larvae, 8–10 FRMFamide immunoreactive neurons of this ganglion (Figure 3.5C) send axons over the crop to the base of the four caeca and adjacent midgut region. Immunoreactive GEC are located only in a narrow central part of the midgut. Adult midgut is devoid of caeca and the GEC are present in the anterior and especially the middle region of the midgut (Figure 3.5D).

3.3 CHEMICAL IDENTITY OF REGULATORY PEPTIDES PRODUCED IN INSECT MIDGUT

The nature of insect enteric hormones was first investigated by immunohistochemistry. Both ENS and GEC react with antibodies to the peptidic hormones of mammals and other animals. Consistent reactions with the antibodies to certain antigens led to the proposition that the enteric endocrine system of insects contains compounds that are antigenically similar to the brain/gut peptides of mammals (De Loof, 1987).

Žitňan *et al.* (1993) examined the distribution of antigens related to mammalian brain/gut peptides in different insect taxa (Table 3.1). They concluded that antibodies to bombesin, neurotensin, motilin, and secretin do not react with any cells or axons in insect gut, whereas antibodies to pancreatic polypeptide (PP), *β*-endorphin, enkephalins, and vasopressin recognize certain GEC and enteric axons of all species. Antibodies to other gastroenteropancreatic hormones of mammals react only in some insects. Insects ENS and GEC also contain some types of antigens that occur in vertebrates exclusively in the CNS, e.g. vasopressin, urotensin I, growth hormone releasing factor, and FMRFamide. Certain other insect gut peptides (allatotropin, prothoracicotropin, diuretic hormone) are not known from any other group of animals. By their structural features and active conformation, peptidic hormones can be arranged into families that presumably embrace peptides of common origin. Certain structurally unrelated hormones are linked by being produced as a single prohormone from which they are derived by enzymatic cleavage. Other prohormones yield a single hormone or multiple copies of an identical or only slightly modified peptide. Some gastrointestinal hormones are characterized by size heterogeneity that is due to enzymatic truncation of the long form. Since the enteric hormones of insects are insufficiently known, their arrangement into groups listed below is only tentative. Structures are given in single-letter

Table 3.1 Reaction of enteric neurons (N) and midgut endocrine cells (E) of representative insects with antibodies against mammalian brain/gut hormones and distribution of the respective peptides in the gastroenteropancreatic system of mammals. (From Žitňan *et al.*, 1993.)

Peptide	Lepisma		Aeschna		*Roaches*		Pyrrhocoris		Calliphora		Galleria		*Mammals*	
	N	*E*	*N*	*E*	*N*	*E*	*N*	*E*	*N*	*E*	*N*	*E*	*N*	*E*
FMRFamide	+	+	+	+	+	+	+	+	+	+	+	+	+	+
Gastrin/CCK	+	+	+	+	+	+	−	−	−	−	+	+	+	+
Substance P	?	?	−	+	−	+	?	?	?	?	?	?	+	+
Vasopressin	+	+	?	?	+	+	+	+	+	+	+	+	−	−
CRF	?	?	?	?	+	+	?	?	?	?	?	?	−	+
PP	+	+	+	+	+	+	+	+	+	+	+	+	−	+
VIP/PHI	+	+	+	+	+	+	−	−	−	−	+	+	+	−
Glucagon	?	?	?	?	−	+	?	?	?	?	−	−	−	+
Glicentin	?	?	?	?	−	+	?	?	?	?	−	−	−	+
Somatostatin	−	−	+	+	+	+	−	−	−	−	−	−	+	+
Enkephalins	+	+	+	+	+	+	+	−	+	−	+	+	+	+
β-endorphin	+	−	?	?	+	+	+	−	+	+	+	−	+	+
ACTH/α-MSH	−	−	?	?	+	+	−	−	−	−	−	−	+	+

+, positive; −, absent; ?, not tested.
Abbreviations of peptide names: CCK, cholecystokinin; CRF, corticotropin releasing factor; PP, pancreatic polypeptide; VIP, vasoactive intestinal peptide; PHI, peptide histidine isoleucine; ACTH, adrenocorticotropin; MSH, melanocyte stimulating hormone.

codes of amino acid residues. Modified residues include pE, pyroglutamate; Ys and Y(s), constitutively or facultatively sulphonated tyrosine; hP, hydroxyproline; a, amidated carboxy terminus.

3.3.1 Myotropic pentapeptides

A number of enteric hormones exert myotropin activities but only Mas-MG-MT 1 (*Manduca sexta* midgut myotropin I) has been isolated with the aid of a myotropic bioassay from the midgut of a caterpillar. It was identified as AEPYTa (Yi *et al.*, 1995). Another myotropic pentapeptide, the proctolin RYLPT, was first prepared from the whole body extracts of a cockroach, but its occurrence in various insects and other arthropods was amply demonstrated (Orchard *et al.*, 1989). It is produced by several neurons of the terminal abdominal ganglion and axonally transported to the posterior ENS (Eckert *et al.*, 1981; O'Shea and Adams, 1986).

3.3.2 FMRFamide

There is ample evidence for the reaction to FMRFa antibodies to insect ENS and GEC, but because the antibodies do not distinguish between different peptides with the -RFa or -RYa carboxy terminus (Veenstra and Schooneveld, 1984), the immunohistochemical data merely indicate that one of the following, or a hitherto unknown but related, member of the peptide family is present: FMRFa-type, myosupressins, 'head peptides', sulphakinins, and PP-type peptides. Occurrence in insects of neuropeptides with the -FMRFa carboxy terminus was proven both by chemical and genomic analyses. In *Drosophila*, a single gene encodes several hexa- and octapeptides with the -FMRFa carboxy terminus (Nambu *et al.*, 1988; Schneider and Taghert, 1988; Chin *et al.*, 1990). Similar peptides were identified in the blowfly, *Calliphora vomitoria* (Duve *et al.*, 1992). The production of identified FMRFa peptides by ENS or GEC, however, has not been demonstrated in any species.

3.3.3 Myosupressins

These neuropeptides, characterized by the -FLRFa carboxy terminus, were isolated from cockroaches (Holman *et al.*, 1986), locusts (Robb *et al.*, 1989) and other insects. Information on *Drosophila* indicates that myosupressins are not encoded by the *FMRFa* gene; this supports the assumption that they represent a separate peptide family. Dromyosupressin TDVDHVFLRFa was isolated from *Drosophila* extracts (Nichols, 1992a) and localized with the aid of an antibody against its TDVDHV portion to the CNS and to limited regions of the ENS (McCormick and Nichols, 1993). An indication that some myosupressins are also produced in GEC comes from a study on *Manduca*. Preliminary data indicate that a form of myosupressin QDVVHSFLRFa, which was isolated from the cephalic part of pharate adult CNS (Kingan *et al.*, 1990), also occurs in larval midgut (Kingan, Žitňan and Beckage, unpublished).

3.3.4 'Head peptides'

Peptides pERPhPSLKTRFa (Aea-HP-I for *Aedes aegypti* Head Peptide I) and TRFa (Aea-HP-II) were isolated from mosquito heads with the aid of an antibody to FMRFa (Matsumoto *et al.*, 1989). Antiserum raised against Aea-HP-1 reacted with GEC of a caterpillar (Crim *et al.*, 1992) and of adult mosquito (Brown *et al.*, 1994). Isolation of a similar peptide ANRSPSLRLRFa from the midgut of the cockroach, *Periplaneta*, provided direct proof that the family of 'head peptides' does occur in the enteric system of insects (Veenstra and Lambrou, 1995).

3.3.5 Sulphakinins

Peptides EQPEDY(s)GHMRFa and pESDDY(s)GHMRFa, which were isolated from the heads of the cockroach, *Leucophaea maderae*, and called leucosulphakinins (Nachman *et al.*, 1986a,b) are insect members of the gastrin/cholecystokinin (CCK) hormone family. Eight similar sulphakinins were identified in other cockroaches, locusts, and flies (for review see Schoofs *et al.*, 1993). Genes encoding putative sulphakinins FDDY(s)GHMRFa and GGDDQFDDY(s)GHMRFa (and a third peptide) were isolated from *Drosophila* (Nichols *et al.*, 1988). Antibodies to these drososulphakinins recognize both ENS and GEC of the fly (Nichols, 1992b). Antisera to vertebrate gastrin (C-terminus: Y(s)GWMDFa) and CCK (C-terminus: YsMGWMDFa) react strongly in the ENS and GEC of many insects (e.g. Andriès and Beauvillain, 1988), but the reacting antigen(s) was not identified. For the blowfly CNS, neurons producing sulphakinins are different from those reacting with the gastrin/CCK antisera (Duve *et al.*, 1994). This proof is lacking for the digestive tract and cross-reactivity of the antisera with the sulphakinins is possible.

3.3.6 Tachykinins

The search for the myotropic neurohormones in *Locusta migratoria* (Schoofs *et al.*, 1990a,b) and subsequently in other insects (Champagne and Ribeiro, 1994; Lundquist *et al.*, 1994a) yielded several peptides of the tachykinin family. Vertebrate hormones of this family, which include substance P from the digestive tract and substance K and neuromedin K from the spinal cord, consist of 11 amino acids with the C-terminus -FXGLMa. Insect tachykinins range in size from 9 to 11 amino acids and contain either the vertebrate-type C-terminus -FX_1GX_2Ma or, less frequently, -FX_1GX_2Ra. Immunohistochemical data indicated the occurrence of compounds related to substance P in insect GEC (Table 3.1). This was confirmed in *Locusta* and *Calliphora* with antisera to the locustatachykinins and callitachykinins that recognized numerous CNS neurons and GEC (Nässel, 1993; Lundquist *et al.*, 1994b). In *Leucophaea*, tachykinin-type antigens were further detected in some neurons of the stomatogastric ganglia and in nerve fibres of the gut musculature (Nässel *et al.*, 1995).

3.3.7 Allatostatins

The name allatostatins was originally given to five peptides isolated from the nervous system of the cockroach, *Diploptera punctata* (Pratt *et al.*, 1989, Woodhead *et al.*, 1989), as inhibitors or juvenile hormone production from explanted corpora allata. Additional allatostatins were identified in *Diploptera* and two other cockroaches, the blowfly, *Calliphora*, and the cricket, *Gryllus bimaculatus* (reviewed by Lorenz *et al.*, 1995). Today, we know of nearly 30 compounds ranging in size from 6 to 18 amino acids and sharing carboxy terminal sequence -YXFGLa or FXFGLa. In *Diploptera*, a cDNA encoding seven allatostatins and six other putative peptides has been isolated (Donly *et al.*, 1993). The gene is expressed in GEC, whereas allatostatins present in the stomatogastric and proctodeal nerves seem to originate in the CNS (Reichwald *et al.*, 1994). The presence of octadecapeptidic allatostatin in the midgut was confirmed by chemical analysis. Quantification of allatostatin with a bioassay revealed that the midgut contains about half the allatostatic activity compared to the brain (Reichwald *et al.*, 1994). The Leu-collatostatin gene of blowflies is also expressed in GEC of the posterior midgut section and not in ENS (East *et al.*, 1995).

3.3.8 Oxytocin/vasopressin family

Two peptides, which were purified from the suboesophageal and thoracic ganglia of *Locusta* with the aid of antibody against mammalian arginine vasopressin, proved to be CLITNCPRGa and the antiparallel dimer of the same sequence (Schooley *et al.*, 1987). The sequence differs from that of arginine vasopressin only by Leu^2 in place of Tyr^2, but locust peptide does not occur as cyclic monomer, which is the functional form of vasopressins and oxytocins. Locust dimer was called 'arginine vasopressin peptide-like insect diuretic hormone' (AVP-like IDH) Proux *et al.*, 1987) but its diuretic function was questioned (Coast *et al.*, 1993).

3.3.9 CRF (corticotropin releasing factor) family

CRF from the hypothalamus of mammals, urotensin I from the urophysis of fish, and sauvagine from frog skin are about 40 amino acid peptides that share large stretches of internal sequence homology. Several shorter peptides of the CRF family were isolated with a diuretic assay from insect CNS: two from *Manduca* (Kataoka *et al.*, 1989b; Blackburn *et al.*, 1991), and one each from *Acheta domesticus* (Kay et al., 1991a), *Locusta* (Kay *et al.*, 1991b; Lehmberg *et al.*, 1991), and *Periplaneta* (Kay *et al.*, 1992). Antiserum to one of the *Manduca* diuretic hormones reacts with the gut innervation and GEC of caterpillars (Žitňan *et al.*, 1993). This observation and the reaction of antibodies to urotensin I in the digestive tracts of *Gryllus* and *Periplaneta* (Iwanaga *et al.*, 1986) indicate that CRF-type peptides occur in insect ENS and GEC, but their sequence structure remains to be elucidated.

3.3.10 *Manduca* allatotropin family

The neurohormone GFKNVEMMTARGFa was identified in *Manduca* as a stimulator of juvenile hormone production (Kataoka *et al.*, 1989a). A homologue [Ala^6, Leu^7, Ser^8] was isolated from the male accessory glands of *Locusta* as a myotropic agent (Paemen *et al.*, 1991). Wide distribution of this type of peptide in insects is suggested by the isolation of an allatotropin-like antigen from *Periplaneta*, where it occurs in the CNS and the retrocerebral gland complex (Veenstra and Hagedorn, 1993). Immunohistochemistry of the larval digestive tract of *Manduca* and *Galleria* with the allatotropin antibody revealed staining in both ENS and GEC (Žitňan *et al.*, 1993).

3.3.11 Insulin family

Insulins *sensu stricto* are dimeric proteins linked by two cysteine bridges. Occurrence of insulin-type proteins in insects was proven by identification of several neurohormones called bombyxins in *Bombyx* (Kawakami *et al.*, 1989) and one neurohormone in *Locusta* (Lagueux *et al.*, 1990). It is not clear, however, whether insects produce an insulin-type compound in their digestive tract. Although radioimmunoassays and metabolic bioassays of midgut extracts indicated the presence of insulin-like protein(s) in several insects (Moreau *et al.*, 1981; Teller *et al.*, 1983; Ben Khay *et al.*, 1987), diverse antisera to mammalian insulins and monoclonal antibody to bombyxin (Mizoguchi *et al.*, 1987), which readily react with certain ENS neurons and corpora cardiaca, failed to react with GEC (Žitňan *et al.*, 1990).

3.3.12 NPY family

Mammalian neurotransmitter neuropeptide Y (NPY), pancreatic polypeptide (PP) from the endocrine cells of pancreas, and peptide YY from the intestinal GEC are homologous peptides 36 amino acids long and sharing C-terminus TRPRYa or TRPQYa. Antibodies to these peptides react in insect ENS and GEC (e.g. Schoofs *et al.*, 1988). Characterization of the *Drosophila* receptor for peptide YY (Li *et al.*, 1992) supports the assumption that insects contain peptides of the NPY family.

3.3.13 Other possible peptides of insect ENS and GEC

Glucagon-related peptides are an important group of vertebrate enteric hormones. Antisera to vasoactive intestinal peptide (VIP), peptide histidine methionine (PHM), and peptide histidine isoleucine (PHI) readily react with both ENS and GEC of various species (Table 3.1), suggesting that this glucagon subfamily is represented in the insects. On the other hand, antigen related to glucagon was reported only in one study (Iwanaga *et al.*, 1981) and antisera to the very similar vertebrate hormone, secretin, failed to react in several insects (Žitňan *et al.*, 1993). The presence in insects of antigens related to glicentin, a structurally different peptide but derived in vertebrates from the same prohormone as glucagon, is also uncertain.

Somatostatin, AGCKNFFWKTFTSC, was originally isolated from mammalian hypothalamus. Somatostatin and its N-terminally extended forms of up to 28 amino acids are also abundant in vertebrate ENS and GEC. In our tests on several insects we did not detect somatostatin-like materials in the digestive tract, but positive reaction was reported for a dragonfly (Andriès and Tramu, 1985) and a cockroach (Iwanaga *et al.*, 1981).

Mammalian opioid peptides are derived from two prohormones and include Met-enkephalin (YGGFM), Leu-enkephalin (YGGFL), and several other peptides extended at the C-terminus. Endorfins, which originate from the propiomelanocortin prohormone (see below) possess YGGFM- as their N-terminus and are up to 31 amino acids long. Immunohistochemical data obtained with antibodies to mammalian peptides (Table 3.1) suggest that insects contain enkephalins and endorfins in both ENS and GEC. Most reliable evidence was obtained in *Calliphora* with a radioimmunoassay specific for the C-terminus of Met-enkephalin-Arg^6-Phe^7 and Met-enkephalin-Arg^6-Gly^7-Leu^8 peptides (Duve *et al.*, 1992).

Immunohistochemical data (Table 3.1) indicate that the enteric endocrine system of insects contains peptides antigenically similar to the products of the vertebrate prohormone propiomelanocortin, such as β-endorphin, adrenocorticotropin (ACTH, 39 amino acids), and α-MSH (melanocyte stimulating hormone, 13 N-terminal amino acids of ACTH). Antibodies to all three peptides react with insect ENS and GEC (in mammals, α-MSH occurs exclusively in CNS and β-endorphin is found in GEC rather exceptionally). Extracts of an ACTH-related diuretic hormone were prepared from the corpora cardiaca of *Locusta* (Rafaeli *et al.*, 1987). Several regulatory peptides, which are believed to be restricted in vertebrates to the CNS, were detected immunohistochemically in insect ENS and GEC. For example, antigens to the hypothalamic growth hormone releasing factor (GHRF) and luteinizing hormone releasing factor (LHRH) occur in GEC of the dragonfly, *Aeschna*, and in the cockroach, *Blaberus* (Andriès and Tramu, 1985; Schols *et al.*, 1987).

Some insect neurohormones are probably confined to the CNS. However, antibody to *Bombyx* prothoracicotropin (Mizoguchi *et al.*, 1990) reacted in the caterpillars of *Galleria* and *Manduca* with neurons of the stomatogastric ganglia (Žitňan *et al.*, 1993). It should be emphasized that, in the brain, this antibody reacts very specifically with only two neurons in each hemisphere.

3.4 FUNCTIONS OF MIDGUT ENDOCRINE PRODUCTS

Widespread occurrence, chemical diversity and structural complexity, which are maintained when the gut is rebuilt at metamorphosis, indicate that both the innervation and the endocrine cells of the gut are of great importance but their physiological roles are very little understood. Most hormones circulating in the haemolymph are probably derived from the CNS, e.g. the elevated levels of FMRFa-like compounds in the flying and running cockroaches seem to be released from the thoracic ganglia (Elia *et al.*, 1995). Regulatory peptides produced in the digestive tract may act as synaptic transmitters and modulators, or parahormones and hormones, but at present we can only register non-specified humoral effects.

3.4.1 Myotropic action

The availability of an easy and reliable myotropic bioassay facilitated the identification of a variety of neuropeptides (reviewed by Schoofs *et al.*, 1993), of which tachykinins, myosupressins and sulphakinins are also produced in the enteric endocrine system. Several peptides may interact in eliciting a myotropic response. For example, Puiroux *et al.* (1992) demonstrated in hindgut cell membranes the presence of a specific proctolin receptor that also binds, albeit with a lower affinity, FMRFamide, Leu-enkephalin and Met-enkephalin-RF. The pentapeptides proctolin and Mas-MG-MT1 possibly act primarily as myotropic agents. Regulation of muscle contractions also seems to be the principal action of the tachykinins (Nässel *et al.*, 1995) but the existence of several structures suggests multifunctional activities (Schoofs *et al.*, 1993; Nässel *et al.*, 1995). A variety of functions, possibly exerted by different members of a peptide family, is characteristic for other types of peptides. For example, the adenotropic allatostatins also antagonize the myotropic effects of proctolin (Hertel and Penzlin, 1992) and slow down gut peristalsis (Lange *et al.*, 1993; Duve and Thorpe, 1994), and the *Manduca* allatotropin exhibits cardioacceleratory activity (Veenstra *et al.*, 1994). Of the six FMRFa-type endogenous nonapeptides of *Calliphora* two exert a stimulatory effect on the heart contractions (Duve *et al.*, 1993), whereas the action of the others is unknown. Myotropic activity in insects was also shown for the mammalian substance P (Penzlin *et al.*, 1980)

3.4.2 Diuresis

The oxytocin/vasopressin and CRF-related peptides were isolated from the central nervous system of insects with the aid of diuretic bioassays. Control of diuresis by the vasopressin-like peptides of locusts was shown with a preparation of a short gut section with attached Malpighian tubules (Proux *et al.*, 1987) but was not confirmed in tests with single Malpighian tubules (Coast *et al.*, 1993). Diuretic action of CRF-like peptides was shown repeatedly in locusts and other insects and it was proposed that the effect is mediated by cyclic AMP (Audsley *et al.*, 1993; Coast and Kay, 1994; Patel *et al.*, 1995). Locust nervous system reportedly contains a third type of diuretic hormone belonging to the ACTH-like peptide family (Rafaeli *et al.*, 1987).

3.4.3 Control of digestion

Involvement of the gut hormones in the control of food processing was proposed by Brown *et al.* (1986), who showed that the number of immunoreactive GEC and their content of PP-like and FMRFa-like materials in the midgut of adult mosquito females decrease after the blood meal. By contrast, in the larval corn earworm, *Heliothis zea*, feeding stimulates FMRFa-immunoreactivity in GEC, and the blood concentration of the immunoreactive agent is higher in the starved animals (Jenkins *et al.*, 1989). The stimulatory effect of some endogenous enkephalin and FMRFa-related peptides of *Calliphora* on the fluid secretion from explanted salivary glands (Duve *et al.*, 1992) may also be interpreted as an effect on digestion.

3.4.4 Control of development and reproduction

In a number of cases, it has been shown that the progress of larval development and initiation of reproduction depend on the food intake. Either a mechanical extension of the gut or ingestion of certain nutrients provide a stimulus by which a cascade of neurohormonal signals is initiated in the brain. In the case of bugs, presumably a nervous signal from the gut causes release of a humoral factor from ventral ganglia, and this factor acts on the brain (Muley and Davey, 1995). In the adult blowfly, *Phormia regina*, however, the midgut releases a humoral stimulus that acts directly on the brain (Yin *et al.*, 1994).

It cannot be excluded that certain gut hormones affect directly corpora allata or prothoracic glands. The occurrence of prothoracicotropin- and allatotropin-immunoreactivity in the midgut (Table 3.1) indicates similar possibilities. The presence of prothoracicotropic activity in the hindgut of caterpillars was shown with a specific bioassay (Gelman *et al.*, 1991), but the active component, possibly derived from ventral nerve cord, has not been identified. In *Diploptera*, the midgut appears to be a source of allatostatins that curb juvenile hormone production (Pratt *et al.*, 1989; Woodhead *et al.*, 1989), but no such effect was seen in the blowfly (Duve and Thorpe, 1994).

3.4.5 Other possible functions

It has been proposed that gut hormones influence the development of parasites and pathogens which are transmitted by the insects (Brown and Lea, 1989). Stefano *et al.* (1989) demonstrated that enkephalins affect migration and adherence of insect haemocytes, indicating that they may be involved in insect's 'immunodefence'. Locustatachykinins III and IV were found to mimic the action of the pheromone biosynthesis stimulating neurohoromone and to induce pheromone production in adult silkworm females (Fonagy *et al.*, 1992). A more direct effect on behaviour was reported for the 'head peptide' Aea-HP-I which suppresses host-seeking in adult females of the *Aedes* mosquito (Brown *et al.*, 1994).

ACKNOWLEDGEMENTS

Preparation for this chapter was supported by grant 305/94/0975 of the Grant Agency of the Czech Republic.

REFERENCES

Al-Yousuf, S. (1990) Neuropeptides in annelids, in *Progress in Comparative Endocrinology* (eds A. Epple, C.G. Scanes and M.H. Stetson), Wiley-Liss, New York, pp. 232–41.

Andrew, A. (1981) APUD cells and paraneurons: embryonic origin. *Adv. Cell Neurobiol.*, **2**, 3–12.

Andriès, J.C. and Beauvillain, J.C. (1988) Ultrastructural study of cholecystokinin-like immunoreactivity in endocrine cells of the insect midgut. *Cell Tissue Res.*, **254**, 75–81.

Andriès, J.C. and Tramu, G. (1985) Distribution pattern of mammalian-like peptide immunoreactive cells in the midgut of *Aeshna cyanea* (Insecta, Odonata). *Experientia*, **41**, 5000–3.

Audsley, N., Coast, G.M. and Schooley, D.A. (1993) The effects of *Manduca sexta* diuretic hormone on fluid transport by the Malpighian tubules and cryptonephric complex of *Manduca sexta*. *J. Exp. Biol.*, **178**, 231–43.

Bayliss, W.M. and Starling, E.H. (1902) The mechanism of gallbladder contraction and evacuation. *J. Physiol. (Lond.)*, **28**, 325–53.

Ben Khay, A., Gordoux, L., Moreau, R. and Dutrieu, J. (1987) Effect of intestinal

insulin-like peptide on glucose catabolism in male adult *Locusta migratoria*. *Arch. Insect Biochem. Physiol.*, **4**, 233–9.
Billingsley, P.F. and Downe, A.E.R. (1986) Nondigestive cell types in the midgut epithelium of *Rhodnius prolixus* (Hemiptera: Reduviidae). *J. Med. Entomol.*, **23**, 212–16.
Blackburn, M.B., Kingan, T.G., Bodnar, W. *et al.* (1991) Isolation and identification of a new diuretic peptide from the tobacco hornworm, *Manduca sexta*. *Biochem. Biophys. Res. Commun.*, **181**, 927.
Brown, B.E. and Starratt, A.N. (1975) Isolation of proctolin, a myotropic peptide, from *Periplaneta americana*. *J. Insect Physiol.*, **21**, 1879–81.
Brown, M.R., Crim, J.W. and Lea, A.O. (1986) FMRFamide- and pancreatic polypeptide-like immunoreactivity of endocrine cells in the midgut of a mosquito. *Tissue Cell*, **18**, 419–28.
Brown, M.R., Klowden, M.J., Crim, J.W. *et al.* (1994) Endogenous regulation of mosquito host-seeking behavior by a neuropeptide. *J. Insect Physiol.*, **40**, 399–406.
Brown, M.R. and Lea, A.O. (1989) Neuroendocrine and midgut endocrine systems in the adult mosquito. *Adv. Dis. Vector Res.*, **6**, 29–58.
Champagne, D.E. and Ribeiro, J.M.C. (1994) Sialokinin-I and sialokinin-II – vasodilatory tachykinins from the yellow-fever mosquito *Aedes aegypti*. *Proc. Natl Acad. Sci. USA*, **91**, 138–42.
Cantera, R., Veenstra, J.A. and Nässel, D.R. (1994) Postembryonic development of corazonin-containing neurons and neurosecretory cells in the blowfly, *Phormia terranovae*. *J. Comp. Neurol.*, **350**, 559–72.
Cassier, P., Albert, J. and Fain-Maurel, M.A. (1972) Sur la présence de cellules de type endocrine dans l'intestin moyen de *Petrobius maritimus* Leach (Insecte aptèrygote, Thysanoure). *C. R. Acad. Sci. Paris*, **281**, 1405–7.
Cassier, P. and Fain-Maurel, M.A. (1977) Sur la présence d'un système endocrine diffus dans le mésenteron de quelques insects. *Arch. Zool. Exp. Gen.*, **118**, 197–209.
Chen, Y.T., Veenstra, J.A., Hagedorn, H. and Davis, N.T. (1994) Leucokinin and diuretic hormone immunoreactivity of neurons in the tobacco hornworm. *Manduca sexta*, and colocalization of this immunoreactivity in lateral neurosecretory cells of abdominal ganglia. *Cell Tissue Res.*, **278**, 493–507.
Chin, A., Reynolds, E. and Scheller, R.H. (1990) Organization and expression of the *Drosophila* FMRFamide-related prohormone gene. *DNA Cell Biol.*, **9**, 263–71.
Coast, G.M. and Kay, I. (1994) The effects of *Acheta* diuretic peptide on isolated Malpighian tubules from the house cricket, *Acheta domesticus*. *J. Exp. Biol.*, **187**, 225–43.
Coast, G.M., Rayne, R.C. Hayes, T.K. *et al.* (1993) A comparison of the effects of two putative diuretic hormones from *Locusta migratoria* on isolated locust Malpighian tubules. *J. Exp. Biol.*, **175**, 1–14.
Copenhaver, P.F. and Taghert, P.H., (1989a) Development of the enteric nervous system in the moth 1. Diversity of cell types and the embryonic expression of FMRFamide-related neuropeptides. *Dev. Biol.*, **131**, 70–84.
Copenhaver, P.F. and Taghert, P.H. (1989b) Development of the enteric nervous system in the moth 2. Stereotyped cell migration precedes the differentiation of embryonic neurons. *Dev. Biol.*, **131**, 85–101.
Copenhaver, P.F. and Taghert, P.H. (1991) Origins of the insect enteric nervous system: differentiation of the enteric ganglia from a neurogenic epithelium. *Development*, **113**, 1115–32.
Crim, J.W., Jenkins, A.C. and Brown, M.R. (1992) A mosquito neuropeptide in a

moth larva (*Helicoverpa zea*): relation to FMRF-amide immunoreactivity. *Tissue Cell*, **24**, 537–45.

De Loof, A. (1987) The impact of the discovery of vertebrate-type steroids and peptide hormone-like substances in insects. *Entomol. Exp. Appl.*, **45**, 105–13.

De Loof, A. and Schoofs, L. (1990) Homologies between the amino acid sequence of some vertebrate pepticide hormones and peptides isolated from invertebrate sources. *Comp. Biochem. Physiol.*, **95B**, 459–68.

de Priester, W. (1971) Ultrastructure of the midgut epithelial cells in the fly *Calliphora erythrocephala*. *J. Ultrastruct. Res.*, **36**, 783–805.

Desbuquois, B. (1990) Gastrointestinal hormones, in *Hormones: from Molecules to Disease* (eds E.-E. Baulieu and P.A. Kelly), Chapman & Hall, London, pp. 539–89.

Dockray, G.J. (1979) Comparative biochemistry and physiology of gut hormones. *Annu. Rev. Physiol.*, **41**., 83–95.

Dockray, G.J. (1988) Regulatory peptides and the neuroendocrinology of gut–brain relations. *Q. J. Exp. Physiol.*, **73**, 703–27.

Donly, B.C., Ding, Q., Tobe, S.S. and Bendena, W.G. (1993) Molecular cloning of the gene for the allatostatin family of neuropeptides from the cockroach *Diploptera punctata*. *Proc. Natl Acad. Sci. USA*, **90**, 8807–11.

Duve, H., Elia, A.J., Orchard, I. *et al.* (1993) The effects of CalliFMRFamides and other FMRFamide-related neuropeptides on the activity of the heart of the blowfly *Calliphora vomitoria*. *J. Insect Physiol.*, **39**, 31–40.

Duve, H., Johnsen, A.H., Sewell, J.C. *et al.* (1992) Isolation, structure, and activity of -Phe-Met-Arg-Phe-NH_2 neuropeptides (designated calliFMRF-amides) from the blowfly *Calliphora vomitoria*. *Proc. Natl Acad. Sci. USA*, **89**, 2326–30.

Duve, F., Rehfeld, J.F., East, P. and Thorpe, A. (1994) Localization of sulfakinin neuronal pathways in the blowfly *Calliphora vomitoria*. *Cell Tissue Res.*, **275**, 177–86.

Duve, H. and Thorpe, A. (1982) The distribution of pancreatic polypeptide in the nervous system and gut of the blowfly, *Calliphora vomitoria* (Diptera). *Cell Tissue Res.*, **227**, 67–77.

Duve, H. and Thorpe, A. (1994) Distribution and functional significance of Leu-callatostatins in the blowfly *Calliphora vomitoria*. *Cell Tissue Res.*, **276**, 367–79.

East, P.D., Thorpe, A. and Duve, H. (1995) Leu-callatostatin gene expression in the blowflies *Calliphora vomitoria* and *Lucilia cuprina* studied by *in situ* hybridization: Comparison with Leu-callatostatin confocal laser scanning immunocytochemistry. *Cell Tissue Res.*, **280**, 355–64.

Eckert, M., Agricola, H. and Penzlin, H. (1981) Immunocytochemical identification of proctolin-like immunoreactivity in the terminal ganglion and hindgut of the cockroach *Periplaneta americana* (L). *Cell Tissue Res.*, **217**, 633–45.

Elia, A.J. Money T.G.A. and Orchard, I. (1995) Flight and running induce elevated levels of FMRFamide-related peptides in the haemolymph of the cockroach, *Periplaneta americana* (L.). *J. Insect Physiol.*, **41**, 565–70.

El-Salhy, M., Abou-El-Ela, R., Falkmer, S. *et al.* (1980) Immunohistochemical evidence of gastro-entero-pancreatic neurohormonal peptides of vertebrate type in the nervous system of the larva of a dipteran insect, the hoverfly *Eristalis aenus*. *Regul. Pept.*, **1**, 187–204.

Endo, Y., Iwanaga, T. and Fujita, T. (1990) Gut endocrine cells of invertebrates, in *Progress in Comparative Endocrinology* (eds A. Epple, C.G. Scanes and M.H. Stetson), Wiley-Liss, New York, pp. 499–503.

Endo, Y., Iwanaga, T., Fujita, T. and Nishiitsutsui Uwo, J. (1982) Localization of

pancreatic polypeptide (PP)-like immunoreactivity in the central and visceral nervous systems of the cockroach *Periplaneta, Cell Tissue Res.*, **227**, 1–9.
Endo, Y. and Nishiitsutsuji-Uwo, J. (1981) Gut endocrine cells in insects: the ultrastructure of the gut endocrine cells of the lepidopterous species. *Biomed. Res.*, **2**, 270–80.
Endo, Y., Sugihara, H., Fujita, S. and Nishiitsutsui-Uwo, J. (1983) Kinetics of columnar and endocrine cells in the cockroach midgut. *Biomed. Res.*, **4**, 51–60.
Fonagy, A., Matsumoto, S., Schoofs, L. *et al.* (1992) *In vivo* and *in vitro* pheromonotropic activity of two locustatachykinin peptides in *Bombyx mori. Biosci. Biotech. Biochem.*, **56**, 1692–3.
Fontaine, J. and LeDouarin, N. (1977) Analysis of endoderm formation in the avian blastoderm by the use of quail-chick chimaeras. The problem of neuroectodermal origin of the cells of the APUD series. *J. Embryol. Exp. Morphol.*, **41**, 209–22.
Fujita, T., Kanno, T. and Kobayashi, S. (1988). *The Paraneuron*, Springer, Heidelberg, 367 pp.
Fujita, T., Yui, R., Iwanaga, T. *et al.* (1981) Evolutionary aspects of 'brain–gut' peptides: an immunohistochemical study. *Peptides*, **2** (Suppl. 2), 123–31.
Gelman, D.B., Thyagaraja, B.S., Kelly, T.J. *et al.* (1991) The insect gut: a new source of ecdysiotropic peptides. *Experientia*, **47**, 77–80.
Greenberg, M.J., Payza, K., Nachman, R.J. *et al.* (1988) Relationships between the FMRFamide-related peptides and other peptide families. *Peptides*, **9** (Suppl. 1), 613–18.
Grimmelikhuijzen, C.J.P., Graff, D. and McFarlane, I.D. (1989) Neurones and neuropeptides in coelenterates. *Arch. Histol. Cytol.*, **52** (Suppl.), 265–76.
Hecker, H., Freyvogel, T.A., Briegel, H. and Steiger, R. (1971) Ultrastructural differentiation of the midgut epithelium in female *Aedes aegypti* L. (Insecta, Diptera) imagines. *Acta Trop.*, **28**, 80–104.
Hertel, W. and Penzlin, H. (1992) Function and modulation of the antennal heart of *Periplaneta americana. Acta Biol. Hung.*, **43**, 113–25.
Holets, V.R., Hokfelt, T., Ude, J. *et al.* (1984) Coexistence of proctolin with TRH and 5-HT in the rat CNS. *Soc. Neurosci.*, **10**, 692 (abstr.).
Holman, G.M., Cook, B.J. and Nachman, R.J. (1986) Primary structure and synthesis of a blocked myotropic neuropeptide isolated from the cockroach, *Leucophaea maderae. Comp. Biochem. Physiol.*, **85C**, 329–33.
Ivanova-Kazas, O.M. (1981) *Comparative Embryology of the Invertebrates. Atelocerata*, Nauka, Moscow, 207 pp. (In Russian).
Iwanaga, T., Fujita, T., Nishiitsutsui-Uwo, J. and Endo, Y. (1981) Immunohistochemical demonstration of PP- somatostatin-, enteroglucagon-, and VIP-like immunoreactivities in the cockroach midgut. *Biomed. Res.*, **2**, 202–7.
Iwanaga, T., Fujita, T., Takeda, N. *et al.* (1986) Urotensin I-like immunoreactivity in the midgut endocrine cells of the insects *Gryllus bimaculatus* and *Periplaneta americana. Cell Tissue Res.*, **244**, 565–8.
Jenkins, A.C., Brown, M.R. and Crim, J.W. (1989) FMRF-amide immunoreactivity in the midgut of the corn earworm (*Heliothis zea*). *J. Exp. Zool.*, **252**, 71–8.
Kataoka, H., Toschi, A., Li, J.P. *et al.* (1989a) Identification of an allatotropin from adult *Manduca sexta, Science*, **243**, 1481–3.
Kataoka, H., Troetschler, R.G., Li, J.P. *et al.* (1989b) Isolation and identification of a diuretic hormone from the tobacco hornworm, *Manduca sexta. Proc. Natl Acad. Sci. USA*, **86**, 2976–80.
Kawakami, A., Iwami, M., Nagasawa, H. *et al.* (1989) Structure and organization

of four clustered genes that encode bombyxin, an insulin-related brain secretory peptide of the silkmoth *Bombyx mori*. *Proc. Natl Acad. Sci. USA*, **86**, 6843–7.

Kay, I., Coast, G.M., Cusinato, O. *et al.* (1991a) Isolation and characterization of a diuretic peptide from *Achaeta domesticus*: evidence for a family of diuretic peptides. *Biol. Chem. Hoppe-Seyler*, **372**, 505–12.

Kay, I., Patel, M., Coast, G.M. *et al.* (1992) Isolation, characterization and biological activity of a CRF-related diuretic peptide from *Periplaneta americana* L. *Regul. Pept.*, **42**, 111–22.

Kay, I., Wheeler, C.H., Coast, G.M. *et al.* (1991b) Characterization of a diuretic peptide from *Locusta migratoria*. *Biol. Chem. Hoppe-Seyler*, **372**, 929–34.

Kingan, T.G., Teplow, D.B., Phillips, J.M. *et al.* (1990) A new peptide of the FMRFamide family isolated from the CNS of the hawkmoth, *Manduca sexta*. *Peptides*, **11**, 849–56.

Kirby, P., Beck, R. and Clarke, K.U. (1984) The stomatogastric nervous system of the house cricket *Acheta domesticus* L. I. The anatomy of the system and the innervation of the gut. *J. Morphol.*, **180**, 81–103.

Kobayashi, M. (1971) Fine structure of the basal granular cell in the mid-gut epithelium of the silkworm, *Bombyx mori*. *J. Sericult. Sci. Jap.*, **40**, 101–6.

Lagueux, M., Lwoff, L., Meister, M. *et al.* (1990) cDNAs from neurosecretory cells of brains of *Locusta migratoria* (Insecta, Orthoptera) encoding a novel member of the superfamily of insulins. *Eur. J. Biochem.*, **187**, 249–54.

Lange, A.B., Chan, K.K. and Stay, B. (1993) Effects of allatostatins and proctolin on antennal pulsatile organ and hindgut muscle in the cockroach. *Diploptera punctata*. *Arch. Insect Biochem. Physiol.*, **24**, 79–92.

Lehmberg, E., Ota, R.B., Furuya, K. *et al.* (1991) Identification of a diuretic hormone of *Locusta migratoria*. *Biochem. Biophys. Res. Commun.*, **179**, 1036–41.

Li, X.-J., Wu, Y.-N., North, A. and Forte, M. (1992) Cloning, functional expression, and developmental regulation of a neuropeptide-Y receptor from *Drosophila melanogaster*. *J. Biol. Chem.*, **267**, 9–12.

Lorenz, M.W., Kellner, R. and Hoffmann, K.H. (1995) Identification of two allatostatins from the cricket, *Gryllus bimaculatus* de Geer (Ensifera, Gryllidae): additional members of a family of neuropeptides inhibiting juvenile hormone biosynthesis. *Regul. Pept.*, **57**, 227–36.

Lundquist, C.T., Clottens, F.L., Holman, G.M. *et al.* (1994a) Callitachykinin-I and callitachykinin-III, 2 novel myotropic peptides isolated from the blowfly, *Calliphora vomitoria*. *Peptides*, **15**, 761–8.

Lundquist, C.T., Clottens, F.L., Holman, G.M. *et al.* (1994b) Locustatachykinin immunoreactivity in the blowfly central-nervous-system and intestine. *J. Comp. Neurol.*, **341**, 225–40.

Matsumoto, S., Brown, M.R., Crim, J.W. *et al.* (1989) Isolation and primary structure of neuropeptides from the mosquito, *Aedes aegypti*, immunoreactive to FMRFamide antiserum. *Insect Biochem.*, **19**, 277–83.

McCormick, J. and Nichols, R. (1993) Spatial and temporal expression identity of dromyosuppressin as a brain–gut peptide in *Drosophila melanogaster*. *J. Comp. Neurol.*, **338**, 278–88.

McGuigan, J.E. (1968) Immunochemical studies with synthetic human gastrin. *Gastroenterology*, **54**, 1005–11.

Mizoguchi, A., Ishizaki, H., Nagasawa, H. *et al.* (1987) A monoclonal antibody against a synthetic fragment of bombyxin (4K-prothoracicotropic hormone) from the silkmoth, *Bombyx mori*: characterization and immunohistochemistry. *Mol. Cell Endocrinol.*, **51**, 227–35.

Mizoguchi, A., Oka, T., Kataoka, H. *et al.* (1990) Immunohistochemical localization of prothoracicotropic hormone-producing neurosecretory cells in the brain of *Bombyx mori*. *Dev. Growth Diff.*, **32**, 579–86.

Moreau, R., Raoelison, C. and Sutter, B.C.J. (1981) An intestinal insulin-like molecule in *Apis mellifera* L. (Hymenoptera). *Comp. Biochem. Physiol.*, **69A**, 79–83.

Muley, H. and Davey, K.G. (1995) The feeding stimulus in *Rhodnius prolixus* is transmitted to the brain by a humoral factor. *J. Exp. Biol.*, **198**, 1087–92.

Nachman, R.J., Holman, G.M., Cook, B.J. *et al.* (1986b) Leucosulphakinin-II, a blocked sulphated insect neuropeptide with homology to cholecystokinin and gastrin. *Biochem. Biophys. Res. Commun.*, **140**, 357–64.

Nachman, R.J., Holman, G.M., Haddon, W.F. and Ling, N. (1986a) Leucosulphakinin, a sulphated insect neuropeptide with homology to gastrin and cholecystokinin. *Science*, **234**, 71–3.

Nambu, J.R., Murphy-Erdosh, C., Andrews, P. *et al.* (1988) Isolation of a *Drosophila* neuropeptide family. *Neuron*, **1**, 55–61.

Nässel, D.R. (1993) Insect myotropic peptides: differential distribution of locustatachykinin and leukokinin-like immunoreactive neurons in the locust brain. *Cell Tissue Res.*, **273**, 1–29.

Nässel, D.R. Karlsson, A., Kim M.-Y. *et al.* (1995) Tachykinin-related neuropeptides in the insect nervous system: structure, distribution and putative functions, in *Insects, Chemical, Physiological and Environmental Aspects* (ed. D. Konopińska), University of Wroclaw, Poland, pp. 242–7.

Nichols, R. (1992a) Isolation and structural characterization of *Drosophila* TDVDHVFLRFamide and FMRFamide-containing neural peptides. *J. Mol. Neurosci.*, **3**, 213–18.

Nichols, R. (1992b) Isolation and expression of the *Drosophila* drosulfakinin neural peptide gene product, DSK-I. *Mol. Cell. Neurosci.*, **3**, 342–7.

Nichols, R., Schneuwly, A. and Dixon J.E. (1988) Identification and characterization of a *Drosophila* homologue to the vertebrate neuropeptide cholecystokinin. *J. Biol. Chem.*, **263**, 12167–70.

Nishiitsutsui-Uwo, J. and Endo, Y. (1981) Gut endocrine cells in insects: the ultrastructure of the endocrine cells in the cockroach midgut. *Biomed. Res.*, **2**, 30–44.

Orchard, I., Belanger, J.H. and Lange, A.B. (1989) Proctolin: a review with emphasis on insects. *J. Neurobiol.*, **20**, 470–96.

O'Shea, M. and Adams, M. (1986) Proctolin: from 'gut factor' to model neuropeptide. *Adv. Insect Physiol.*, **19**, 1–28.

Paemen, L., Tips, A., Schoofs, L. *et al.* (1991) A novel myotropic peptide from the male accessory glands of *Locusta migratoria*. *Peptides*, **12**, 7–10.

Patel, N., Hayes, T.K. and Coast, G.M. (1995) Evidence for the hormonal function of a CRF-related diuretic peptide (*Locusta*-DP) in *Locusta migratoria*. *J. Exp. Biol.*, **198**, 793–804.

Pearse, A.G.E. (1969) The cytochemistry and ultrastructure of polypeptide hormone-producing cells of the Apud series and the embryologic, physiologic and pathologic implications of the concept. *J. Histochem. Cytochem.*, **17**, 303–13.

Penzlin, H. (1985) Stomatogastric nervous system, in *Comprehensive Insect Physiology, Biochemistry and Pharmacology* Vol. 5 (eds G.A. Kerkut and L.I. Gilbert), Pergamon Press, Oxford, pp. 371–406.

Penzlin, H., Wieduwilt, I. and Hertel, W. (1989) Evidence for a myotropic effect of substance P in *Periplaneta americana* L. *Gen. Comp. Endocrinol.*, **75**, 88–95.

Pratt, G.E., Farnsworth, D.E., Siegel, N.R. *et al.* (1989) Identification of an allatostatin from adult *Diploptera punctata*. *Biochem. Biophys. Res. Commun.*, **163**, 1243–7.

Price, D. and Greenberg, M. (1977) Structure of a molluscan cardioexcitatory neuropeptide. *Science*, **189**, 670–1.
Proux, J., Miller, C.A., Li, J.P. *et al.* (1987) Identification of an arginine-vasopressin-like diuretic hormone from *Locusta migratoria. Biochem. Biophys. Res. Commun.*, **149**, 180–6.
Puiroux, J., Pedelaborde, A. and Loughton, B.G. (1992) Characterization of a proctolin binding site on locust hindgut membranes. *Insect Biochem. Mol. Biol.*, **22**, 547–51.
Rafaeli, A., Moshitzky, P., Applebaum, S.W. (1987) Functional similarity and immunological cross-reactivity of locust diuretic hormone and corticotropin. *Gen. Comp. Endocrinol.*, **67**, 1–6.
Raffa, R.B. (1988) The action of FMRFamide (Phe-Met-Arg-Phe-NH_2) and related peptides on mammals. *Peptides*, **9**, 915–22.
Rawdon, B.B. and Andrew, A. (1990) Vertebrate gut endocrine cells: comparative and developmental aspects, in *Progess in Comparative Endocrinology* (eds A. Epple, C.G. Scanes and H.H. Stetson), Wiley-Liss, New York, pp. 504–9.
Reichwald, K., Unnithan, G.C., Davis, N.T. *et al.* (1994) Expression of the allatostatin gene in endocrine cells of the cockroach midgut. *Proc. Natl Acad. Sci. USA*, **91**, 11894–8.
Reinecke, J.P., Cook, B.J. and Adams, T.J. (1973) Larval hindgut of *Manduca sexta* (L.) (Lepidoptera: Sphingidae). *Int. J. Insect Morphol. Embryol.*, **2**, 277–90.
Reuter, M. and Gustafsson, M. (1989) 'Neuroendocrine cells' in flatworms – progenitors to metazoan neurones? *Arch. Histol. Cytol.*, **52**, 253–63.
Robb, S., Packman, L.C. and Evans, P.D. (1989) Isolation, primary structure and bioactivity of schistoFLRG-amide, a FMRF-amide-like neuropeptide from the locust, *Schistocerca gregaria. Biochem. Biophys. Res. Commun.*, **160**, 850–6.
Schneider, L.E. and Taghert, P.H. (1988) Isolation and characterization of a *Drosophila* gene that encodes multiple neuropeptides related to Phe-Met-Arg-Phe-NH_2 (FMRFamide). *Proc. Natl Acad. Sci. USA*, **85**, 1993–7.
Schols, D., Verhaert, P., Huybrechts, R. *et al.* (1987) Immunocytochemical demonstration of propiomelanocortin- and other opioid-related substances and a CRF-like peptide in the gut of the American cockroach. *Periplaneta americana* L. *Histochemistry*, **86**, 345–51.
Schoofs, L., Danger, J.M., Jegou, S. *et al.* (1988) NPY-like peptides occur in the nervous system and midgut of the migratory locust, *Locusta migratoria* and in the brain of the grey fleshfly, *Sarcophaga bullata. Peptides*, **9**, 1027–36.
Schoofs, L., Holman, G.M., Hayes, T.K. *et al.* (1990a) Locustatachykinins I and II, two novel insect neuropeptides with homology to peptides of the vertebrate tachykinin family. *FEBS Lett.*, **261**, 397–401.
Schoofs, L., Holman, G.M., Hayes, T.K. *et al.* (1990b) Locustatachykinin III and IV: two additional insect neuropeptides with homology to peptides of the vertebrate tachykinin family. *Regul. Pept.*, **31** 199–212.
Schoofs, L., Van den Broeck, J. and De Loof, A. (1993) The myotropic peptides of *Locusta migratoria:* structures, distribution, functions and receptors. *Insect Biochem. Mol. Biol.*, **23**, 859–81.
Schooley, D.A., Miller, C.A. and Proux, J.P. (1987) Isolation of two arginine vasopressin-like factors from ganglia of *Locusta migratoria. Arch Insect Biochem. Physiol.*, **5**, 157–66.
Sehnal, F. and Žitňan, D. (1990) Endocrines of insect gut, in *Progess in Comparative Endocrinology* (eds A. Epple, C.G. Scanes and M.H. Stetson), Wiley-Liss, New York, pp. 510–15.
Stefano, G.B., Leung, M.K., Zhao, X. and Scharrer, B. (1989) Evidence for the

involvement of opioid neuropeptides in the adherence and migration of immunocompetent invetebrate hemocytes. *Proc. Natl Acad. Sci. USA*, **86**, 626–30.

Teller, J.K., Rosiński, G., Pilc, L. *et al.* (1983) The presence of insulin-like hormone in head and midguts of *Tenebrio molitor* L. (Coleoptera) larvae. *Comp. Biochem. Physiol.*, **74A**, 463–5.

Truman, J.W. and Copenhaver, P.F. (1989) The larval eclosion hormone neurones in *Manduca sexta:* identification of the brain–proctodeal neurosecretory system. *J. Exp. Bio.*, **147**, 457–70.

Veenstra, J.A. and Hagedorn, H.H. (1993) Sensitive enzyme immunoassay for *Manduca* allatotropin and the existence of an allatotropin-immunoreactive peptide in *Periplaneta americana. Arch. Insect Biochem. Physiol.*, **23**, 99–109.

Veenstra, J.A. and Lambrou, G. (1995) Isolation of a novel RFamide peptide from the midgut of the American cockroach, *Periplaneta americana. Biochem. Biophys. Res. Commun.*, **213**, 519–24.

Veenstra, J.A., Lehman, H.K. and Davis, N.T. (1994) Allatotropin is a cardioacceleratory peptide in *Manduca sexta. J. Exp. Biol.*, **188**, 347–54.

Veenstra, J.A. and Schooneveld, H. (1984) Immunocytochemical localization of neurons in the nervous system of the colorado beetle with antisera against FMRFamide and bovine pancreatic polypeptide. *Cell Tissue Res.*, **235**, 303–8.

Westfall, J.A. (1973) Ultrastructural evidence for a granule-containing sensory-motor interneuron in *Hydra litoralia. J. Ultrastruct. Res.*, **42**, 268–82.

Woodhead, A.P., Stay, B., Seidel, S.L. *et al.* (1989) Primary structure of four allatostatins: neuropeptide inhibitors of juvenile hormone synthesis. *Proc. Natl Acad. Sci. USA*, **86**, 5997–6001.

Yi, S.-X., Tirry, L., Bai, C. *et al.* (1995) Isolation, identification, and synthesis of Mas-MG-MT I, a novel peptide from the larval midgut of *Manduca sexta* (Lepidoptera: Sphingidae). *Arch. Insect Biochem. Physiol.*, **28**, 159–71.

Yin, C.M., Zou, B.X., Li, M.F. and Stoffolano, J.G. (1994) Discovery of a midgut peptide hormone which activates the endocrine cascade leading to oogenesis in *Phormia regina* (Meigen). *J. Insect Physiol.*, **40**, 283–92.

Žitňan, D., Endo, Y. and Sehnal, F. (1989) Stomatogastric nervous system of *Galleria mellonella* L. (Lepidoptera: Pyralidae); changes during metamorphosis with special reference to FMRFamide neurons. *Int. J. Insect Morphol. Embryol.*, **18**, 227–37.

Žitňan, D., Kingan, T.G. and Beckage, N.E. (1995) Parasitism-induced accumulation of FMRFamide-like peptides in the gut innervation and endocrine cells of *Manduca sexta. Insect Biochem. Mol. Biol.*, **25**, 669–78.

Žitňan, D., Šauman, I. and Sehnal, F. (1993) Peptidergic innervation and endocrine cells of insect midgut. *Arch. Insect Biochem. Physiol.*, **22**, 113–32.

Žitňan, D., Sehnal, F., Mizoguchi, A. *et al.* (1990) Developmental changes in the bombyxin- and insulin-like immunoreactive neurosecretory system in the wax moth, *Galleria mellonella. Dev. Growth Diff.*, **32**, 637–45.

4

The peritrophic matrix

R.L. Tellam

4.1 INTRODUCTION

Two historical motivations for the study of insects involves their competition with human beings for the same food supply and their influence on human health. The insect midgut has pivotal roles in both of these aspects of the relationship between humans and insects (see also Chapters 6–8 and 16). In particular, the functions of the peritrophic matrix (PM) which lines the gut of most insects, is intimately associated with the digestive process in insects and the cycle of invasion and transmission of many insect-borne pathogens. Despite its central importance in these events, there is relatively little known of the detailed molecular structure and functions of the PM. In part, this deficiency is caused by the wide variety of insect PM structures and the multiple functions of this matrix. Because of this variety, it is impractical to comprehensively describe all PM structures. Rather, examples will be cited which illustrate general aspects of the PM structure, recent advances in the understanding of its molecular structure and unifying concepts relating to the functions of this matrix.

4.1.1 Definition of peritrophic matrix

First observed last century in insects, the PM was described as a 'membranous sac which directly surrounds the food in the lumen of the intestine' (Balbiani, 1890). Since then the name has also included any

Biology of the Insect Midgut.
Edited by M.J. Lehane and P.F. Billingsley.
Published in 1996 by Chapman & Hall, London.
ISBN 0 412 61670 X.

membranous layer which lines the insect gut (e.g. Figure 4.1). Peters (1992) emphasized the point that most insect midguts contain a number of discrete PMs and suggested that these membranes collectively should be called a peritrophic envelope. There has been criticism of the term peritrophic membrane because of the inference that it is related to the cytoplasmic membrane of cells whereas the PM is acellular being principally composed of chitin, glycosaminoglycans and proteins. Consequently, the term peritrophic matrix is used here. Richards and Richards (1977) listed a number of features which are indicative of a PM:

1. a positive test for chitin;
2. a membrane(s) forming a tube or sac around ingested food and capable of being physically separated from the rest of the gut;
3. a line separating the food from the digestive epithelia in histological sections;
4. any recognizable layer around the food;
5. any membranous or filamentous secretion of midgut cells whether or not it is concerned with food.

Peters (1992) has remarked on the great variety of PM types and their lack of absolute conformity to these characteristics. The semipermeable nature of the matrix may be a common functional characteristic which could be added to the above list. In most cases, the original description given by Balbiani (1890) is adequate with the provision that the PM may have evolved to include many divergent structures and functions.

4.1.2 Species range

Although the PM is often thought to be only associated with insects, Peters (1992) in a comprehensive review, described the presence of PM-like structures in many arthropod classes as well as several other phyla. He also noted the presence of a mucoid-type layer in the gut of many animals including vertebrates and suggested that this layer may be functionally analogous to a PM. The majority of the Insecta contain PMs, at least at some stage of their life cycle. There is a minority of insects which do not contain a PM in any of their life stages, e.g. some lice species (Phthiraptera) (Waterhouse, 1953a; Peters, 1992). The PM is developmentally expressed in different life stages of many insects. For example, the adult flea (Siphonaptera) does not contain a PM whereas its larval stage does (Peters, 1992). The structural form and number of PMs can also be very different between the various life stages of the same insect. Some PMs are continually expressed whereas others are induced by the ingestion of a meal. One of the best examples of the latter is the production of PMs in adult female mosquitoes which occurs after ingestion of a blood meal. Consequently, the establishment of the

(a)

(b)

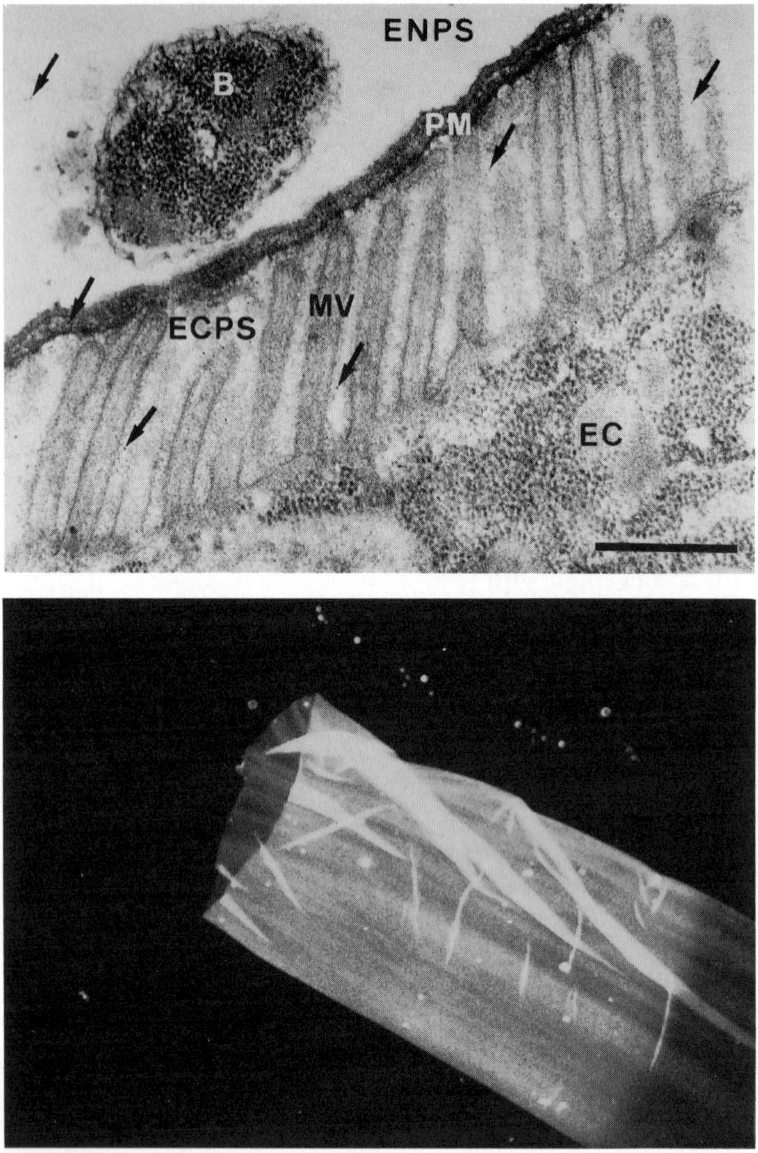

Figure 4.1 Peritrophic matrix from higher Diptera. (a) Electron micrograph of peritrophic membrane from larva of *Lucilia cuprina*; PM, peritrophic membrane; MV, microvilli, ECPS, ectoperitrophic matrix space; ENPS, endoperitrophic matrix space; EC, midgut epithelial cells; B, bacterium. The scale bar represents 500 nm. The arrows denote 6 nm colloidal gold particles which freely diffuse from the gut lumen through the peritrophic matrix to the ectoperitrophic matrix space (courtesy of C. Eisemann). (b) Peritrophic matrix dissected from adult *Haematobia exigua irritans* and labelled with FITC-wheatgerm lectin (courtesy of D. Kemp and P. Allingham).

presence or absence of a PM in an insect species requires careful examination of each life stage of the insect and an awareness that the PM may be presented in many different structural forms.

The operational definition used to define the presence of PMs in an insect species often places considerable weight on the detection of chitin. It is therefore important to consider the specificity and sensitivity of different methods used to detect chitin. One of the most widely used tests for the detection of chitin is the chitosan reaction. Several authors have commented on the uncertainty of this technique which is attributed to a number of factors including the presence of other sugar types and proteins (Peters, 1992). A second widely used method involves the interaction of fluorescently labelled wheatgerm lectin (or gold-labelled wheatgerm lectin) with the linear polymers of *N*-acetyl-D-glucosamine which make chitin. It is assumed that wheatgerm lectin binds to chitin embedded in the PM. An alternative possibility is that this lectin binds to oligosaccharides with terminal *N*-acetyl-D-glucosamine or *N*-acetylneuraminic acid residues on glycoproteins present in or on the PM. Indeed, an integral PM glycoprotein (peritrophin-95) from the larvae of *Lucilia cuprina* has been isolated and shown to bind specifically wheatgerm lectin (Tellam, unpublished result). Consequently, considerable care should be exercised in the interpretation of data obtained by wheatgerm lectin localizations. Peters and Latka (1986) suggested that the binding of wheatgerm lectin to chitin can be differentiated from that to glycoproteins on the basis of the greater strength of the former interaction as determined by the ability of competitive sugars to inhibit the interaction. However, it is more appropriate that a combination of technical approaches is used for the identification of chitin in a PM. Peters (1992) has listed a number of additional techniques which can also detect chitin.

4.2 PERITROPHIC MATRIX STRUCTURE AND SYNTHESIS

4.2.1 Types of peritrophic matrix

As there is a great diversity in the structures of insect PMs (Peters, 1992), the definition of strict structural classes of PMs is difficult. The most widely accepted system involves the categorization of PMs into two classes which are differentiated on the basis of their mode of synthesis (Wigglesworth, 1930). Type I PM is synthesized by the entire midgut digestive epithelium, often in response to distension of the midgut during feeding. In many instances this PM forms a sac completely surrounding the ingested food. Type I PMs are the most widespread in insects and are particularly prevalent in lepidopterans (Spence, 1991;

and Ryerse *et al.*, 1992 for an alternative view). Type II PM is produced as a continuous tube-like structure from a small belt of specialized cells at the anterior end of the midgut, i.e. the cardia. This class of PM is present before the ingestion of food by the insect. The larvae of all Diptera produce type II PMs (Peters, 1992). Both types of PM can be synthesized by different developmental stages of the same species. The larvae of mosquitoes produce a type II PM whereas the adult produces a type I PM after a blood meal (Freyvogel and Stäubli, 1965; Richards and Richards, 1971; Richardson and Romoser, 1972; Houk *et al.*, 1979; Berner *et al.*, 1983). Like all systems of categorization there are some insects which produce PMs which do not easily fit into these two classes (Spence, 1991). These insects include the weevils *Cionus scrophulariae* and *C. pulchelles* which secrete a PM from the posterior end of the midgut and the beetle *Ptinus tectus* which synthesizes PM from specialized cells in the middle of their midgut (Spence, 1991).

4.2.2 Peritrophic matrix structure

The PM structure is probably most highly developed in the larvae of the Diptera which typically produce a single type 2 PM characterized by distinct and regular electron-lucent and electron-dense layers (Binnington, 1988; Peters, 1992). Often the electron-dense layer is associated with the luminal surface of the PM and is thought to correspond to proteoglycans present in the PM (Peters, 1992). Adult Diptera, in contrast, typically produce multiple tubular PMs, each arising from distinct regions in the cardia (Peters, 1992). For example, the blowfly *Lucilia cuprina* produces a single PM in the larval stage and three PMs in the adult stage (Binnington, 1988) whereas the adult stable fly (*Stomoxys calcitrans*) produces two individual PMs which unite in the cardia to form a single composite structure in the midgut (Lehane, 1976). The striking feature of all Dipteran PMs is their regular lamellar structure and constant thickness. This structural regularity probably reflects an ordered molecular structure of the components within these PMs and their highly defined mode of synthesis from the cardia. The thickness of type 2 PMs from the Diptera are typically in the range of approximately 100–2000 nm (Spence, 1991; Peters, 1992). The minimum PM thickness is 10 times greater than the thickness of a cellular cytoplasmic membrane. Thus, PMs are a considerable physical barrier between the contents of the insect gut lumen and the underlying digestive epithelial cells of the insect midgut.

Type I PMs generally do not contain discrete layered structures. In some insects they are intimately associated with the microvilli of the digestive epithelial cells lining the insect midgut. The microvilli are thought to synthesize the chitin fibrils and other molecular components of the PM and act as a 'template' for the construction of type I PM (Mercer and Day, 1952). Often the regular arrangement of the microvilli is imprinted on the PM structure giving rise to hexagonal (or honeycomb, i.e. chitin fibres arranged at 60° to each other) and orthogonal (or grid-like, i.e. chitin fibres arranged at 90° to each other) textures (Peters, 1992). Similar structures are not found in other chitinous structures, such as insect cuticle. The majority of insect species, however, contain PMs characterized by a random (felt-like) distribution of the chitin microfibrils. The formation of type I PMs is thought to occur in three steps: first, the secretion of chitin fibril precursors from the microvilli of the midgut digestive epithelial cells; second, the maturation of this material into a cross-linked network and; third, the attachment of a matrix of material (probably protein and proteoglycans) which interlocks the chitin fibrils (Reid and Lehane, 1984; Blackburn *et al.*, 1988; Martin and Kirkham, 1989; Weaver and Scott, 1990; Spence, 1991; Walters *et al.*, 1993; Ryerse *et al.*, 1994). Thus, the PM is a structure which has a number of maturation phases.

4.2.3 The production of peritrophic matrix

In many insects the PM is produced at a very rapid rate. In some instances, such as in the larvae of the blowflies *Calliphora fallax* and *C. augur*, this rate can be as fast as 5–10 mm/h (Waterhouse, 1954; Becker, 1978a). Further, these larvae, like many insects, are capable of sustained production of their type II PM over relatively long periods of time. Ultimately the PM is typically shed from the anus of the larvae into the external environment from which the PM can, in some instances, be harvested (East *et al.*, 1993). Blowfly larvae are focusing a great deal of energy into the production of PM presumably as an adjunct to their voracious appetite which underlies their rapid growth at this stage. What is not clear is why the PM is apparently produced at a rate which exceeds the physical growth requirements of the larvae. Continual renewal of the PM may be required to replace PM which has been modified by proteases in the harsh digestive environment of the midgut or PM which has lost its semipermeable character due to the clogging of the PM with debris from digested material.

In some insect groups the formation of the PM is induced by the ingestion of food. The PM in these instances is typically a sac-like structure completely enveloping the ingested food in the gut lumen. Multiple PMs are often layered around the food bolus (Waterhouse, 1953b). The trigger for the formation of the PM in many of these insects is the distension of the gut because suspensions of latex particles, saline solutions and even air can be effective (Richards and Richards, 1977; Spence, 1991; Billingsley and Rudin, 1992).

Cardia from adult blowflies (*Calliphora erythrocephala*) have been cultured *in vitro* and shown to produce small quantities of PM which has enabled study of factors influencing PM growth and morphology (Zimmermann *et al.*, 1973). The osmolarity, pH and temperature of the culture medium affected the rate of growth of cultured PM (Becker *et al.*, 1975). The optimum temperature (~28°C), optimum pH (~7.0) and optimum osmolarity (340 mOsmol) were broadly consistent with the biological context of the cardia in the anterior midgut region. Significant PM growth occurred over a broad range of these parameters although irregularities in the structure of the PM were apparent at the extremes of these parameters.

The expression of different types of PM in the various developmental stages of many insects suggests that production of PMs could be primarily under the control of hormones such as ecdysone and juvenile hormone which are interacting regulators of the development of many insects. Using the *in vitro* cardia culture system, Becker (1978b) demonstrated that 20-hydroxyecdysone increased adult *C. erythrocephala* PM formation whereas *Cercopia* juvenile hormone inhibited PM production. Both hormones altered the structure of the matrix. A number of other agents were shown to affect either the rate of production and/or the structure of the PM. Some of these agents are insecticides, e.g. the chitin synthetase inhibitors diflubenzuron and polyoxin D (Clarke *et al.*, 1977; Becker, 1978b, 1980; Zimmermann and Peters, 1987), DDT (Abedi and Brown, 1961; Richards and Richards, 1977) and cyromazine (Friedel *et al.*, 1988). The significance of these observations in terms of the primary mechanism of action of these insecticides is unclear.

One of the first points of direct contact between some insect biocontrol agents and the insect is at the PM. The endotoxin from *Bacillus thuringiensis*, which is used for the control of some insect species, has been shown to alter markedly the quantity of five glycoproteins in the PM from larval *Manduca sexta* (Rupp and Spence, 1985). The toxin resulted in the production of very fragile PMs. This may allow easier access of the toxin to midgut cells, the primary target of the toxin. In addition, the protein composition of the PM from *Trichoplusia ni* (cabbage looper) is altered by a factor secreted by specific baculoviruses which presumably gain entry into the tissues of this insect by the altered PM (Derksen and Granados, 1988). Baculoviruses are also used as insect biocontrol agents (Christian and Oakeshott, 1989). Thus, the structure of the PM may be an important factor in the determination of the efficacy of these biocontrol strategies. Indeed, it may be predicted that resistance to these biocontrol agents may be possible in insects which develop thicker or structurally altered PMs.

4.3 THE COMPOSITION OF THE PERITROPHIC MATRIX

It is generally agreed that PMs are primarily composed of chitin, proteins and proteoglycans (Spence, 1991; Peters, 1992). The relative amounts of these components may vary between different species and life stages as well as between the different maturation states of the PM in one life stage. Assessments of the composition of the PM should be considered in relation to the difficulty of ensuring that there is no contamination of the PM by ingested food or gut cells. Possible approaches which avoid this issue include the collection of PMs from tissue culture *in vitro* or larval culture and the dissection and removal of PM from artificially distended guts of insects (using air or saline to distend the gut).

4.3.1 Chitin

Chitin (poly-β-(1,4)-*N*-acetyl-D-glucosamine) is a characteristic constituent of PMs. Values of the chitin content ranging between 3.7 and 13% of the total mass of the PMs from a variety of insects have been reported (de Mets and Jeuniaux, 1962; Peters, 1992). The insect cuticle, in contrast, contains a much higher content of chitin (~25–40%; Richards, 1978). Chitin can form three types of conformational structures: α-chitin composed of antiparallel chains; β-chitin, composed of parallel chains; and γ-chitin composed of three chitin chains arranged in different orientations. The α- and γ-chitins predominate in the very few analyses of insect PMs which have been performed (Kenchington, 1976; Peters, 1992). The hydrated chitin fibrils in the PM probably contribute to the strength of the PM, particularly its ability to resist stretching and compression.

4.3.2 Proteoglycans

Proteoglycans are characterized by a core protein to which are attached polysaccharide components particularly glycosaminoglycans. The latter are a family of diverse macromolecules composed of linear polymers of a disaccharide unit consisting of hexosamine and uronic acid. The hexosamine is usually highly charged as a result of the attachment of carboxylic or sulphate ester groups. Proteoglycans (or acid mucopolysaccharides) are present in most of the PMs that have been examined (Mello *et al.*, 1971; Lehane, 1976; Stamm *et al.*, 1978; Dimitriadis, 1985; Peters, 1992). The uronic acid content of the PM from larvae of *Bombyx mori* represents 1.6% of the total mass of the PM (de Mets and Jeuniaux, 1962). In most insect PMs the proteoglycans are evenly distributed. However, in the more differentiated PMs of the higher Diptera, proteoglycans are concentrated into electron-dense layers which give the characteristic lamellar appearance to these PMs (Peters, 1976, 1979, 1992; Miller and Lehane, 1993b). Proteoglygans in vertebrate connective tissue absorb water and swell as a consequence of a Donnan osmotic effect caused by the high density of anionic charges on these molecules. The hydration of the proteoglycans may aid in the formation of a gel-like meshwork in the PM which could contribute to the strength of the PM. Lehane (1976) suggested that proteoglycans are involved in determining the permeability of the PM (see below).

4.3.3 Proteins

Protein is the major component of the PM in most of the insect species that have been examined (Ono and Kato, 1968; Zimmermann *et al.*, 1973, 1975; Richards and Richards, 1977; Peters, 1992). Values of the protein content of PMs range between 21 and 55.5% of the total PM mass. Richards and Richards (1977) observed that the sum of the percentages of protein, proteoglycans and chitin is often substantially less than 100%. One explanation for this deficit is that there is another component(s) in the PM which was not measured. Another possibility is that this result is a reflection of the technical limitations of the procedures used to measure each of these components in the PM (Peters, 1992). The amino acid compositions of several insect PMs have been determined but are not particularly informative presumably because they reflect a weight-averaged composition obtained from a number of proteins (Ono and Kato, 1968; Zimmermann *et al.*, 1975; Adang and Spence, 1982; Peters, 1992).

The proteins bound to the PM can be divided into four classes based on the ease of their release from the membrane. First, there are the proteins that are covalently attached to the PM and which therefore cannot be extracted even with strong denaturants such as 8 M urea or 6 M guanidine hydrochloride. There is little or no evidence that this group of proteins represents a significant proportion of the total protein content of the PM (Tellam, unpublished observations). Second, there is a select group of proteins which are bound non-covalently, but very strongly, to the PM. These proteins are released by the denaturants described above and have been called 'peritrophins' because of the structural and functional similarities of the small number of these proteins which have been characterized (see below). The peritrophins are a major constituent of the proteins present in PMs from a number of higher Diptera and are probably intimately involved in the determination of the structure and function of these PMs. The third group of PM proteins can be removed from the PM by mild detergent treatment (e.g. 2% Triton X-100 or 2% Zwittergent 3–14). These proteins are the peripheral PM proteins. Fourth, there is a small group of proteins which are very loosely bound to the PM and can be extracted with very mild treatments such as low or high ionic strength buffers. It is unclear whether these proteins are bound to the PM or simply entrapped within the fibrillar network of the PM.

(a) Range of proteins present in peritrophic matrices

There have been only very limited characterizations of the proteins present in PMs. Usually these analyses were performed by analytical sodium dodecyl sulphate–polyacrylamide gel electrophoresis (SDS–PAGE) and in some cases the corresponding glycoprotein profiles were determined. The best characterized PMs are those from the larvae of the higher Diptera (representative of type 2 PMs) and from the Lepidoptera (Stamm *et al.*, 1978; Adang and Spence, 1982; Rupp and Spence, 1985; Dörner and Peters, 1988; East *et al.*, 1993; Lehane *et al.*, 1996). In general, the proteins isolated from these PMs are representative of the integral PM proteins or these proteins in combination with the peripheral PM proteins. Comparison of the SDS–PAGE protein profiles of larval and adult PMs from the blowfly, *Calliphora erythrocephala*, and the fleshfly, *Sarcophaga barbata*, demonstrated some similarities but also substantial differences. Presumably, the differing protein profile reflects the structural differences between adult and larval PMs in these insects. There are two general characteristics which are common to all these analyses. First, there is only a very limited repertoire of proteins (6–12) present in the PMs. The sizes of these proteins are in the range 15–220 kDa with predominance of individual proteins in the 30–60 kDa and 15–20 kDa ranges. Second, where it has been analysed, there is evidence that the majority of these proteins are glycosylated, typically with oligosaccharides with terminal mannose or *N*-acetyl-D-galactosamine residues (Stamm *et al.*, 1978; Adang and Spence, 1982; Rupp and Spence, 1985; Dörner and Peters, 1988; East *et al.*, 1993). Labelling of intact PM from a variety of Dipteran larvae and assorted other insects with gold-labelled lectins indicated the presence of glycoproteins or glyco-conjugates with predominantly terminal *N*-acetyl-D-glucosamine, α-mannose, *N*-acetyl-D-galactosamine or fucose residues (Dörner and Peters, 1988; Rudin and Hecker, 1989; Lehane *et al.*, 1996). The detailed distribution of these oligosaccharides on PMs varies considerably between different insects and between the different PMs present in insects with multiple PMs (Peters, 1992; Lehane *et al.*, 1996). Particular oligosaccharides often showed a polarized distribution on either the ecto- or endo-PM surface, suggesting a polarized distribution of some of these glycoproteins in the PM (Dörner and Peters, 1988; Lehane *et al.*, 1996).

There is much less known about the proteins present in type I PMs. A study of adult black flies (*Simulium vittatum*) demonstrated a close association between the appearance of the type I PM induced by the feeding process in this insect and the presence of two proteins (66 and 61 kDa) (Ramos *et al.*, 1994). The pattern of proteins present in this PM was relatively simple compared to that in the type I PM from mosquitoes (Ramos *et al.*, 1994). The protein profile present in the PM from a lepidopteran (cabbage looper; *Trichoplusia ni*) was analysed by Derksen and Granados (1988), who showed a restricted number of PM proteins (~10 major proteins) with a particular abundance of proteins in the region between 30 and 70 kDa.

(b) Molecular structure of intrinsic peritrophic matrix proteins

The detailed molecular structure of three intrinsic PM glycoproteins (peritrophin-44, peritrophin-48 and peritrophin-95) from the larvae of *Lucilia cuprina* has been determined (Tellam *et al.*, 1992, 1994a,b; Elvin *et al.*, 1996). These proteins were purified from PM obtained by larval culture using a combination of differential extraction procedures, size exclusion chromatography and anion-exchange chromatography. Strong denaturants such as 6 M urea or 6 M guanidine hydrochloride were required for the initial solubilization of these proteins from the PM. Consequently, these proteins are representative of the intrinsic PM proteins or peritrophins. The amino acid sequences of these proteins have been determined by cloning and sequencing the cDNAs coding for these proteins. Figure 4.2 shows a schematic representation of the domain structures of these proteins. There is some limited homology between these sequences but more particularly each protein contains five copies of ~65 amino acid domain characterized by a specific register of six cysteines. The cysteines are probably involved in intradomain disulphide bonds. It was demonstrated that peritrophin-44 is extremely stable to proteolysis and that this stability is mediated by the presence of extensive disulphide bonds. The stability of this protein may have particular relevance in the strong proteolytic environment in contact with the PM in the gut of larvae. All of the deduced amino acid sequences contained potential N-linked glycosylation sites but at different relative positions. Peritrophin-95 contains an additional carboxy-terminal domain of approximately 100 amino acids. This domain is characterized by the complete absence of cysteine residues and an abundance of proline and threonine residues which reflect multiple non-identical copies of a 10 amino acid structure which may be involved in O-linked glycosylation.

Searches of the protein sequence databases at first indicated that there were no proteins which were similar to these peritrophins. However, a

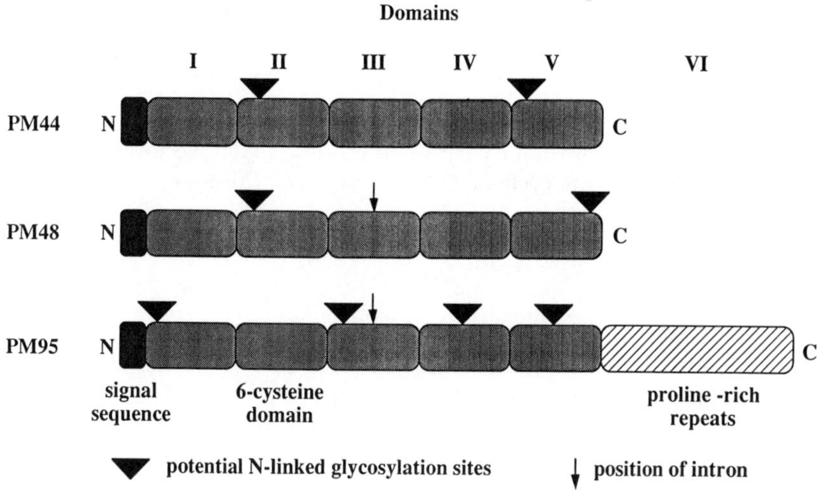

Figure 4.2 Schematic representation of the domain structures of three intrinsic peritrophic matrix proteins, peritrophin-95 (PM95), peritrophin-48 (PM48) and peritrophin-44 (PM44) from larvae of *Lucilia cuprina*.

subsequent search performed with a consensus sequence derived from the characteristic spacing of the cysteine residues in each cysteine-rich domain, discovered two significant matches. Both of these matches were the single six-cysteine domains found at the carboxy-terminal ends of two animal chitinases. The function of these domains in the chitinases is not clear as they are not in the catalytic domain but by analogy with the structure of plant chitinases, they may represent chitin-binding domains. Indeed, it has now been demonstrated that peritrophin-44 binds specifically to reacetylated chitin *in vitro*. This is probably the mechanism whereby these proteins bind to chitin within the PM. The strength of attachment of the peritrophins to the PM may be a reflection of the multiple chitin-binding domains present in these proteins.

Interestingly, there is significant amino acid sequence similarity between the proline-rich domain from peritrophin-95 and the protein procyclin (Figure 4.3). The latter is a surface coat protein from the tsetse fly gut stage of the trypanosome parasite. The significance of this observation is not clear. There is also an overall structural similarity between this region in peritrophin-95 and various mucins. The latter proteins are extensively O-glycosylated and involved in the lubrication and protection of mucous membranes in many animals (Allen, 1983). The proline-rich domain in peritrophin-95 may perform a similar function in the gut of *L. cuprina* larvae. The mRNA coding for peritrophin-44 was restricted to the cardia thereby confirming the

```
PM95pro     EPEETKPTETEPEKTTPATTEPEETKPTETEPEKTTPATTEPEPVETTL
parpAα      EPEETGPEETGPEETGPEETGPEETGPEETGPEETEP   EPEPGAATL
            ::::: : :: :: : :  : :::: : :: :: : :   ::::   ::
```

Figure 4.3 Amino acid sequence similarity between the proline-rich domain of peritrophin-95 (PM95pro) and parpAα, the acidic repetitive protein from the procyclic stage of the life cycle of *Trypanosoma brucei* (Mowatt *et al.*, 1989).

importance of the cardia in the synthesis of type II PMs. It is known that proteins which are immunologically related to peritrophin-44 and peritrophin-95 are present in the type II PMs from larvae of the flies *Chrysomya rufifacies* and *Haematobia exigua irritans* but are not present in the PMs from corresponding adults of these species (Tellam and Wijffels, unpublished observations). Thus, one fundamental question relates to the reasons for the stage-specific expression of these peritrophins when both the larval and adult forms of these insects produce a type II PM (i.e. from cardia). Presumably, this result reflects significant structural and functional differences between the PMs of the larvae and adults. But why is this necessary? Perhaps the differential expression of these peritrophins is somehow related to the changing roles of the larvae and adults. The former is primarily concerned with rapid growth whereas the latter is primarily concerned with reproduction. The relevance of these studies to other insect PM proteins, particularly proteins present in type I PMs, is not yet clear.

(c) Protein interactions and biological roles

The potential for multiple chitin-binding domains in the *L. cuprina* peritrophins may give them the capability of cross-linking chitin fibrils in the PM thereby interweaving a protein network throughout the chitin fibrils. Consistent with this idea is the uniform distribution of these peritrophins throughout the PM (Figure 4.4). It is noteworthy that the chitin-binding domains in these peritrophins may also have the potential to interact with *N*-acetyl-D-glucosamine expressed as a terminal sugar on oligosaccharides attached to peritrophins. This could also lead to extensive cross-linking of these glycoproteins. The potential for multiple modes of cross-linking of the peritrophins within the PM may give strength and elasticity to the PM. Indeed, Zimmermann and Peters (1987) concluded that PM proteins are necessary for the maintenance of the structural integrity of the PM. In addition, potential cross-linking functions of the intrinsic PM proteins may play a key role in dictating the permeability characteristics of the PM. Antibodies to the peritrophins

Figure 4.4 Immunofluorescence localization of peritrophin-44 on the peritrophic membrane from *Lucilia cuprina* larva.

ingested by *L. cuprina* larvae block the movement across the PM of small 6 nm colloidal gold particles (Willadsen *et al.*, 1993). These gold particles normally freely diffuse across the PM. This result is consistent with the involvement of these proteins in the determination of the porosity of the PM in this insect. As discussed below, the permeability of PMs in general is highly regulated. However, it is not readily apparent how a peritrophin-mediated chitin cross-linking process could regulate the highly defined porosity present in most PMs.

4.4 PERITROPHIC MATRIX FUNCTIONS

The PM separates the contents of the gut lumen from the underlying digestive epithelial cells. In so doing, the PM partitions molecules between the endo- and ecto-PM spaces. The latter space lies between the PM and the digestive epithelial cells lining the gut. The PM has multiple functions which are intimately associated with the ability of the PM to compartmentalize the gut. These functions of the PM can be broadly grouped into its roles in digestion and protection of the midgut epithelium.

4.4.1 Digestion

(a) Semipermeability

The PM is a semipermeable membrane which allows the selective movement of small molecules from the gut lumen to the digestive epithelial cells and the retention in the gut lumen of larger molecules as well as bacteria, viruses and parasites. The porosity of the PM is probably also an important determinant regulating the movement of newly synthesized digestive enzymes from the surface of the midgut epithelial cells into the gut lumen. The 'pore' sizes in PMs from a number of insect species have been measured using a variety of techniques and generally range between 4 and 10 nm (Zimmermann and Mehlan, 1976; Richards and Richards, 1977; Peters and Wiese, 1986; Santos and Terra, 1986; Miller and Lehane, 1990; Terra, 1990; Spence, 1991; Peters, 1992; Ferreira *et al.*, 1994). This size range is ideal for discrimination between different-sized proteins, e.g. the tsetse fly *Glossina palpalis* PM did not allow haemoglobin (68 kDa) to pass across its PM but did allow myoglobin (17 kDa) (Peters and Wiese, 1986; Miller and Lehane, 1990; Peters, 1992; for an alternative view see Barbenhenn and Martin, 1995). The actual process of movement of molecules across the PM may be considerably more complex than that expected for a simple semipermeable membrane as evidenced by a few isolated experiments. Zhuzhikov (1964) demonstrated that there was directional polarity in the ability of human salivary amylase to move across the PMs of the flies *Musca domestica* and *Calliphora erythrocephala*. This enzyme could pass freely from the ecto-PM space to the endo-PM space but not vice versa. Lehane (1976) suggested that the permeability of PMs may be effected by pH and ionic concentration. Indeed, the flux of a 70 kDa protein through the PM of the cockroach *Periplaneta americana* was 10 times greater in saline than in water (Rippon, 1987). More recently, Miller and Lehane (1993b) demonstrated that the permeability of PM isolated from the tsetse fly *Glossina morsitans morsitans* was increased in the presence of calcium ions. It was proposed that the calcium ions neutralize anionic charges of glycosaminoglycans present in the PM which then alters the porosity of the membrane. Lehane *et al.* (1996) established that sulphated glycosaminoglycans are an integral component of at least one of the three PMs present in the tsetse fly and that there is a regular distribution of the anionic charges on this PM. It was suggested that changes in the midgut pH of this fly could also alter the density of anionic charges associated with the glycosaminoglycans thereby altering the permeability of the PM. This proposal could be particularly imporant in several insect species where different regions of the gut are characterized by markedly different pH values (Chapter 8).

Spence and Kawata (1993) demonstrated that the permeability of the PM from *Manduca sexta* larvae was decreased at the posterior end compared with the anterior end and also decreased with the age of the insect.

The detailed molecular mechanisms determining the porosity of the PM are not clear. Lehane (1976) suggested that mucopolysaccharides may be involved in determining the permeability of the PM. In particular, the packing properties of these molecules may make them responsive to changes in pH and ionic composition. As already discussed (section 4.3.3(c)), there is indirect evidence that the peritrophins may be involved in determining the porosity of the PM from *Lucilia cuprina* larvae (Willadsen *et al.*, 1993). The lectins, wheatgerm lectin, lentil lectin and Con A, also inhibit the free movement of 6 nm gold particles across the PM of *L. cuprina* (Eisemann *et al.*, 1994). The mechanism of this inhibition involves the formation of an impervious gel-like layer of undefined composition on the gut lumen side of the PM. Interestingly, Con A and lentil lectin bind to all three characterized peritrophins whereas wheatgerm lectin binds to peritrophin-95. An apparently identical impervious layer is formed on the PM when antibodies to these peritrophins are fed to *L. cuprina* larvae (Willadsen *et al.*, 1993). This suggests a common mode of action of the antibodies and the lectins mediated by their binding to the same (or similar) target proteins on the PM.

(b) Partitioning of digestive enzymes

The diets of most insects typically contain considerable quantities of high-molecular-weight polymeric molecules such as cellulose and protein. The digestion of these polymers to a level where their constituents are able to be utilized by the insect involves a number of sequential steps. The PM plays a major role in this process by the partitioning of digestive enzymes and semi-digested polymers between the endo- and ecto-PM spaces (Terra and Ferreira, 1983; Santos and Terra, 1986; van Handel and Romoser, 1987; Ferreira *et al.*, 1994). In this manner, the PM regulates the sequential degradation of ingested polymeric molecules (Terra, 1990; Terra and Ferreira, 1994). Digestive enzymes are synthesized largely by the gut epithelial cells in most insects (Chapter 8). Some of these enzymes such as the endoproteinases and endoglycosidases pass across the PM into the gut where they degrade the ingested polymeric macromolecules into smaller fragments which subsequently move into the ecto-PM space. Here a new group of digestive enzymes (particularly exoproteinases and exoglycosidases) is present which specializes in the degradation of these polymer fragments. Presumably, the sequential degradation of these polymeric molecules increases the efficiency of the digestive process. The mechanisms whereby some of these digestive enzymes are restricted to the ecto-PM space are not entirely clear. The larger sizes of many enzymes in this group may prevent them from crossing the PM. In addition, some of these enzymes are anchored to the plasma membrane of the digestive epithelial cells lining the gut (Terra and Ferreira, 1983). The digestive enzymes secreted into the gut lumen are likely to be relatively small and soluble and may also be present in the ecto-PM space. Further discussion of the role of the PM in digestion in the insect midgut can be found in Chapters 6–8.

(c) Counter-current flow

Terra and Ferreira (1981) proposed an hypothesis for the 'recycling' of digestive enzymes in the endo-PM space. It was suggested that these enzymes are 'recovered' in the posterior midgut region by their movement across the PM into the ecto-PM space. A counter-current water flow (i.e. in the anterior direction) in the ecto-PM space is then postulated to move these enzymes back toward the anterior midgut region where they again pass across the PM into the endo-PM space (Ferreira *et al.*, 1981). This process would greatly improve the efficiency of use of digestive proteases.

(d) The peritrophic matrix as a matrix for enzyme immobilization

The PM may act as a solid phase matrix for the immobilization of proteins particularly enzymes (Peters and Kalnins, 1985; Ramos *et al.*, 1993). There may be a number of biological advantages for these immobilized proteins: minimization of losses of these proteins in the posterior flow of waste products through the gut; enhanced resistance to inactivation either due to proteolysis or unfolding; improved efficiency of the overall digestive process; the localization of proteins which are involved in the defence of the insect from invasion by viruses, bacteria and parasites. There is only very limited direct evidence in support of these possibilities in the context of proteins bound to the PM.

A leucine aminopeptidase activity was localized to the PM from larvae of *Drosophila melanogaster* using histochemical staining procedures (Walker *et al.*, 1980), and Peters and Kalnins (1985) using light and electron microscopic techniques demonstrated that aminopeptidases were located on the PM from a number of insects. Biochemical techniques were also used to demonstrate the presence of proteolytic activity associated with the PM (Eguchi and Iwamoto, 1976; Eguchi *et al.*, 1982). However, it is not clear whether these proteases are actually bound to the PM or simply entrapped within the PM (Ferreira *et al.*, 1994). The presence of considerable quantities of proteases in the midgut of most insects and the necessity that these proteases reach the gut lumen via movement across the PMs suggest that at any point in time there may be significant quantities of proteases entrapped within the PM. One experiment which may differentiate between these two possibilities is the localization of residual protease activity after extensive washing of the PM (Peters and Kalnins, 1985). In a functional sense, there may be no need to differentiate between these two possibilities. Terra (1990) has suggested that proteases bound to the PM are particularly concerned with preventing the non-specific clogging of the pores of the PM by partially digested material. Recently, a sucrase activity has also been located in the PM isolated from silkworm *Bombyx mori* (Sumida *et al.*, 1994). Peters *et al.* (1983) demonstrated the presence of a mannose-specific lectin on the gut lumen side of the PM from *Calliphora erythrocephala*. It was suggested that this lectin could be involved in interactions with the pili of gut bacteria. Possibly, this lectin has a defensive role on the PM or may be responsible for the immobilization of symbiotic bacteria.

(e) Role of the peritrophic matrix in water and ion movements

Proteoglycans in the PM are probably substantially hydrated as a consequence of osmotic forces associated with their highly charged glycosaminoglycans. These hydrated matrices can exert considerable osmotic force which could play an important role in regulating the movement of water across the PM. The charge associated with the glycosaminoglycans on the PM could also play a role in regulating the movement of ions across the PM. Miller and Lehane (1993b) have highlighted a number of similarities between the vertebrate renal glomerular basement membrane and the PM from the tsetse fly and suggest that these similarities are an example of convergent evolution.

4.4.2 The peritrophic matrix as a protective barrier

The thickness and resilience of most insect PMs attest to their potential to act as a protective barrier between the contents of the gut lumen and insect tissues. The PM has particular relevance in insects which transmit viruses and parasites to human beings. These pathogens are often present in the blood meal of the insect vector and some undergo a specific developmental phase of their life cycle in the tissues of the insect. One of the earliest events in this process is the movement of the pathogen from the insect gut into insect tissues. The PM plays an important role in this process. Peters (1992) has compiled a comprehensive list of organisms which either interact with, pass through or avoid the PM during the course of an insect infection. (Chapter 16 also describes in some detail the interactions between the insect midgut and a number of parasites.)

(a) Abrasion

Earlier views of the primary function of the PM were strongly dominated by the idea that the PM protected the digestive epithelial cells lining the midgut from damage caused by abrasive particles ingested by the insect. These particles can take the form of plant debris or crystals of haemoglobin in the case of haematophagous insects (Berner *et al.*, 1983; Peters, 1992). This is a significant function of the PM in some insects (Sudha and Muthu, 1988). However, the view that this is the primary function of the PM is not consistent with the presence of PMs in some fluid-feeding insects (Peters, 1992). Thus, the presence of a PM is not strongly associated with the dietary habits of insects. Mucin-like proteins and mucopolysaccharides on the surface of the PM may be responsible for lubrication of the gut thereby facilitating the passage of ingested food along the insect gut. The PM is also thought to protect digestive epithelial cells from the potentially detrimental effects of tannic acid polymers ingested by some insects (Barbehenn and Martin, 1992).

(b) Bacteria

The PM acts as a barrier which helps prevent invasion of insects by viruses, bacteria and parasites. The general lack of microbial colonization of insect tissues may attest to the protective efficacy of the PM. The pore sizes in the PM are generally too small to allow passive movement of these micro-organisms into the ecto-PM space. Some of the bacteria in the insect gut lumen are probably in a symbiotic relationship with the insect. In this instance, the PM may help to maintain the localization of these bacteria in the insect gut where they can aid in the digestion of ingested food. Bacteria such as *Bacillus thuringiensis* which produce insecticidal endotoxins are being used as insect biocontrol agents (Chapter 13). These endotoxins usually exert their effects by disrupting the cell membranes of digestive epithelial cells which line the insect midgut. The refractory nature of some insects to these bacterial endotoxins may be related to the inability of these toxins to traverse the PM (Yunovitz *et al.*, 1986).

(c) Viruses

Most viruses are introduced into an insect by ingestion and subsequent infection through the gut. There is evidence that the PM can act as a physical barrier to viral infections. Stoltz and Summers (1971) demonstrated that the PM from *Aedes taeniorhynchus* larvae prevented infection by mosquito iridescent virus, granulosis virus and nuclear polyhedrosis virus. However, a number of viruses ingested by insects do gain access to the insect via the PM. The restrictive pore sizes in the PM suggest that such an invasion process must be an active process involving modification of the PM. Indeed, it has been demonstrated that baculoviruses express a 'viral enhancement factor' (VEF) which alters the protein composition of the PM of lepidopterans and is associated with enhanced infectivity of the insect with the virus presumably by facilitating movement of the virus across the PM (Derksen and Granados, 1988). Alternatively, viral infections of the insect midgut can take place before the PM is fully formed especially in insects which produce PMs in response to feeding.

(d) Parasites

The PM has recently received considerable attention because of its role in the propagation of parasites which are ultimately transmitted to man by haematophagous insects. Presumably, the PM is one of the relatively more vulnerable areas for direct attack by a parasite because of the presence of tough, relatively impenetrable cuticle layers covering the foregut, hindgut and external surface of insects. In addition, some parasites may gain entry into an insect using windows of opportunity that arise before the PM is fully synthesized especially in insects producing type I PMs in response to feeding (Miller and Lehane, 1993a). Other parasites may utilize weak defensive points in the insect at the junction of the PM with the cuticle of the hindgut or foregut. Thus, parasites transmitted by insects in general have developed a range of infection strategies which either circumvent or minimize the protective role of the insect PM. A few examples of these strategies are described below.

Mosquitoes produce a type I PM in direct response to ingestion of a blood meal. The formation of the PM in mosquitoes can take as long as 1–1.5 days (Miller and Lehane, 1993a). Ingested malaria parasites such as *Plasmodium berghei* and *P. falciparum* ookinetes may have sufficient time during this period to mature fully and gain entry into the insect. However, maturation times of the PM and parasite are similar and slight variations in either could have considerable influence on infectivity rates. Indeed, different mosquito species form their PMs at different rates (Perrone and Spielman, 1988) and it has been suggested that this influences the capability of these mosquitoes to become infected with different malaria species (Ponnudurai *et al.*, 1988; Miller and Lehane, 1993a). It has also been shown that the thickness of the PM and therefore possibly the maturation state of the PM influences the rate of infection of the mosquito *Aedes aegypti* with *Plasmodium gallinaceum* (Stohler, 1957; Billingsley and Rudin, 1992). However, there is also substantial evidence that this parasite can directly penetrate the PM, possibly by using a secreted chitinase (Meis and Ponnudurai, 1987; Huber *et al.*, 1991; Sieber *et al.*, 1991; Billingsley and Rudin, 1992; Miller and Lehane, 1993a; Shahabuddin, 1995). Thus, the PM plays a role in limiting rather than preventing *Plasmodium* infection of the mosquito (Peters, 1992; Billingsley, 1994).

The relationship between the PM of blackflies (e.g. *Simulium damnosum* and *S. vittatum*) and their infection with microfilariae is more clear. The PM is formed relatively rapidly after a blood meal (beginning approximately 15 min after engorgement and fully matured after several hours; Peters 1992; Ramos *et al.*, 1994). Once formed, any remaining ingested microfilariae are encased in the peritrophic envelope from which very few can escape and subsequently infest the insect. Thus, infection of the insect occurs primarily in the relatively brief period before the PM is formed (Reid and Lehane, 1984; Peters, 1992). Ingestion of microfilariae by the insect appears to induce increased PM production leading ultimately to a thicker PM, which is an even greater barrier to this parasite (Bain *et al.*, 1976).

The *Leishmania* parasite ingested by the sandfly *Lutzomyia longipalpis* has an unusual mechanism for avoiding the PM. After feeding on blood containing the parasite, the sandfly produces a PM which completely surrounds the blood meal and the ingested parasite. The enveloped parasites transform from the ingested amastigotes into promastigotes which then multiply and attach themselves to or within the PM. The eventual disintegration of the PM 3–4 days later allows the parasite to colonize the posterior midgut epithelium subsequently leading to the infestation of the sandfly (Peters, 1992; Miller and Lehane, 1993a). A chitinase secreted by the parasite is thought to aid in the disintegration of the PM (Schlein *et al.*, 1991).

The African trypanosomes cause considerable morbidity and mortality in human beings and livestock animals. This parasite is transmitted by the bite of the tsetse fly (*Glossina*). The tsetse fly contains a type II PM which is produced prior to feeding. Consequently, the presence of the PM should limit the direct invasion of the fly by the ingested parasite. It was thought that the parasite enters the ecto-PM space at the junction of the PM with the cuticle of the hindgut (Peters, 1992). In this region, spines on the hindgut damage the PM providing a unique opportunity for entry into the ecto-PM space where parasite growth and development occurs. Ample evidence, however, demonstrates that the parasite can enter the ecto-PM space at other sites which may include directly traversing the PM in the midgut or entry in the region of the proventriculus (Ellis and Evans, 1977; Evans *et al.*, 1979; Evans and Ellis, 1983; Peters, 1992; Miller and Lehane, 1993a).

4.4.3 Faecal pellets and cocoons

Insects often produce faecal pellets which are enclosed by a membrane thought to originate from the midgut region (Peters, 1992). This membrane has several characteristics which are consistent with its description as a modified PM. In some species of insects potentially harmful ingested material (e.g. tannins and plant toxins) is enclosed in the gut by the PM and extruded from the insect within a faecal pellet (Bernays and Chamberlain, 1980; Bernays *et al.*, 1980). In a few instances the PM has been adapted for use in the production of insect cocoons (e.g. the Coleoptera; Rudall and Kenchington, 1971, 1973; Kenchington, 1976; Peters, 1992). The material used for these cocoons contains chitin and is synthesized in the midgut region.

4.5 PERITROPHIC MATRIX: A TARGET FOR THE CONTROL OF INSECTS

It has been demonstrated that isolated peritrophins from *Lucilia cuprina* larval PM, when injected into sheep, induced a humoral immune response which inhibited the growth of larvae which subsequently fed on those sheep or their sera (Tellam *et al.*, 1992; 1994a,b; East *et al.*, 1993; Willadsen *et al.*, 1993). This effect was shown to be mediated by specific antibody which survived in sufficient quantities in the gut to bind to the PM. The binding of these antibodies led to the production of a gel-like layer of undefined composition on the gut lumen-side of the PM which probably prevents nutrients from moving across the PM to the underlying digestive epithelial cells, i.e. the larvae feeding on these antibodies are starved. This leads to inhibition of the growth of these larvae. The magnitude of the larval growth inhibition probably depends on the delicate balance between the quantity of ingested antibody, the rate of antibody digestion in the gut and the rate of synthesis of PM. The larval growth inhibition translates into larval mortality at sufficiently high antibody concentrations. It is interesting to note that similar effects can be obtained when *L. cuprina* larvae are fed on very low quantities of the lectins, wheatgerm lectin, Con A and lentil lectin (Eisemann *et al.*, 1994). Presumably, these lectins are binding to glycosylated PM proteins in a manner analogous to the binding of specific antibody and in both cases this leads to the clogging of the PM. The observation that these lectins have a detrimental effect on a wide range of insect groups suggests that the clogging of the PM may have potential as a general means of insect control. Indeed, this may be one of the reasons why many plants contain relatively high concentrations of lectins in their tissues. Lectins are considered by many as excellent candidates for expression in transgenic plants as a means of protection against insect attack (Gatehouse *et al.*, 1992; Boulter, 1993). In an analogous manner, recombinant antibodies to an insect's PM proteins could be produced in transgenic plants or delivered to plants by appropriately engineered viruses to effect control of plant-eating insects.

4.6 CONCLUSIONS

Insect PMs have a multitude of structures and functions. This versatility may be the reason for the success of this structure in insects. Future characterizations of the detailed molecular structure of PMs may have important outcomes. First, these studies may identify PM molecules which are pivotal to the success of infection of some insect vectors by specific parasites. Many of these parasites are ultimately transmitted to human beings and livestock animals. Disruption of the parasite–insect relationship at the insect PM by immunological or chemical means may break the life cycle of these parasites and help in their control. Second, it may be possible to control specific insect populations which are directly competing for the same food supply as human beings by the use of new strategies which disrupt the functions of the PM in these insects. Transgenic plants expressing lectins, chitinases and recombinant antibodies to PM proteins are currently being produced. These molecules, when ingested by insects, have the potential to interfere with the movement of nutrients across the PM. Consequently, transgenic plants of this type may have the capability of reducing damage to food crops caused by feeding insects.

ACKNOWLEDGEMENT

I am grateful to Susan Briscoe who helped locate a number of references.

REFERENCES

Abedi, Z.H. and Brown, A.W.A. (1961) Peritrophic membrane as a vehicle for DDT and DDE excretion in *Aedes aegypti* larvae. *Ann. Entomol. Soc. Am.*, **54**, 539–42.

Adang, M.J. and Spence, K.D. (1982) Biochemical comparisons of the peritrophic membrane of the lepidopterans *Orgyia pseudotsugata* and *Manduca sexta*. *Comp. Biochem. Physiol.*, **73B**, 645–9.

Allen, A. (1983) Mucus – a protective secretion of complexity. *Trends Biochem. Sci.*, **8**, 169–73.

Bain, O., Philippon, B., Séchan, Y. and Cassone, J. (1976) Corrélation entre le nombre de microfilaires ingérées et l'épaisseur de la membrane péritrophique du vecteur dans l'Onchocercose de savane africain. *C. R. Acad. Sci. Paris D*, **283**, 391–2.

Balbiani, E.G. (1890) Études anatomiques et histologiques sur le tube digestif des *Cryptops*. *Arch. Zool. Exp. Gen.*, **8**, 1–82.

Barbehenn, R.V. and Martin, M.M. (1992) The protective role of the peritrophic membrane in the tannin-tolerant larvae of *Orgyia leucostigma* (Lepidoptera). *J. Insect Physiol.*, **38**, 973–80.

Barbehenn, R.V. and Martin, M.M. (1995) Peritrophic envelope permeability in herbivorous insects. *J. Insect Physiol.*, **41**, 303–11.

Becker, B. (1978a) Determination of the formation rate of peritrophic membranes in some Diptera. *J. Insect Physiol.*, **24**, 529–33.

Becker, B. (1978b) Effects of 20-hydroxy-ecdysone, juvenile hormone, Dimilin, and Captan on *in vitro* synthesis of peritrophic membranes in *Calliphora erythrocephala*. *J. Insect Physiol.*, **24**, 699–705.

Becker, B. (1980) Effects of Polyoxin D on *in vitro* synthesis of peritrophic membranes in *Calliphora erythrocephala*. *Insect Biochem.*, **10**, 101–6.

Becker, B., Peters, W. and Zimmermann, U. (1975) *In vitro* synthesis of peritrophic membranes of the blowfly, *Calliphora erythrocephala*. *J. Insect Physiol.*, **21**, 1463–70.

Bernays, E.A. and Chamberlain, D.J. (1980) Tolerance of ingested tannin in *Schistocerca gregaria*. *J. Insect Physiol.*, **26**, 415–20.

Bernays, E.A., Chamberlain, D.J. and McCarthy, P. (1980) The differential effects of ingested tannic acid on different species of Acridoidea. *Entomol. Exp. Appl.*, **28**, 158–66.

Berner, R., Rudin, W. and Hecker, H. (1983) Peritrophic membranes and protease activity in the midgut of the malaria mosquito, *Anopheles stephensi* (Liston) (Insecta: Diptera) under normal and experimental conditions. *J. Ultrastruct. Res.*, **83**, 195–204.

Billingsley, P.F. (1994) Vector–parasite interactions for vaccine development. *Int. J. Parasitol.*, **24**, 53–8.

Billingsley, P.F. and Rudin, W. (1992) The role of the mosquito peritrophic membrane in blood meal digestion and infectivity of *Plasmodium* species. *J. Parasitol.*, **78**, 430–40.

Binnington, K.C. (1988) Ultrastructure of the peritrophic membrane-secreting cells in the cardia of the blowfly, *Lucilia cuprina*. *Tissue Cell*, **20**, 269–81.

Blackburn, K., Wallbanks, K.R., Molyneux, D.H. *et al.* (1988) The peritrophic membrane of the female sandfly *Phlebotomus papatasi*. *Ann. Trop. Med. Parasitol.*, **82**, 613–19.

Boulter, D. (1993) Insect pest control by copying nature using genetically engineered crops. *Phytochemistry*, **34**, 1453–66.

Christian, P.D. and Oakeshott, J.G. (1989) The potential of genetically engineered baculoviruses for insect pest control. *Aust. J. Biotechnol.*, **3**, 264–6.

Clarke, L., Temple, G.H.R. and Vincent, J.F.V. (1977) The effects of a chitin inhibitor – Dimilin – on the production of peritrophic membrane in the locust, *Locusta migratoria*. *J. Insect Physiol.*, **23**, 241–6.

de Mets, R., and Jeuniaux, C. (1962) Sur les substances organiques constituant la membrane péritrophique des insectes. *Arch. Int. Physiol. Biochim.*, **70**, 93–6.

Derksen, A.C.G. and Granados, R.R. (1988) Alteration of a lepidopteran peritrophic membrane by baculoviruses and enhancement of viral infectivity. *Virology*, **167**, 242–50.

Dimitriadis, V.K. (1985) Ultrastructural analysis of peritrophic membrane function of *Drosophila auraria* larvae. *J. Submicrosc. Cytol.*, **17**, 293–7.

Dörner, R. and Peters, W. (1988) Localization of sugar components of glycoproteins in peritrophic membranes of larvae of Diptera (Culicidae, Simuliidae). *Entomol. Gen.*, **14**, 11–24.

East, I.J., Fitzgerald, C.J., Pearson, R.D. *et al.* (1993) *Lucilia cuprina*: inhibition of larval growth induced by immunization of host sheep with extracts of larval peritrophic membrane. *Int. J. Parasitol.*, **23**, 221–9.

Eguchi, M. and Iwamoto, A. (1976) Alkaline proteases in the midgut tissue and digestive fluid of the silkworm *Bombyx mori*. *Insect Biochem.*, **6**, 491–6.

Eguchi, M., Iwamoto, A. and Yamauchi, K. (1982) Interrelation of proteases from the midgut lumen, epithelia and peritrophic membrane of the silkworm, *Bombyx mori* L. *Comp. Biochem. Physiol.*, **72A**, 359–63.

Eisemann, C.H., Donaldson, R.A., Pearson, R.D. *et al.* (1994) Larvicidal activity of lectins on *Lucilia cuprina*: mechanism of action. *Entomol. Exp. Appl.*, **72**, 1–10.

Ellis, D.S. and Evans, D.A. (1977) Passage of *Trypanosoma brucei rhodesciense* through the peritrophic membrane of *Glossina morsitans morsitans*. *Nature (London)*, **267**, 834–5.

Elvin, C.M., Vuocolo, T., Pearson, R.D. *et al.* (1996) Characterization of a major peritrophic membrane-protein, peritrophin-44, from the larvae of *Lucila cuprina* – cDNA and deduced amino acid sequences. *J. Biol. Chem.*, **271**, 8925–35.

Evans, D.A. and Ellis, D.S. (1983) Recent observations on the behaviour of certain trypanosomes within their insect hosts. *Adv. Parasitol.*, **22**, 1–42.

Evans, D.A., Ellis, D.S. and Stamford, S. (1979) Ultrastructural studies on certain aspects of the development of *Trypanosoma congolense* in *Glossina morsitans morsitans*. *J. Protozool.*, **26**, 557–63.

Ferreira, C., Capella, A.N., Sitnik, R. and Terra, W.R. (1994) Properties of the digestive enzymes and the permeability of the peritrophic membrane of *Spodoptera frugiperda* (Lepidoptera) larvae. *Comp. Biochem. Physiol.*, **107a**, 631–40.

Ferreira, C., Ribeiro, A.F. and Terra, W.R. (1981) Fine structure of the larval midgut of the fly *Rhynchgosciara americana* and its physiological implications. *J. Insect Physiol.*, **27**, 559–70.

Freyvogel, T. and Stäubli, W. (1965) The formation of peritrophic membranes in Culicidae. *Acta Trop.*, **22**, 118–47.

Friedel, T., Hales, D.F. and Birch, D. (1988) Cyromazine-induced effects on the larval cuticle of the sheep blowfly, *Lucilia cuprina*: ultrastructural evidence for a possible mode of action. *Pesticide Biochem. Physiol.*, **31**, 99–107.

Gatehouse, A.M.R., Hilder, V.A., Powell, K. *et al.* (1992) Potential of plant-derived genes in the genetic manipulation of crops for insect resistance. *Proceedings of the 8th Symposium of Insect–Plant Relationships*, Kluwer Academic, Dordrecht, pp. 221–33.

Houk, E.J., Obie, F. and Hardy, J.L. (1979) Peritrophic membrane formation and the midgut barrier to arboviral infection in the mosquito, *Culex tarsalis* Coquillet (Insecta, Diptera). *Acta Trop.*, **36**, 39–45.

Huber, M., Cabib, E. and Miller, L.H. (1991) Malaria parasite chitinase and penetration of the mosquito peritrophic membrane. *Proc. Natl Acad. Sci. USA*, **88**, 2807–10.

Kenchington, W. (1976) Adaptation of insect peritrophic membranes to form cocoon fabrics, in *The Insect Integument* (ed. H.R. Hepburn), Elsevier, Amsterdam, pp. 497–513.

Lehane, M.J. (1976) Formation and histochemical structure of the peritrophic membrane in the stablefly, *Stomoxys calcitrans*. *J. Insect Physiol.*, **22**, 1551–7.
Lehane, M.J., Allingham, P. and Weglicki, P. (1996) Peritrophic matrix composition of the tsetse fly, *Glossina morsitans morsitans*. *Cell Tissue Res.*, **283**, 375–84.
Màrtin, J.S. and Kirkham, J.B. (1989) Dynamic role of microvilli in peritrophic membrane formation. *Tissue Cell*, **21**, 627–38.
Meis, J.F.G.M. and Ponnudurai, T. (1987) Ultrastructural studies on the interaction of *Plasmodium falciparum* ookinetes. *Parasitol. Res.*, **73**, 500–6.
Mello, M.L., Vidal, B.C. and Valdrighi, L. (1971) The larval peritrophic membrane of *Melipona quadrifasciata* (Hymenoptera: Apoidae). *Protoplasma*, **73**, 349–65.
Mercer, E.H. and Day, M.F. (1952) The fine structure of the peritrophic membrane of certain insects. *Biol. Bull.*, **103**, 384–94.
Miller, N. and Lehane, M.J. (1990) *In vitro* perfusion studies on the peritrophic membrane of the tsetse fly *Glossina morsitans morsitans* (Diptera, Glossinidae). *J. Insect Physiol.*, **36**, 813–18.
Miller, N. and Lehane, M.J. (1993a) Peritrophic membranes, cell surface molecules and parasite tropisms within arthropod vectors. *Parasitol. Today*, **9**, 45–50.
Miller, N. and Lehane, M.J. (1993b) Ionic environment and the permeability properties of the peritrophic membrane of *Glossina morsitans morsitans*. *J. Insect Physiol.*, **39**, 139–44.
Mowatt, M.R., Wisdom, G.S. and Clayton, C.E. (1989) Variation of tandem repeats in the developmentally regulated procyclic acidic repetitive proteins of *Trypanosoma brucei*. *Mol. Cell. Biol.*, **9**, 1332–5.
Ono, M. and Kato, S. (1968) Amino acid composition of the peritrophic membrane of the silkworm, *Bombyx mori* L. *Bull. Sericultural Exp. Station Jap.*, **23**, 1–8.
Perrone, J.B. and Spielman, A. (1988) Time and site of assembly of the peritrophic membrane of the mosquito *Aedes aegypti*. *Cell Tissue Res.*, **252**, 473–8.
Peters, W. (1976) Investigations on the peritrophic membranes of Diptera, in *The Insect Integument* (ed. H.R. Hepburn), Elsevier, Amsterdam, pp. 515–43.
Peters, W. (1979) The fine structure of peritrophic membranes of mosquito and blackfly larvae of the genera *Aedes*, *Anopheles*, *Culex* and *Odagmia* (Diptera: Culicidae/Simuliidae). *Entomol. Gen.*, **5**, 289–99.
Peters, W. (1992) *Peritrophic Membranes*, Zoophysiology Series Vol. 30, Springer, Berlin, 238 pp.
Peters, W. and Kalnins, M. (1985) Aminopeptidases as immobilized enzymes on the peritrophic membranes of insects. *Entomol. Gen.*, **11**, 25–32.
Peters, W., Kolb, H. and Kolb-Bachofen, V. (1983) Evidence for a sugar receptor (lectin) in the peritrophic membrane of the blowfly larva, *Calliphora erythrocephala* Mg. (Diptera). *J. Insect Physiol.*, **29**, 275–80.
Peters, W. and Latka, I. (1986) Electron microscopic localization of chitin using colloidal gold labelled with wheat germ agglutinin. *Histochemistry*, **84**, 155–60.
Peters, W. and Wiese, B. (1986) Permeability of the peritrophic membranes of some Diptera to labelled dextrans. *J. Insect Physiol.*, **32**, 43–9.
Ponnudurai, T., Billingsley, P.F. and Rudin, W. (1988) Differential infectivity of *Plasmodium* for mosquitoes. *Parasitol. Today*, **4**, 319–21.
Ramos, A., Mahowald, A. and Jacobs-Lorena, M. (1993) Gut-specific genes from

the black fly *Simulium vittatum* encoding trypsin-like and carboxy-peptidase-like proteins. *Insect Mol. Biol.*, **1**, 149–63.

Ramos, A., Mahowald, A. and Jacobs-Lorena, M. (1994) Peritrophic matrix of the black fly *Simulium vittatum*: formation, structure and analysis of its protein components. *J. Exp. Zool.*, **268**, 269–81.

Reid, G.D.P. and Lehane, M.J. (1984) Peritrophic membrane formation in three temperature simuliids, *Simulium ornatum*, *S. equinum* and *S. lineatum*, with respect to the migration of onchocercal microfilariae. *Ann. Trop. Med. Parasitol.*, **78**, 527–39.

Richards, A.G. (1978) The chemistry of insect cuticle, in *Biochemistry of Insects* (ed. M. Rockstein), Academic Press, New York, pp. 205–32.

Richards, A.G. and Richards, P.A. (1971) Origin and composition of the peritrophic membrane of the mosquito, *Aedes aegypti*. *J. Insect Physiol.*, **17**, 2253–75.

Richards, A.G. and Richards, P.A. (1977) The peritrophic membranes of insects. *Annu. Rev. Entomol.*, **22**, 219–40.

Richardson, M. and Romoser, W.S. (1972) The formation of the peritrophic membrane in adult *Aedes triseriatus* (Say) (Diptera: Culicidae). *J. Med. Entomol.*, **9**, 495–500.

Rippon, G.D. (1987) An investigation of the structure and function of the peritrophic membrane of the American cockroach, *Periplaneta americana*, with special reference to the possible effects of tannins on water movement and permeability of dyes through the peritrophic membrane. PhD Thesis, University of Adelaide, South Australia, Australia.

Rudall, K.M. and Kenchington, W. (1971) Arthropod silks: the problem of fibrous proteins in animal tissues. *Annu. Rev. Entomol.*, **16**, 73–96.

Rudall, K.M. and Kenchington, W. (1973) The chitin system. *Biol. Rev.*, **48**, 597–636.

Rudin, W. and Hecker, H. (1989) Lectin-binding sites in the midgut of the mosquitoes *Anopheles stephensi* Liston and *Aedes aegypti* L. (Diptera: Culicidae). *Parasitol. Res.*, **75**, 268–79.

Rupp, R.A. and Spence, K.D. (1985) Protein alterations in *Manduca sexta* midgut and haemolymph following treatment with a sublethal dose of *Bacillus thuringiensis* crystal endotoxin. *Insect Biochem.*, **15**, 147–54.

Ryerse, J.S., Purcell, J.P. and Sammons, R.D. (1994) Structure and formation of the peritrophic membrane in the larva of the southern corn rootworm, *Diabrotica undecimpunctata*. *Tissue Cell*, **26**, 431–7.

Ryerse, J.S., Purcell, J.P., Sammons, R.D. and Lavrik, P.B. (1992) Peritrophic membrane structure and formation in the larva of a moth, *Heliothis*. *Tissue Cell*, **24**, 751–71.

Santos, D.C. and Terra, W.R. (1986) Distribution and characterization of oligomeric digestive enzymes from *Erinnyis ello* larvae and inferences concerning secretory mechanisms and the permeability of the peritrophic membrane. *Insect Biochem.*, **16**, 691–700.

Schlein, Y., Jacobson, R.L. and Schlomai, J. (1991) Chitinase secreted by *Leishmania* functions in the sandfly vector. *Proc. R. Soc. London*, **245**, 121–6.

Shahabuddin, M. (1995) Chitinase as a vaccine. *Parasitol. Today*, **11**, 46–7.

Sieber, K.-P., Huber, M., Kaslow, D. *et al.* (1991) The peritrophic membrane as a barrier: its penetration by *Plasmodium gallinaceum* and the effect of a monoclonal antibody to ookinetes. *Exp. Parasitol.*, **72**, 145–56.

Spence, K.D. (1991) Structure and physiology of the peritrophic membrane, in *Physiology of the Insect Epidermis* (eds K. Binnington and A. Retnakaran), Inkata Press, Melbourne, pp. 77–93.

Spence, K.D. and Kawata, M.Y. (1993) Permeability characteristics of the peritrophic membranes of *Manduca sexta* larvae. *J. Insect Physiol.*, **39**, 785–90.

Stamm, B., D'Haese, J. and Peters, W. (1978) SDS gel electrophoresis of proteins and glycoproteins from peritrophic membranes of some Diptera. *J. Insect Physiol.*, **24**, 1–8.

Stohler, H. (1957) Analyse des Infektionsverlaufes von *Plasmodium gallinaceum* im Darme von *Aedes aegypti*. *Acta Trop.*, **14**, 302–52.

Stoltz, D.B. and Summers, M.D. (1971) Pathway of infection of mosquito iridescent virus. I. Preliminary observations on the fate of ingested virus. *J. Virol.*, **8**, 900–9.

Sudha, P.M. and Muthu, S.P. (1988) Damage to the midgut epithelium caused by food in the absence of peritrophic membrane. *Curr. Sci.*, **57**, 624–5.

Sumida, M., Yuan, X.L. and Matsubara, F. (1994) Sucrose activity and its kinetic properties in peritrophic membrane, and in membrane-bound and soluble fractions of midgut in the silkworm, *Bombyx mori*. *Comp. Biochem. Physiol.*, **108b**, 255–64.

Tellam, R.L., Casu, R.E. and Eisemann, C.H. (1994a) Recombinant blowfly strike antigen. Australian Provisional Patent Application No. PM5235.

Tellam, R.L., Eisemann, C.H., East, I. and Elvin, C. (1992) Flystrike antigen and vaccine and method for preparation. Australian Patent Application No. 29716/92.

Tellam, R.L., Schorderet, S. and Eisemann, C. (1994b) Antigen for inclusion in a vaccine against blowfly strike. Australian Provisional Patent Application No. PM8452.

Terra, W.R. (1990) Evolution of digestive systems of insects. *Annu. Rev. Entomol.*, **35**, 181–200.

Terra, W.R. and Ferreira, C. (1981) The physiological role of the peritrophic membrane and trehalase: digestive enzymes in the midgut and excreta of starved larvae of *Rhynchosciara*. *J. Insect Physiol., 27L*, 325–31.

Terra, W.R. and Ferreira, C. (1983) Further evidence that enzymes involved in the final stages of digestion by *Rhynchosciara* do not enter the endoperitrophic space. *Insect Biochem.*, **13**, 143–50.

Terra, W.R. and Ferreira, C. (1994) Insect digestive enzymes – properties, compartmentalisation and function. *Comp. Biochem. Physiol. [B]*, **109**, 1–62.

van Handel, E. and Romoser, W.S. (1987) Proteolytic activity in the ectoperitrophic fluid of blood-fed *Culex nigripalpus*. *Med. Vet. Entomol.*, **1**, 251–5.

Walker, V.K., Geer, B.W. and Williamson, J.H. (1980) Dietary modulation and histochemical localization of leucine aminopeptidase activity in *Drosophila melanogaster* larvae. *Insect Biochem.*, **10**, 543–8.

Walters, L.L., Irons, K.P., Guzman, H. and Tesh, R.B. (1993) Formation and composition of the peritrophic membrane in the sand fly, *Phlebotomus perniciosus* (Diptera: Psychodidae). *J. Med. Entomol.*, **30**, 179–98.

Waterhouse, D.F. (1953a) Studies on the digestion of wool by insects. IX. Some features of digestion in chewing lice (Mallophaga) from bird and mammalian hosts. *Aust. J. Biol. Sci.*, **6**, 257–75.

Waterhouse, D.F. (1953b) Occurrence and endodermal origin of the peritrophic membrane in some insects. *Nature (London)*, **172**, 676.

Waterhouse, D.F. (1954) The rate of production of the peritrophic membrane in some insects. *Aust. J. Biol. Sci.*, **7**, 59–72.

Weaver, S.C. and Scott, T.W. (1990) Peritrophic membrane formation and cellular turnover in the midgut of *Culiseta melanura* (Diptera: Culicidae). *J. Med. Entomol.*, **27**, 864–73.

Wigglesworth, V.B. (1930) The formation of the peritrophic membrane in insects, with special reference to the larvae of mosquitoes. *Q. J. Microsc. Sci.*, **73**, 593–616.

Willadsen, P., Eisemann, C.H. and Tellam, R.L. (1993) 'Concealed' antigens: expanding the range of immunological targets. *Parasitol. Today*, **9**, 132–4.

Yunovitz, H., Sneh, B., Schuster, S. *et al.* (1986) A new sensitive method for determining the toxicity of a highly purified fraction from d-endotoxin produced by *Bacillus thuringiensis* var. *entomocidus* on isolated larval midgut of *Spodoptera littoralis* (Lepidoptera, Noctuidae). *J. Invert. Pathol.*, **48**, 223–31.

Zhuzhikov, D.P. (1964) Function of the peritrophic membrane in *Musca domestica* L. and *Calliphora erythrocephala* Meig. *J. Insect Physiol.*, **10**, 273–8.

Zimmermann, D. and Peters, W. (1987) Fine structure and permeability of peritrophic membranes of *Calliphora erythrocephala* (Meigen) (Insecta: Diptera) after inhibition of chitin and protein synthesis. *Comp. Biochem. Physiol.*, **86b**, 353–60.

Zimmermann, U. and Mehlan, D. (1976) Water transport across peritrophic membranes of *Calliphora erythrocephala*. *Comp. Biochem. Physiol.*, **55A**, 119–26.

Zimmermann, U., Mehlan, D. and Peters, W. (1973) Investigations on the transport function and structure of peritrophic membranes. III. Periodic incorporation of glucose, methionine, and cysteine into the peritrophic membranes of the blowfly *Calliphora erythrocephala* Mg. *in vivo* and *in vitro*. *Comp. Biochem. Physiol.*, **45B**, 683–93.

Zimmermann, U., Mehlan, D. and Peters, W. (1975) Investigations on the transport function and structure of peritrophic membranes. V. Amino acid analysis and electron microscopic investigations of the peritrophic membranes of the blowfly *Calliphora erythrocephala* Mg. *Comp. Biochem. Physiol.*, **51B**, 181–6.

5

Structural macromolecules of the cell membranes and the extracellular matrices of the insect midgut

N.J. Lane, R. Dallai and D.E. Ashhurst

5.1 MODIFICATIONS OF THE CELL MEMBRANE

5.1.1 Organization of the intercellular junctions

Insect midgut cells are joined to each other on their lateral borders by junctional complexes whose function can be mechanical (smooth septate junctions, *fasciae adhaerentes,* hemi-adhering junctions) or for intercellular communication (gap junctions). The most prominent intercellular associations are the apical smooth septate junctions, which form a circumferential belt around the luminal borders of the midgut cells. Beneath this region, gap junctions are usually found. In the deep part of the lateral border the two adjacent cell membranes may form strip or spot-like adhering junctions, *fasciae adhaerentes,* whereas the basal plasma membrane that is in contact with the underlying extracellular matrix of the basal lamella may be modified to form hemi-adhering junctions. The general appearance of intercellular junctions in the insect midgut epithelium is thus markedly different from that of the lower chordates and higher vertebrates. The latter have a junctional complex consisting of *zonulae occludentes* (tight junctions (TJ)) at the luminal

Biology of the Insect Midgut.
Edited by M.J. Lehane and P.F. Billingsley.
Published in 1996 by Chapman & Hall, London.
ISBN 0 412 61670 X.

edge, *zonulae adhaerentes* (ZA or intermediate junctions) beneath, with *maculae adhaerentes* (MA, desmosomes) and gap junctions on the more basal part of the lateral borders (Stevenson and Paul, 1989). Other invertebrates may also exhibit differences from insects, in that their midgut cells may have a pronounced apical *zonula adhaerens* beneath which a long septate junction, of different type to the smooth septate, may be found. Similarly, in the insect oesophagus and rectum, which are cuticle lined, in contrast to the midgut, extensive apical zonular adhering structures are found, with underlying pleated septate junctions.

Most of the above types of cell junction serve as an attachment for one or more cytoskeletal elements so that they form an integrated system with these cell components. For example, in some cases a striking fibrous belt-like structure, bearing some similarities to the terminal web of vertebrate epithelia, is found in the apical cytoplasm of insect midgut cells and makes distinct associations with the smooth septate junctions (Lane and Flores, 1988; Dallai *et al.*, 1993) apparently by inserting into their membranes (Figure 5.6).

The structural molecules of the insect midgut thus include the molecules associated with (a) the junctional transmembranous modifications, (b) the extracellular structures in the intercellular cleft associated with the junctions, (c) the cytoskeletal components that link into the junctional membranous modifications from the cytoplasm and (d) those transmembrane molecules that extend into the extracellular matrix at non-junctional regions. Although it will be seen from the following that the fine structural details of these modifications have been well established, the detailed molecular biology of most of them is still poorly understood.

5.1.2 Adhering junctions

In the epithelial tissues of vertebrates the function of maintaining cell-to-cell adhesion and providing mechanical support is attributed to two types of intercellular adherens junctions: the spot desmosomes or *maculae adhaerentes* and the so-called belt desmosomes or intermediate junctions, the *zonulae adhaerentes*. The former are characterized by the insertion of intermediate filaments (IF), the latter by actin microfilaments. A different situation is found in insect gut. In general, adhering junctions are in the apical region of epithelia in fore- and hindgut. The insect midgut lacks cuticle as well as adhering junctions in the apical-most regions; those that exist are present in more basal areas. In many reports, insect adhering junctions have been termed desmosomes but they cannot genuinely be so, since arthropods lack IFs, apart from the nuclear lamin. In insects the 'desmosomal' plaques seem to be associated with underlying cytoplasmic microtubules (Ashhurst, 1970), rather than with IFs but the molecular nature of this association has not been investigated. Such a situation seems peculiar to the arthropods.

In thin sections of insect midgut, adhering junctions are found beneath the septate junctions. Here the two adjacent membranes lie parallel to one another and are separated by a 15–25 nm intercellular space (Figure 5.1). This space is filled with filamentous material but it is narrower than that of the vertebrate spot desmosome interspace; no central stratum is apparent in the midline. On the cytoplasmic sides, dense mats of microfilaments, clearly not IFs, are closely apposed to the membranes and microtubules are often found beyond them.

In insect tissues, the group of adherens junctions in which microfilaments are associated with membrane specializations have often been referred to as *fasciae* or *puncta adhaerentes*. The first of these are discontinuous strip-like junctions whereas the second of these are small cell–cell contact areas; they are extremely common in insect gut. They often also occur as half junctions, or focal contacts, similar to hemidesmosomes (but lacking IFs), on membranes juxtaposed to the basal lamina (Figure 5.2); there is no consistent freeze-fracture profile associated with any of these junctions.

The vertebrate ZA and desmosomes are multiprotein complexes organized around transmembrane cell adhesion molecules. They are quite distinct biochemically and each junctional complex contains several junction-specific components (Anderson *et al.*, 1993). In ZA, the protein α-catenin may mediate anchoring of actin filaments, and α-actinin and vinculin may also be present in the adjacent cytoplasm (Geiger, 1989). In desmosomes, the protein desmoplakin appears to be involved in anchoring the IFs, mainly cytokeratins, to the membranes, and cadherins, desmoglein and desmocollin are also to be found (Garrod, 1993). Hemidesmosomes link IFs into the membrane via desmoplakin to the transmembrane integrins that are associated with the adjacent extracellular matrix molecules such as fibronectin. Recently, a molecular model for the adherens junction in *Drosophila* has been proposed and it is extremely similar to that for the vertebrates (Peifer, 1995). The integrity of the junction is maintained by the cell–cell interaction through D-E-cadherin. On its cytoplasmic side, D-E-cadherin organizes a multiprotein complex where the protein Armadillo interacts directly with the cadherin, the cytoskeletal elements, and perhaps with one other protein in the complex. This represents the best model to date for the constructions associated with insect adherens junctions, but it is not known if such an organization exists in the midgut. Rather little cytochemistry has been done on this tissue, although Reinhardt and Hecker (1973) found that the so-called '*maculae adhaerentes*' and 'hemidesmosomes' of the mosquito midgut did not stain with periodic acid–TCH–silver–proteinate (specific for polysaccharide) but that the cytoplasmic mat stained strongly with phosphotungstic acid (specific for basic amino-acid-rich protein).

5.1.3 Smooth septate junctions

Smooth septate junctions, found only in endodermal-derived tissues and previously termed continuous junctions or *zonula continua* by Noirot and Noirot-Timothée (1967) represent a variation of the pleated septate junction that is found in ectodermal tissues. In conventional thin sections there is essentially no difference between the two junctions,

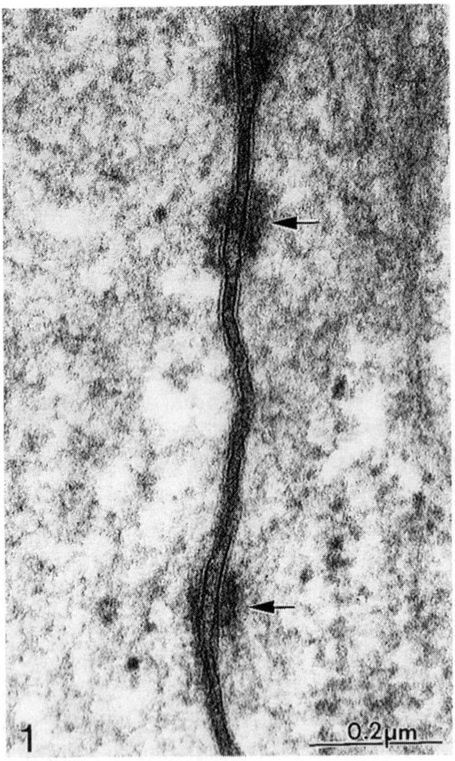

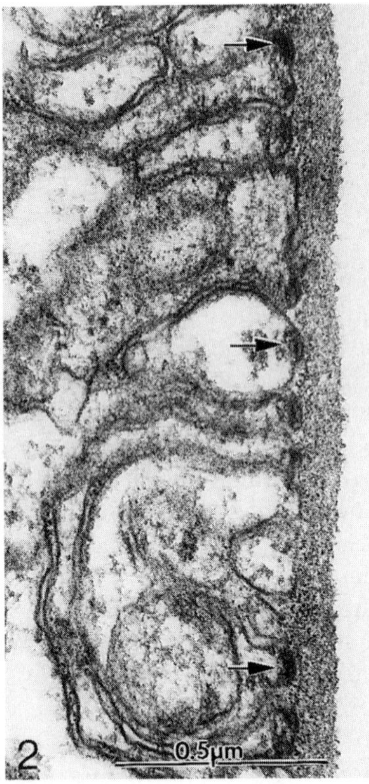

Figure 5.1 Lateral borders between adjacent midgut cells from the cockroach *Periplaneta americana*. Note the adhering junctions to be seen at intervals (arrows), characterized by the insertion of cytoskeletal elements.

Figure 5.2 The basal processes of the insect *Rhodnius prolixus*, where midgut cells abut against the extracellular matrix. Here they exhibit adhering junctions, looking rather like hemidesmosomes (arrows) but which are associated with actin, not intermediate, filaments.

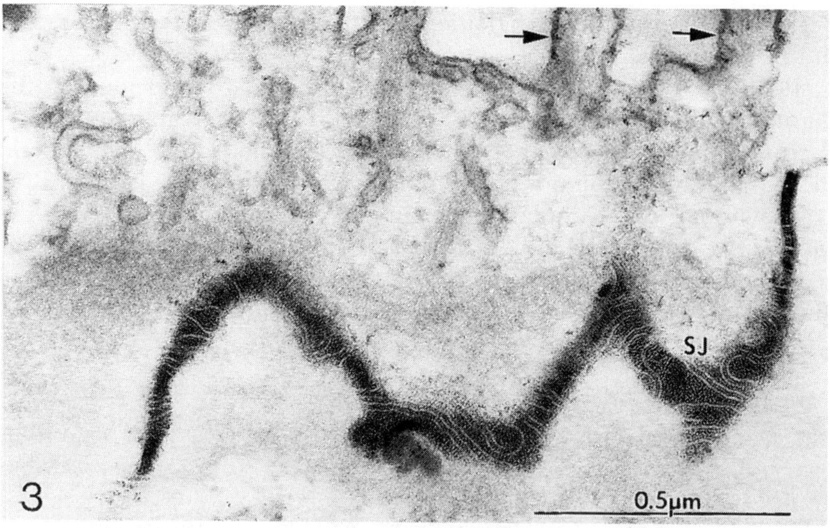

Figure 5.3 Smooth septate junctions (SJ) lying between component cells of the midgut in the Collembolan springtail, *Isotomurus palustris*. The microvilli (arrows) lie above these junctions and here the cells have been incubated in lanthanum and cut tangentially to reveal the smooth septate junctional ribbons. Note that in this organism they often run in double parallel rows.

except for a greater density of intercellular material that almost obscures the septa which straddle the intercellular cleft in the smooth septate junction. The regular 15–20 nm intercellular space is crossed by indistinct septa so that they often appear as cross striations. After lanthanum treatment, however, septal ribbons become evident and in tangential sections the ultrastructural appearance is distinct. Pleated septate junctions show intercellular septa as electron-transparent ribbons running in a zig-zag fashion between the membranes, whereas the smooth septate junctions have septa in the shape of undulating ribbons (Figure 5.3) with interseptal columns that appear *en face* as small electron-transparent dots. The septa themselves show a particulate substructure of several units (as in Lane and Harrison, 1978) and in both types the septal ribbons tend to run in parallel arrays.

The appearance of the smooth septate junctions after lanthanum treatment is very similar in the midgut cells of all arthropods and in Onychophora (Dallai and Giusti, 1979; Lane *et al.*, 1994). Small variations in this pattern have been observed in the proturans, in the sea-spider picnogonids and also in the horseshoe crab, *Limulus*, where the intercellular septa appear preferentially as double rather than single rows, after tracer infiltration (Lane and Harrison, 1978; Green, 1981; Xué and Dallai, 1992).

The ribbons seen in lanthanum-treated material correspond to the alignment of intramembrane particles (IMPs) observable in freeze-fracture replicas (Figure 5.4). With this technique one may obtain additional information on the intimate organization of the junctions. In fixed material, 10–12 nm IMPs remain associated with the P-face of the membrane, whereas complementary grooves are visible on the E-face (Figures 5.4 and 5.5). IMPs often fuse to form short cylinders. Replicas show that the smooth septate junctions start at the luminal surface, with

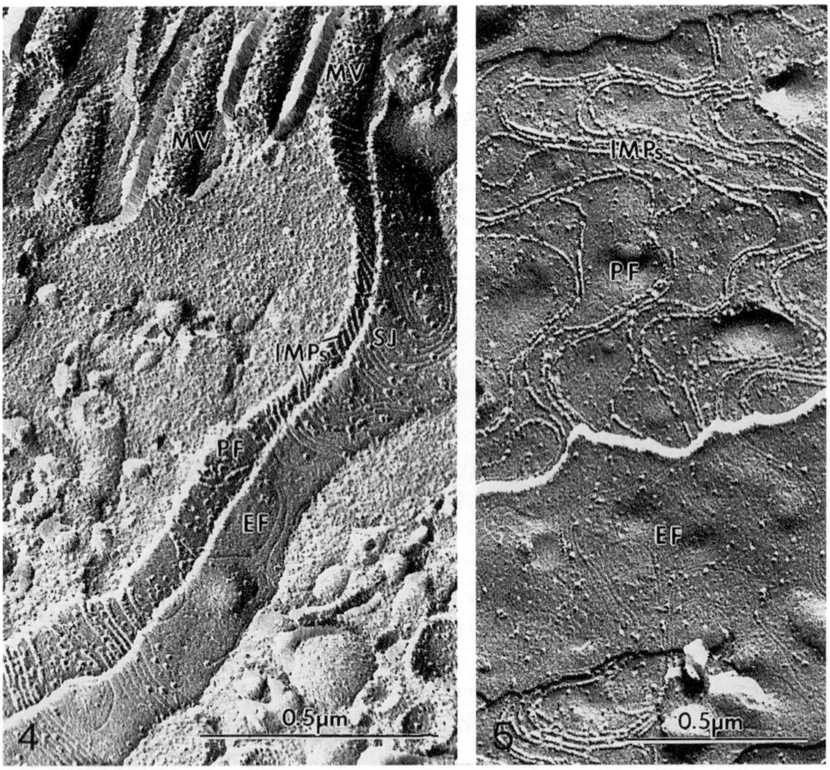

Figures 5.4 and 5.5 Freeze-fracture preparations from the midgut of the dipteran medfly, *Ceratitis capitata*; the replicas are made from glutaraldehyde-fixed material. The smooth septate junctions (SJ) feature rows of intramembranous particles (IMPs) on the P-face (PF) which are in register with complementary grooves on the E-face (EF). The junctions continue up to the base of the microvilli (MV), with no adhering structures apical to them (Figure 5.4). Along the lateral surfaces the particle rows (PF) and grooves (EF) present an undulating pattern (Figure 5.5).

ridges of particles evident at the base of the microvilli (Figure 5.4); this confirms the absence of adhering junctions in this apical-most region. In unfixed tissues, on the other hand, the ridges of IMPs preferentially cleave on the E-face, (as in Figure 5.10) so that a reversed array of IMPs occurs (Flower and Filshie, 1975). In rapidly frozen preparations, the ridges of IMPs appear to be particulate (Kachar *et al.*, 1986). Preliminary isolation of smooth septate junctions suggested that they were composed of both glycoprotein and proteins (Green *et al.*, 1983). A protein of 23–24 kDa has since been found, associated possibly with the integral membrane component of the smooth septate junction (Lane and Dilworth, 1989). This protein is a little larger than that of the pleated septate junction (22–23 kDa) and may be correlated with the larger size of the IMPs of the smooth septate junction. Three-dimensional models of septate junctions can be constructed correlating the intercellular septal ribbons seen in thin section with the IMP ridges visible in freeze-fracture replicas (Lane, 1984); it is assumed that the septal ribbons insert into the apical junctional membranes in the regions of the intra-membranous particles and ridges.

There is also a protein found in the dlg mutant of *Drosophila* that has sequence homology with other proteins such as tight junctional proteins ZO-1 and ZO-2, P55 protein and postsynaptic densities (Woods and Bryant, 1991, 1993). This protein is associated with the septate junction region of insect epithelia and is thought to be involved in cell signalling. Recently, there have been a number of interesting developments with regard to junctions and cell adhesion (Gumbiner, 1993; Peifer, 1995). The idea is emerging that certain cell adhesion molecules may function as tumour suppressors (Tsukita *et al.*, 1993) whereby the expression of their products results in the suppression of dedifferentiation, invasion and metastasis. These cell adhesion molecules are intimately associated with the cytoskeleton at specialized junctional regions and may also play a role in signal transduction from the adhesion receptors or junctional components. ZO-1 (from vertebrate tight junctions) is a member of this family of proteins (Willot *et al.*, 1993) as is the discs-large (dlg) tumour suppressor protein of septate junctions (Woods and Bryant, 1991) and submembrane junctional plaque proteins (Tsukita *et al.*, 1993). Reviews considering these developments (Anderson *et al.*, 1993; Citi, 1993; Woods and Bryant, 1993) discuss the relationships between ZO-1, the 160 kDa protein from ZO-2, the 130 kDa protein (ZO-3?) unrelated to cingulin (140 kDa), the 155 kDa protein called the 7H6 antigen and occludin (Furuse *et al.*, 1993). It is noteworthy that a 4.1 kDa protein encoded by the *coracle* gene, is a component of *Drosophila* septate junctions, where it could be involved in signal transduction. The colinar distribution of homologous domains within ZO-1 and the dlg tumour suppressor gene product suggests a common evolutionary origin and a similarity between the tight junctions of vertebrates and the septate junctions of the invertebrates. Their possible role in the transduction of membrane signals may be related to the regulation of paracellular permeability as well as intercellular junction assembly. The extent to which similar homologies may be found among other junctional proteins has yet to be established.

The general pattern of the smooth septate junctions may be altered in certain groups of insects. For example, in the collembolan *Isotomurus palustris*, the junctions are associated with a distinct terminal web-like structure (Figure 5.6). This fibrous belt, which surprisingly does not contain actin (Dallai *et al.*, 1993), receives the microfilaments descending from microvilli and is anchored to the lateral plasma membrane at the junction. This feature is clearly revealed after quick-freezing and deep-etching preparations, which enhance the three-dimensional arrangements of the structure (Figure 5.7). IMPs of the junctional membrane in this region of the terminal web are randomly arranged, unlike the regular rows of those in the typical smooth septate junction that occurs below the web.

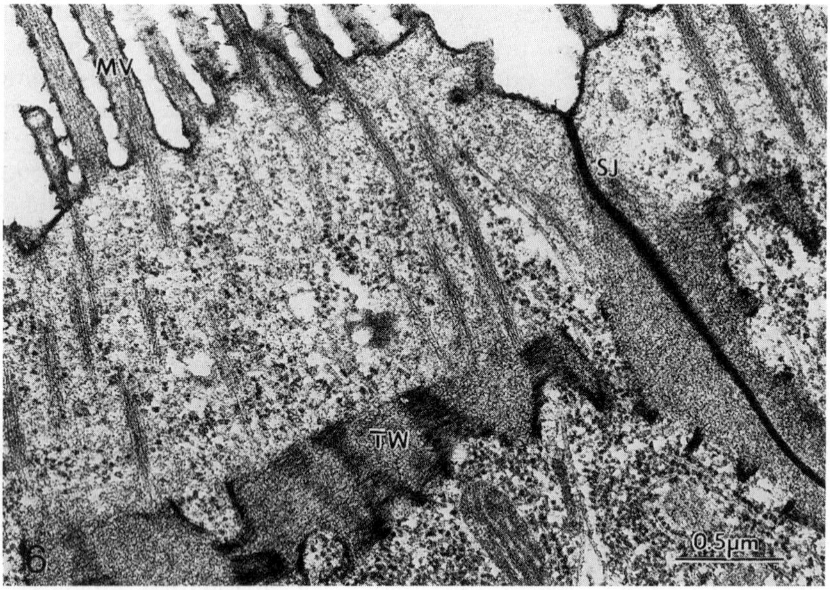

Figure 5.6 Section through the midgut of the springtail, *Isotomurus palustris*, showing the terminal web (TW) anchored to the actin bundles from the microvilli (MV) and by cytoskeletal links to the smooth septate junctions (SJ) on the lateral borders.

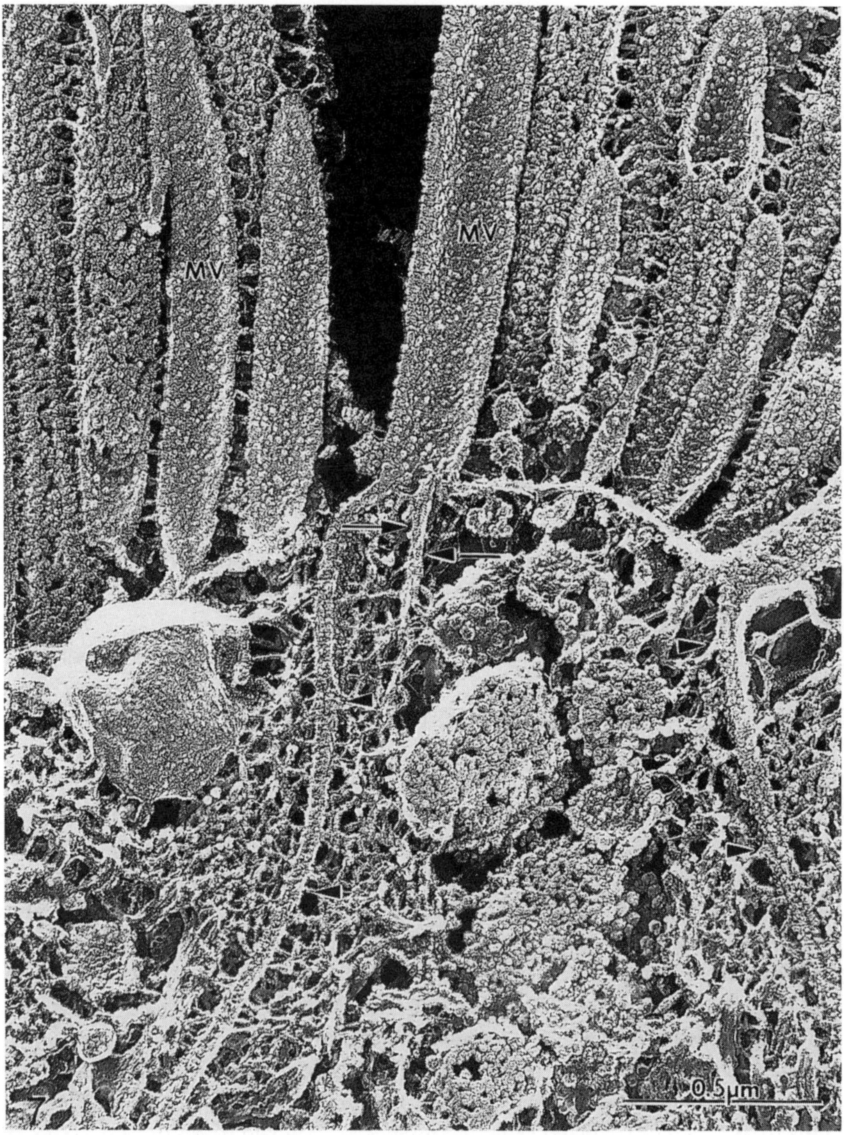

Figure 5.7 Midgut of the springtail visualized by quick-freezing, followed by the deep-etch and rotary replication technique, after having physically broken the frozen cells open. Note the cytoskeletal elements (presumably actin fibrils) anchored at the septate junctional level (arrowheads). Clusters of microfilaments (actin) (arrows) are to be seen emerging from the microvilli (MV) and descending to make contact with the junctional-associated cytoskeleton.

The association of cytoskeletal elements to the cytoplasmic face of the septate junction has already been described in other insects (Lane and Flores, 1988, 1990; Lane and Dilworth, 1989; Colombo *et al.*, 1993) and would seem to play an important role in the transmembrane lateral movement and assembly of IMPs into their mature junctional configuration (Lane and Swales, 1982). They also function in maintaining the structural integrity of the junction, as treatment with cytochalasin, an actin depolymerizer, leads to disorganization of the intramembrane components of the junction (Lane and Flores, 1988). In some cases, actin is apparently tethered onto microtubules in the underlying cytoplasm (Lane *et al.*, 1987), so they, too, may be involved in junctional assembly. This hypothesis is supported by the results of experiments on insect gut of treatments with colchicine and nocodazole which interfere with microtubular assembly and structure; these drugs have an impact not only on microtubule integrity, but also on their associated actin fibrils (Lane *et al.*, 1987; Lane and Flores, 1990). The first sign of assembly of the smooth septate junctions appears in embyronic midgut cells with individual IMPs becoming aligned into short rows. These subsequently become more extensive and arranged in parallel fashion and finally take on the mature configuration (Lane and Swales, 1982) characteristic of the fully assembled junction.

Clearly the septate junctions are involved in the adherence between adjacent cells; they may also form a partial permeability barrier (Lane, 1986). Septate junctions have often been considered to be analogous to the *zonula occludentes* of vertebrates on the basis of a similar apical location in epithelial cells (Noirot-Timothée and Noirot, 1980). Recent evidence, based on homologous sequences encoding a guanylate kinase homologue in different proteins associated with both tight and septate junctions, has strengthened this hypothesis (Woods and Bryant, 1991; Willot *et al.*, 1993). Similar to the function of tight junctions, which prevent fluid flowing paracellularly from the luminal to the basal compartment, the septate junctions have been thought to slow down the diffusion of molecules from the lumen into the intercellular space. The junctional matrix is charged and may bind certain ions, thus preventing the further entry of other molecules (Swales and Lane, 1985). However, it has been shown that compounds of different molecular weight can actually move fairly rapidly through the intercellular space of the septate junction (Skaer *et al.*, 1987). In some midgut cells and particularly in their associated mesenteric caeca (Flores and Lane, 1990), there may also be a movement of materials from the basal or lateral parts of the intercellular clefts into a system of intracytoplasmic cisternae. This is clearly shown by tracer infiltration (Figure 5.8), where the continuity of the extracellular space with a plasma membrane reticular system is apparent. This differs from the tubulocisternal endoplasmic reticulum

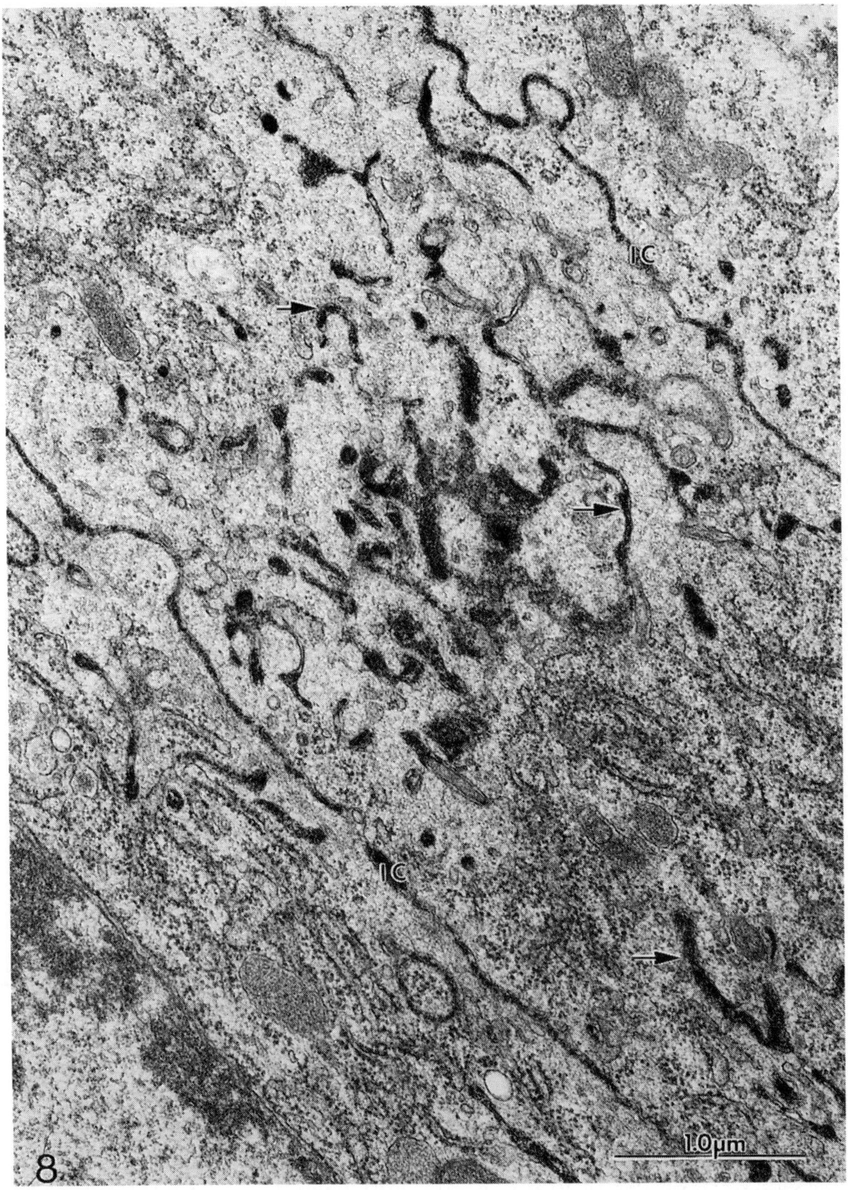

Figure 5.8 Basal part of the midgut of the locust, *Schistocerca gregaria* after incubation in lanthanum. The tracer has penetrated, not only into the intercellular clefts (IC), but also into the system of intracytoplasmic cisternae (arrows).

that terminates in subsurface cisternae in the apical cytoplasm, and may be the morphological basis of an insect midgut transcellular transport system.

5.1.4 Gap junctions

In insects the epithelial cells of the midgut frequently communicate with one another by gap junctions; these occur beneath the septate junctions and are randomly distributed as macular associations. Gap junctions are intercellular associations where the cleft between the adjacent cells is reduced from the conventional 15–20 nm to 2–3 nm. These close appositions often show a striated appearance across the space and in *en face* view, as is evident in tangential sections, the component connexons can sometimes be seen (Figure 5.9). Connexons made up of six hexameric connexin subunits, straddle the intercellular cleft and form a transmembrane hemichannel which, when juxtaposed to and aligned with a similar hemichannel in the membrane of the adjacent cell, form a channel or pore between the two cells. The channels of such connexons, arranged in clusters or plaques, form the morphological basis of the 'coupling' that is observed in cells linked by gap junctions (Musil, 1994); exchange of ions or small molecules ($\geqslant$1.4 kDa) is thought to take place between cells, leading to ionic and metabolic communication. These channels in insects are larger than those in vertebrate tissues; this is consistent with the passage of larger molecules between coupled cells in insect tissues (Chapters 2 and 9). The structure of gap junctions is similar in thin sections of midgut cells in different insects. In freeze-fracture the connexons measure about 13 nm in diameter and the preferential plane of cleavage of these gap junctional subunits is onto the E-face (Figure 5.10), leaving complementary PF pits, in contrast to that of the vertebrates and other invertebrate groups, where smaller (8–10 nm) connexons are observed, which cleave onto the P-face.

A 16–18 kDa highly conserved protein has also been found in gap junctions isolated from many arthropods, including *Manduca* midgut (Lane and Finbow, 1988; Finbow *et al.*, 1990). This protein, originally found in arthropod tissues (Finbow *et al.*, 1984), has now been found in vertebrates, usually as a 16 kDa protein, and is quite distinct from connexin with which it exhibits no sequence homology. The vertebrate and the arthropod 16 kDa proteins are homologous with one another as well as with both the 16 kDa subunit of the vacuolar H^+-ATPase, and the F_0 proton channel of the F_0 F_1 ATP synthase of mitochondria and chloroplasts. This protein, called ductin, has been isolated and characterized from the midgut of *Manduca sexta* (Lane and Finbow, 1988) and *Drosophila* (Finbow *et al.*, 1994), and forms the gap junction channels. In its other form, originally found in endomembrane systems (such as the

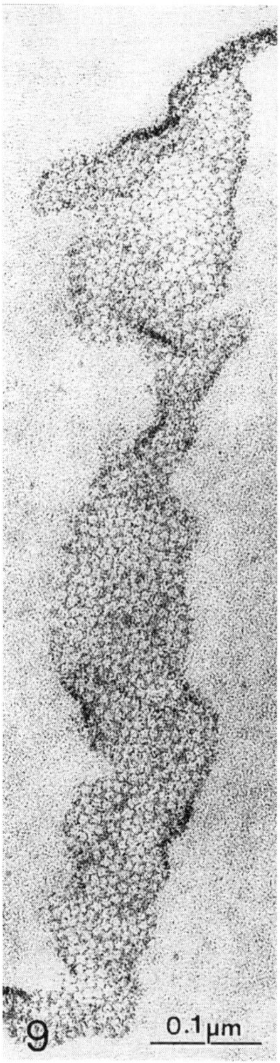

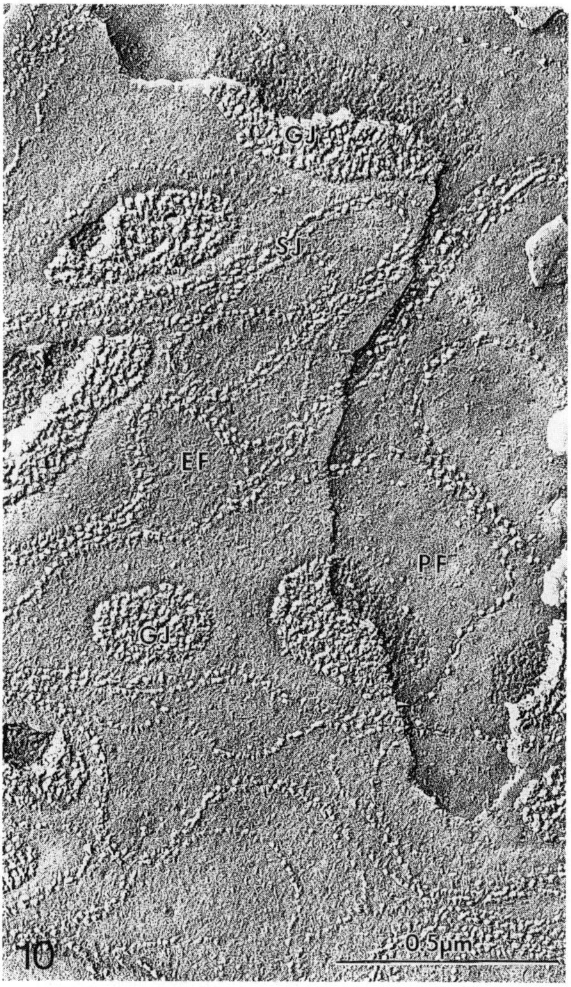

Figure 5.9 Thin section of lanthanum-infiltrated gap-junctional region of the midgut of *Isotomurus alticola*. Note that the connexons straddling the cleft are unstained whereas the intercellular cleft between them is laden with tracer, so that the connexons are in negative contrast.

Figure 5.10 Freeze-fracture replica of the midgut of the grasshopper *Pezotettix giornai*; this tissue was unfixed and features gap junctions (GJ), with the component connexon IMPs on the E-face (EF), and complementary pits on the P-face (PF). In circumstances such as this, with no fixative, the component IMPs of the septate junctions (SJ) atypically fracture onto the E-face.

lysosomes and Golgi), it is now implicated in epithelial ion transport in both vertebrates and arthropods. Antibodies to ductin have recently been found to inhibit gap junction-based cell–cell communication between cells in *Drosophila* ovarian follicles (Bohrmann, 1993). The precise role of the gap junctions in insect midgut, apart from presumably aiding in integrating cellular activities, has not yet been elucidated, since extensive studies with antibodies have not yet been performed.

5.1.5 Glycocalyx

The luminal plasma membrane of the midgut epithelium is organized into regular folds of microvilli (Chapter 1) which increase the surface area of the cells available for absorption. In the insect midgut, the microvilli are typically covered on the outer surface of the luminal plasma membrane by an electron-dense, fuzzy coat termed the glycocalyx (Figures 5.11 and 5.12) (Del Bene *et al.*, 1991; Kitajima, 1995). In vertebrates the glycocalyx is formed from oligosaccharide side chains of integral membrane glycoproteins, proteoglycans and glycolipids plus glycoproteins and proteoglycans absorbed onto the cell surface. The layer is typically relatively thin and is usually continuous along the epithelium. Although the glycocalyx has rarely been examined in its own right, the carbohydrate components are almost certainly derived from such cell surface molecules as membrane-associated enzymes on the microvilli (Chapter 8) and ion transporters (Chapter 9). Staining with charged colloid has demonstrated the presence of glycosaminoglycans on the glycocalyx of the mosquito (Houk *et al.*, 1986a,b), and which in other species are rich in *N*-acetylglucosamine and *N*-acetylgalactosamine (Rudin and Hecker, 1989). It has been suggested that, in contrast to vertebrate gut, the midgut glycocalyx of insects is either neutral or weakly anionic (Dimitriadis and Pirpasopoulou, 1992), though this may depend on the local conditions of the midgut. Studies of membrane glycoproteins of another mosquito, *Anopheles stephensi*, indicate that the variation in cell surface glycosylation can be quite restricted (Wilkins and Billingsley, 1996) (Chapter 12). However, although the function of the glycocalyx *per se* remains unexamined, its most obvious role may be in protecting the midgut cells against damage by abrasion, by proteases secreted into the midgut lumen, and against extremes of pH. An alternative glycocalyx arrangement is sometimes observed where the microvilli are covered by a filamentous layer (Smith *et al.*, 1969), which can be PAS-positive (Gouranton and Maillet, 1965) and have anionic groups (Noirot and Noirot-Timothée, 1972).

In the Hemiptera, however, an electron-dense glycocalyx-like layer separates the plasma membrane from an outer or perimicrovillar membrane (Figure 5.13) (Marshall and Cheung, 1970; Reger, 1971;

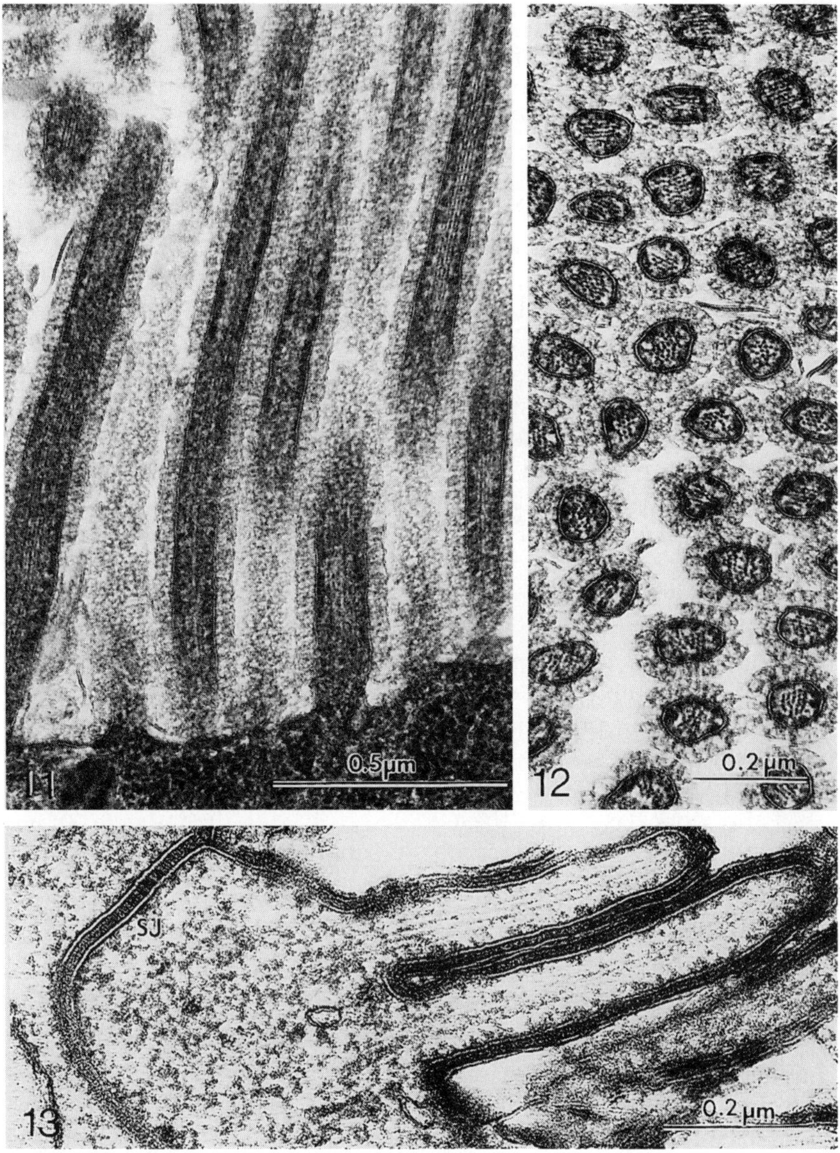

Figures 5.11, 5.12 and 5.13 Thin sections through the midgut of the Western flower thrips, *Frankliniella occidentalis* (Figures 5.11 and 5.12) and the hemipteran bug, *Lygaeus saxatilis* (Figure 5.13). These demonstrate the extensive glycocalyx seen in both longitudinal (Figure 5.11) and transverse (Figure 5.12) section and the 'double' membrane on the microvilli found in the midgut of hemipterans and some other bugs (Figure 5.13). SJ, septate junction.

Guitiérrez and Burgos, 1978; Lane and Harrison, 1979). The spacing between the inner (I) and outer (O) membranes is a regular 10 nm (Figure 5.14), and the perimicrovillar membrane is produced continuously into the midgut lumen (Billingsley and Downe, 1983), possibly from secretory, lysosome-like vesicles (Andries and Torpier, 1982). Lanthanum impregnation reveals the presence of columns inclined at an angle within the I–O space. Both microvillar membranes fracture into faces containing either PF particles or EF pits arranged as spiral ridges or grooves around the sides and across the tip of each microvillus (Lane and Harrison, 1979). These could be the insertion sites of one or both of the I–O columns (Figure 5.14). Ironically, the function of this replacement glycocalyx is better understood, as it compartmentalizes the enzymes involved in terminal sugar and protein digestion (Billingsley and Downe, 1985; Ferreira *et al.*, 1988 and Chapter 8).

In addition to the double membrane described above, other arrangements have been observed in the insect gut where intercellular columns provide the support for a junction-like structure. In the midgut of lepidopteran larvae, apical cytoplasmic projections are often associated with each other by 'septa' which can be seen by lanthanum impregnation to be due to double rows of small particles or rods (Flower and Filshie, 1976). Parts of some of the cytoplasmic microvillus-like projections in these midgut cells are differentiated in that they possess linear arrays of particles with complementary grooves (Flower, 1974; Flower and Filshie, 1976). In addition, projections called 'portasomes' have been observed on the extracellular surface of the lumen of the goblet cells in some insect midguts (Anderson and Harvey, 1966; Harvey, 1980, 1992; Harvey *et al.* 1981) and work has shown these to be involved in the transport of ions across the inner goblet cell membane (Chapter 9).

5.2 THE EXTRACELLULAR MATRICES

5.2.1 The basement membranes and associated connective tissues

In common with other insect organs, the midgut is surrounded by connective tissue which separates the cells from the haemolymph. The wide variation in these connective tissues, even among closely related species, makes it necessary to describe them briefly. The midgut epithelium of the locusts, *Locusta migratoria* and *Schistocerca gregaria*, lies on a typical basement membrane with an electron-lucent lamina rara and a lamina densa of amorphous material (Anderson and Cochrane, 1978; Nasiruddin and Mordue, 1993). This is surrounded by a layer of muscle cells, tracheoles and fibroblasts in an extracellular matrix containing banded collagen fibrils (Figure 5.15). Other insects in which a similar arrangement of cells and connective tissue around the midgut

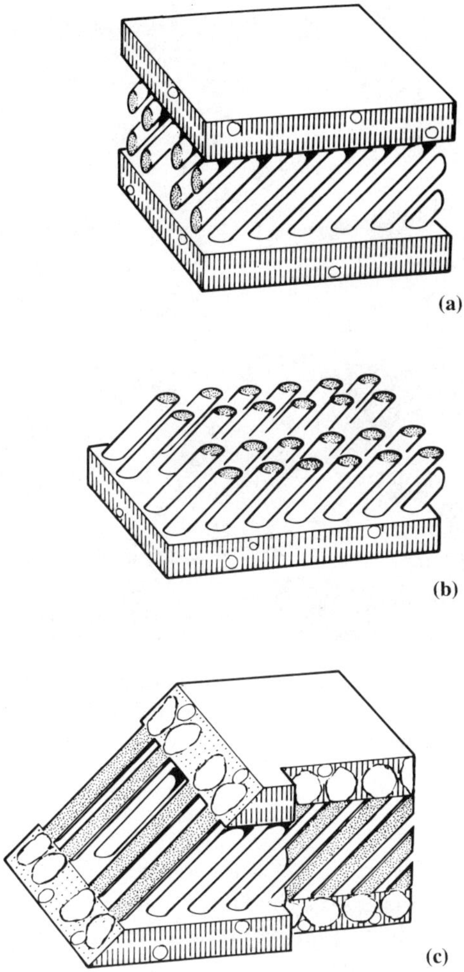

(a)

(b)

(c)

Figure 5.14 Diagram illustrating a possible interpretation, in microvillar 'double' membranes, of the structure of the I–O space between the two plasma membranes that lie around the tip of the microvilli in the midgut of haematophagous insects, such as *Rhodnius prolixus*. Analyses of lanthanum-impregnated thin sections suggests that pegs or columns exist between the two membranes; it appears that these are not set at right angles to the membranes, but obliquely. Hence, in transverse sections of the microvilli several columns are cut through (a); in longitudinal sections, they are cut through (b); only in oblique sections (c) can they be seen in their entirety. Freeze-fracture replicas show rows of intramembranous particles that cleave onto the P-face, which may also lie in double rows, as indicated in (c); these could represent the insertion or attachment points of the intercellular columns. Dotted areas correspond to the regions which appear non-opaque against the dense lanthanum background in tracer-impregnated sections.

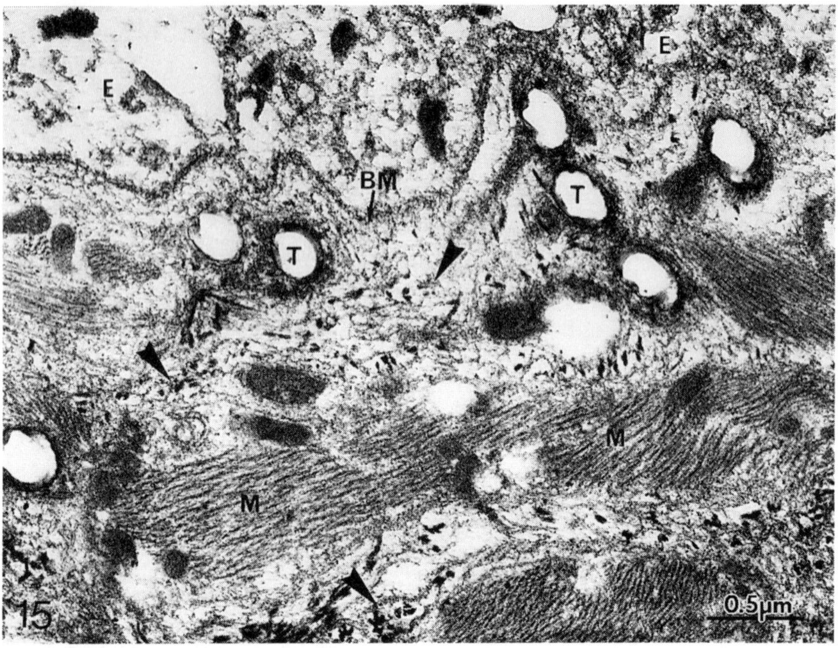

Figure 5.15 The basal cell membrane of the midgut epithelial cells (E) of *Locusta migratoria* lies on a thin basement membrane (BM). The underlying connective tissue containing collagen fibrils (arrowheads) supports the trachea (T) and muscle (M).

has been described include the cockroach, *Periplaneta americana* (Dictyoptera) (François, 1978), the flies, *Calliphora erythrocephala*, *Simulium slossonae* and *S. congareenarum* (Diptera) (De Priester, 1971; Steele *et al.*, 1992), *Cephalobaena tetrapoda*, *Raillietiella boulengeri* and *Reighardi sternae* (Pentastomida) (Thomas and Böckeler, 1992) and several blood-feeding insects, including *Anopheles* sp. (Billingsley, 1990; Syafruddin *et al.*, 1991; Meis *et al.*, 1992). It should be noted that the collagen fibrils of several orders, including the Coleoptera, Diptera, Lepidoptera and Protura, are thin and indistinctly banded and the collagen has not been characterized (Ashhurst, 1985).

Very thick, finely granular basement membranes are found under the midgut epithelium of adult *Rhodnius prolixus* (Pacheco, 1970), of larvae of *Ephestia kühniella* (Smith *et al.*, 1969) and of *Cantharis fusca* (Figure 5.16) (Holter, 1970). In some insects, particularly members of the Coleoptera, the midgut basement membrane has a more elaborate structure. The simplest, in *Coccinella septempunctata* is a thick layer with a network of fine filaments (Holter, 1970). There are thick basement membranes containing granules 18–50 nm in diameter under the epithelium of *Dermestes haemorrhoidalis*, *Galerucella nymphaeae*, *Ilybius fenestratus*, *Nebria brevicollis* and *Tenebrio molitor* (Figure 5.17) (Holter, 1970; Tano *et al.*, 1987).

Further variants are complex, layered basement membranes incorporating rods or plates arranged in lattices (Holter, 1970). A simple form is found in *Harpalus rufipes*; in longitudinal sections the basement membrane consists of a parallel arrangement of electron-dense rods perpendicular to the basal cell membrane (Figure 5.18). In transverse section, the rods appear square and are of two sizes, one with sides of 22 nm, the other of 30 nm; they are arranged in a square lattice (Figure 5.19). An hexagonal lattice of two elements, one thick and one thin, is found in *Aphodius rufipes*, *Bledius spectabilis* and *Quedius fuliginosus* (Figures 5.20, 5.21 and 5.22). In *Oryctes nasicornis* and *O. rhinocerus* (Hess and Pinnock, 1975; Bayon and François, 1976), there is a similar layer next to the basal cell membrane, but there are only three thin rods for each thick one. Outside this layer, there are two further layers in which, while the arrangement of the elements remains the same, the size of the larger rods increases so that in the outer layer they are plate-like (Figures 5.23 and 5.24). All these elements are held together by thin septa.

Similar lattices are found in other orders. The fleas, *Echidnophaga gallinacea*, *Tunga penetrans* and *Xenopsylla cheopis* (Siphonaptera) have a midgut basement membrane that consists of three elements: a large (95 × 70 nm) ovoid element, and two round (25 nm and 20 nm) elements in a lattice (Reinhardt *et al.*, 1972). Another flea, *Ctenophthalmus* sp., has a basal layer consisting of only two elements in an hexagonal array (Richards and Richards, 1968). The adult heteropteran midgut (*Nepa cinerea*, *Ranata linearis*) has a series of electron-dense, elongated plates, approximately 0.2 by 0.4 μm in size and linked together by many thin filaments under the epithelial cells (Gouranton, 1970). There is a square lattice of filaments with 'holes' at the intersections in the midgut of the female mosquitoes, *Aedes aegypti* and *A. dorsalis* (Terzakis, 1967; Reinhardt and Hecker, 1973; Houk *et al.*, 1980). After feeding, some distortion of the lattice was observed in *A. aegypti*, but not in *A. dorsalis*. In *Culex tarsalis*, a similar structure is split into layers and no changes were observed after feeding (Houk, 1977).

5.2.2 Matrix macromolecules

The major constituent of all extracellular matrices is the fibrous protein, collagen (for review see Van der Rest and Garrone, 1991). There are few detailed descriptions of the connective tissue matrix that surrounds the midgut (see above), but in all instances, collagen fibrils, which in some instances are very thin, are present (Ashhurst, 1985). Most insect

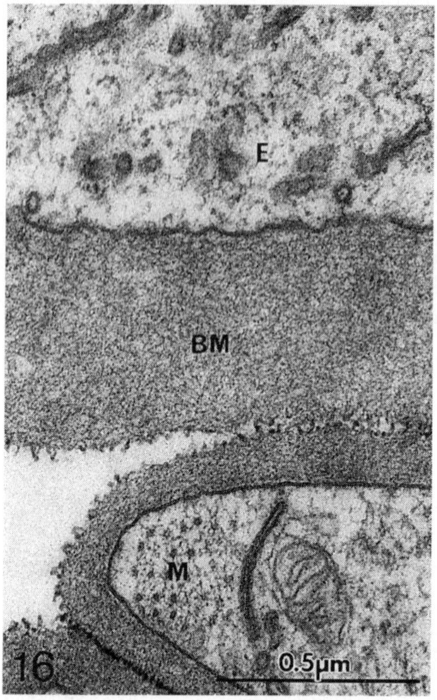
E
BM
M
16
0.5μm

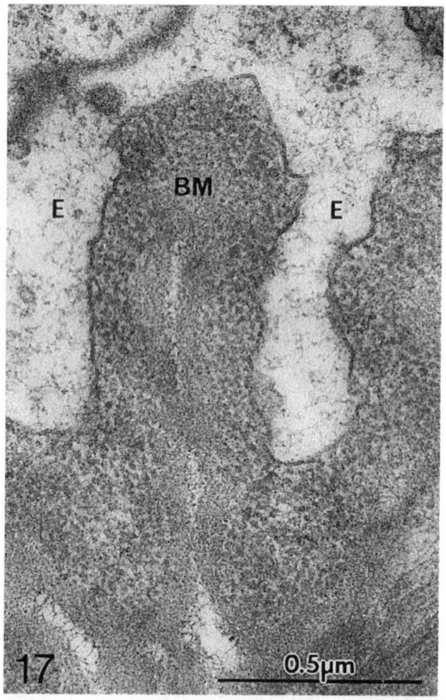
E
BM
E
17
0.5μm

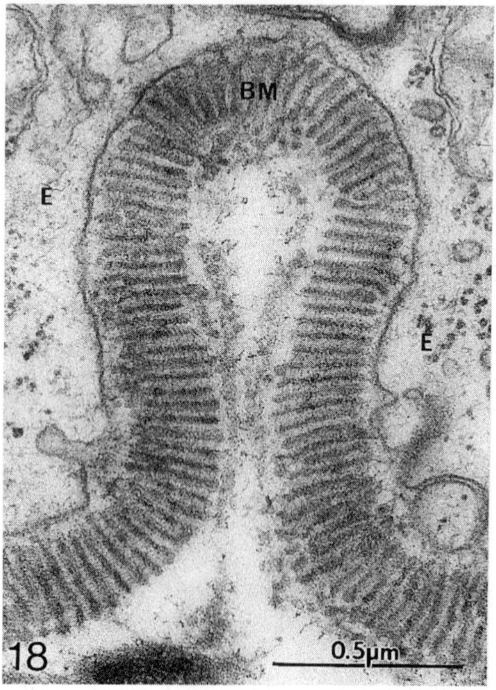
BM
E
E
18
0.5μm

19
0.5μm

collagen fibrils are banded and have a repeating periodicity of 67 nm. The appearance of fibrils reconstituted from solubilized insect fibrillar collagens is identical to those of reconstituted rat tail tendon, that is mammalian type I collagen fibrils (Figures 5.25 and 5.26) (Ashhurst and Bailey, 1980; François *et al.*, 1980; François, 1985). Similarly, the segment-long-spacing crystallites are identical in both banding pattern and length. It follows from these observations that the molecules of insects and mammals are of the same length (about 300 nm) and that the distribution of polar amino acids along them is very similar.

All collagen molecules have three α-chains that form a triple helical region. In the helical regions every third amino acid is glycine and there are many proline and hydroxyproline residues. The molecules of the fibrous collagens have a central helical region of over 1000 amino acids and globular N- and C-terminal domains. Locust, cockroach and *Tenebrio* fibrillar collagens share these properties with the mammalian fibrillar collagens, especially type I collagen (Ashhurst and Bailey, 1980; François *et al.*, 1980; François, 1985). The insect collagen molecules are trimers of three identical α-chains, whereas the mammalian type I collagen molecule has two identical α1(I) chains and one α2(I) chain. Trimers of α1(I) occur in tissues, such as skin (Uitto, 1979).

The main structural component of basement membranes is type IV collagen which was the first non-fibrillar collagen to be identified and characterized. Six different α-chains have been identified in mammalian tissues and the resulting molecules contain at least two different α-chains. The molecules self-assemble into networks. The amino-terminal segments of four molecules are held together by covalent bonding to form a four-armed structure with the carboxy-terminals at their ends. Two carboxy-terminal (NC1) domains are joined by disulphide bonding and further covalent interactions occur between helical regions (Ayad *et al.*, 1994; Yurchenco and O'Rear, 1994; Kühn, 1995). A typical type IV collagen was extracted from *Drosophila* cell cultures. The triple helical regions of the molecules are similar in both amino acid sequence and position along the molecule to those of mammalian type IV collagen, but

---

Figure 5.16 The thick basement membrane (BM) under the midgut epithelial cells (E) of *Cantharius fusca* appears amorphous. A thinner, similar basement membrane surrounds the muscle (M).

Figure 5.17 The basement (BM) under the midgut epithelial cells (E) of *Ilybius fenestratus* contains electron-dense granules.

Figures 5.18 and 5.19 In *Harpalus rufipes*, the basement membrane (BM) of the midgut epithelial cells (E) consists of short rods vertical to the cell membrane (Figure 5.18). In transverse section (Figure 5.19), they are seen to be of two sizes, square and arranged in a square lattice.

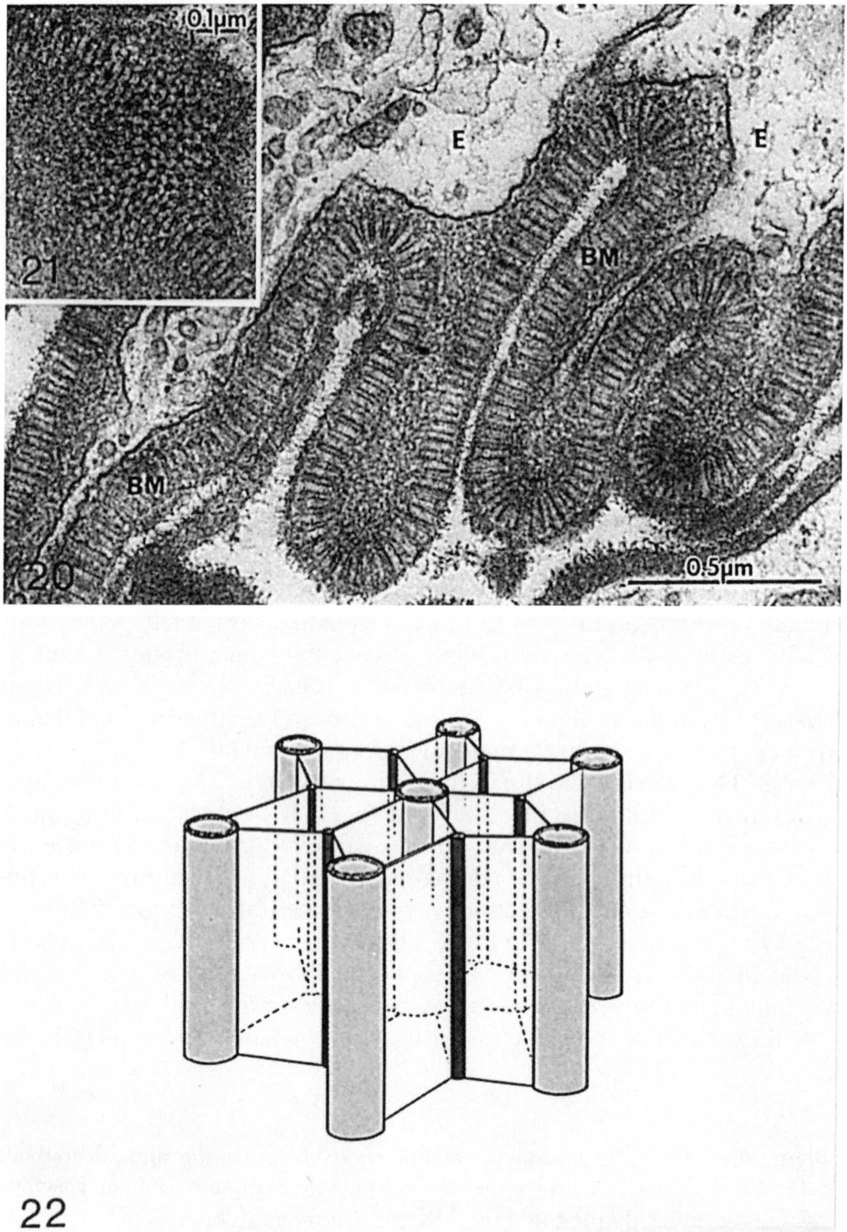

Figures 5.20, 5.21 and 5.22 The basement membrane (BM) under the midgut epithelial cells (E) of *Quedius fuliginosus* consists of thick and thin rods perpendicular to the cell membrane and forming an hexagonal lattice (Figure 5.21). The rods are held together by thin septa shown diagrammatically in Figure 5.22.

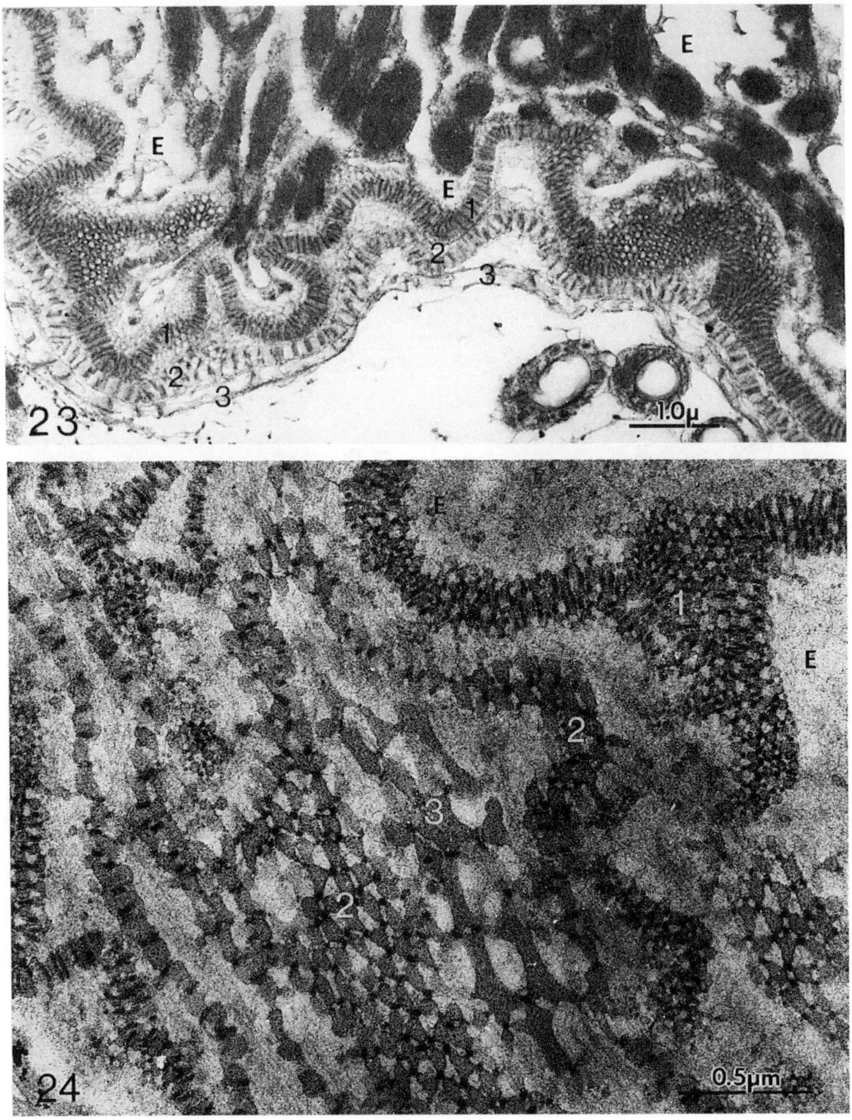

Figures 5.23 and 5.24 The basement membrane of the midgut epithelial cells (E) of *Oryctes nasicornis* consists of three layers (1, 2 and 3). In oblique section (Figure 5.24), the layers are seen to consist of two elements, large and small rods. The small rods are of similar size in the three layers. The larger rods increase in size from layers 1 to 3. In layers 2 and 3 they appear triangular and are attached to the thin rods at the angles.

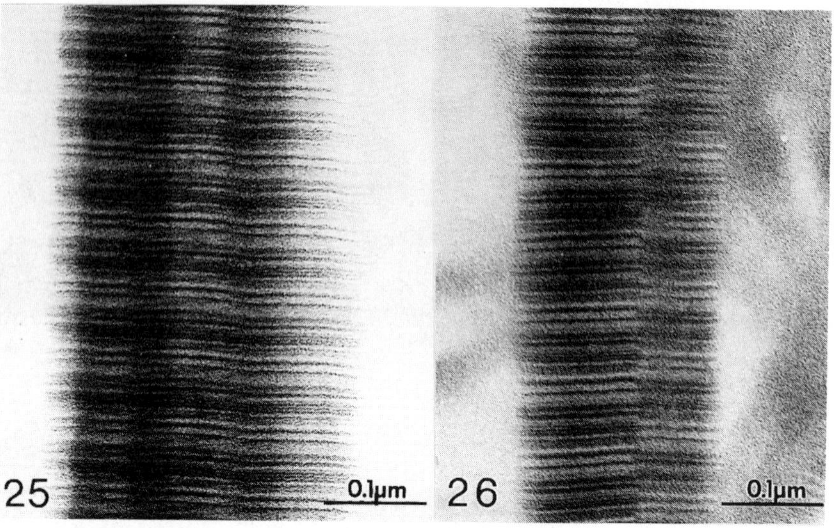

Figures 5.25 and 5.26 Reconstituted collagen fibrils of *Locusta migratoria* and rat tail tendon, respectively. The banding patterns are identical.

the non-helical regions are less conserved (Blumberg *et al.*, 1988; Lunstrum *et al.*, 1988; Fessler and Fessler, 1989).

Laminin is another universal constituent of mammalian basement membranes. Electron micrographs of isolated molecules of laminin isoforms (laminins 1 to 4) typically show a cruciform molecule with three arms of equal length and one which is longer, but in some isoforms (laminins 5 to 7) one of the short arms is truncated (*note,* laminin 2 = merosin, 3 = s-laminin, 4 = s-merosin, 5 = kalinin/nicein, 6 and 7 = k-laminin) (Timpl and Brown, 1994). Laminin-1 self-assembles by interactions of the amino-terminal domains of the three short arms to form a network; the long arms remain free to take part in cell and glycosaminoglycan interactions (Yurchenco and O'Rear, 1994). A molecule, similar in structure, chain sequence and properties to mammalian laminins, was isolated from *Drosophila* cell cultures (Fessler *et al.*, 1987; Chi and Hui, 1989; Garrison *et al.*, 1991). Using antibodies to *Drosophila* laminin, it was located in the basement membranes of several organs, including the midgut, during development (Fessler *et al.*, 1987; Kusche-Gullberg *et al.*, 1992).

A fibronectin-like molecule has been detected in the haemolymph throughout the *Drosophila* life cycle and in extracts of mosquito midgut and cell cultures (Figure 5.27) (Gratecos *et al.*, 1988; Billingsley *et al.*, unpublished observations).

Elastin has not been reported in insects and other invertebrates (Sage and Gray, 1979, 1980; Ashhurst, 1985). A recent observation, suggestive of the presence of elastin in mosquitoes, is that a molecule in reduced extracts of midgut tissue contains an epitope recognized by a monoclonal antibody to mammalian elastin (Figure 5.27) (Billingsley *et al.*, unpublished observations). Whether the mosquito molecule is elastin has still to be demonstrated.

Extracellular matrices contain proteoglycans which are large molecules that typically consist of a core protein to which glycosaminoglycan (GAG) chains are attached. GAGs are long polymers of repeating dimeric units, each of which consists of a uronic acid and an aminosugar. These may be carboxylated and sulphated. The proteoglycan is, therefore, a large polyanionic molecule. There are six GAGs in vertebrates, hyaluronan, chondroitin-4- and chondroitin-6-sulphate, dermatan sulphate, heparan sulphate and keratan sulphate; hyaluronan is peculiar in that it has only -COOH groups and is not attached to a core protein. Heparan sulphate is a component of vertebrate basement membranes, whereas the other GAGs are found in other extracellular matrices.

It has been suggested that all the GAGs are found in insects, but the methods for their identification are not so rigorous as those used with vertebrate tissues. Both carboxylated and highly sulphated substances have been located in connective tissues of cockroaches, locusts and

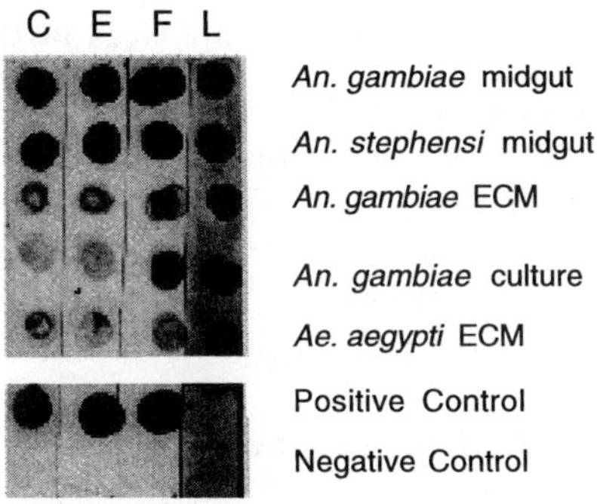

Figure 5.27 An immunoblot of extracellular matrix molecules in mosquito extract. Commercial antibodies against mammalian type IV collagen (C), elastin (E), cellular fibronectin (F) and EHS laminin (L) were used. The extracts probed were from midguts of *Anopheles gambiae* and *An. stephensi*, crude preparations of cell cultures and scrapings of deposited matrix (ECM) from cultures of *A. gambiae* or *Aedes egypti* cell lines. Cross-reacting molecules are present in the midgut extracts, whereas laminin and fibronectin appear to be produced preferentially by the cell lines.

other insects using Alcian Blue and other basic dyes (Ashhurst and Costin, 1971a,b,c; François 1978, 1989). On the basis of their lability to testicular hyaluronidase and chondroitinase ABC, it is assumed that these molecules are very similar to the mammalian GAGs, chondroitin, dermatan and keratan sulphates and hyaluronan. GAGs have been extracted both from whole insects and the midguts of several species including *Calliphora erythocephala*, *Ceratitus capitata*, *Periplaneta americana*, *Rhodnius prolixus* and *Tenebrio molitor* and their electrophoretic mobility was similar to that of known samples of mammalian GAGs, including heparan sulphate (Höglund, 1976; Cassaro and Dietrich, 1977; Theocharis *et al.*, 1985; Garcia *et al.*, 1986; François, 1989). Using more rigorous techniques, a heparan-sulphate-like molecule has been identified in *Drosophila* and localized to several embryonic tissues, including the midgut (Spring *et al.*, 1994). A proteoglycan-like molecule, papilin, in which O-linked carbohydrate chains are attached to a core protein was identified in *Drosophila* cell culture medium (Campbell *et al.*, 1987).

Glutactin is an acidic, sulphated glycoprotein that was isolated from *Drosophila* cell cultures (Fessler and Fessler, 1989; Olson *et al.*, 1990); it is found in embryonic basement membranes, particularly those of the developing nervous system and muscles. *Drosophila* cell cultures have yielded two further matrix proteins tiggrin and peroxidasin (Fogerty *et al.*, 1994; Nelson *et al.*, 1994). Tiggrin is primarily associated with muscle cells and their attachments to apodemes where it binds to integrins and the extracellular matrix.

5.2.3 Constituents of the extracellular matrices of the midgut

The thick layers of connective tissue that surround the midgut of cockroaches and *Tenebrio molitor* contain collagen fibrils that are of a collagen similar to the mammalian type I trimer (François *et al.*, 1980; François, 1985). It also contains some proteoglycan-like, highly acidic molecules (François, 1989). Nothing is known of the collagen in Diptera and other orders in which the collagen fibrils are thin and unbanded (Ashhurst, 1985).

It may be assumed that the morphologically typical basement membranes contain a type IV collagen-like molecule, but in no instance has the actual collagen of a midgut epithelial basement membrane been characterized. It is very unlikely that these basement membranes do not contain such a molecule. Antibodies to the *Drosophila* type IV collagen are bound around the base of *Drosophila* midgut cells (Lunstrum *et al.*, 1988; Fessler and Fessler, 1989). Attempts to use these antibodies on tissues of other species of insects have so far proved unsuccessful (Ashhurst, unpublished observations). Commercially available antibodies to mammalian type IV collagen are bound by the midgut basement membrane of *Anopheles stephensi* and this collagen has been identified in extracts of midguts and cell cultures of several *Anopheles* spp. by immunoblotting (Figure 5.27) (Billingsley *et al.*, unpublished observations).

Laminin is found in all basement membranes and was located in *Drosophila* midgut basement membranes using species-specific antibodies (Fessler *et al.*, 1987; Fessler and Fessler, 1989), but these did not cross-react with proteins in the basement membranes of other insects. A rabbit anti-mouse EHS sarcoma laminin antibody cross-reacts with the midgut basement membranes of *Calliphora erythrocephala* and *D. melanogaster* (Dennis *et al.*, 1991), whereas commercially available antibodies to vertebrate laminin cross-react with the basement membrane and extracts of the midgut of *Anopheles stephensi* and other mosquitoes (Figure 5.27) (Billingsley *et al.*, unpublished observations).

Sulphated GAGs have been located in the basement membrane of the midgut of *Periplaneta* using histochemical methods (François, 1978). A heparan sulphate proteoglycan has since been identified in *Drosophila* midgut basement membrane (Spring *et al.*, 1994). Another proteoglycan-like molecule, papilin, also found in *Drosophila* basement membranes, may fulfil a similar function (Campbell *et al.*, 1987).

The nature of the proteins that comprise the complex layered basement membranes remains unknown. Gouranton (1970) reported that the dense plaques in *Ranatra* are digested by collagenase. The use of enzymes, such as bacterial and mammalian collagenases, to identify the presence of a collagenous protein is not appropriate because bacterial collagenase breaks down many proteins in addition to collagen, and mammalian collagenase does not normally cleave type IV collagen; all commercial samples contain large amounts of non-specific proteases. The layers of all species tested are periodic acid–Schiff-positive, that is, glycoproteins are present, but only the basement membrane in *Oryctes* is stained by Alcian Blue and hence possibly contains some GAGs (Gouranton, 1970; Holter, 1970; Bayon and François, 1976). Thus, all that can be said with certainty is that these structures are composed of glycoproteins.

5.2.4 Functional significance of the connective tissues

The primary function of connective tissues is mechanical support and this is obviously true of the layers of fibrous tissue and basement membranes around the midgut which anchor the surrounding nerves, muscles and tracheoles into their appropriate positions. All the molecules that pass between the epithelial cells and haemolymph have to negotiate the basement membrane and other connective tissue and it is known from the vertebrates that they can act as charged molecular sieves. No work has been done on the functions of insect connective tissues, but the similarities between the component molecules, particularly of the basement membranes, in insects and mammals, permits the assumption that the functions of the molecules will be similar in these diverse animals.

The network of type IV collagen is the main structural component and it provides the scaffold to which the other molecules are attached. It is elastic and provides a strong support for the epithelial cells (Welling *et al.*, 1995). Laminin also forms a network which is attached to the type IV collagen network, but it has more varied functions (Engel, 1992). It interacts with integrins, heparan sulphate proteoglycans and receptors on the cell surface. Both laminin and tiggrin act as ligands for integrins during *Drosophila* muscle and wing morphogenesis (Gotwals *et al.*, 1994a,b). Amounts of laminin increase in *Drosophila* midgut basement membrane during morphogenesis and it is essential for the development of the epithelium (Kusche-Gullberg *et al.*, 1992; Yarnitzky and Volk, 1995). This latter observation may be significant because exposure to laminin induced an increase of brush border enzymes on enterocytes of a human cell line, that is, differentiation and expression of the genes for the enzymes was increased by contact with laminin (Vachon and Beaulieu, 1995). Heparan sulphate is bound by laminin and in turn it has binding sites for cytokines and growth factors; this ensures appropriate concentrations of these factors near epithelial cells. Presumably heparan sulphate and papilin have similar functions in *Drosophila* (Campbell *et al.*, 1987; Spring *et al.*, 1994). The filtration or sieve-like properties of basement membranes have been studied most extensively in the glomerular basement membrane of the mammalian kidney (Williams, 1994). Here the basement membrane is a barrier to molecules with a diffusion radius greater than 10 nm (Orloff and Berliner, 1973). The negative charges of the proteoglycans also mean that the passage of charged molecules may be regulated at the basement membrane. This may be important for the further passage of ions and molecules that have been transported through the midgut epithelial cells into large extracellular spaces created by the invagination of the basal plasma membrane and which have only a few small openings to the underlying basement membrane (Berridge, 1970); sodium pumps are located in these membranes (MacVicker, 1993). The typical basement membranes under the midgut are likely to perform all these functions.

Drosophila has also provided a tool for investigating insect integrins. To date, most of those identified are associated with the basement membranes of the developing wings and muscle–apodeme junctions, but some have been located in the midgut. Position specific (PS) integrins are found in association with the midgut during early embryonic development (Roote and Zusman, 1995) and a specific integrin subunit, βv, is confined to the midgut of embryos and larvae (Yee and Hynes, 1993). Integrins may mediate signals, which influence differentiation and proliferation of the cells, from the basement membrane and other connective tissues to the cytoskeleton (Paulsson, 1992). Epitopes associated with molecules involved in cell–matrix interactions in vertebrates have been located in *Drosophila* basement membranes, but not those of the midgut (Dennis *et al.*, 1991).

The function of the complex layered basement membranes is not clear and cannot be deduced from their composition, which remains a mystery. Their occurrence cannot be correlated with feeding habits (Holter, 1970), and indeed, there may be variation from a simple, to a very complex, basement membrane within one family of the Coleoptera. They have also been described in some Crustacea (see for example, François and Graf, 1988). Whether similar basement membranes occur under the epithelium of other insect organs is not clear. Until more is known of their chemical composition and distribution, the functional significance of these basement membranes will remain an enigma.

5.3 CONCLUDING REMARKS

The morphology and ultrastructure of the insect midgut have been well analysed by a variety of methods, but the details of most of the macromolecules making up the component structures of the cell junctions and extracellular matrices have yet to be clarified. The paucity of material present in each insect precludes extensive biochemical studies. The relative infrequency of the junctions and the very thin layers of connective tissues means that vast numbers of insects are required for any analysis of constituent proteins. Progress on the nature of cell junctional proteins may be more rapid in the future because of the availability of mutants in *Drosophila*, from which gene products can be isolated and characterized, antisera raised, and their localization then determined by immunocytochemical techniques. Such methods have already transformed our understanding of the relationship of septate to tight junctions by virtue of their common involvement in signal transduction (section 5.1.3). A number of matrix macromolecules have been isolated from *Drosophila* cell cultures in sufficient quantities to enable them to be characterized biochemically. Antibodies have been raised to these molecules, but the observed lack of cross-species reactivity emphasizes the diversity of insect matrix macromolecules.

ACKNOWLEDGEMENTS

Figure 5.15 is reproduced by courtesy of Dr M. Nasiruddin, Figures 5.16–5.22 by courtesy of Dr P. Holter and Figures 5.23 and 5.24 by courtesy of Dr J. François. The immunoblot (Figure 5.27) was kindly provided by Dr P.F. Billingsley and Mr D. Gare. We are grateful to Dr Pietro Lupetti for her helpful technical collaboration with the fast freezing work. One of us (NJL) is indebted to the Wellcome Trust for financial support during the preparation of this report (032970/1.4U and 032970/Z/90/B) and RD thanks MURST for grants to support his laboratory.

REFERENCES

Anderson, E. and Harvey, W.R. (1966) Active transport by the cecropia midgut. II. Fine structure of the midgut epithelium. *J. Cell Biol.*, **31**, 107–34.

Anderson, J.M., Balda, M.S. and Fanning, A.S. (1993) The structure and regulation of tight junctions. *Curr. Opin. Cell Biol.*, **5**, 772–8.

Anderson, M. and Cochrane, D.G. (1978) Studies on the midgut of the desert locust *Schistocerca gregaria*. II. Ultrastructure of the muscle coat and its innervation. *J. Morphol.*, **156**, 257–78.

Andries, J.C. and Torpier, G. (1982) An extracellular brush border coat of lipid membranes in the midgut of *Nepa cinerea* (Insecta, Heteroptera): ultrastructure and genesis. *Biol. Cell.*, **46**, 195–202.

Ashhurst, D.E. (1970) An insect desmosome. *J. Cell Biol.*, **46**, 421–5.

Ashhurst, D.E. (1985) Connective tissues, in *Comprehensive Insect Physiology, Biochemistry and Pharmacology* Vol. 3 (eds G.A. Kerkut and L.I. Gilbert), Pergamon, Oxford, pp. 249–87.

Ashhurst, D.E. and Bailey, A.J. (1980) Insect collagen, morphological and biochemical characterization. *Eur. J. Biochem.*, **103**, 75–83.

Ashhurst, D.E. and Costin, N.M. (1971a) Insect mucosubstances. I. The mucosubstances of developing connective tissue in the locust, *Locusta migratoria*. *Histochem. J.*, **3**, 279–95.

Ashhurst, D.E. and Costin, N.M. (1971b) Insect mucosubstances II. The mucosubstances of the central nervous system. *Histochem. J.*, **3**, 297–310.

Ashhurst, D.E. and Costin, N.M. (1971c) Insect mucosubstances III. Some mucosubstances of the nervous system of the wax moth (*Galleria mellonella*) and the stick insect (*Carausius morosus*). *Histochem. J.*, **3**, 379–87.

Ayad, S., Boot-Handford, R., Humphries, M.J. *et al.* (1994) *The Extracellular Matrix Factsbook*, Academic Press, London, 163 pp.

Bayon, C. and François, J. (1976) Ultrastructure de la lame basale due mesenteron chez la larve d'*Oryctes nasicornis* L. (Coleoptera: Scarabaeidae). *Int. J. Insect Morphol. Embryol.*, **5**, 205–17.

Berridge, M.J. (1970) A structural analysis of intestinal absorption. *Symp. R. Entomol. Soc., Lond.*, **5**, 135–51.

Billingsley, P.F. (1990) The midgut ultrastructure of hematophagous insects. *Annu. Rev. Entomol.*, **35**, 219–48.

Billingsley, P.F. and Downe, A.E.R. (1983) Ultrastructural changes in posterior midgut cells associated with feeding in adult female *Rhodnius prolixus* Stål (Hemiptera: Reduviidae). *Can. J. Zool.*, **61**, 2574–86.

Billingsley, P.F. and Downe, A.E.R. (1985) Cellular localization of aminopeptidase in the midgut of *Rhodnius prolixus* Stål (Hemiptera: Reduviidae) during blood digestion. *Cell Tissue Res.*, **241**, 421–8.

Blumberg, B., MacKrell, A.J. and Fessler, J.H. (1988) *Drosophila* basement membrane procollagen α1(IV). *J. Biol. Chem.*, **263**, 18328–37.

Bohrmann, J. (1993) Antisera against a channel-forming 16 kDa protein inhibit dye-coupling and bind to cell membranes in *Drosophila* ovarian follicles. *J. Cell Sci.*, **105**, 513–18.

Burgos, M.H. and Guitiérrez, L.S. (1976) The intestine of *Triatoma infestans*. I. Cytology of the midgut. *J. Ultrastruct. Res.*, **57**, 1–9.

Campbell, A.G., Fessler, L.I., Salo, T. and Fessler, J.H. (1987) Papilin: a *Drosophila* proteoglycan-like sulfated glycoprotein from basement membranes. *J. Biol. Chem.*, **262**, 17605–12.

Cassaro, C.M.F. and Dietrich, C.P. (1977) Distribution of sulfated mucopolysaccharides in invertebrates. *J. Biol. Chem.*, **252**, 2254–61.

Chi, H.C. and Hui, C.F. (1989) Primary structure of the *Drosophila* laminin B2 chain and composition with human, mouse, and *Drosophila* laminin B1 and B2 chains. *J. Biol. Chem.*, **264**, 1543–50.

Citi, S. (1993) The molecular organization of tight junctions. *J. Cell Biol.*, **121**, 485–90.

Colombo, A., Bonfanti, P. and Camatini, M. (1993) Actin, α-actin, and vinculin are associated with septate junctions in Insecta. *Cell Motil. Cytoskel.*, **26**, 205–13.

Dallai, R. and Giusti, F. (1979) The epithelial cell junctions in *Onychophora* in *Myriapod Biology* (ed. M. Camatini), Academic Press, London, pp. 433–43.

Dallai, R., Trastullo, E., Lupetti, P. and Mencarelli, C. (1993) Unusual cytoskeletal association with the intercellular septate junction in the midgut of *Collembola* (Insecta: Apterygota). *Int. J. Insect Morphol. Embryol.*, **22**, 473–86.

Del Bene, G., Dallai, R. and Marchini, D. (1991) Ultrastructure of the midgut and the adhering tubular salivary glands of *Frankliniella occidentalis* (Pergande) (Thysanoptera: Tripidae). *Int. J. Insect Morphol. Embryol.*, **20**, 15–24.

Dennis, R.D., Marini, R. and Schachner, M. (1991) Expression of carbohydrate epitopes L2/HNK-1 and L3 in the larva and imago of *Drosophila melanogaster* and *Calliphora vicina*. *Cell Tissue Res.*, **265**, 589–600.

De Priester, W. (1971) Ultrastructure of the midgut epithelial cells in the fly, *Calliphora erythrocephala*. *J. Ultrastruct. Res.*, **36**, 783–805.

Dimitriadis, V.K. and Pirpasopoulou, A. (1992) Complex carbohydrates present in the gut and Malpighian tubules of *Drosophila auraria* larvae (Insecta, Diptera): a cytochemical study. *Cytobios*, **70**, 159–70.

Engel, J. (1992) Laminins and other strange proteins. *Biochemistry*, **31**, 10643–51.

Ferreira, C., Ribeiro, A.F., Garcia, E.S. and Terra, W.R. (1988) Digestive enzymes trapped between and associated with the double plasma membranes of *Rhodnius prolixus* posterior midgut cells. *Insect Biochem.*, **18**, 521–30.

Fessler, L.I., Campbell, A.G., Duncan, K.G. and Fessler, J.H. (1987) *Drosophila* laminin: characterization and localization. *J. Cell Biol.*, **105**, 2383–91.

Fessler, J.H. and Fessler, L.I. (1989) *Drosophila* extracellular matrix. *Annu. Rev. Cell Biol.*, **5**, 309–39.

Finbow, M.E., Buultjens, T.E.J., Lane, N.J. *et al.* (1984) Isolation and characterization of arthropod gap junctions. *EMBO J.*, **3**, 2271–8.

Finbow, M.E., Goodwin, S., Meagher, L. *et al.* (1994) Evidence that the 16 kDa proteolipid (Subunit C) of the vacuolar H^+-ATPase and ductin from gap junctions are the same polypeptide in *Drosophila* and *Manduca*: molecular cloning of the Vha 16 k gene from *Drosophila*. *J. Cell Sci.*, **107**, 1817–24.

Finbow, M.E., Lane, N.J., Meagher, L. and Findlay, J. (1990) Isolation of the 16 kD vacuolar proton channel from insects in a gap-junction-like state. *J. Cell Biol.*, **111**, 277A.

Flores, V. and Lane, N.J. (1990) Evidence for a transcellular cisternal route across the caecal epithelium of an insect. *Cell Tissue Res.*, **261**, 347–54.
Flower, N.E. (1974) Plasma membrane differentiation in the midgut of a lepidopteran larva, *Ephestia kuhniella*, in *Proc. 8th Int. Congr. Electron Microsc.* Vol. 2 (ed. J.V. Sanders and D.J. Goodchild), Australian Academy of Sciences, Canberra, pp. 224–5.
Flower, N.E. and Filshie, B.K. (1975) Junctional structures in the midgut cells of lepidopteran caterpillars. *J. Cell Sci.*, **17**, 221–39.
Flower, N.E. and Filshie, B.K. (1976) Goblet cell membrane differentiations in the midgut of a lepidopteran larva. *J. Cell Sci.*, **20**, 357–75.
Fogerty, F.J., Fessler, L.I., Bunch, T.A. *et al.* (1994) Tiggrin, a novel *Drosophila* extracellular matrix protein that functions as a ligand for *Drosophila* alpha PS2 beta PS integrins. *Development*, **120**, 1747–58.
François, J. (1978) The ultrastructure and histochemistry of the mesenteric connective tissue of the cockroach *Periplaneta americana* L. (Insecta, Dictyoptera). *Cell Tissue Res.*, **189**, 91–107.
François, J. (1985) Caractérisation biochimique et biophysique du collagène de l'insecte *Tenebrio molitor* L. (Coleoptera, Tenebrionidae). *Biochimie*, **67**, 1035–42.
François, J. (1989) The glycosaminoglycans of midgut connective sheath during development of the mealworm *Tenebrio molitor* L. *Comp. Biochem. Physiol.*, **93B**, 93–8.
François, J. and Graf, F. (1988) Unusual basement layer in the midgut of gammaridean *Niphargus virel chevreux* (Crustacea, Amphipoda). *Hitochemistry*, **89**, 379–83.
François, J., Herbage, D. and Junqua, S. (1980) Cockroach collagen: isolation, biochemical and biophysical characterization. *Eur. J. Biochem.*, **112**, 389–96.
Furuse, M., Hirare, T., Itoh, M. *et al.* (1993) Occludin: a novel integral membrane protein localization at tight junctions. *J. Cell Biol.*, **123**, 1777–88.
Garcia, E.S., de Azambuja, P., Nader, H.B. and Dietrich, C.P. (1986) Biosynthesis of sulfated glycosaminoglycans in the Hemipteran *Rhodnius prolixus*. *Insect Biochem.*, **16**, 347–52.
Garrison, K., MacKrell, A.J. and Fessler, J.H. (1991) *Drosophila* laminin A chain sequence, interspecies comparison, and domain structure of a major carboxyl portion. *J. Biol. Chem.*, **266**, 22899–904.
Garrod, D.R. (1993) Desmosomes and hemi-desmosomes. *Curr. Opin. Cell Biol.*, **5**, 30–40.
Geiger, B. (1989) Cytoskeleton-associated cell contacts. *Curr. Opin Cell Biol.*, **1**, 103–9.
Gotwals, P.J., Fessler, L.I., Wehrli, M. and Hynes, R.O. (1994a) *Drosophila* PS1 integrin is a laminin receptor and differs in ligand specificity from PS2. *Proc. Natl Acad. Sci. USA*, **91**, 11447–51.
Gotwals, P.J., Paine-Saunders, S.E., Stark, K.A. and Hynes, R.O. (1994b) *Drosophila* integrins and their ligands. *Curr. Opin. Cell Biol.*, **6**, 734–9.
Gouranton, J. (1970) Etude d'une lame basale présentant une structure d'un type nouveau. *J. Microsc.*, **9**, 1029–40.
Gouranton, J. and Maillet, P.L. (1965) Sur l'existence d'une membrane péritrophique chez un insecte succur de sève, *Cicadella viridis* L. (Homoptera, Jassidae). *C. r. hebd. Séanc., Acad. Sci., Paris*, **261**, 1102–5.
Gratecos, D., Naidet, C., Astier, M. *et al.* (1988) *Drosophila* fibronectin: a protein that shares properties similar to those of its mammalian homologue. *EMBO J.*, **7**, 215–23.

Green, C.R. (1981) Septate junctions of the phylum Hemichordata. *J. Ultrastruct. Res.*, **75**, 1–10.

Green, C.R., Noirot-Timothée, C. and Noirot, C. (1983) Isolation and characterization of invertebrate smooth septate junctions. *J. Cell Sci.*, **62**, 351–70.

Guitiérrez, L.S. and Burgos, M.H. (1978) The intestine of *Triatoma infestans*. II. The surface coat of the midgut. *J. Ultrastruct. Res.*, **63**, 244–51.

Gumbiner, B. (1993) Breaking through the tight junction barrier. *J. Cell Biol.*, **123**, 1631–3.

Harvey, W.R. (1980) Water and ions in the gut, in *Insect Biology in the Future 'VBW 80'* (eds M. Locke and D.S. Smith), Academic Press, New York, pp. 105–25.

Harvey, W.R. (1992) Physiology of V-ATPases. *J. Exp. Biol.*, **172**, 1–17.

Harvey, W.R., Cioffi, M. and Wolfersberger, M.G. (1981) Portasomes as coupling factors in active ion transport and oxidative phosphorylation. *Am. Zool.*, **21**, 775–91.

Hess, R.T. and Pinnock, D.E. (1975) The ultrastructure of a complex basal lamina in the midgut of larvae of *Oryctes rhinoceros* L. (Coleoptera, Scarabaeidae). *Z. Morphol. Tiere*, **80**, 277–85.

Höglund, L. (1976) The comparative biochemistry of invertebrate mucopolysaccharides. V. Insecta (*Calliphora erythrocephala*). *Comp. Biochem. Physiol.*, **53B**, 9–14.

Holter, P. (1970) Regular grid-like substructures in the midgut epithelial basement membrane of some Coleoptera. *Z. Zellforsch.*, **110**, 373–85.

Houk, E.J. (1977) Midgut ultrastructure of *Culex tarsalis* (Diptera: Culcidae) before and after a blood meal. *Tissue Cell*, **9**, 103–18.

Houk, E.J., Chiles, R.E. and Hardy, J.L. (1980) Unique midgut basal lamina in the mosquito, *Aedes dorsalis* (Meigen) (Insecta: Diptera). *Int. J. Insect Morphol. Embryol.*, **9**, 161–4.

Houk, E.J., Hardy, J.L. and Chiles, R.E. (1986a) Histochemical staining of the complex carbohydrates of the midgut of the mosquito, *Culex tarsalis* Coquillet. *Insect Biochem.*, **16**, 667–75.

Houk, E.J., Hardy, J.L. and Chiles, R.E. (1986b) Mesenteronal surface charge of the mosquito, *Culex tarsalis* Coquillet. Binding of colloidal iron hydroxide, native ferritin and cationized ferritin. *J. Submicrosc. Cytol.*, **18**, 385–96.

Kachar, B., Christakis, N.A., Reese, T.S. and Lane, N.J. (1986) The intramembrane structure of septate junctions based on direct freezing. *J. Cell Sci.*, **80**, 13–28.

Kitajima, E.W. (1995) A peculiar type of glycocalyx on the microvilli of the midgut epithelial cells of the thrips *Frankliniella* sp. (Thysanoptera, Tripidae). *Cytobiologie*, **11**, 299–303.

Kühn, K. (1995) Basement membrane (Type IV) collagen. *Matrix Biol.*, **14**, 439–45.

Kusche-Gullberg, M., Garrison, K., Mackrell, A.J. *et al.* (1992) Laminin A chain: expression during *Drosophila* development and genomic sequence. *EMBO J.*, **11**, 4519–27.

Lane, N.J. (1984) A comparison of the construction of intercellular junctions in the CNS of vertebrate and invertebrates. *Trends Neurosci.*, **7**, 95–9.

Lane, N.J. (1986) Arthropod fine structure: towards an understanding of the intricacies of intercellular junctions. *Micron Micros. Acta*, **17**, 137–47.

Lane, N.J., Dallai, R., Martinucci, G. and Burighel, P. (1994) Electron microscopic structure and evolution of epithelial junctions, in *Molecular Mechanisms of Epithelial Cell Junctions: from Development to Disease* (ed. S. Citi), R.G. Landes, Austin, Texas, pp. 23–43.

Lane, N.J. and Dilworth, S. (1989) Isolation and biochemical characterization of septate junctions. Differences between the proteins in smooth and pleated varieties. *J. Cell Sci.*, **93**, 123–31.

Lane, N.J. and Finbow, M.E. (1988) Isolation of gap and septate junctions from arthropod tissues. *J. Cell Biol.*, **107**, 793A.

Lane, N.J. and Flores, V. (1988) Actin filaments are associated with the septate junctions of invertebrates. *Tissue Cell*, **20**, 211–17.

Lane, N.J. and Flores, V. (1990) The role of cytoskeletal components in the maintenance of intercellular junctions in an insect. *Cell Tissue Res.*, **262**, 373–85.

Lane, N.J., Flores, V., Harrison, J.B. and Lee, W.M. (1987) Septate junction-associated actin is anchored onto microtubules. *Cell Biol. Int. Rep.*, **11**, 250.

Lane, N.J. and Harrison, J.B. (1978) An unusual type of continuous junction in Limulus. *J. Ultrastruct. Res.*, **64**, 85–97.

Lane, N.J. and Harrison, J.B. (1979) An unusual cell surface modification: a double plasma membrane. *J. Cell Sci.*, **39**, 353–72.

Lane, N.J. and Swales, L.S. (1982) Stages in the assembly of pleated and smooth septate junctions in developing insect embryos. *J. Cell Sci.*, **56**, 245–62.

Lunstrum, G.P., Bachinger, H.-P., Fessler, L.I. *et al.* (1988) *Drosophila* basement membrane procollagen IV. *J. Biol. Chem.*, **263**, 18318–27.

MacVicker, J.A.K. (1993) Na^+/K^+-ATPases in the midguts of haematophagous insects; biochemical and immunochemical studies. PhD Thesis, University of London.

Marshall, A.T. and Cheung, W.W.K. (1970) Ultrastructure and cytochemistry of an extensive plexiform surface coat on the midgut cells of a Fulgorid insect. *J. Ultrastruct. Res.*, **33**, 161–72.

Meis, J.F.G.M., Wismans, P.G.P., Jap. P.H.K. *et al.* (1992) A scanning electron microscopic study of the sporogonic development of *Plasmodium falciparum* in *Anopheles stephensi*. *Acta Trop.*, **50**, 227–36.

Musil, L.S. (1994) Structure and assembly of gap junctions, in *Molecular Mechanisms of Epithelial Cell Junctions* Vol. 10, CRC Press, Boca Raton, pp. 173–94.

Nasiruddin, M. and Mordue, A.J. (1993) The effect of azadirachtin on the midgut histology of the locusts, *Schistocerca gregaria* and *Locusta migratoria*. *Tissue Cell*, **25**, 875–84.

Nelson, R.E., Fessler, L.I., Takagi, Y. *et al.* (1994) Peroxidasin: a novel enzyme-matrix protein of *Drosophila* development. *EMBO J.*, **13**, 3438–47.

Noirot, C. and Noirot-Timothée, C. (1967) Un nouveau type de jonction intercellulaire (zonula continua) dans l'intestin moyen des insects. *C. r. hebd. Séanc. Acad. Sci., Paris*, **264**, 2796–8.

Noirot, C. and Noirot-Timothée, C. (1972) Structure fine de la bordure en brosse de l'intestine moyen chez les insectes. *J. Microsc.*, **13**, 85–96.

Noirot-Timothée, C. and Noirot, C. (1980) Septate and scalariform junctions in arthropods. *Int. Rev. Cytol.*, **63**, 97–140.

Olson, P.F., Fessler, L.I., Nelson, R.E. *et al.* (1990) Glutactin, a novel *Drosophila* basement membrane-related glycoprotein with sequence similarity to serine esterases. *EMBO J.*, **9**, 1219–27.

Orloff, J. and Berliner, R.W. (1973) *Handbook of Physiology*. Section 8: Renal Physiology. American Physiological Society, Washington, DC.

Pacheco, J. (1970) Ultrastructura del piloro de *Rhodnius prolixus* (Hemiptera, Reduviidae). *Acta Biol. Venez.*, **7**, 41–70.

Paulsson, M. (1992) Basement membrane proteins: structure, assembly, and cellular interactions. *Crit. Rev. Biochem. Mol. Biol.*, **27**, 93–127.

Peifer, M. (1995) Cell adhesion and signal transduction: the Armadillo connection. *Trends Cell Biol.*, **5**, 224–9.

Reger, J.F. (1971) Fine structure of the surface coat of midgut epithelial cells in the homopteran *Phylloselis atra* (Fulgorid). *J. Submicrosc. Cytol.*, **3**, 353–8.

Reinhardt, C. and Hecker, H. (1973) Structure and function of the basal lamina and of the cell junctions in the midgut epithelium (stomach) of female *Aedes aegypti* L. (Insecta, Diptera). *Acta Trop.*, **30**, 213–36.

Reinhardt, C., Schulz, U., Hecker, H. and Freyvogel, T.A. (1972) Zur Ultrastruktur des Mitteldarmepithels bei Flöhen (Insecta, Siphonaptera). *Rev. Suisse Zool.*, **79**, 130–7.

Richards, A.G. and Richards, P.A. (1968) Flea *Ctenophthalmus*: heterogeneous hexagonally organized layer in the midgut. *Science*, **160**, 423–4.

Roote, C.E. and Zusman, S. (1995) Function for PS integrins in tissue adhesion, migration, and shape changes during early embryonic development in *Drosophila*. *Dev. Biol.*, **169**, 322–36.

Rudin, W. and Hecker, H. (1989) Lectin binding-sites in the midgut of the mosquitoes, *Anopheles stephensi* Liston and *Aedes aegypti* L. (Diptera: Culicidae). *Parasitol. Res.*, **75**, 268–79.

Sage, H. and Gray, W.R. (1979) Studies on the evolution of elastin. I. Phylogenetic distribution. *Comp. Biochem. Physiol.*, **64B**, 313–27.

Sage, H. and Gray. W.R. (1980) Studies on the evolution of elastin. II. Histology. *Comp. Biochem. Physiol.*, **66B**, 13–22.

Skaer, H. le B., Maddrell, S.H.P. and Harrison, J.B. (1987) The permeability properties of septate junctions in Malpighian tubules of *Rhodnius*. *J. Cell Sci.*, **88**, 251–65.

Smith, D.S., Compher, K., Janners, M. *et al.* (1969) Cellular organization and ferritin uptake in the midgut epithelium of a moth, *Ephestia kuhniella*. *J. Morphol.*, **127**, 41–72.

Spring, J., Paine-Saunders, S.E., Hynes, R.O. and Bernfield, M. (1994) *Drosophila syndecan*: conservation of a cell-surface heparan sulfate proteoglycan. *Proc. Natl Acad. Sci. USA*, **91**, 3334–8.

Steele, E.J., Noblet, G.P. and Noblet, R. (1992) Sporogonic development of *Leucocytozoon smithi*. *J. Protozool.*, **39**, 690–9.

Stevenson, B.R. and Paul, D.L. (1989) The molecular constituents of intercellular junctions. *Curr. Opin. Cell Biol.*, **1**, 884–91.

Swales, L.S. and Lane, N.J. (1985) Embryonic development of glial cells and their junctions in the locust CNS. *J. Neurosci.*, **5**, 117–27.

Syafruddin, R.A., Kamimura, K. and Kawamoto, F. (1991) Penetration of the mosquito midgut wall by the ookinetes of *Plasmodium yoelii nigeriensis*. *Parasitol. Res.*, **77**, 230–6.

Tano, Y., François, J. and Noirot-Timothée, C. (1987) Métamorphose de l'intestin antérieur et de l'intestin moyen chez *Tenebrio molitor* L. (Insecte, Coleoptére). *Can. J. Zool.*, **65**, 1923–34.

Terzakis, J.A. (1967) Substructure in an epithelial basal lamina (basement membrane). *J. Cell Biol.*, **35**, 273–8.

Theocharis, D.A., Anagnostides, S.T. and Tsiganos, C.P. (1985) Distribution and changes of glycosaminoglycans in the three stages of development of the insect *Ceratitis capitata*. *Comp. Biochem. Physiol.*, **81B**, 933–7.

Thomas, G. and Böckeler, W. (1992) Light and electron microscopical investigations of the midgut epithelium of different Cephalobaenida (Pentastomida) during digestion. *Parasitol. Res.*, **78**, 587–93.

Timpl, R. and Brown, J.C. (1994) The laminins. *Matrix Biol.*, **14**, 275–81.

Tsukita, Sh., Itoh, M., Nagafuchi, A. *et al.* (1993) Submembranous junctional plaque proteins include potential tumor suppressor molecules. *J. Cell Biol.*, **123**, 1049–53.

Uitto, J. (1979) Collagen polymorphism: isolation and partial characterization of $\alpha 1$(I)-trimer molecules in normal human skin. *Arch. Biochem. Biophys.*, **192**, 371–9.

Vachon, P.H. and Beaulieu, J.-F. (1995) Extracellular heterotrimeric laminin promotes differentiation in human enterocytes. *Am. J. Physiol.*, **268**, G857–G867.

Van der Rest, M. and Garrone, R. (1991) Collagen family of proteins. *FASEB J.*, **5**, 2814–23.

Welling, L.W., Zupka, M.T. and Welling, D.J. (1995) Mechanical properties of basement membrane. *News Physiol. Sci.*, **10**, 30–5.

Wilkins, S. and Billingsley, P.F. (1996) Oligosaccharides on midgut microvillar glycoproteins of the mosquito, *Anopheles stephensi* histon (submitted).

Williams, J.C. (1994) Permeability of basement membranes to macromolecules. *Proc. Soc. Exp. Biol. Med.*, **207**, 13–19.

Willot, E., Balda, M.S., Fanning, A.S. *et al.* (1993) The tight junction protein ZO-1 is homologous to the *Drosophila* discs-large tumor suppressor protein of septate junctions. *Proc. Natl Acad. Sci. USA*, **90**, 7834–8.

Woods, D.F. and Bryant, P.J. (1991) The discs-large tumor suppressor gene of *Drosophila* encodes a guanylate kinase homolog localized at septate junctions. *Cell*, **66**, 451–64.

Woods, D.F. and Bryant, P.J. (1993) Apical junctions and cell signalling in epithelia. *J. Cell Sci.* (Suppl. **17**), 171–81.

Yarnitzky, T. and Volk, T. (1995) Laminin is required for heart, somatic muscles, and gut development in the *Drosophila* embryo. *Dev. Biol.*, **169**, 609–18.

Yee, G.H. and Hynes, R.O. (1993) A novel, tissue specific integrin subunit, *ß*v, expressed in the midgut of *Drosophila melanogaster*. *Development*, **118**, 845–58.

Yurchenco, P.D. and O'Rear, J.J. (1994) Basal lamina assembly. *Curr. Opin. Cell Biol.*, **6**, 674–81.

Xue, L. and Dallai, R. (1992) Cell junctions in the gut of *Protura*. *Tissue Cell*, **24**, 51–9.

Part Two

Digestion and Transport

6

Digestive enzymes

W.R. Terra, C. Ferreira, B.P. Jordão and R.J. Dillon

6.1 INTRODUCTION

It has been estimated that nearly two thirds of the world's species are in the class Insecta. The reasons for the phenomenal success of the Insecta have attracted the attention of researchers for centuries (Wigglesworth, 1972). Perhaps one significant factor for their success is the ability of insects to exploit a wide range of organic material in their nutrition. These materials extend from refractory wood, humus, wool and wax to the more digestible micro-organisms, plant and animal tissues and finally nutrient imbalanced diets such as plant sap or blood. The digestive capacity depends on the enzymes present and how they are compartmentalized in the insect's gut.

In addition to the ability to digest refractory materials, insects have exploited food sources not available to other animals because of their toxicity. Some species of phytophagous insects have circumvented the effects of protease inhibitors and secondary plant compounds which become toxic after digestive enzyme action (Harborne, 1993). Thus the insects have expanded into ecological niches unexploited by other organisms and this offers an unrivalled opportunity for enzymologists to study the digestive physiology of organisms associated with the widest range of diets. The ability of insect species to utilize different materials as a food source has inevitably led to some being classified as pests by man in the areas of agriculture, food storage, the building industry and human health. This has contributed to a need for knowledge of insect digestion as a prerequisite for developing methods of control that act via the gut, such as the use of transgenic plants to

Biology of the Insect Midgut.
Edited by M.J. Lehane and P.F. Billingsley.
Published in 1996 by Chapman & Hall, London.
ISBN 0 412 61670 X.

control phytophagous insects. Further interest is guaranteed through the knowledge that digestive enzymes are implicated in resistance to the transmission of medically important pathogens. Transmission blocking vaccines to protect mammalian hosts against pathogens transmitted by insects also require knowledge of digestive physiology.

This chapter describes the properties and function of insect digestive enzymes. Enzyme compartmentalization is discussed in Chapter 8. The present chapter emphasizes digestive enzymes for which molecular properties are known, although the occurrence of less common digestive enzymes is not disregarded. Because overall patterns of digestion, midgut morphologies, and digestive properties correlate well with the phylogenetic position of the insect (see following sections and Chapter 8), all insect species mentioned are classified up to the order and family levels (see discussion on insect phylogeny in Chapter 8).

6.2 CHARACTERISTICS OF DIGESTIVE ENZYMES

In view of the varied background of the potential readers of this book, we will survey in this section the most important characteristics of enzymes. In the following sections data from many insect digestive enzymes are described and the reader is directed to representative studies which may be used as models of enzyme characterization.

Enzyme kinetic parameters are meaningless, unless assays are performed in conditions in which enzymes are stable. The easiest way to certify that enzymes are stable in a given condition is to develop a rectilinear plot of product formation (or substrate disappearance) versus time. Activities (velocities) calculated from this plot are reliable parameters. According to the International Union of Biochemistry and Molecular Biology, the assay temperature should be 30°C, except when the enzyme is unstable at this temperature or for specific purposes. Owing to partial inactivation, the optimum temperature usually decreases as assay times increase. Hence optimum temperature is not a true property of enzymes and therefore should not be included in the characterization. Enzyme pH optimum should be determined using different buffers to discount the effects of chemical constituents of the buffers and their ionic strength on enzyme activity. The number of molecular forms of a given enzyme should be evaluated by submitting the enzyme preparation to a separation process (gel permeation, ion-exchange chromatography, electrophoresis, gradient ultracentrifugation, etc.), followed by assays of the resulting fractions. Substrate specificity of each molecular form of a given enzyme should be evaluated and substrate preference quantified by determining V_{max}/K_m ratios for each substrate. Substrate preference expressed as the percentage activity towards a given substrate in relation to the activity upon a reference substance may be misleading because, in this condition, enzyme activities are determined at different substrate saturations. The isoelectric points of many enzymes can be determined after staining with specific substrates after separation of the native enzymes on isoelectrofocusing gels.

If enzyme characterization is performed as part of a digestive physiology study, emphasis should be given to enzyme compartmentalization, substrate specificity and substrate preference, in order to disclose the sequential action of enzymes during the digestive process. Knowledge of the effect of pH on enzyme activity is useful in evaluating enzyme action in gut compartments with different pH values. Finally, the determination of digestive enzyme M_r values, associated with the identification of those soluble enzymes able to pass through the peritrophic matrix, allows estimation of the peritrophic matrix pore sizes. M_r values determined in gentle conditions are preferred, since in these conditions the enzymes should maintain their *in vivo* aggregation states. The method of choice in this case is gradient ultracentrifugation.

Complete enzymological characterization requires purification to homogeneity and analysis of amino acid composition and sequence. Furthermore, details of the catalytic mechanism, including involvement of amino acid residues in catalysis, should be known. This kind of study permits the classification of insect digestive enzymes into catalytic families and will enable us to establish evolutionary relationships with enzymes from other living organisms.

6.3 CLASSIFICATION OF DIGESTIVE ENZYMES

Digestive enzymes are hydrolases. The enzyme classification and numbering system used here is that recommended by the Nomenclature Committee of the International Union of Biochemistry and Molecular Biology (Nomenclature Committee, 1992).

Peptidases (peptide hydrolases, EC 3.4) are enzymes acting on peptide bonds and include the proteinases (endopeptidases, EC 3.4.21–24) and the exopeptidases (EC 3.2.4.11–19). Proteinases are divided into subclasses on the basis of catalytic mechanism, as shown with specific reagents or effect of pH. Specificity is only used to identify individual enzymes within subclasses. Serine proteinases (EC 3.4.21) have a serine and a histidine in the active site. Cysteine proteinases (EC 3.4.22) possess a cysteine in the active site and are inhibited by mercurial compounds. Aspartic proteinases (EC 3.4.23) have a pH optimum below 5, due to the involvement of a carboxyl residue in catalysis. Metalloproteinases (EC 2.3.24) need a metal ion in the catalytic process. Exopeptidases include enzymes which hydrolyse single amino acids from the N-terminus (aminopeptidases, EC 3.4.11) or from the C-terminus (carboxypeptidases, EC 3.4.16–18) of the peptide chain and those enzymes specific for dipeptides (dipeptide hydrolases, EC 3.4.13).

Glycosidases (EC 3.2) are classified according to their substrate specificities but for the purpose of this review, we will distinguish two broad categories. One includes the depolymerases, that is, enzymes which cleave internal bonds in polysaccharides, and that are usually named from their substrates, exemplified by amylase, cellulase, pectinase and chitinase. The second category includes enzymes that hydrolyse oligosaccharides and disaccharides. Oligosacharidases and disaccharidases are usually named based on the monosaccharide that gives its reducing group to the glycosidic bond and on the configuration (α or β) of this bond.

Lipids consist of a large and heterogeneous group of substances that are relatively insoluble in water but readily soluble in apolar solvents. Some contain fatty acids (fats, phospholipids, glycolipids, waxes) and others lack them (terpenes, steroids and carotenoids). Ester bonds are hydrolysed in lipids containing fatty acids before they are absorbed. The enzymes which hydrolyse ester bonds comprise: (a) carboxylic ester hydrolases (EC 3.1.1), exemplified by lipases, esterases and phospholipases A and B; (b) phosphoric monoester hydrolases (EC 3.1.3), which are the phosphatases; (c) phosphoric diester hydrolases (EC 3.1.4), which include phospholipases C and D.

6.4 PEPTIDASES

6.4.1 Trypsins

Trypsins (EC 3.4.21.4) are serine proteinases that preferentially cleave protein chains on the carboxyl side of basic L-amino acids such as arginine or lysine. Digestive trypsin-like activity has been reported in most examined insect species. Important exceptions are Hemipteran species and species belonging to the series Cucujiformia of Coleoptera (section 6.4.3).

Table 6.1 summarizes the properties of the best known insect midgut trypsin-like enzymes. Most trypsin M_r values are in the range 20 000 to 35 000 and pI values are variable (most of them in the range 4–5). pH optima are always alkaline (most between 8 and 9), irrespective of the pH prevailing in midguts from which the trypsins were isolated (see pH of gut contents of representative families of the major insect orders, Chapter 8). Nevertheless, trypsins isolated from lepidopteran insects have higher pH optima, corresponding to the higher pH values found in their midguts. Cleavage specificity against polypeptides was studied in trypsins from several insects (Terra and Ferreira, 1994). Results showed that specificities of these enzymes are similar (but not identical) to that of

Table 6.1 Properties of trypsin-like enzymes from insect midguts

Order *Species (Family)*	*Stage*	*Degree of purification*	M_r	*pI*	pH_o	*Reference*
Thysanura						
Thermobia domestica (Lepismatidae)	Adult	Partial	21 000 (P)	4.95	9.0	Zinkler and Polzer (1992)
Orthoptera						
Locusta migratoria (Acrididae)	Adult	Homogeneous	24 000 (SP)	3.5	9.0	Sakal *et al.* (1989)
Coleoptera						
Pterostichus melanarius (Carabidae)	Adult	Partial	16 000 (F)	<7.0	8.0	Gooding and Huang (1969)
Tenebrio molitor (Tenebrionidae)	Larva	Homogeneous	24 300 (SP)	8.0	8.0	Levinsky *et al.* (1977)
Costelytra zealandica (Scarabaeidae)	Larva	Partial	23 000 (SP)	–	9.5	Christeller *et al.* (1989)
Attagenus megatoma (Dermestidae)	Larva	Partial	21 000 (F)	–	8.5	Baker (1981b)
Hymenoptera						
Vespa crabo (Vespidae)	Larva	Homogeneous	28 000 (F)	–	8.0	Jany *et al.* (1978)
Apis mellifera (Apidae)	Adult	Homogeneous	21 000(F)	–	–	Giebel *et al.* (1971)
Diptera						
Aedes aegypti (Culicidae)	Adult	Homogeneous	27 000 (SP)	4.0	8.4	Graf and Briegel (1985)
Glossina morsitans (Glossinidae)	Adult	Partial	20 000 (F)	–	8.5	Gooding and Rolseth (1976)
Musca domestica (Muscidae)	Larva	Homogeneous	33 000 (SP)	–	8.5	Lemos and Terra (1992)
Hypoderma lineatum (Oestridae)	Larva	Homogeneous	27 000 (SP)	–	7.8	Tong *et al.* (1981)
Lepidoptera						
Bombyx mori (Bombycidae)	Larva	Homogeneous	22 000 (SP)	–	10	Sasaki and Suzuki (1982)
Choristoneura fumiferana (Tortricidae)	Larva	Homogeneous	25 000 (SP)	10.3	9.5	Milne and Kaplan (1993)

Enzymes included are representative of the more thoroughly characterized enzymes of each major taxa. References chosen are those which describe most properties of each enzyme. Where isoenzymes occur, only the major or, if there is not one, that of higher M_r value is described. M_r values were determined using gel filtration (F), SDS–PAGE (SP), polyacrylamide gel electrophoresis under native conditions (P) or ultracentrifugation (U). pH optima (pH_o) were determined using synthetic substrates.

vertebrate trypsins. Nevertheless, some properties of insect trypsins contrast with those of vertebrate trypsins. Thus, insect trypsins are not activated or stabilized by calcium ions (e.g. Levinsky *et al.*, 1977; Jany *et al.*, 1978; Lemos and Terra, 1992), in most cases are unstable in acid pH (e.g. Sakal *et al.*, 1989) and have different sensitivities to natural trypsin inhibitors (Purcell *et al.*, 1992 and references therein).

Barillas-Mury *et al.* (1991) sequenced what seems to be the precursor of midgut trypsin in *Aedes aegypti*. The sequence shows significant differences from the vertebrate trypsin precursors in the region of the activation peptide. Similar results were found with a putative trypsinogen from *Simulium vittatum* (Diptera: Simuliidae) (Ramos *et al.* 1993). These differences suggest that the processing of precursors of insect trypsins may be different from that of vertebrates. In *Erinnyis ello* (Santos *et al.*, 1986) and in *Musca domestica* (Lemos and Terra 1992; Jordão *et al.*, 1995) trypsin is synthesized in midgut cells in an active form, but is associated with the membranes of vesicles. These vesicles then migrate to the cell apex and trypsin precursors are processed to a soluble form before being secreted. Secretory granules isolated from the opaque zone cells from *Stomoxys calcitrans* (Diptera: Muscidae) adults contain trypsin precursor which is also different from that found in vertebrates (Moffat and Lehane, 1990).

6.4.2 Chymotrypsins

Chymotrypsins (EC 3.4.21.1) are serine proteinases that preferentially cleave protein chains on the carboxyl side of aromatic amino acids. It seems that the distribution of chymotrypsin-like enzymes among insect taxa is similar to that of trypsin (Applebaum, 1985). Most insect chymotrypsin M_r values are in the range 20 000–30 000 and pH optima in the range 8–9, irrespective of the pH prevailing in the midguts from which the chymotrypsins were isolated (Table 6.2). The sequences of the chymotrypsin-like proteinases were determined from *Vespa orientalis* and *Lucilia cuprina* and are similar to vertebrate chymotrypsins (Jany *et al.*, 1983; Casu *et al.*, 1994). Also, insect chymotrypsins act on glucagon and β-chain of oxidized insulin in a manner similar to vertebrate chymotrypsins. However, some properties of insect chymotrypsins contrast to those of vertebrate chymotrypsins, such as their instability at acid pH and their strong inhibition by soybean trypsin inhibitor.

6.4.3 Cysteine proteinases

Gooding (1969) was the first to suggest that enzymes similar to mammalian cathepsins may be present in insect midguts. Now it is evident that cysteine proteinases are common in midguts of hemipteran

Table 6.2 Properties of chymotrypsin-like enzymes isolated from insect midguts

Order *Species (Family)*	*Stage*	*Degree of purification*	M_r	*pI*	*pH$_o$*	*Reference*
Orthoptera						
Locusta migratoria (Acrididae)	Adult	Homogeneous	24 000 (SP)	10.1	8.7	Sakal *et al.* (1988)
Coleoptera						
Tenebrio molitor (Tenebrionidae)	Larva	Homogeneous	23 400 (SP)	–	–	Garty quoted in Applebaum (1985)
Hymenoptera						
Vespa orientalis (Vespidae)	Larva	Homogeneous	23 000 (SP)	–	8.2	Jany and Pfleiderer (1974)
Diptera						
Glossina morsitans (Glossinidae)	Adult	Partial	35 000 (F)	–	8.5	Gooding and Rolseth (1976)
Lepidoptera						
Pieris brassicae (Pieridae)	Larva	Partial	32 000 (U)	–	10	Lecadet and Dedonder (1966)

Details as in Table 6.1.

Heteroptera (Houseman and Downe, 1980; Terra *et al.*, 1988; Terra and Ferreira, 1994).

The predominant digestive proteinase in beetles (Coleoptera) of the family Bruchidae is a cysteine proteinase (Wieman and Nielsen, 1988; Campos *et al.*, 1989). A similar enzyme was found in the following Coleoptera families: Tenebrionidae, Chrysomelidae, Coccinelidae, Curculionidae, Meloidae and Silphidae (Wolfson and Murdock, 1990). With the exception of Silphidae, whose proteolytic activity is not activated by cysteine, although it is inhibited by *p*-mercuribenzoate, all the mentioned families belong to series Cucujiformia. In contrast, there is no cysteine proteinase in Dermestidae, Scarabaeidae, Elateridae and Carabidae (Terra and Ferreira, 1994). Thus the data suggest that the Cucujiformia ancestor was a beetle adapted to ingest seeds rich in naturally-occurring trypsin inhibitors by using a cysteine proteinase, instead of a trypsin-like proteinase, for protein digestion. Despite the obvious importance of insect digestive cysteine proteinases, which may derive from lysosomes, characterization of these enzymes has not progressed very far (Table 6.3).

6.4.4 Aspartic proteinases

Aspartic proteinases are active at acid pH, hydrolyse internal peptide bonds in proteins, and some also attack synthetic substrates. The first report of aspartic proteinase in insects was made by Greenberg and Paretsky (1955), who found a strong proteolytic activity at pH 2.5–3.0 in homogenates of whole bodies of *Musca domestica*. They hypothesized that this activity may be due to a pepsin-like enzyme. Lemos and Terra (1991a) were able to demonstrate that the enzyme is cathepsin-D-like. Sequence studies have shown that pepsin may have evolved from the same archetypical gene as cathepsin D in vertebrates. A similar evolutionary trend seems to have occurred in cyclorrhaphous Diptera, which apparently use cathepsin D as a digestive enzyme in the acid zone of their midguts. An aspartic proteinase similar to cathepsin D was found in several families of Hemiptera. These enzymes were poorly characterized (Table 6.3).

Aspartic proteinases were found in the following families of Coleoptera: Meloidae, Chrysomelidae, Coccinelidae (Wolfson and Murdock, 1990) and Bruchidae (Silva and Xavier-Filho, 1991), but not in Tenebrionidae (Terra and Ferreira, 1994). All these families belong to the series Cucujiformia. Thus, it is possible that aspartic proteinases occur together with cysteine proteinases in most Coleoptera. The aspartic proteinase isolated from *Callosobruchus maculatus* (Table 6.3) was partially purified and shown to be similar to cathepsin D (Silva and Xavier-Filho, 1991).

Table 6.3 Properties of cysteine, aspartic and metalloproteinases isolated from insect midguts

Enzyme class / *Species (Order)*	*Stage*	*Degree of purification*	M_r	*pI*	pH_o	*Reference*
Cysteine proteinase						
Thermobia domestica (Thysanura)	Nymph	Partial	42 000 (P)	3.5	6.0	Zinkler and Poltzer (1992)
Phymata wolffi (Hemiptera)	Adult	Crude	40 000 (F)	–	5.5	Houseman *et al.* (1985)
Rhodnius prolixus (Hemiptera)	Nymph	Crude	40 000 (U)	–	5.5	Terra *et al.* (1988)
Acanthoscelides obtectus (Coleoptera)	Larva	Homogeneous	24 000 (F)	–	6.0	Wieman and Nielsen (1988)
Callosobruchus maculatus (Coleoptera)	Larva	Partial	25 000 (F)	–	6.0	Campos *et al.* (1989)
Aspartic proteinase						
Rhodnius prolixus (Hemiptera)	Nymph	Crude	88 000 (U)	–	3.5	Terra *et al.* (1988)
Callosobruchus maculatus (Coleoptera)	Larva	Partial	62 000 (F)	–	3.3	Silva and Xavier-Filho (1991)
Musca domestica (Diptera)	Larva	Crude	80 000 (U)	5.5	3.5	Lemos and Terra (1991a)
Metalloproteinase						
Tineola bisselliella (Lepidoptera)	Larva	Partial	24 000 (F)	–	9.4	Ward (1975a)

pH optima were determined using synthetic (cysteine proteinases) or protein (aspartic and metallo proteinases) substrates. Other details as in Table 6.1.

6.4.5 Miscellaneous proteinases

Christeller *et al.* (1990) partially purified an enzyme similar to the serine proteinase elastase (EC 3.4.21.36) from *Teleogryllus commodus* (Orthoptera:Gryllidae) midguts. An elastase-like enzyme was also found in midguts of Coleoptera Scarabaeidae and of several families of Lepidoptera (Christeller *et al.*, 1992). Nevertheless, given the similiarities of chymotrypsin and elastase, as recognized by Christeller *et al.* (1992), the relationship betwen the two activities in insects needs clarification.

6.4.6 Aminopeptidases

Aminopeptidases (EC 3.4.11.) hydrolyse single amino acids from the N-terminus of the peptide chain and are classified on the basis of their dependence on metal ions (usually Zn^{2+} or Mn^{2+} and substrate specificity.

The best characterized aminopeptidases found in insect midguts are described in Table 6.4. Insect aminopeptidases have alkaline pH optima (range 7.2–9.0), irrespective of the pH of the midgut lumen from which they come. K_m values are similar (range 0.13–0.78 mM). Exceptions are the lower K_m values in Isoptera and the higher K_m values in *Rhynchosciara americana* membrane-bound enzymes, which are probably related to the specificity of these enzymes (see below). The M_r values of most aminopeptidases are in the range 90 000–130 000. Higher M_r values may result from aggregation, as demonstrated with *R. americana* aminopeptidases (Ferreira and Terra, 1985, 1986a,b). *Tineola bisselliella* larval digestive aminopeptidases were partially purified (Ward 1975b,c). Although it is possible that those aminopeptidases are products of several different genes (as observed with lysozyme in *Drosophila melanogaster*, Kylsten *et al.*, 1992), it is more probable that the aminopeptidases derive from two precursors by limited proteolysis.

Aminopeptidases are usually metalloenzymes. Nevertheless, there have been few attempts to study the role of metal ions in catalysis by insect midgut enzymes. Two subsites were proposed to occur in the active centre of the microvillar *R. americana* aminopeptidase: a hydrophobic subsite, to which isoamyl alcohol binds exposing the catalytic metal ion, and a polar subsite, to which hydroxylamine binds. Exposure of the metal ion after isoamyl alcohol binding may be analogous to the situation that results when part of the substrate occupies the hydrophobic subsite, and to be related to conformational changes associated with the catalytic step (Ferreira and Terra, 1986b).

Detailed substrate specificity studies are known for the M_r 115 700 soluble and the two membrane-bound aminopeptidases of *R. americana*, and for the midgut aminopeptidases of *Attagenus megatoma* and *T. bisselliella* (see Table 6.4 for references). All these enzymes have a broad

Table 6.4 Properties of aminopeptidases isolated from insect midguts

Order *Species (Family)*	*Stage*	*Source*	*Degree of purification*	M_r	pH_o	K_m (*mM*)	*Reference*
Orthoptera							
Teleogryllus commodus (Gryllidae)	Adult	S	Partial	94 000 (F)	8.0	0.53	Christeller *et al.* (1990)
Isoptera							
Trinervitermes trinervoides (Termitidae)	Adult	S	Homogeneous	88 700 (F)	8.5	0.004	Van der Westhuizen *et al.* (1981)
Hemiptera							
Rhodnius prolixus (Reduviidae)	Nymph	S	Crude	61 000 (U)	8.0	0.36	Ferreira *et al.* (1988)
Coleoptera							
Costelytra zealandica (Scarabaeidae)	Larva	S	Partial	96 400 (F)	8.0	0.78	Christeller *et al.* (1989)
Attagenus megatoma (Dermestidae)	Larva	S	Partial	126 000 (F)	7.5	0.32	Baker and Woo (1981)
Hymenoptera							
Scaptotrigona bipunctata (Apidae)	Larva	S	Crude	190 000 (U)	7.5	–	Schumaker *et al.* (1993)
Diptera							
Rhynchosciara americana (Sciaridae)	Larva	S	Partial	115 700 (U)	8.0	0.5	Ferreira and Terra (1984)
Rhynchosciara americana (Sciaridae)	Larva	S	Partial	117 000 (F)	7.5	0.8	Klinkowstrom *et al.* (1994)
Rhynchosciara americana (Sciaridae)	Larva	M	Partial	169 000 (U)	7.2	1.3	Ferreira and Terra (1986a,b)
Rhynchosciara americana (Sciaridae)	Larva	M	Partial	107 000 (U)	7.2	1.8	Ferreira and Terra (1985)
Glossina morsitans (Glossinidae)	Adult	S	Partial	163 700 (F)	8.0	–	Cheeseman and Gooding (1985)
Drosophila melanogaster (Drosophilidae)	Larva	M	Partial	280 000 (F)	7.6	0.057	Walker *et al.* (1981)
Tineola bisselliella (Tineidae)	Larva	S	Partial	94 000 (F)	7.7	0.17	Ward (1975b)
Tineola bisselliella (Tineidae)	Larva	S	Partial	240 000 (F)	8.4	0.3	Ward (1975c)

The source of enzyme was the soluble (S) or membrane (M) fraction of midgut homogenates, pH optima and K_m values were determined using LpNA (most) or L-leucine-β-naphthylamide (*D. melanogaster* and *T. biselliella*) as substrate. Other details as in Table 6.1.

specificity. V_{max}/K_m ratios show that the soluble *R. americana* amino-peptidase has a substrate specificity similar to that of the high electrophoretic migrating enzyme from *T. bisselliella*. Both enzymes hydrolyse L-leucine-ß-naphthylamide (LeußNA) > L-arginine-ß-naphthylamide (ArgßNA) > DL-methionine-ß-naphthylamide (MetßNA) > L-proline-ß-naphthylamide (ProßNA) and do not significantly hydrolyse L-glutamic acid-ß-naphthylamide (GlußNA) or act on di- and tripeptides. The amino-peptidase from *A. megatoma* is similar to the *T. bisselliella* aminopeptidase of intermediate electrophoretic migration in being very active on MetßNA. The two *R. americana* membrane-bound midgut aminopeptidases have a specificity toward synthetic substrates similar to the soluble enzyme. The M_r 107 000 membrane-bound aminopeptidase prefers tetrapeptides instead of tripeptides, as observed for the soluble enzyme, the contrary being true for the M_r 169 000 membrane-bound aminopeptidase. In spite of these differences, all these enzymes resemble vertebrate digestive aminopeptidase N (EC 3.4.11.2).

The recently described (Klinkowstrom *et al.*, 1994) soluble, glycocalyx-associated M_r 117 000) aminopeptidase from *R. americana* removes N-terminal aspartic or glutamic residues from peptides that are not efficiently attacked by the other aminopeptidases. This enzyme resembles vertebrate digestive aminopeptidase A (EC 3.4.11.7).

6.4.7 Carboxypeptidases

Carboxypeptidases (EC 3.4.16–18) hydrolyse single amino acids from the C-terminus of the peptide chain and are divided into three classes on the basis of their catalytic mechanism. Serine carboxypeptidases (EC 3.4.16) are most active in the acid range and have a serine in their active site. Metallocarboxypeptidases (EC 3.4.17) require bivalent cations, usually Zn^{2+}, for activity. Cysteine carboxypeptidases (EC 3.4.18) are inhibited by thiol-blocking reagents.

Insect digestive carboxypeptidases have been classified as carboxypeptidase A or B (metallocarboxypeptidases) depending on activity against *N*-carbobenzoxy-glycyl-L-phenylalanine (ZGlyPhe) (or hippuryl ß-phenyl-lactic acid, HPLA) or *N*-carbobenzoxy-glycyl-L-arginine (ZGlyArg) (or hippuryl-L-arginine), respectively. Properties of the best known carboxypeptidase-A-like enzymes from insects are described in Table 6.5. The enzyme from *Attagenus megatoma* is a true carboxypeptidase A, based on inhibition data and substrate specificity. The enzyme from *T. bisselliella* cannot be clearly classified under any carboxypeptidase class, although it seems to be closer to carboxypeptidase A than to any other carboxypeptidase. All other carboxypeptidases described in Table 6.5 lack inhibition data to support an unambiguous classification. Most insect carboxypeptidase-A-like

Table 6.5 Properties of carboxypeptidase-A-like enzymes from insect midguts

Order *Species (Family)*	*Stage*	*Source*	*Degree of purification*	M_r	pH_o	K_m (mM)	*Substrate*	*Reference*
Orthoptera								
Teleogryllus commodus (Gryllidae)	Adult	S	Partial	39 700 (F)	8.5	0.06	HPLA	Christeller *et al.* (1990)
Coleoptera								
Costelytra zealandica (Scarabaeidae)	Larva	S	Partial	42 800 (F)	8.5	0.028	HPLA	Christeller *et al.* (1989)
Attagenus megatoma (Dermestidae)	Larva	S	Partial	35 500 (F)	8.0	–	HPLA	Baker (1981a)
Diptera								
Glossina morsitans (Glossinidae)	Adult	S	Partial	30 200 (F)	8.0	–	HPLA	Gooding and Rolseth (1976)
Musca domestica (Muscidae)	Larva	S	Crude	45 000 (U)	8.0	1.2	ZGlyPhe	Jordão and Terra (1989)
Musca domestica (Muscidae)	Larva	M	Crude	58 000 (U)	7.5	2.3	ZGlyPhe	Jordão and Terra (1989)
Lepidoptera								
Tineola bisselliella (Tineidae)	Larva	S	Partial	72 000 (F)	7.6	0.44	ZGlyPhe	Ward (1976)

The source of enzymes was the soluble (S) or membrane (M) fraction of midgut homogenates. Other details as in Table 6.1.

enzymes have M_r values in the range 20 000–50 000 (Table 6.5). Those presenting higher M_r values are probably composed of subunits, or may be overestimated due to associated detergent molecules (Jordão and Terra, 1989). K_m values are varied and seem to be lower with HPLA than with ZGlyPhe.

Carboxypeptidases classified as B because of their alkaline pH optima and activity on HA were partially characterized in Diptera. There are data for adult *Glossina morsitans* (pH optimum 7.8, M_r 22 000; Gooding and Rolseth, 1976).

Cysteine carboxypeptidases seem to occur in Hemiptera (Houseman and Downe, 1981). These enzymes may have a lysosomal origin, as previously discussed for cysteine proteinases.

6.4.8 Dipeptidases

Dipeptidases (EC 3.4.13) hydrolyse dipeptides and are classified according to their substrate specificities.

Chromatographic data showed the occurrence of a true dipeptidase in *Rhodnius prolixus* (Hemiptera: Reduviidae) midguts (Garcia and Guimarães, 1979). Dipeptidase activity in *Rhynchosciara americana* (Diptera: Sciaridae) is found mainly in caeca (Klinkowstrom *et al.*, 1995), the major site of terminal digestion. The caecal dipeptidase activity (substrate: Gly-Leu) is partly soluble and partly membrane bound. The soluble activity is resolved by gel filtration into two enzymes (M_r 63 000 and 73 000), that hydrolyse both Gly-Leu and Pro-Gly, although with different efficiency. The membrane-bound dipeptidase (pH optimum 8.2) was solubilized by Triton X-100 and papain. Gel filtration data suggest that there is only one detergent form (M_r 86 000), of the membrane-bound dipeptidase. No activity upon Gly-Pro was found in *R. americana* midguts, whereas the activity on β-Ala-His (carnosine), although occurring in both soluble and membrane-bound fractions, is independent of enzymes hydrolysing Gly-Leu and pro-Gly. *R. americana* dipeptidases that hydrolyse Gly-Leu resemble dipeptide hydrolase (dipeptidase, EC 3.4.13.18) although, in contrast to the mammalian enzyme (Norén *et al.*, 1973), they are very active on Pro-Gly. *R. americana* also seems to have an aminoacyl-histidine dipeptidase (carnosinase, EC 3.4.13.3).

6.5 GLYCOSIDASES

6.5.1 α-Amylases

From the enzymes known to act preferentially on long α-1,4-glucan chains such as native starch or glycogen, only α-amylases have been found in insects.

The best characterized insect amylases are described in Table 6.6. Most amylase M_r values are in the range 48 000–68 000. Amylase pI values are acidic (most in the range 3.5–4.0) and K_m values with soluble starch are about 0.1%, with a few amylases displaying K_m values between 0.2 and 0.4%. pH optima generally correspond to the pH prevailing in midguts from which the amylases were isolated. Action patterns of insect amylases against starch vary. *R. americana* amylase has a degree of multiple attack between that of the amylase of *Bacillus subtilis* and porcine pancreas (Terra *et al.*, 1977). Amylases from larvae and adults of *Sitophilus oryzae* (Baker, 1987) and larvae of *Bombyx mori* (Kanekatsu, 1978) have action patterns similar to that of porcine pancreas amylase. Calcium ions activate (*Calosobruchus chinensis*, Podoler and Applebaum, 1971b), afford thermal stability (*Bombyx mori*, Kanekatsu, 1978) or protect amylases against inactivation during dialysis against buffer (*T. molitor*, Buonocore *et al.*, 1976) or EDTA (*R. americana*, Terra *et al.*, 1977). Thus insect amylases seem to be calcium-dependent enzymes.

Activation of insect amylases by chloride and other anions, and displacements of the pH optima in the presence of these ions were first described in Hemiptera (Hori, 1972). *R. americana* midgut amylase binds chloride (K_D = 1.6 mM at 37°C) resulting in a 34-fold increase in V_{max}, without affecting K_m, but causing a shift in the pH optimum from 6.8 to 8.0. Activation also occurs with anions other than chloride, such as bromide and nitrate, and it seems to be dependent on the ionic size (Terra *et al.*, 1977). The similarity of these findings to those reported for mammalian amylases suggests that insect and mammalian amylases are activated by chloride by a similar mechanism. However, there are insect amylases that are not activated by chloride (Table 6.6). Lack of chloride activation in amylases displaying a very high pH optimum (Table 6.6) may result from the fact that these enzymes must necessarily have an essential protonated group with a high pK_a, which perhaps cannot be increased further. The absence of chloride activation in amylases with neutral to acidic pH optima (Table 6.6) deserves further investigation.

Amylase coding sequences and circular dichroism data suggest that insect amylases have three-dimensional structures similar to that determined for porcine pancreatic α-amylase (Chen *et al.*, 1992). However, in spite of the many resemblances between insect and mammalian amylases, there are sufficient differences between these two groups and even among insect amylases (Table 6.6) to justify more studies on insect amylases. Related to these differences are findings that amylases from different insect species are distinctly affected by a wide range of protein inhibitors (Chen *et al.*, 1992).

Table 6.6 Properties of α-amylases isolated (or thought to be originated) from insect midguts

Order *Species (Family)*	*Stage*	*Source*	*Degree of purification*	M_r	*pI*	*pH$_o$*	K_m (%)	*Chloride activation*	*Reference*
Orthoptera									
Locusta migratoria (Acrididae)	Adult	G	Crude	68 000 (F)	–	6.0	0.089	–	Droste and Zebe (1974)
Coleoptera									
Tenebrio molitor (Tenebrionidae)	Larva	WB	Homogeneous	68 000 (F)	4.0	5.8	0.13	Yes	Buonocore *et al.* (1976)
Callosobruchus chinensis (Bruchidae)	Larva	WB	Homogeneous	–	–	5.7	0.23	No	Podoler and Applebaum (1971a,b)
Sitophylus oryzae Curculionidae)	Adult	WB	Homogeneous	56 000 (SP)	3.7	4.8	0.25	Yes	Baker and Woo (1985); Baker (1987)
Sitophylus oryzae (Curculionidae)	Adult	WB	Homogeneous	56 000 (SP)	3.8	4.8	0.07	Yes	Baker and Woo (1985); Baker (1987)
Hymenoptera									
Scaptotrigona bipunctata (Apidae)	Larva	G	Crude	68 000 (U)	–	5.5	–	No	Schumaker *et al.* (1993)
Diptera									
Rhynchosciara americana (Sciaridae)	Larva	G	Crude	64 500 (U)	–	8.0	0.14	Yes	Terra *et al.* (1977) Terra and Ferreira (1983)
Lepidoptera									
Bombyx mori (Bombycidae)	Larva	G	Homogeneous	47 000 (F)	–	9.3	–	–	Kanekatsu (1978)
Anagasta kuhniella (Pyralidae)	Larva	WB	Partial	56 000 (P)	4.2	9.0	0.037	No	Baker (1989)
Antheraea mylitta (Saturniidae)	Larva	G	Homogeneous	58 000 (SP)	<6.5	9.5	0.113	No	Nagarajn and Abraham (1995)

The source of enzymes was the soluble fraction of homogenates from midgut (G) or whole bodies (WB). pH optima and K_m values were determined using soluble starch as substrate. Other details as in Table 6.1.

6.5.2 Cellulases

Cellulose digestion occurs in several insects that have, as a rule, nutritionally poor diets (for review see Martin, 1991 and section 6.8). Although it is widely believed that insects are dependent on symbionts for cellulose digestion (Martin, 1991), there is increasing evidence that insects secrete enzymes able to hydrolyse crystalline cellulose (Slaytor, 1992). The role of symbiotic organisms in insects which digest cellulose is becoming controversial (Slaytor, 1992), although symbiotic nitrogen-fixing organisms are certainly involved in increasing the nutritive value of diets of many insects (for review see Terra, 1990).

The cellulase system of *Ergates faber* (Coleoptera: Cerambycidae) larvae was resolved into three homogeneous enzymes. Enzyme 1 M_r 25 000; pH optimum 4.5) is a typical exo-β-1,4-glucanase, whereas enzyme 2 (M_r 57 000; pH optimum 5.2) and enzyme 3 (M_r 70 000; pH optimum 5.2) are active against cellobiose, carboxymethyl cellulose and crystalline cellulose (Chararas *et al.*, 1983). An exo-β-1,4-glucanase was purified to homogeneity from the fungus-growing termite *Macrotermes mülleri* (Rouland *et al.*, 1988a). This enzyme has M_r 52 000 and pH optimum of 4.4 and is also very active on carboxymethylcellulose. Endo-β-1,4-glucanases have been found in several insects (for review see Martin, 1991), and two enzymes acting on crystalline cellulose were described in the higher termite *Nasutitermes walkeri*: one enzyme liberates glucose whereas the other hydrolyses internal bonds (Schultz *et al.*, 1986). β-1,4-Glucosidases play roles other than cellulolysis and are reviewed below.

6.5.3 Hemicellulases

Hemicellulose is a mixture of polysaccharides and is associated with cellulose in plant cell walls. The major polysaccharides in the hemicellulose fraction are those based on β-1,4-linked D-glycan backbones made up predominantly of glucose (glucans), xylose (xylans), mannose (mannans) and mannose and glucose (glucomannans) units.

Hemicellulases include all the enzymes which hydrolyse hemicellulose polysaccharides and must be widespread, as nutritional studies have shown that hemicelluloses may be utilized by insects (e.g. Terra *et al.*, 1987).

Xylanase (endo-β-1,4-xylanase, EC 3.2.1.8) has been found in grasshoppers (Orthoptera), stone-flies (Plecoptera) and caddisflies (Trichoptera) (Vonk and Western, 1984) but it has been purified only from termites (Isoptera) (Rouland *et al.*, 1988b). The purified enzyme proved to be from fungus ingested by the termites rather than the termite itself.

Laminaran is a β-glucan consisting chiefly of β-1,3-linked glucose residues. Laminaranases (laminarinases) have been found in species of Orthoptera, Coleoptera, Trichoptera, Diptera and Isoptera (Vonk and Western, 1984). Two laminaranases were secreted by the larval midgut of *Rhagium inquisitor* (Coleoptera: Cerambycidae). Laminaranase 1, which has been partially purified and characterized (Chipoulet and Chararas, 1984) has M_r 95 000–100 000, pH optimum of 5.5 and, based upon products of hydrolysis of laminaran, it was classified as an endo-β-1,3-glucanase. The enzyme is inactive on lichenan (β-1,3–1,4-glucan). A similar enzyme has been found in the grasshopper *Abracris flavolineata* (Orthoptera; Acrididae) (Ferreira, Marana, Silva and Terra, unpublished results). Insect laminaranases are supposed to digest cell walls of fungi (Martin *et al.*, 1981), or the small amounts of callose present in phloem (Chipoulet and Chararas, 1984).

Lichenases (endo-β-1,3; 1,4-glucanase) (EC 3.2.1.73) hydrolyse lichenan and cereal β-glucans, but are inactive on laminaran. Electrophoretical evidence suggests that this enzyme occurs in *R. inquisitor* (Chipoulet and Chararas, 1985) and in *A. flavolineata* (Ferreira, Marana, Silva and Terra, unpublished results). The putative physiological role of this enzyme is to hydrolyse β-1,3–1,4-glucans which are abundant in grasses.

6.5.4 Pectic enzymes

There are no records of pectic enzymes other than pectinases in insects. Pectinases (polygalacturonases, EC 3.2.1.15) seem to occur in Orthoptera, Hemiptera, Coleoptera, Diptera and Trichoptera, although they are apparently absent in Dictyoptera and in some Orthoptera and Coleoptera (Vonk and Western, 1984). No insect pectinases have been characterized.

6.5.5 Chitinases and β-*N*-acetylglucosaminidases

Chitin is a 1,4-β homopolymer of *N*-acetylglucosamine. Chitinolytic enzymes include: chitinase (EC 3.2.1.14), which catalyses the random hydrolysis of internal bonds in chitin forming smaller oligosaccharides, and β-*N*-acetyl-D-glucosaminidase (EC 3.2.1.52), which liberates *N*-acetylglucosamine from the non-reducing end of oligosaccharides. Lysozyme, as mentioned below, also has some chitinase activity, whereas chitinase has no lysozyme activity. Chitinase is found in the midguts of several insects (Vonk and Western, 1984), but no insect digestive chitinases have been characterized.

An enzyme related to chitinolytic enzymes is *β-N*-acetyl-D-hexosaminidase (EC 3.2.1.52), which differs from *β-N*-acetyl-D-glucosaminidase in having a rather wide substrate specificity. The enzyme is found in the glycocalyx of midgut cells and in the ectoperitrophic space of *Rhynchosciara americana* (Diptera, Sciaridae) larvae (Terra and Ferreira, 1983). The physiological role of the *R. americana* enzyme is probably the hydrolysis of *N*-acetylglucosamine *β*-linked compounds such as glycoproteins diffusing out of the endo-peritrophic space. This enzyme has a pH optimum of 6.2, M_r 141 000 and a K_m for *p*-nitrophenyl-*N*-acetyl-*β*-D-glucosaminide of 0.56 mM. Larvae of *Erinnyis ello* (Lepidoptera, Sphingidae) and *Spodoptera frugiperda* (Lepidoptera, Noctuidae) have similar enzymes (Terra and Ferreira, 1994).

6.5.6 Lysozyme

Lysozyme (EC 3.2.1.17) catalyses the hydrolysis of the 1,4-*β*-glycosidic linkage between *N*-acetylmuramic acid and *N*-acetylglucosamine of the peptidoglycan present in the cell wall of many bacteria. Lysozyme is widespread throughout nature as part of the defence mechanism against bacteria and is also involved in the midgut digestion of bacteria in some organisms, such as ruminants, that harbour a bacterial culture in their foreguts. The larvae of *Musca domestica* (Diptera, Cyclorrhapha, Muscidae) feed on decomposing organic material and bacteria (their main food) are lysed in the middle region of their midguts through the combined action of low pH, lysozyme and a cathepsin-D-like proteinase. The capacity to digest bacteria in the midgut seems to be an ancestral trait of Diptera Cyclorrhapha (Lemos and Terra, 1991b).

Two lysozymes were purified to homogeneity from *M. domestica* larval midguts (Lemos *et al.*, 1993). Lysozymes 1 and 2 have M_r 17 000 and pI values of 7.9 and 8.2, respectively. Lysozymes 1 and 2 have identical kinetic properties that include a decrease in activity, displacement of the pH optimum toward acidic values and an increase in K_m as the ionic strength becomes higher. Lysozymes are resistant to a cathepsin-D-like proteinase present in *M. domestica* midgut, and display a chitinase activity that is six-fold higher than that of chicken lysozyme.

6.5.7 α-Glucosidases

α-Glucosidases (EC 3.2.1.20) catalyse the hydrolysis of terminal, non-reducing α-1,4-linked glucose residues from aryl-glucosides (as *p*-nitrophenyl-α-D-glucoside), disaccharides or oligosaccharides with varied efficiency. Insect α-glucosidases best characterized are described in Table 6.7.

Table 6.7 Properties of α-glucosidases isolated (or thought to be originated) from insect guts

Order / *Species (Family)*	*Stage*	*Source*	*Degree of purification*	M_r	*pI*	*pH_o*	K_m (*mM*)	*Reference*
Hemiptera								
Dysdercus peruvianus (Pyrrhocoridae)	Adult	M,G	Homogeneous	138 000 (SP)	–	5.0	26 (Mal)	Silva and Terra (1995)
Coleoptera								
Sitophilus zeamais (Circulionidae)	Adult	G,S	Partial	130 000 (U)	4.9	5.5	13 (Mal)	Baker (1991)
Hymenoptera								
Apis mellifera (Apidae)	Adult	WB,S	Homogeneous	82 000 (F)	6.5	6.5	4 (NPG)	Huber and Mathison (1976)
Apis mellifera (Apidae)	Adult	WB,S	Homogeneous	100 000 (F)	7.3	6.5	0.5 (NPG)	Huber (1975)
Diptera								
Drosophila melanogaster (Drosophillidae)	Adult	WB,S	Homogeneous	56 000 (F)	–	5.0	8.9 (Suc)	Tanimura *et al.* (1979)
Drosophila melanogaster (Drosophillidae)	Adult	WB,S	Homogeneous	200 000 (F)	–	6.0	13 (Suc)	Tanimura *et al.* (1979)
Musca domestica (Muscidae)	Larva	G,S	Partial	72 700 (U)	–	6.3	5.2 (Mal)	Jordão and Terra (1989)
Musca domestica (Muscidae)	Larva	G,S	Partial	330 000 (U)	–	6.1	4.6 (Mal)	Jordão and Terra (1989)
Musca domestica (Muscidae)	Larva	G,M	Partial	240 000 (U)	–	6.6	1.0 (Mal)	Jordão and Terra (1989)
Lepidoptera								
Thaumetopoea pityocampa (Notodontidae)	Larva	WB,S	Homogeneous	190 000 (F)	3.7	6.0	0.5 (NPG)	Pratviel-Sosa *et al.* (1986)

The source of enzymes was soluble (S) or membrane (M) fraction of homogenates from midgut (G) or whole bodies (WB). pH optima were determined using maltose (Mal), *p*-nitrophenyl-*β*-D-glucoside (NPG) or sucrose (Suc). Other details as in Table 6.1.

Insect soluble α-glucosidases were purified to homogeneity only from whole bodies. The *Apis mellifera* M_r 82 000 α-glucosidase (Table 6.7) has a transglucosidase activity and is strongly inhibited by Tris (Huber and Mathison, 1976). This enzyme is probably synthesized in hypopharyngeal glands and, after ingestion, converts nectar sucrose into glucose and fructose in the crop. The *A. mellifera* M_r 100 000 α-glucosidase (Table 6.7) also has a transglucosidase activity and is inhibited by Tris. This enzyme is found in significant amounts only in abdomens (Huber, 1975). Thus, its physiological role should be hydrolysis of α-glucosides in the midgut. The same function is proposed for the enzymes purified from the whole bodies of *D. melanogaster* and *Thaumetopoea pityocampa* (Table 6.7).

A membrane-bound α-glucosidase was purified to homogeneity from the midgut perimicrovillar membranes of *Dysdercus peruvianus* (Table 6.7). The enzyme has only one subunit with M_r 138 000 and catalyses the hydrolysis of different ingested α-glucosides at the same active site, which seem to have five subsites.

Insect midgut α-glucosidases have a number of common properties, for example their inhibition by Tris (Table 6.7). M_r values are in the range 60 000–80 000 or are multiples of these values, suggesting the occurrence of oligomeric enzymes. Subunits were found in the M_r 200 000 α-glucosidase from *D. melanogaster* (Tanimura *et al.*, 1979) and in the enzymes from *Sitophilus zeamais* (Baker, 1991). Insect α-glucosidases have acidic pI values, except for *A. mellifera*, which show nearly neutral pI values (Table 6.7). An interesting feature of insect midgut α-glucosidases is their pH optima, which are in the range 5–6.5 irrespective of the corresponding midgut pH value (see midgut pH values in Chapter 8).

The α-glucosidases from *S. zeamais* (Baker, 1991), *T. pityocampa* (Pratviel-Sosa *et al.*, 1986), *M. domestica* (Jordão and Terra, 1991), and *D. peruvianus* (Silva and Terra, 1995) are able to hydrolyse oligosaccharides up to at least maltopentaose. Since the *M. domestica* α-glucosidases occur at different sites in the midgut, and have different specificities, they provide an efficient stepwise hydrolysis of α-chains formed by the action of amylase on starch (Jordão and Terra, 1991).

Understanding of the molecular basis of substrate specificity depends to a large extent on protein-sequencing studies. Only the *A. aegypti* salivary α-glucosidase has been studied thus far. This enzyme, which may be intracellular and not present in the saliva, has a domain whose sequence is similar to that of a yeast maltase which is supposed to provide the glucosidase functions (Marinotti and James, 1990).

6.5.8 β-Glucosidases and β-galactosidases

β-Glycosidases catalyse the hydrolysis of terminal, non-reducing β-1,4-linked monosaccharide residues from the corresponding glycoside. Depending on the monosaccharide which is removed, the β-glycosidase is named β-glucosidase (glucose), β-galactosidase (galactose), β-xylosidase (xylose), and so on. For the purpose of this review it is convenient to define two activities: glycosyl β-glycosidase, which is the activity upon disaccharides or oligosaccharides, and aryl (or alkyl) β-glycosidase, which is the activity upon a monosaccharide linked to a non-sugar residue (aglycone).

Table 6.8 describes some properties of the best characterized insect β-glycosidases. M_r and pI values are variable, whereas K_m values are found in the range of 0.2–2 mM, with only one exception (8.1 mM). pH optima are in the range 4.5–6.2, irrespective of the luminal pH of the midguts from which they were isolated. Nevertheless, enzymes found in Lepidoptera seem to have pH optima somewhat higher (6–7) than other insects.

Insect β-glucosidases may be divided into three classes based on their substrate specificities (Terra and Ferreira, 1994). Class 1 includes the enzymes with glycosyl β-glycosidase and aryl β-glycosidase activity, that is, enzymes active on cellobiose, *p*-nitrophenyl-β-D-glucoside (NPβ Glu), *p*-nitrophenyl-β-D-galactopyranoside (NPβGal), *p*-nitrophenyl-β-D-fucoside (NPβFuc) and lactose, etc. Class 2 includes enzymes with only glycosyl β-glucosidase activity, i.e. those that only hydrolyse cellobiose and lactose efficiently. Finally, Class 3 is composed of enzymes with only aryl (or alkyl) β-glycosidase activity, that is, those that have significant activity only on NPβGlu and similar substrates. Some of these latter enzymes are also active on NPβGal.

The best-known enzymes of Class 1 are the β-glucosidases from *E. ello* and *R. americana* (Table 6.8). These enzymes hydrolyse β-D-glucosides, β-D-galactosides and β-D-fucosides at the same active site. Class 2 β-glucosidases have been partially characterized in *Abracris flavolineata* and *Sitophilus oryzae* (Table 6.8), whereas Class 3 β-glucosidases were described in *A. flavolineata*, *S. oryzae* and*T. pityocampa* (Table 6.8).

Although having different substrate specificities, Class 1 and 2 β-glucosidases hydrolyse cellobiose and other disaccharides and are thus typical β-glucosidases (EC 3.2.1.21). These enzymes may hydrolyse cellobiose that results from the degradation of cellulose in the few insects that have a cellulose (section 6.5.2). However, in most insects it is probably responsible for finishing hemicellulose digestion (section 6.5.3) and/or cleaving the carbohydrate moieties of glycoproteins. Because Class 3 β-glucosidases preferentially hydrolyse monosaccharides linked to a hydrophobic aglycone, it is probable they are glycosylceramidases

Table 6.8 Properties of β-glucosidases isolated (or thought to be originated) from insect guts

Order Species (Family)	Stage	Source	Degree of purification	M_r	pI	pH_o	K_m (mM)	Substrates: Good	Substrates: Poor	Reference
Orthoptera										
Abracris flavolineata (Acrididae)	Adult	G,S	Partial	82 000 (F)	7.2	4.8	–	NPG, OGL	–	Marana *et al.* (1995)
Abracris flavolineata (Acrididae)	Adult	G,S	Partial	82 000 (F)	6.8	5.5	–	Cel	NPG	Marana *et al.* (1995)
Hemiptera										
Rhodnius prolixus (Reduviidae)	Adult	G,S	Crude	64 000 (U)	–	4.5	–	–	–	Terra *et al.* (1988)
Coleoptera										
Phoracantha semipunctata (Cerambycidae)	Larva	WB,S	Homogeneous	100 000 (P)	–	5.3	2.2 (NPG)	Cel, NPG, Sal	–	Chararas and Chipoulet (1982)
Sitophilus oryzae (Curculionidae)	Adult	WB,S	Partial	170 000 (F)	–	5.5	8.1 (NPG)	Cel	NPG	Baker and Woo (1992)
Sitophilus oryzae (Curculionidae)	Adult	WB,S	Partial	140 000 (F)	–	5.5	1.9 (NPG)	NPG	Cel	Baker and Woo (1992)
Diptera										
Rhynchosciara americana (Sciaridae)	Larva	G,M	Partial	106 000 (U)	5.4	6.2	0.53 (NPG)	Cel, NPG, NPF	–	Ferreira and Terra (1983)
Lepidoptera										
Thaumetopoea pityocampa (Notodontidae)	Larva	WB,S	Homogeneous	67 000 (P)	3.6	6.0	0.39 (NPG)	NPG, PGA, Lam	Cel	Pratviel-Sosa *et al.* (1987)
Erinnyis ello (Sphingidae)	Larva	G,S	Partial	129 000 (U)	6.8	6.5	0.18 (NPG)	Cel, NPG, PGA	–	Santos and Terra (1985)

The source of enzymes was soluble (S) or membrane (M) fraction of homogenates from midgut (G) or whole bodies (WB). pH optima and K_m were determined using cellobiose (Cel) or *p*-nitrophenyl-β-D-glucoside (NPG). Substrates tested included octyl β-glucoside (OGL), salicin (Sal), *p*-nitrophenyl-β-D-fucoside (NPF), *p*-nitrophenyl-β-D-galactoside (PGA) or laminaribiose (Lam). Other details as in Table 6.1.

(EC 3.2.1.62). Class 3 β-glucosidases may also hydrolyse monogalactosyldiglycerides (see below).

Insects lacking Class 1 β-glucosidase have both Class 2 and Class 3 β-glucosidases, as observed in *A. flavolineata* (Marana *et al.*, 1995) and *S. oryzae* (Baker and Woo, 1992). This condition may be an adaptation of ancestors of different taxa to diets containing plant glycosides. Most plant glycosides have a hydrophobic aglycone and are β-linked *o*-glycosyl compounds. Thus, they are readily hydrolysed by insects with a glycosylceramidase-like activity. Since aglycones are usually more toxic than the glycosides themselves (Yu, 1989), detoxification may result from lower glycosylceramidase activity. Such control of expression would be more easily accomplished if β-glucosidase and glycosylceramidase activities occur in separate enzymes. Any decrease in the synthesis of glycosylceramidase would not necessarily affect the synthesis of β-glucosidase, and subsequently would not affect the final digestion of hemicellulose, glycoproteins and cellulose.

Midgut β-galactosidase activity may be entirely associated with the β-glucosidase, as in *E. ello* larvae (Santos and Terra, 1985) and perhaps other insects described in Table 6.8. It may also be partly associated with specific aryl β-galactosidases such as those found in adults of *L. migratoria* (Morgan, 1975) and *A. flavolineata* (Marana, Terra and Ferreira, unpublished results) and in *T. molitor* larvae (Ferreira, Torres and Terra, unpublished results). The enzyme from *L. migratoria* has a pH optimum of 5.5 and M_r 65 000 (gel filtration), whereas that of *A. flavolineata* has a pH optimum of 4.5, K_m 0.52 mM (NPβGal) and M_r 72 000 (gel filtration). These enzymes actively hydrolyse NPβGal and are nearly inactive on lactose and on β-glucosides. Thus, the natural substrates of these aryl β-galactosidases are probably galactolipids such as galactosyldiglycerides. 2,3-Diacyl-1-β-D-galactopyranosyl-D-glycerol (monogalactosyldiglyceride) and 2,3-diacyl-1-(α-D-galactopyranosyl (1→6)-β-D-galactopyranosyl-D-glycerol (digalactosyldiglyceride) are the major lipids in photosynthetic tissues and are, consequently, probably the world's commonest lipids (Harwood, 1980). These lipids are readily hydrolysed in the midgut lumen of several insects, the resulting galactose and fatty acids being absorbed and utilized (Turunen, 1992 and references therein, see Chapter 11). Digalactosyldiglyceride may be converted into monogalactosyldiglyceride (the presumed natural substrate of aryl β-galactosidase) by the action of a α-galactosidase.

6.5.9 β-Fructosidase

Sucrose hydrolysis is catalysed by enzymes that are specific for the α-glucosyl (α-glucosidase, EC. 3.2.1.20, see above) or for the β-fructosyl residue (β-fructosidase, EC 3.2.1.26) of the substrate. In insect midguts sucrose hydrolysis generally occurs by action of the conspicuous α-glucosidase. Probably because insect *β*-fructosidase is relatively uncommon, only a small number of reports verify its presence and there has been only one (Santos and Terra, 1986) attempt at characterization. Larvae and adults of *Erinnyis ello* (Lepidoptera : Sphingidae) have a midgut *β*-fructosidase with pH optimum 6.0, K_m 30 mM (sucrose), pI 5.2 and M_r 78 000 (Santos and Terra, 1986). The physiological role of *E. ello* midgut *β*-fructosidase is to hydrolyse the major leaf (larvae) or nectar (adults) carbohydrate sucrose which is not efficiently digested by *E. ello* midgut α-glucosidase (Terra *et al.*, 1987).

6.5.10 Trehalase

Trehalase (EC 3.2.1.28) hydrolyses α,α′-trehalose into two glucose molecules and is one of the most widespread carbohydrases in insects. The best characterized insect gut trehalases are described in Table 6.9. Trehalase M_r values are in the range 60 000–140 000, K_m values vary from 0.4 to 1.1 mM and Tris is inhibitory with K_1 values around 60 mM. As discussed above for α-glucosidases, trehalase pH optima are in the range 4.8–6.0, irrespective of the luminal pH of the midguts from which they were isolated.

Only the soluble trehalases from *B. mori* adults (Sumida and Yamashita, 1983) and *L. dispar* larvae (Valaitis and Bowers, 1993) were purified to homogeneity. Although only partially purified, the soluble trehalase from *Rhynchosciara americana* larvae is the best-known insect midgut trehalase (Terra *et al.*, 1978, 1979b, 1983). This enzyme catalyses the hydrolysis of trehalose according to rapid equilibrium kinetics. During the reaction the trehalose glycosyl oxygen is protonated, followed by the release of a glucose molecule, leading to development of a carbonium ion, which is attacked by water. Both the protonated group that attacks the glycosyl oxygen, as well as the deprotonated group that stabilizes the developing carbonium ion seem to be carboxyl groups. It is possible, however, that the protonated carboxyl group participates in the reaction through another amino acid residue, such as histidine (Terra *et al.*, 1979b).

Apical and basal trehalases can be distinguished in insect midguts. The apical trehalase may be soluble (glycocalyx-associated or secreted into the midgut lumen) or microvillar, whereas the midgut basal trehalase is an integral protein of the basal plasma membrane (Terra *et al.*, 1979a, 1985; Azuma and Yamashita, 1985; Santos *et al.*, 1986; Espinoza-Fuentes *et al.*, 1987). The apical midgut trehalase is a true digestive enzyme. The midgut basal trehalase probably plays a role in the midgut utilization of haemolymph trehalose (Azuma and Yamashita, 1985).

Table 6.9 Properties of trehalases isolated from insect guts

Order *Species (Family)*	*Stage*	*Source*	*Degree of purification*	M_r	*pI*	pH_o	K_m (mM)	*Tris* K_i (mM)	*Reference*
Dictyoptera									
Blaberus discoidales (Blaberidae)	Adult	S	Crude	70 000 (F)	–	4.8	0.5	–	Gilby *et al.* (1967)
Hymenoptera									
Apis mellifera (Apidae)	Adult	S	Partial	135 000 (F)	–	6.0	1.1	50	Talbot and Huber (1975)
Coleoptera									
Tenebrio molitor (Tenebrionidae)	Larva	S	Crude	60 000 (U)	4.0	5.0	0.4	–	Terra *et al.* (1985)
Diptera									
Rhynchosciara americana (Sciaridae)	Larva	S	Partial	122 000 (P)	4.6	6.0	0.67	74	Terra *et al.* (1978)
Lepidoptera									
Bombyx mori (Bombycidae)	Adult	S	Homogeneous	138 000 (F)	–	5.4	0.46	47	Sumida and Yamashita (1983)
Lymantria dispar (Lymantriidae)	Larva	S	Homogeneous	60 000 (SP)	4.6	5.7	0.4	–	Valaitis and Bowers (1993)

pH optima and K_m values were determined using trehalose as substrate. Other details as in Table 6.1.

6.5.11 Miscellaneous glycosidases

Insect midgut α-galactosidases, α-mannosidases and β-mannosidases were studied only in midgut homogenates (Ferreira *et al.*, 1988; Chipoulet and Chararas, 1985). M_r values are high (except for *R. inquisitor* α-galactosidase) and variable. pH optima are acidic and are irrespective of the pH found in the corresponding midgut lumen. The three enzymes are thought to be involved in cleaving the carbohydrate moieties of glycoproteins and in terminal digestion of hemicelluloses. In addition to these roles, α-galactosidase may hydrolyse digalactosyldiglyceride (section 6.5.8) and the sugars melibiose and raffinose which occur in plant tissue.

6.6 LIPASES, PHOSPHOLIPASES AND ESTERASES

Lipids that contain fatty acids comprise storage lipids and membrane lipids. Storage lipids, such as oils present in seeds and fats in adipose tissues of animals, are triacylglycerols (triglycerides). Membrane lipids include phospholipids, glycolipids and sulpholipids.

6.6.1 Triacylglycerol lipase, acylglycerol lipase and galactolipase

Triacylglycerol lipases (EC 3.1.1.3) are enzymes that preferentially hydrolyse the outer ester links of triacylglycerols and act only on the water–lipid interface. The activity of the lipase is increased as the interface becomes larger due to lipid emulsification caused by emulsifiers (surfactants). Insects lack emulsifiers comparable to the bile salts of vertebrates, but surfactant phospholipids, including lysolecithin, occur in their midguts in sufficient concentration to alter the surface tension of midgut contents (De Veau and Schultz, 1992). Lysolecithin, and other surfactants, may be formed by the action of phospholipase A on ingested phospholipids (see below).

Insect midgut triacylglycerol lipases have been studied in few insects and only in crude preparations. The data suggest that the enzyme preferentially releases fatty acids from the α-positions (Bollade *et al.*, 1970; Hoffman and Downer, 1979), prefers unsaturated fatty acids (Weintraub and Tietz, 1973), and is activated by calcium ions (Gilbert *et al.*, 1965), thus resembling the action of mammalian pancreatic lipase. The resulting 2-monoacylglycerol may be absorbed, as in *Periplaneta americana* (Dictyoptera: Blattidae) (Bollade *et al.*, 1970), or further hydrolysed, as seen in *Locusta migratoria* (Orthoptera: Acrididae) (Weintraub and Tietz, 1973) and *Manduca sexta* (Lepidoptera: Sphingidae) (Tsuchida and Wells, 1988). Hydrolysis of 2-monoacylglycerol may be accomplished by the triacylglycerol lipase, following migration of the fatty acid to the 1-position, which seems to be favoured by the alkaline midgut pH in *Manduca sexta* (Tsuchida and Wells, 1988). Alternatively, 2-monoacylglycerol hydrolysis may be catalysed by an acylglycerol lipase (EC 3.1.1.27), although the presence of this enzyme in insects is unknown.

A combination of α- and β-galactosidases may remove galactose residues from mono- and digalactosylglyceride (section 6.5.7) to leave a diacylglycerol which may be hydrolysed by a triacylglycerol lipase. Alternatively, galactosyldiglycerides may be hydrolysed by galactolipase (EC 3.1.1.26), yielding free fatty acids and galactosyl glycerol. Although it is known that galactose, glycerol and the fatty acids from galactosyldiglycerides are utilized by insects (Turunen, 1992), no data exist on the digestive enzymes involved.

6.6.2 Phospholipases

Phospholipase A_2 (EC 3.1.1.4) and phospholipase A_1 (EC 3.1.1.32) remove from phosphatides the fatty acid attached to the 2-position and 1-position, respectively, resulting in a lysophosphatide. Lysophosphatide is more stable in micellar aggregates than on cell membranes. Thus, the action of phospholipase A on membrane phosphatides causes the solubilization of cell membranes, rendering the cell contents free to be acted on by the appropriate digestive enzymes. Phospholipase A was found in the midgut of several species of Lepidoptera (Sommerville and Pockett, 1976; Turunen and Kastari, 1979) and is probably widespread among insects. The phospholipase A_2 partially purified from the midgut of adult *Cincindella circumpicta* (Coleoptera: Cincindelidae) is M_r 22 000, calcium dependent, has a pH optimum of 9.0 and is inhibited by the site-specific inhibitor oleyoxyethyl phosphorylcholine. Unfed beetles did not express the phospholipase in the midgut contents (Uscian *et al.*, 1995). Although lysophosphatide may be further hydrolysed by a lysophospholipase (phospholipase B, EC 3.1.1.5), evidence suggests it is absorbed intact by insects (Turunen and Kastari, 1979). Phosphatides may also be hydrolysed by phospholipase C (EC 3.1.4.3) yielding the phosphoryl base moiety and diacylglycerol, or phospholipase D (EC 3.1.4.4), resulting in phosphatidate and the base. Both enzymes have been found in insect midguts (Turunen, 1993).

6.6.3 Esterases

Esterases, which are usually named the carboxylesterases (ali-esterases, EC 3.1.1.1) catalyse the hydrolysis of carboxyl ester into alcohol and carboxylate. This enzyme, in contrast to lipases, attacks molecules which are completely dissolved in water. It also hydrolyses water-insoluble long-chain fatty acid esters in the presence of surfactants, but at a rate much slower than that of triacylglycerol lipase. Insect midgut esterases are better known from a molecular perspective than the other lipid-hydrolysing enzymes. Esterases with a broad specificity have been purified to apparent homogeneity from midguts of insects from different orders including Dictyoptera (Hipps and Nelson, 1974), Isoptera (Sreerama and Veerabhadrappa, 1991) and Diptera (Whyard *et al.*, 1994), and their molecular weights were found to differ widely. A role for esterases in digestion is unclear, although they may be responsible for the hydrolysis of cholesterol and vitamin esters. Nevertheless, midgut esterases play a role in insect resistance to insecticide (Whyard *et al.*, 1994) and to allelochemicals (Lindroth, 1989).

6.7 OTHER DIGESTIVE ENZYMES

Phosphate moieties need to be removed from phosphorylated compounds prior to absorption. This is accomplished by non-specific phosphatases. The phosphatases may be active in an alkaline (alkaline phosphatase, EC 3.1.3.1) or acid (acid phosphatase, EC 3.1.3.2) medium. Alkaline phosphatase is usually a midgut microvillar membrane marker in dipteran and lepidopteran species, although they may also occur in midgut basolateral membranes and even as a secretory enzyme. Acid phosphatase is usually soluble in the cytosol of midgut cells in many insects and may also appear in midgut contents or be found membrane bound in midgut cells (Terra and Ferreira, 1994).

The best-known alkaline phosphatases are those from *Bombyx mori* (Lepidoptera: Bombycidae) larval midguts. The major membrane-bound and the minor soluble alkaline phosphatases were purified and shown to be monomeric enzymes with the following properties: (a) soluble enzyme, M_r 61 000, pH optimum 9.8, K_m for *p*-nitrophenyl phosphate (NPP) = 1.5 mM; (b) membrane-bound enzyme, M_r 58 000, pH optimum 10.9, K_m for NPP = 2 mM. Both enzymes present wide substrate specificity and are inhibited by cysteine (Okada *et al.*, 1989). The membrane-bound alkaline phosphatase occurs in the microvillar membranes of columnar cells, whereas the soluble enzyme is loosely attached to the goblet cell apical membrane facing the cell cavity (Azuma and Eguchi, 1989). The determination of the complete sequence of the membrane-bound alkaline phosphatase led to the finding of putative regions for phosphatidylinositol anchoring, zinc site but not for N-glycosylation, despite the fact that the enzyme contains N-linked oligosaccharides (Itoh *et al.*, 1991).

Acid phosphatases have been characterized in some detail only in *Rhodnius prolixus* (Hemiptera: Reduviidae). The major enzyme activity is soluble and has the following properties: wide specificity, M_r 82 000, K_m for NPP = 0.7 mM, and inhibited by fluoride, tartrate and molybdate. The minor enzyme activity is membrane-bound and is resolved into two enzymes (M_r 123 000 and M_r 164 000) which are resistant to fluoride and tartrate (Terra *et al.*, 1988).

6.8 DIGESTIVE ENZYMES DERIVED FROM MICRO-ORGANISMS

Most insects harbour a substantial microbiota including bacteria, yeasts and protozoa. Micro-organisms might be symbiotic or fortuitous contaminants from the external environment (Chapter 15). Micro-organisms are found in the lumen, adhering to the peritrophic matrix, attached to the midgut surface or intracellular. Micro-organisms produce and secrete their own hydrolases and cell death will result in the release of enzymes into the intestinal milieu. Any consideration of the spectrum of hydrolase activity in the midgut must include the possibility that some activities are derived from micro-organisms.

A number of factors combine to cause difficulties in drawing conclusions about the presence of microbially derived enzymes. Micro-organisms may be growing undetected in the gut because detection often depends on the use of highly selective culture media which do not allow the growth of all microbes found in the gut. In fact it is now thought that the majority of microscopically visualized cells cannot be cultivated. Enzyme assays with micro-organisms cultured from the gut can be misleading because the induction of enzymes will depend on the presence of different substrates and culture conditions. Lack of enzyme production in culture could simply be the result of catabolite repression.

Conclusions based on the use of antibiotics to establish the presence of microbially derived hydrolases should also be interpreted with care. Antibiotics have selective actions, will cause overgrowth by non-susceptible micro-organisms and some (antimycotics such as cycloheximide) interfere with protein synthesis. Studies with germ-free insects offer an alternative method for establishing the source of the enzymes but there have been few studies of this type. A study of germ-free locusts *Schistocerca gregaria* showed that although there were no differences in β-glucosidase or β-1:3 glucanase activities, there were significantly higher α-glucosidase activities in conventional insects (with a gut microbiota) than their germ-free counterparts (Charnley *et al.*, 1985). Measurements for a range of hydrolases failed to reveal any enzymes present in the conventional insect that were absent in the germ-free locust (Dillon, Hunt and Charnley, unpublished).

An approach which compares the characteristics of the hydrolases from the gut with those produced by cultured symbionts can give useful information on the origin of the enzymes. Supernatants and sonicates of the cultured symbiont *Nocardia rhodnii* found within the gut of *Rhodnius prolixus* were examined for hydrolase activity (Terra *et al.*, 1988). Although several hydrolases were detected in the cultures, only amylase was thought to be derived from the symbiont in the midgut. Symbiont aminopeptidase differed from the *R. prolixus* enzyme in being particle bound and showing different relative activities towards a range of aminopeptidase substrates. Other symbiont hydrolases had much lower specific activities than in the posterior midgut. Amylase from the midgut was not activated by chloride ions: a characteristic of bacterial and plant amylases. Both symbiont and midgut amylase isolates were particle bound, further evidence that the enzyme was from the symbiont.

The contribution of the insect gut microbiota to cellulose digestion has been the subject of much debate (see reviews by Martin, 1987, 1991; Slaytor, 1992). There is no involvement of protozoa in cellulase production in the non-fungus cultivating higher termites of the Nasutitermitinae (Slaytor, 1992) and the enzymes with activity towards crystalline cellulose are secreted from the midgut epithelium (section 6.5).

In the fungus-cultivating higher termites the cellulase components appear to originate from the termite and possibly from fungi (*Termitomyces* sp.). The extent of the contribution of the fungi is not clear. Martin (1987) suggested that exo-ß-1,4-glucanase (EC 3.2.1.91) is acquired from the fungus in *Macrotermes natalenses* based on isoelectrofocusing experiments showing identical pI values for the proteins isolated from the gut and fungal nodules. Further purification and characterization is required to determine the extent of the contribution for the acquired enzymes.

Rouland *et al.* (1988a) suggested that the exo-ß-1,4-glucanase in *Macrotermes mülleri* is of termite origin (section 6.5), but the endoglucanase is acquired from the *Termitomyces* sp. A comprehensive comparison of the physical and kinetic parameters of the single endo-ß-1,4-glucanase (EC 3.2.1.4) purified from fungal culture and the termite gut suggested the enzymes were identifical. The fungal endoglucanase was a relatively thermostable monomeric glycoprotein of M_r 34 000, displayed a pH optimum of 4.4, a K_m against carboxymethylcellulose of 0.75% and was inhibited by Mn^{2+} and Ag^+. In contrast, calculations suggest that fungal endo-ß-1,4-glucanase does not provide a substantial contribution to the cellulose digestion in *Macrotermes michaelseni* (Veivers *et al.*, 1991). Differences in conclusions about the importance of fungal cellulases might result in part from studies of different species and castes of termites combined with a lack of information about the stability of fungal enzymes in the gut of the termite.

Cellulolytic protozoa are the main agents for cellulase activity in the phylogenetically lower termites, e.g. *Coptotermes lacteus* (Hogan *et al.*,

1988). Five enzymes from the protozoa in the paunch were resolved on Superose 6B column as two β-1,4-glucosidases completely inhibited by 30 mM δ-gluconolactone, two endo-β-1,4 glucanases and an exo-β-1,4-glucosidase which is not inhibited by δ-gluconolactone. Cellulases of termite origin were also isolated from the upper regions of the gut. Tetracycline treatment to remove gut bacteria in the cockroach (*Panesthia cribrata*), did not result in reduced cellulase (endo-β-1,4-glucanase and β-1,4-glucosidase) activity (Scrivener *et al.*, 1989). However, conclusions based on the use of antibiotics, as stated above, must be treated with caution.

Acquired fungal enzymes are also thought to participate in cellulose digestion in larvae of the siricid wasp, *Sirex cyaneus* (reviewed by Martin, 1987). Larval and fungal extracts compared after ion-exchange and gel chromatography yielded identical separations of endo-1,3-β-xylanase (EC 3.2.1.32) (pI 5.7, 5.3 and 4.5) and two major endo-β-1,4-glucanases (pI 4.9 and <4.0). An endo-1,3-β-xylanase purified from the termite *Macrotermes mülleri* was suggested to be identical to a xylanase from the symbiotic fungus *Termitomyces* sp. (Rouland *et al.*, 1988b). The xylanase was a glycoprotein with M_r 28 500, pH optimum 5.2 and K_m against carboxymethylcellulose of 0.5×10^{-3}%.

There are relatively few studies which show an unambiguous contribution of microbial hydrolases towards digestion in insects. Many studies are confined to a record of the specific enzymes activities obtained comparing crude enzyme preparations from the insect gut and microbial cultures, the microbial contribution is inferred. When the hydrolases are known to be microbially derived, the quantitative significance to digestion should then be assessed.

6.9 CONCLUDING REMARKS

Until recent, the study of digestive enzymes concentrated on correlations between diet composition and the types of hydrolases present in the insect. It is now clear, at least from the species studied, that most species possess the full array of hydrolases. Our survey of the literature reveals that good comparative data are only available for a limited number of insect species. Inclusion of earlier published accounts of insect digestive enzymes is hindered by the often poor understanding of enzymological methods; the use of crude homogenates with unoptimized assays on undefined substrates. Even the data of those species that have been studied may be restricted in terms of the stages of life cycle.

Studies with α-amylase and serine proteases show that insect digestive enzymes possess similar amino acid sequences to the mammalian enzymes although they differ somewhat in substrate specifity and behaviour in the presence of protein inhibitors. More data are necessary before a clear picture emerges on their catalytic properties and evolutionary relationships among insects and other organisms.

It is likely that many of the digestive enzymes are glycoproteins, although few are currently identified as such. This might explain the presence of multiple forms of some enzymes, due to sugar heterogeneity, and serve to conserve enzymes by reducing hydrolysis by proteases.

Cysteine and aspartic proteinases are major digestive enzymes in coleopteran families belonging to series Cucujiformia and among Hemiptera Heteroptera. Despite the obvious importance of these enzymes, which may derive from lysosomes, characterization of insect digestive cysteine and aspartic proteinases has not progressed very far.

Digestive enzymes involved in the final stages of protein digestion require further study; molecular data are almost completely restricted to aminopeptidase N.

In spite of the fact that a requirement for essential fatty acids is probably universal in insects, progress has been limited in the study of lipid digestion. Presumably the lack of comparatively simple, sensitive assays and the complexities of digestion related to lipid solubilization have hindered work in this areas. Another reason to study enzymes associated with lipid digestion is that they might be important in limiting the development of pathogens and parasites. Hydrolysis of membrane lipids might cause cellular lysis and fatty acid products of digestion possess antibiotic effects.

An increasing number of papers have appeared on oligosaccharidases such as α- and β-glucosidases. β-Glucosidases were, until recently, thought to be vestigial because most insects do not digest cellulose nor have an important role in terminal hemicellulose digestion and in removing monosaccharides from glycoconjugates. β-Glucosidases are responsible for the release of toxic compounds by hydrolysis of plant glycosides thus explaining the resistance of some plant species to insect attack.

The discovery that *Bacillus thuringiensis* δ-endotoxin utilizes microvillar-associated enzymes as binding sites (Chapter 13) suggests that these enzymes could be important in interactions between insects and micro-organisms. This provides a further incentive for the study of digestive enzymes as an aid to the development of novel methods of insect control.

ACKNOWLEDGEMENTS

We are grateful to those authors who supplied us with reprints. Work in Brazil was supported by Brazilian research agencies FAPESP and CNPq.

W.R. Terra and C. Ferreira are staff members of the Biochemistry Department and research fellows of CNPq. B.P. Jordão is a post-doctoral fellow of CNPq.

REFERENCES

Applebaum, S.W. (1985) Biochemistry of digestion, in *Comprehensive Insect Physiology, Biochemistry, and Pharmacology* Vol. 4 (eds G.A. Kerkut and L.I. Gilbert), Pergamon Press, New York, pp. 279–311.

Azuma, M. and Eguchi, M. (1989) Discrete localization of distinct alkaline phosphatase isoenzymes in the cell surface of silkworm midgut epithelium. *J. Exp. Zool.*, **251**, 108–12.

Azuma, M. and Yamashita, O. (1985) Cellular localization and proposed function of midgut trehalase in the silkworm larva, *Bombyx mori. Tissue Cell*, **17**, 539–51.

Baker, J.E. (1981a) Isolation and properties of digestive carboxypeptidases from midguts of larvae of the black carpet beetle *Attagenus megatoma. Insect Biochem.*, **11**, 583–91.

Baker, J.E. (1981b) Resolution and partial characterization of the digestive proteinases from larvae of the black carpet beetle, in *Current Topics in Insect Endocrinology and Nutrition* (eds G. Bhaskaran, S. Friedman and J.G. Rodriguez), Plenum Press, New York, pp. 283–315.

Baker, J.E. (1987) Purification of isoamylases from the rice weevil, *Sitophilus oryzae* (L.) (Coleoptera: Curculionidae), by high-performance liquid chromatography and their interaction with partially-purified amylase inhibitors from wheat. *Insect Biochem.*, **17**, 37–44.

Baker, J.E. (1989) Interaction of partially-purified amylases from larval *Anagasta kuehniella* (Lepidoptera: Pyralidae) with amylase inhibitors from wheat. *Comp. Biochem. Physiol.*, **93B**, 239–46.

Baker, J.E. (1991) Properties of glycosidases from the maize weevil, *Sitophilus zeamais. Insect Biochem.*, **21**, 615–21.

Baker, J.E. and Woo, S.M. (1981) Properties and specificities of a digestive aminopeptidase from larvae of *Attagenus megatoma* (Coleoptera: Dermestidae). *Comp. Biochem. Physiol.*, **69B**, 189–93.

Baker, J.E. and Woo, S.M. (1985) Purification, partial characterization and postembryonic levels of amylases from *Sitophilus oryzae* and *Sitophilus granarius. Arch. Insect Biochem. Physiol.*, **2**, 415–28.

Baker, J.E. and Woo, S.M. (1992) β-Glucosidases in the rice weevil, *Sitophilus oryzae*: purification, properties, and activity levels in wheat and legume-feeding strains. *Insect Biochem. Mol. Biol.*, **22**, 495–504.

Barillas-Mury, C., Graf, R., Hagedorn, H.H. and Wells, M.A. (1991) cDNA and deduced amino acid sequence of a blood meal-induced trypsin from the mosquito, *Aedes aegypti. Insect Biochem.*, **21**, 825–31.

Bollade, D., Paris, R. and Moulins, M. (1970) Origine et mode d'action de la lipase intestinale chez les blattes. *J. Insect Physiol.*, **16**, 45–53.

Buonocore, V., Poerio, E., Silano, V. and Tomasi, M. (1976) Physical and catalytic properties of α-amylase from *Tenebrio molitor* L. larvae. *Biochem. J.*, **153**, 621–5.

Campos, F.A.P., Xavier-Filho, J., Silva, C.P. and Ary, M.B. (1989) Resolution and partial characterization of proteinases and α-amylases from midguts of larvae of the bruchid beetle *Callosobruchus maculatus* (F.). *Comp. Biochem. Physiol.*, **92B**, 51–7.

Casu, R.E., Pearson, R.D., Sarmey, J.M. *et al.* (1994) Excretory/secretory chymotrypsin from *Lucilia cuprina:* purification, enzymatic specificity and amino acid sequence deduced from mRNA. *Insec Mol. Biol.*, **3**, 201–11.

Chararas, C. and Chipoulet, J.M. (1982) Purification by chromatography and properties of a *β*-glucosidase from the larvae of *Phoracantha semipunctata. Comp. Biochem. Physiol.*, **72B**, 559–64.

Chararas, C., Eberhard, R., Courtois, J.E. and Petek, F. (1983) Purification of three cellulases from the xylophagous larvae of *Ergates faber* (Coleoptera: Cerambycidae). *Insect Biochem.*, **13**, 213–18.

Charnley, A.K., Hunt, J. and Dillon, R. (1985) The germ free culture of desert locusts, *Schistocerca gregaria. J. Insect Physiol.*, **31**, 477–85.

Cheeseman, M.T. and Gooding, R.H. (1985) Proteolytic enzymes from tsetse flies, *Glossina morsitans* and *Glossina palpalis* (Diptera: Glossinidae). *Insect Biochem.*, **15**, 677–80.

Chen, M.S., Feng, G., Zen, K.C. *et al.* (1992) α-Amylases from three species of stored grain Coleoptera and their inhibition by wheat and corn proteinaceous inhibitors. *Insect Biochem. Mol. Biol.*, **22**, 261–8.

Chipoulet, J.M. and Chararas, C. (1984) Purification and partial characterization of a laminarinase from the larvae of *Rhagium inquisitor. Comp. Biochem. Physiol.*, **77B**, 699–706.

Chipoulet, J.M. and Chararas, C. (1985) Survey and electrophoretical separation of the glycosidases of *Rhagium inquisitor* (Coleoptera: Cerambycidae) larvae. *Comp. Biochem. Physiol.*, **80B**, 241–6.

Christeller, J.T., Laing, W.A., Markwick, N.P. and Burgess, E.P.J. (1992) Midgut protease activities in 12 phytophagous lepidopteran larvae: dietary and protease inhibitor interactions. *Insect Biochem. Mol. Biol.*, **22**, 735–46.

Christeller, J.T., Laing, W.A., Shaw, B.D. and Burgess, E.P.J. (1990) Characterization and partial purification of the digestive proteases of the black field cricket, *Teleogryllus commodus* (Walker): elastase is a major component. *Insect Biochem.*, **20**, 157–64.

Christeller, J.T., Shaw, B.D., Gardiner, S.E. and Dymock, J. (1989) Partial purification and characterization of the major midgut proteases of grass grub larvae (*Costelytra zealandica*). Coleoptera: Scarabaeidae. *Insect Biochem.*, **19**, 221–31.

De Veau, E.J.I. and Schultz, J.C. (1992) Reassessment of interaction between gut detergents and tannins in Lepidoptera and significance for gypsy moth larvae. *J. Chem. Ecol.*, **18**, 1437–53.

Droste, H.J. and Zebe, E. (1974) Carbohydrasen und Kohlenhydratverdauung im Darmtrakt von *Locusta migratoria. J. Insect Physiol.*, **20**, 1639–57.

Espinoza-Fuentes, F.P., Ribeiro, A.F. and Terra, W.R. (1987) Microvillar and secreted digestive enzymes from *Musca domestica* larvae. Subcellular fractionation of midgut cells with electron microscopy monitoring. *Insect Biochem.*, **17**, 819–27.

Ferreira, C. and Terra, W.R. (1983) Physical and kinetic properties of a plasma-membrane-bound *β*-D-glucosidase (cellobiase) from midgut cells of an insect (*Rhynchosciara americana* larva). *Biochem. J.*, **213**, 43–51.

Ferreira, C. and Terra, W.R. (1984) Soluble aminopeptidases from cytosol and luminal contents of *Rhynchosciara americana* midgut caeca. Properties and phenanthroline inhibition. *Insect Biochem.*, **14**, 145–50.

Ferreira, C. and Terra, W.R. (1985) Minor aminopeptidases purified from the plasma membrane of midgut caeca cells of an insect (*Rhynchosciara americana*) larva. *Insect Biochem.*, **15**, 619–25.

Ferreira, C. and Terra, W.R. (1986a) The detergent form of the major aminopeptidase from the plasma membrane of midgut caeca cells of *Rhynchosciara americana* (Diptera) larva. *Comp. Biochem. Physiol.*, **84B**, 373–6.

Ferreira, C. and Terra, W.R. (1986b) Substrate specificity and binding loci for inhibitors in an aminopeptidase purified from the plasma membrane of midgut cells of an insect (*Rhynchosciara americana*) larva. *Arch. Biochem. Biophys.*, **244**, 478–85.

Ferreira, C., Ribeiro, A.F., Garcia, E.S. and Terra, W.R. (1988) Digestive enzymes trapped between and associated with the double plasma membranes of *Rhodnius prolixus* posterior midgut cells. *Insect Biochem.*, **18**, 521–30.

Garcia, E.S. and Guimarães, J.A. (1979) Proteolytic enzymes in the *Rhodnius prolixus* midgut. *Experientia*, **35**, 305–6.

Giebel, W., Zwilling, R. and Pfleiderer, G. (1971) The evolution of endopeptidases-XII. The proteolytic enzymes of the honeybee (*Apis mellifica* L.). *Comp Biochem. Physiol.*, **38B**, 197–210.

Gilbert, L.I., Chino, H. and Domroese, K.A. (1965) Lipolytic activity of insect tissues and its significance in lipid transport. *J. Insect Physiol.*, **11**, 1057–70.

Gilby, A.R., Wyatt, S.S. and Wyatt, G.R. (1967) Trehalases from the cockroach, *Blaberus discoidalis*: activation, solubilization and properties of the muscle enzymes and some properties of the intestinal enzyme. *Acta Biochim. Pol.*, **14**, 83–100.

Gooding, R.H. (1969) Studies on proteinases from some blood-sucking insects. *Proc. Entomol. Soc. Ont.*, **100**, 139–45.

Gooding, R.H. and Huang, C.T. (1969) Trypsin and chymotrypsin from the beetle *Pterostichus melanarius*. *J. Insect Physiol.*, **15**, 325–39.

Gooding, R.H. and Rolseth, B.M. (1976) Digestive processes of haematophagous insects. XI. Partial purification and some properties of six proteolytic enzymes from the tsetse fly *Glossina morsitans morsitans* Westwood (Diptera: Glossinidae). *Can. J. Zool.*, **54**, 1950–9.

Graf, R. and Briegel, H. (1985) Isolation of trypsin isozymes from the mosquito *Aedes aegypti* (L.). *Insect Biochem.*, **15**, 611–18.

Greenberg, B. and Paretsky, D. (1955) Proteolytic enzymes in the house fly, *Musca domestica* (L.). *Ann. Entomol. Soc. Am.*, **48**, 46–50.

Harborne, J.B. (1993) *Introduction to Ecological Biochemistry*, 4th edn, Academic Press, London.

Harwood, J.L. (1980) Plant acyl lipids. Structure, distribution and analysis, in *The Biochemistry of Plants: A Comprehensive Treatise* Vol. 4. *Lipids: Structure and Function* (ed. P.K. Stumpf), Academic Press, New York, pp. 1–56.

Hipps, P.P. and Nelson, D.R. (1974) Esterases from the midgut and gastric caecum of the American cockroach, *Periplaneta americana* (L.). Isolation and characterization. *Biochim. Biophys. Acta*, **341,** 421–36.

Hoffman, A.G.D. and Downer, R.G.H. (1979) End product specificity of triacylglycerol lipases from intestine, fat body, muscle and haemolymph of the American cockroach, *Periplaneta americana* L. *Lipids*, **14**, 893–9.

Hogan, M.E., Schultz, M.W., Slaytor, M. *et al.* (1988) Components of termite and protozoal cellulases from the lower termite, *Coptotermes lacteus* Frogatt. *Insect Biochem.*, **18**, 45–51.

Hori, K. (1972) Comparative study of a property of salivary amylase among various heteropterous insects. *Comp. Biochem. Physiol.*, **42B**, 501–8.

Houseman, J.G. and Downe, A.E.R. (1980) Endoproteinase activity in the posterior midgut of the *Rhodnius prolixus* Stal (Hemiptera: Reduviidae). *Insect Biochem.*, **10**, 363–6.

Houseman, J.G. and Downe, A.E.R. (1981) Exoproteinase activity in the posterior midgut of *Rhodnius prolixus* Stal (Hemiptera: Reduviidae). *Insect Biochem.*, **11**, 579–82.

Houseman, J.G., Morrison, P.E. and Downe, A.E.R. (1985) Cathepsin B and aminopeptidase in the posterior midgut of *Phymata wolffi* (Hemiptera: Phymatidae). *Can. J. Zool.*, **63**, 1288–91.

Huber, R.E. (1975) The purification and study of a honey bee abdominal sucrase exhibiting unusual solubility and kinetic properties. *Arch. Biochem. Biophys.*, **168**, 198–209.

Huber, R.E. and Mathison, R.D. (1976) Physical, chemical, and enzymatic studies on the major sucrase of honey bee (*Apis mellifera*). *Can. J. Biochem.*, **54**, 153–64.

Itoh, M., Takeda, S., Yamamoto, H. *et al.* (1991) Cloning and sequence analysis of membrane-bound alkaline phosphatase cDNA or silkworm, *Bombyx mori. Biochim. Biophys. Acta*, **1129**, 135–8.

Jany, K.D., Bekelaer, K., Pfleiderer, G. and Ishay, J. (1983) Amino acid sequence of an insect chymotrypsin from the larvae of the hornet, *Vespa orientalis. Biochem. Biophys. Res. Commun.*, **110**, 1–7.

Jany, K.D., Haug, H. and Ishay, J. (1978) Trypsin-like endopeptidases from the midguts of the larvae from the hornets of *Vespa orientalis* and *Vespa crabro. Insect Biochem.*, **8**, 221–30.

Jany, K.D. and Pfleiderer, G. (1974) Purification and some physical properties of a chymotrypsin-like protease of the larva of the hornet, *Vespa orientalis. Eur. J. Biochem.*, **42**, 419–28.

Jordão, B.P. and Terra, W.R. (1989) Distribution, properties, and functions of midgut carboxypeptidases and dipeptidases from *Musca domestica* larvae. *Arch. Insect Biochem. Physiol.*, **11**, 231–44.

Jordão, B.P. and Terra, W.R. (1991) Regional distribution and substrate specificity of digestive enzymes involved in terminal digestion in *Musca domestica* hind-midguts. *Arch. Insect Biochem. Physiol.*, **17**, 157–68.

Jordão, B.P., Terra, W.R., Ribeiro, A.F. *et al.* (1996) Trypsin secretion in *Musca domestica* larval midguts. A biochemical and immunocytochemical study. *Insect Biochem. Mol. Biol.* (in press).

Kanekatsu, R. (1978) Studies on further properties for an alkaline amylase in the digestive juice of the silkworm, *Bombyx mori. J. Fac. Text. Sci. Technol.*, **76** (Series E no. 9), 1–21.

Klinkowstrom, A.M., Terra, W.R. and Ferreira, C. (1994) Aminopeptidase A from *Rhynchosciara americana* (Diptera) larval midguts. Properties and midgut distribution. *Arch. Insect Biochem. Physiol.*, **27**, 301–15.

Klinkowstrom, A.M., Terra, W.R. and Ferreira, C. (1995) Midgut dipeptidases from *Rhynchosciara americana* (Diptera) larvae. Properties of soluble and membrane-bound forms. *Insect Biochem. Mol. Biol.*, **25**, 303–10.

Kylsten, P., Kimbrell, D.A., Daffre, S. *et al.* (1992) The lysozyme locus in *Drosophila melanogaster*. Different genes are expressed in midgut and salivary glands. *Mol. Gen. Genet.*, **232**, 335–43.

Lecadet, M.M. and Dedonder, R. (1966) Les protéases de *Pieris brassicae*. I. Purification et propriétés. *Bull. Soc. Chim. Biol.*, **48**, 631–60.

Lemos, F.J.A., Ribeiro, A.F. and Terra, W.R. (1993) A bacteria-digesting midgut-lysozyme from *Musca domestica* (Diptera) larvae. Purification, properties and secretory mechanism. *Insect Biochem. Mol. Biol.*, **23**, 533–41.

Lemos, F.J.A. and Terra, W.R. (1991a) Properties and intracellular distribution of a cathepsin D-like proteinase active at the acid region of *Musca domestica* midgut. *Insect Biochem.*, **21**, 457–65.

Lemos, F.J.A. and Terra, W.R. (1991b) Digestion of bacteria and the role of midgut lysozyme in some insect larvae. *Comp. Biochem. Physiol.*, **100B**, 265–8.

Lemos, F.J.A. and Terra, W.R. (1992) Soluble and membrane-bound forms of trypsin-like enzymes in *Musca domestica* larval midguts. *Insect Biochem. Mol. Biol.*, **22**, 613–19.

Levinsky, H., Birk, Y. and Applebaum, S.W. (1977) Isolation and characterization of a new trypsin-like enzyme from *Tenebrio molitor* L. larvae. *Int. J. Pept. Protein Res.*, **10**, 252–64.

Lindroth, R.L. (1989) Biochemical detoxication: mechanism of differential tiger swallowtail tolerance to phenolic glycosides. *Oecologia*, **81**, 219–24.

Marana, S.R., Terra, W.R. and Ferreira, C. (1995) Midgut β-D-glucosidases from *Abracris flavolineata* (Orthoptera: Acrididae). Physical properties, substrate specificities and function. *Insect Biochem. Mol. Biol.*, **25**, 835–43.

Marinotti, O. and James, A.A. (1990) An α-glucosidase in the salivary glands of the vector mosquito, *Aedes aegypti. Insect Biochem.*, **20**, 619–23.

Martin, M.M. (1987) *Invertebrate–Microbial Interactions: Ingested Fungal Enzymes in Arthropod Biology*, Cornell, Ithaca.

Martin, M.M. (1991) The evolution of cellulose digestion in insects. *Philos. Trans. R. Soc. Lond. B*, **333**, 281–8.

Martin, M.M., Kukor, J.J., Martin, J.S. *et al.* (1981) Digestive enzymes of fungus-feeding beetles. *Physiol. Zool.*, **54**, 137–45.

Milne, R. and Kaplan, H. (1993) Purification and characterization of a trypsin-like digestive enzyme from spruce budworm (*Choristoneura fumiferana*) responsible for the activation of δ-endotoxin from *Bacillus thuringiensis. Insect Biochem. Mol. Biol.*, **23**, 663–73.

Moffatt, M.R. and Lehane, M.J. (1990) Trypsin is stored as an inactive zymogen in the midgut of *Stomoxys calcitrans. Insect Biochem.*, **20**, 719–723.

Morgan, M.R.J. (1975) Relationship between gut cellobiase, lactase, aryl β-glucosidase, and aryl β-galactosidase activities of *Locusta migratoria. Insect Biochem.*, **5**, 609–17.

Nagarajn, J. and Abraham, E.G. (1995) Purification and characterization of digestive amylase from the tasar silkworm, *Antheraea mylitta* (Lepidoptera: Saturniidae). *Comp. Biochem. Physiol.*, **110B**, 201–9.

Nomenclature Committee of The International Union of Biochemistry and Molecular Biology (1992), Enzyme Nomenclature 1992, Academic Press, New York.

Norén, O., Sjoström, H. and Josefsson, L. (1973) Studies on a soluble dipeptidase from pig intestinal mucosa. I. Purification and specificity. *Biochim. Biophys. Acta*, **327**, 446–56.

Okada, N., Azuma, M. and Eguchi, M. (1989) Alkaline phosphatase isoenzymes in the midgut of silkworm: purification of high pH-stable in microvillus and labile cytosolic enzymes. *J. Comp. Physiol. B.*, **159**, 123–30.

Podoler, H. and Applebaum, S.W. (1971a) The α-amylase of the beetle *Callosobruchus chinensis*. Purification and action pattern. *Biochem. J.*, **121**, 317–20.

Podoler, H. and Applebaum, S.W. (1971b) The α-amylase of the beetle *Callosobruchus chinensis*: properties. *Biochem. J.*, **121**, 321–5.

Pratviel-Sosa, F., Clermont, S., Percheron, F. and Chararas, C. (1986) Studies on glycosidases and glucanases in *Thaumetopoea pityocampa* larvae. Part 1. Purification and some properties of an α-glucosidase. *Comp. Biochem. Physiol.*, **84B**, 77–81.

Pratviel-Sosa, F., Clermont, S., Percheron, F. and Chararas, C. (1987) Studies on glycosidases and glucanases in *Thaumetopoea pityocampa* larvae. II. Purification and some properties of a broad specificity β-D-glucosidase. *Comp. Biochem. Physiol.*, **86B**, 173–8.

Purcell, J.P., Greenplate, J.T. and Sammons, R.D. (1992) Examination of midgut luminal proteinase activities in six economically important insects. *Insect Biochem. Mol. Biol.*, **22**, 41–7.

Ramos, A., Mahowald, A. and Jacobs-Lorena, M. (1993) Gut-specific genes from the black-fly *Simulium vittatum* encoding trypsin-like and carboxypeptidase-like proteins. *Insect Mol. Biol.*, **1**, 149–63.

Rouland, C., Civas, A., Renoux, J. and Petek, F. (1988a) Purification and properties of cellulases from the termite *Macrotermes mulleri* (Termitidae, Macrotermitinae) and its symbiotic fungus *Termitomyces* sp. *Comp. Biochem. Physiol.*, **91B**, 449–58.

Rouland, C., Renoux, J. and Petek, F. (1988b) Purification and properties of two xylanases from *Macrotermes mulleri* (Termitidae, Macrotermitinae) and its symbiotic fungus *Termitomyces* sp. *Insect Biochem.*, **18**, 709–15.

Sakal, E., Applebaum, S.W. and Birk, Y. (1988) Purification and characterization of *Locusta migratoria* chymotrypsin. *Int. J. Pept. Protein Res.*, **32**, 590–8.

Sakal, E., Applebaum, S.W. and Birk, Y. (1989) Purification and characterization of trypsins from the digestive tract of *Locusta migratoria. Int. J. Pept. Protein. Res.*, **34**, 498–505.

Santos, C.D., Ribeiro, A.F. and Terra, W.R. (1986) Differential centrifugation, calcium precipitation and ultrasonic disruption of midgut cells of *Erinnyis ello* caterpillars. Purification of cell microvilli and inferences concerning secretory mechanisms. *Can. J. Zool.*, **64**, 490–500.

Santos, C.D. and Terra, W.R. (1985) Physical properties substrate specificities and a probable mechanism for a β-D-glucosidase (cellobiase) from midgut cells of the cassava hornworm (*Erinnyis ello*). Biochim. Biophys. Acta, **831**, 179–85.

Santos, C.D. and Terra, W.R. (1986) Midgut α-glucosidase and β-fructosidase from *Erinnyis ello* larvae and imagoes. Physical and kinetic properties. *Insect Biochem.*, **16**, 819–24.

Sasaki, T. and Suzuki, Y. (1982) Alkaline proteases in digestive juice of the silkworm, *Bombyx mori. Biochim. Biophys. Acta*, **703**, 1–10.

Schulz, M.W., Slaytor, M., Hogan, M. and O'Brien, R.W. (1986) Components of cellulase from the higher termite, *Nasutitermes walkeri. Insect Biochem.*, **16**, 929–32.

Schumaker, T.T.S., Cristofoletti, P.T. and Terra, W.R. (1993) Properties and compartmentalization of digestive carbohydrases and proteases in *Scaptotrigona bipunctata* (Apidae: Meliponinae) larvae. *Apidologie*, **24**, 3–17.

Scrivener, A.M., Slaytor, M. and Rose, H.A. (1989) Symbiont-independent digestion of cellulose and starch in *Panesthia cribrata* Saussure, an Australian wood-eating cockroach. *J. Insect Physiol.*, **35**, 935–41.

Silva, C.P. and Terra, W.R. (1995) An α-glucosidase from the perimicrovillar membranes of *Dysdercus peruvianus* (Hemiptera) midgut cells. Purification and properties. *Insect Biochem. Mol. Biol.*, **25**, 487–94.

Silva, C.P. and Xavier-Filho, J. (1991) Comparison between the levels of aspartic and cysteine proteinases of the larval midguts of *Callosobruchus maculatus* (F.)

and *Zabrotes subfasciatus* (Boh.) (Coleoptera: Bruchidae). *Comp. Biochem. Physiol.*, **99B**, 529–33.

Slaytor, M. (1992) Cellulose digestion in termites and cockroaches: what role do symbionts play? *Comp. Biochem. Physiol.*, **103B**, 775–84.

Sommerville, H.J. and Pockett, H.V. (1976) Phospholipase activity in gut juice of lepidopterous larvae. *Insect Biochem.*, **6**, 351–3.

Sreerama, L. and Veerabhadrappa, P.S. (1991) Purification and properties of carboxylesterases from the mid-gut of the termite *Odentotermes horni* W. *Insect Biochem.*, **21**, 833–44.

Sumida, M. and Yamashita, O. (1983) Purification and some properties of soluble trehalase from midgut of pharate adult of the silkworm, *Bombyx mori. Insect Biochem.*, **13**, 257–65.

Talbot, B.G. and Huber, R.E. (1975) Partial purification, stabilization and characterization of adult honey bee midgut trehalase and a new trehalase specific disc gel stain method. *Insect Biochem.*, **5**, 337–47.

Tenimura, T., Kitamura, K., Fukuda, T. and Kikuchi, T. (1979) Purification and partial characterization of three forms of α-glucosidase from the fruit fly *Drosophila melanogaster. J. Biochem.*, **85**, 123–30.

Terra, W.R. (1990) Evolution of digestive systems of insects. *Annu. Rev. Entomol.*, **35**, 181–200.

Terra, W.R. and Ferreira, C. (1994) Insect digestive enzymes: properties, compartmentalization and function. *Comp. Biochem. Physiol.*, **109B**, 1–62.

Terra, W.R. and Ferreira, C. (1983) Further evidence that enzymes involved in the final stages of digestion by *Rhynchosciara* do not enter the endoperitrophic space. *Insect Biochem.*, **13**, 143–50.

Terra, W.R., Ferreira, C. and Bastos, F. (1985) Phylogenetic considerations of insect digestion. Disaccharidases and the spatial organization of digestion in the *Tenebrio molitor* larvae. *Insect Biochem.*, **15**, 443–9.

Terra, W.R., Ferreira, C. and De Bianchi, A.G. (1977) Action pattern, kinetical properties and electrophoretical studies of an alpha-amylase present in midgut homogenates from *Rhynchosciara americana* (Diptera) larvae. *Comp. Biochem. Physiol.*, **56B**, 201–9.

Terra, W.R., Ferreira, C. and De Bianchi, A.G. (1978) Physical properties and Tris inhibition of an insect trehalase and a thermodynamic approach to the nature of its active site. *Biochim. Biophys. Acta*, **524**, 131–41.

Terra, W.R., Ferreira, C. and De Bianchi, A.G. (1979a) Distribution of digestive enzymes among the endo- and ectoperitrophic spaces and midgut cells of *Rhynchosciara* and its physiological significance. *J. Insect Physiol.*, **25**, 486–94.

Terra, W.R., Ferreira, C. and Garcia, E.S. (1988) Origin, distribution, properties and functions of the major *Rhodnius prolixus* midgut hydrolases. *Insect Biochem.*, **18**, 423–34.

Terra, W.R., Terra, I.C.M. and Ferreira, C. (1983) Inhibition of an insect midgut trehalase by dioxane and δ-gluconolactone: enzyme pKa values and geometric relationships at the active site. *Int. J. Biochem.*, **15**, 143–6.

Terra, W.R., Terra, I.C.M., Ferreira, C. and De Bianchi, A.G. (1979b) Carbodiimide-reactive carboxyl groups at the active site of an insect midgut trehalase. *Biochim. Biophys. Acta*, **571**, 79–85.

Terra, W.R., Valentin, A. and Santos, C.D. (1987) Utilization of sugars, hemicellulose, starch, protein, fat and minerals by *Erinnyis ello* larvae and the digestive role of their midgut hydrolases. *Insect Biochem.*, **17**, 1143–7.

Tong, N.T., Imhoff, J.M., Lecroisey, A. and Keil, B. (1981) Hypodermin A, a trypsin-like neutral proteinase from the insect *Hypoderma lineatum. Biochim. Biophys. Acta*, **658**, 209–19.

Tsuchida, K. and Wells, M.A. (1988) Digestion, absorption, transport and storage of fat during the last larval stadium of *Manduca sexta*. Changes in the role of lipophorin in the delivery of dietary lipids to the fat body. *Insect Biochem.*, **18**, 263–8.

Tururen, S. (1992) Efficient use of dietary galactose in *Pieris brassicae. J. Insect Physiol.*, **38**, 503–9.

Turunen, S. (1993) Metabolic pathways in the midgut epithelium of *Pieris brassicae* during carbohydrate and lipid assimilation. *Insect Biochem. Mol. Biol.*, **23**, 681–9.

Turunen, S. and Kastari, T. (1979) Digestion and absorption of lecithin in larvae of the cabbage butterfly, *Pieris brassicae. Comp. Biochem. Physiol.*, **62A**, 933–7.

Uscian, J.M., Miller, J.S., Sarath, G. and Stanley-Samuelson, D.W. (1995) A digestive phospholipase A_2 in the tiger beetle *Cincindella circumpicta. J. Insect Physiol.*, **41**, 135–41.

Valaitis, A.P. and Bowers, D.F. (1993) Purification and properties of the soluble midgut trehalase from the gypsy moth, *Lymantria dispar. Insect Biochem. Mol. Biol.*, **23**, 599–606.

Van der Westhuizen, M.C., Hewitt, P.H. and du Toit, P.J. (1981) Aminopeptidase from the harvester termite *Trinervitermes trinervoides* (Sjostedt): distribution, purification, physical and chemical properties. *Insect Biochem.*, **11**, 311–21.

Veivers, P.C., Muhlemann, R., Slaytor, M. *et al.* (1991) Digestion, diet and polyethism in two fungus growing termites: *Macrotermes subhyalinus* Rambur and *M. michaelseni* Sjostedt. *J. Insect Physiol.*, **37**, 675–82.

Vonk, H.J. and Western, J.R.H. (1984) *Comparative Biochemistry and Physiology of Enzymatic Digestion*, Academic Press, New York.

Walker, V.K., Williamson, J.H. and Church, R.B. (1981) Differential characterization of two leucine aminopeptidases in *Drosophila melanogaster. Biochem. Genet.*, **19**, 47–60.

Ward, C.W. (1975a) Properties and specificity of the major metal-chelator-sensitive proteinase in the keratinolytic larvae of the webbing clothes moth. *Biochim. Biophys. Acta*, **384**, 215–27.

Ward, C.W. (1975b) Aminopeptidases in webbing clothes moth larvae. Properties and specificities of the enzymes of intermediate electrophoretic mobility. *Biochim. Biophys. Acta*, **410**, 361–9.

Ward, C.W. (1975c) Aminopeptidases in webbing clothes moth larvae. Properties and specificities of enzymes of highest electrophoretic mobility. *Aust. J. Biol. Sci.*, **28**, 447–55.

Ward, C.W. (1976) Properties of the major carboxypeptidase in the larvae of the webbing clothes moth. *Tineolla bisselliella. Biochim. Biophys. Acta*, **429**, 564–72.

Weintraub, H. and Tietz, A. (1973) Triglyceride digestion and absorption in the locust, *Locusta migratoria. Biochim. Biophys. Acta*, **306**, 31–41.

Whyard, S., Downe, A.E.R. and Walker, V.K. (1994) Isolation of an esterase conferring insecticide resistance in the mosquito *Culex tarsalis. Insect Biochem. Mol. Biol.*, **24**, 819–27.

Wieman, K.F. and Nielsen, S.S. (1988) Isolation and partial characterization of a major gut proteinase from larvel *Acanthoscelides obtectus* Say (Coleoptera, Bruchidae). *Comp. Biochem. Physiol.*, **89B**, 419–26.

Wigglesworth, V.B. (1972) *The Principles of Insect Physiology*, 7th edn, Methuen, London.

Wolfson, J.L. and Murdock, L.L. (1990) Diversity in digestive proteinase activity among insects. *J. Chem. Ecol.*, **16**, 1089–102.

Yu, S.J. (1989) β-Glucosidase in four phytophagous Lepidoptera. *Insect Biochem.*, **19**, 103–8.

Zinkler, D. and Polzer, M. (1992) Identification and characterization of digestive proteinases from the firebrat. *Thermobia domestica. Comp. Biochem. Physiol.*, **103B**, 669–73.

7

Mechanisms controlling the synthesis and secretion of digestive enzymes in insects

M.J. Lehane, H.M. Müller and A. Crisanti

Insects can be loosely divided into two categories according to their feeding behaviour. Lepidopteran and dipteran larvae and sap feeders feed continuously and a continuous stream of food is passing through the intestine. Carnivores or haematophagous insects have periods of digestive activity and periods when the gut is empty. Continuous and discontinuous feeding present the insects with distinct problems in the control of digestive enzyme secretion which we are only beginning to address. Increasing experimental evidence indicates that such control mechanisms are complex and that there may be separate control over secretion and synthesis (Blakemore *et al.*, 1995). Secretory cells themselves can also be divided into two categories depending on the fate of the newly synthesized secretory product. Constitutively secreting cells release secretory material shortly after it is synthesized and do not accumulate large numbers of secretory granules. In contrast, cells undertaking regulated secretion can store considerable quantities of secretory product that are released in response to a specific physiological signal. The ultrastructural investigations on insect gut cells carried out to date indicate that most secretion in insects probably occurs through the constitutive route (Lehane, 1996) and that most examples of regulated secretion appear in discontinuously feeding insects. During the digestion process the levels of digestive enzymes in the midgut

Biology of the Insect Midgut.
Edited by M.J. Lehane and P.F. Billingsley.
Published in 1996 by Chapman & Hall, London.
ISBN 0 412 61670 X.

change considerably in virtually all insects studied. This, together with the ultrastructural evidence of stored secretory products in the gut cells, demonstrates that control mechanisms regulate the synthesis and secretion of digestive enzymes. It has been suggested that these control mechanisms may depend on nerve action, hormone release, paracrine activity or prandial mechanisms (in which elements of the meal interact directly with the secretory cells effecting control, Lehane *et al.*, 1995) and here we will consider the evidence for each.

Detailed investigations suggest that the innervation of the insect midgut controls only the activity of the musculature (Day and Powning, 1949; Garcia and Garcia, 1977; Anderson and Cochrane, 1978; Zitnan *et al.*, 1993). Accordingly, trypsin synthesis could be induced in isolated midguts (Graf and Briegel, 1989) and trypsin secretion could be stimulated in *in vitro* preparations of *Stomoxys calcitrans* (Blakemore *et al.*, 1993, 1995). These results argue against a direct involvement of the nervous system in the regulation of digestive enzyme levels.

Hormones play a crucial role in regulating key physiological events, such as general protein metabolism, which are linked to digestive enzyme production. To date, evidence suggesting a role of hormones in regulating digestive enzyme levels is inconclusive. This may be due to the difficulty of separating the direct from the incidental influences of the hormonal system under investigation. For example, Englemann (1969) showed food passage rates through the gut were inhibited by neck ligation which is a technique which has often been used to investigate the effect of brain hormones on digestive enzyme secretion. Similarly, median neurosecretory cell (MNC) ablation in some insects (Engelmann and Wilkens, 1969; Muraleedharan and Prabhu, 1979) significantly reduced food intake levels thus making it difficult to discriminate between direct hormonal control of digestive enzyme levels and effects due to reduced food uptake that will also influence these levels. Such confounders need to be appreciated in the design of experiments aimed at investigating the direct influence of hormones on digestive enzyme levels.

Bearing in mind the difficulties in interpreting many of the available data, the mosquitoes provide the best body of evidence for the control of digestive enzyme production by hormonal means. Briegal and Lea (1979) showed that ovariectomy or MNC ablation caused a halving of midgut trypsin activity in *Aedes aegypti* and that this effect could be reversed by re-implanting an ovary. Their results suggest that the ovary or MNC cells have a modulating effect rather than exerting absolute control over enzyme synthesis and secretion as trypsin levels in operated mosquitoes were still sufficient for normal digestion of the meal. As the mosquito eggs reach maturity the insect becomes incapable of digesting a blood meal (Detinova, 1962), and it has been suggested that the decapeptide trypsin modulating oostatic factor (TMOF) isolated from the ovaries of *A. aegypti* (Borovsky *et al.*, 1990, 1993) may be involved. However, micromolar concentrations of TMOF and its analogues are needed in order to reduce trypsin production to 10% of its control level; these concentrations are far above those at which hormones normally operate. In addition, it is not clear if TMOF is acting directly on the midgut cells synthesizing and secreting digestive enzymes or if it exerts a more general effect, for example on protein metabolism. The recent observation that TMOF has no effect on protein-stimulated secretion from *in vitro* preparations of opaque zone cells of *S. calcitrans* (Millett and Lehane, unpublished) would argue against a direct effect of this hormone on enzyme secretion.

The experimental separation of prandial and paracrine mechanisms is difficult. As described in detail in Chapter 3, ultrastructural studies have shown that the midgut epithelium of all insects studied to date contains large numbers of putative endocrine cells (Endo *et al.*, 1990). In support of an endocrine role for these cells the contents of their secretory granules have been shown immunocytochemically to share epitopes with vertebrate neuropeptides (Endo *et al.*, 1990; Sehnal and Zitnan, 1990; Zitnan *et al.*, 1993). Also feeding has been shown to cause quantitative changes in the levels of these putative 'peptide hormones' in mosquitoes (Brown *et al.*, 1986; Jenkins *et al.*, 1989). As in vertebrates (Fujita *et al.*, 1988), these midgut endocrine cells are likely to play a role in the regulation of intestinal activities. However, no clear evidence has yet been gathered to show what role these putative endocrine cells play in control of midgut events and this is due to the difficulty of designing suitable experiments. The evidence for paracrine control currently available is weak (Rounds, 1968; Sreekumar and Prabhu, 1988).

It is well documented that the ingestion of some nutrients will stimulate *de novo* synthesis of digestive enzymes (e.g. Champlain and Fisk, 1956; Engelmann, 1969; Engelmann and Wilkens, 1969; Garcia and Garcia, 1977; Muraleedharan and Prabhu, 1979; Hori *et al.*, 1982; Houseman *et al.*, 1988; Jimenez and Gilliam, 1989; Felix *et al.*, 1991; Stiles *et al.*, 1991; Blakemore *et al.*, 1995). There is evidence in some insects that such stimulation is dependent on the concentration of the ingested nutrient and not on the size of the meal (Briegel and Lea, 1975) which suggests stretch receptors are not involved. This control could operate through hormonal mechanisms linked to foregut receptors but, arguing against this, in some insects the induction of digestive enzymes can be achieved by feeding them the specific nutrients by enema (Briegel and Lea, 1975, 1979). Also, in *Stomoxys calcitrans* nutrients cause a dose-dependent secretion from isolated *in vitro* preparations (Blakemore *et al.*, 1995), suggesting that the nutrients are effecting control through paracrine or prandial mechanisms in the midgut itself.

Proteins have been most widely studied as nutrients regulating digestive enzyme activity. In *Aedes aegypti* the ingestion of egg albumin and blood plasma induced proteolytic activity (Briegel and Lea, 1975) but ingestion of insoluble proteins did not (Felix *et al.*, 1991). This effect was independent of the molecular weight of the protein in mosquitoes (Felix *et al.*, 1991) and *S. calcitrans* (Blakemore *et al.*, 1995). Amino acids and polyamino acids did not stimulate secretion (Engelmann, 1969; Blakemore *et al.*, 1995). Blakemore *et al.* (1995) showed that all soluble proteins tested could stimulate trypsin secretion from their *in vitro* preparation and that the ability was not correlated with any obvious property of the protein other than its being a protein. This reflects the situation reported in several *in vivo* experiments. These observations raise several questions concerning the mechanisms allowing so many disparate molecules to initiate a cascade of events leading to the synthesis and secretion of the digestive enzymes which are discussed by Lehane *et al.* (1995).

Blakemore *et al.* (1995) have shown in *S. calcitrans* a dose-dependent secretion in response to protein in an *in vitro* culture preparation in which the gut segment had been slit lengthways to make a flat sheet of epithelium. Numerous other studies have shown a strong correlation between gut protein content and luminal proteinase levels in several insects (e.g. Gooding, 1977; Lehane, 1977; Houseman *et al.*, 1985; Billingsley and Hecker, 1991) although no such correlation is seen for epithelium-bound aminopeptidase (Houseman and Downe, 1983; Schneider *et al.*, 1987; Billingsley and Hecker, 1991) or α-glucosidase activity (Billingsley and Hecker, 1991). A simple explanation for this might be that protein above a threshold level initiates new synthesis whereas the level of trypsin is determined translationally by the quantity of amino acids produced during digestion of the meal (Lehane, 1977). Correlations between levels of extracellular proteolytic enzymes have also been demonstrated on several occasions. For example, in *S. calcitrans* trypsin, chymotrypsin and carboxypeptidase A and B activities all correlated during the digestive cycle (Schneider *et al.*, 1987). These authors suggested this might be explained neatly if all extracellular proteinases were under the control of the same 'operon'. But the observation that the relationship only holds true after certain types of protein meal (Gooding, 1977) suggests the controlling mechanisms are likely to be more complex and recent studies at the molecular level confirm such a view.

The new tools of cell and molecular biology have provided new mechanisms for the study of regulatory mechanisms operating in the insect midgut. At the forefront is work being carried out on digestive serine proteases in mosquitoes which has illustrated that the patterns of expression of digestive enzymes are more complex than can be easily studied with traditional methods. Mosquito trypsins have been described as *early* and *late* trypsins according to their expression pattern over time. Early trypsins are produced in small amounts within the first few hours after uptake of a blood meal and disappear shortly afterwards. Late trypsins are produced in large amounts 24 h after the blood meal and account for most of the proteolytic activity detected in the mosquito gut during the digestion process. More detailed information on the expression pattern, the activation process, and the function of mosquito trypsins has been provided in recent years by the cloning of both *Aedes aegypti* and *Anopheles gambiae* trypsin genes. In the mosquito *Ae. aegypti* the complete coding sequences of three trypsin genes have been reported (Barillas-Mury *et al.*, 1991; Kalhok *et al.*, 1993). Two of them were shown to be induced in the mosquito midgut after blood feeding (late trypsins) whereas one was constitutively transcribed (early trypsin) in the gut of female mosquitoes. Genomic Southern analysis indicates that additional trypsin related genes may exist in *Ae. aegypti* (Kalhok *et al.*, 1993). In *Anopheles gambiae* trypsin genes are arranged as a tightly clustered gene family consisting of seven related coding sequences, *Antryp1–7*. Two members of this family, *Antryp1* and *Antryp2*, are induced in the mosquito gut upon blood feeding. The other members (*Antryp3, 4, 5, 6* and *7*) are all constitutively expressed in female mosquitoes (Müller *et al.*, 1995). As in *Aedes*, the expression level of the early *Anopheles* trypsin genes appears to be down-regulated shortly after blood feeding (Müller *et al.*, 1995). Blood feeding was also shown to induce chymotrypsin activity (Hörler and Briegel, 1995) and chymotrypsin gene expression (Müller *et al.*, 1995) in *An. albimanus* and *An. gambiae*, respectively. The expression of digestive enzyme in a time- and tissue-specific regulated manner has also been reported in other haematophagous insect species. In the black fly *Simulium vittatum* genes encoding trypsin-like and carboxypeptidase-like proteins were induced in the gut after ingestion of the blood meal (Ramos *et al.*, 1993).

The biochemical mechanisms that regulate the expression and secretion of different serine proteases in the mosquito gut in a temporal co-ordinated manner have not yet been fully elucidated. Most of the information available has originated from experiments carried out in *Aedes* and *Anopheles* mosquitoes. In *Ae. aegypti* early trypsin mRNA was detected in the gut of female mosquitoes at 24 h after emergence and reached its peak level at an adult age of 4–7 days (Noriega *et al.*, 1996). Earlier studies (Graf and Briegel, 1989) failed to detect trypsin activity in the gut of unfed *Ae. aegypti* thus implying that early trypsin synthesis is controlled by a mechanism acting at the mRNA translation level. This conclusion is supported by the finding that translation inhibitors such as puromycin and cycloheximide inhibited the induction of trypsin activity after feeding (Felix *et al.*, 1991). Accordingly, stretching of the mosquito midgut or osmotic changes of its content due to the ingestion of a blood meal (Graf and Briegel, 1989) are believed to start a chain of events leading to the specific translation of early trypsin transcripts. Increasing experimental evidence indicates that in *Anopheles* mosquitoes the mechanisms controlling early trypsin synthesis differ from those of *Ae. aegypti*. In *An. albimanus* trypsin activity could be detected in the gut of mosquitoes that had never taken a blood meal (Hörler and Briegel, 1995). Immunoblot experiments indicate that the early *An. gambiae* trypsin *Antryp4* is synthesized in the gut of female mosquitoes from 1 day after emergence onward (Müller *et al.*, 1995). Two hours after membrane feeding on saline most of the *Antryp4* is recovered in the gut lumen and shows a shift in M_r due to the removal of the activation peptide upon secretion (Müller *et al.*, 1995). These findings indicated that in *Anopheles* mosquitoes early trypsins are stored as zymogen in the gut epithelium before feeding and that their secretion and activation is largely independent of the protein content of the ingested meal.

It has been speculated that early trypsin activity is part of the signal transduction pathway that activates the transcription of late trypsin genes in the midgut of *Ae. aegypti* (Barillas-Mury and Wells, 1993). This hypothesis is supported by the observation that the addition of soybean trypsin inhibitor (SBTI) to the blood meal prevented the transcriptional activation of late trypsin genes in the mosquito gut (Barillas-Mury *et al.*, 1995). This inhibitory effect can be overcome by feeding a meal digested *ex vivo* with bovine trypsin. These results are consistent with the observation that a 5 kDa trypsin digestion product of bovine serum albumin is a potent promoter of trypsin and carboxypeptidase production in *Glossinia morsitans morsitans* (Gooding, 1977). On the basis of these findings a model has been proposed to explain the regulation of trypsin synthesis following a blood meal (Barillas-Mury *et al.*, 1995). According to this model early trypsins in combination with other proteases generate free amino acids and small peptides from the constituents of the blood meal. These products then represent the initial signals for the induction of transcription of the late trypsin genes. An essentially similar hypothesis has also been proposed by Blakemore *et al.* (1995) as an explanation for the control of secretion in *S. calcitrans*. In *Ae. aegypti* the transcription of the late trypsin genes would then be sustained by a positive feedback mechanism mediated by free amino acids and small peptides generated by the action of the late trypsin proteases on the blood meal constituents (Barillas-Mury *et al.*, 1995). In this way early trypsin activity would generate signals that reflect the quality of the blood meal in terms of protein content (Lehane, 1977). This mechanism may have developed to prevent the expression of high levels of late trypsin that could result in self digestion in the absence of protein meal substrates. Recent experimental evidence suggests that in *Anopheles* mosquitoes such early trypsin activity may be important for starting a cascade of events leading to blood meal digestion (Müller *et al.*, 1995).

The promoter enhancer elements of the trypsin and chymotrypsin genes represent the final links in the chain of control signals that regulate the transcription of early and late serine proteases in the mosquito gut. The identification of the DNA regions that function as promoter elements would provide invaluable information for dissecting the molecular mechanisms that link the biochemical signals (hormones, digestion products, etc.) with the tissue specific and time co-ordinated transcription of the trypsin and chymotrypsin genes. Moreover, it has been proposed that transgenic mosquitoes showing a modified vectorial phenotype may prove useful in the genetic control of mosquito vector populations. Thus, DNA sequences controlling the transcription of the trypsin genes could be used to generate transgenic mosquitoes expressing antiparasitic or antiviral agents in the mosquito gut.

Information has rapidly accumulated on the localization and the structure of the promoter enhancer elements that control the transcription of trypsin genes in the mosquito gut. The sequence of the putative transcription regulatory region upstream to one of the *Ae. aegypti* late trypsin genes was the first described (Barillas-Mury and Wells, 1993). The analysis of the DNA region upstream to the translation start site of this trypsin gene revealed the presence of several DNA binding sites for known transcription regulatory proteins. A typical TATA box was found upstream to the translation start codon together with sequences matching the DNA binding sequence of the GCN4 protein (Barillas-Mury and Wells, 1993). The presence of GCN4 binding sites may be of functional significance as this protein was shown in yeast to promote the transcription of several genes in response to different loads of amino acids. This finding indirectly supports the suggestion that in the mosquito gut the early trypsin activity may generate signals acting directly on specific transcription factors to activate the transcription of the late trypsin genes (Barillas-Mury *et al.*, 1995).

In *An. gambiae* the trypsin genes cluster together in a single genetic locus of approximately 11 000 nucleotides. Sequencing of the entire gene cluster (Müller *et al.*, 1993, 1995) revealed that the distance between the coding regions varies between 300 and 1500 nucleotides. The genetic organization of the locus suggests that the putative regulatory sequences are most likely placed within the short non-coding sequences upstream and downstream of the trypsin genes. The analysis of the non-coding DNA sequences upstream to the early and the late trypsin genes failed to detect typical binding sites for known transcription control factors. Moreover, the DNA sequences upstream to the *Anopheles* trypsin genes did not show any similarity with the putative regulatory sequences of the *Ae. aegypti* late trypsin gene. Sequence alignment revealed the presence of a conserved sequence motif upstream of all *Anopheles* trypsin genes. The position of this motif with respect to the transcription start site of the trypsin genes ranged between nucleotide −57 and −112. Blood induced and constitutive trypsin genes differed in the presence of a palindromic sequence upstream of this motif (Müller *et al.*, 1995; Skavdis *et al.*, 1996). The transcriptional activity of the DNA sequences upstream and downstream of the late (*Antryp1* and *Antryp2*) and of the most strongly expressed early trypsin gene, *Antryp4*, were investigated both *in vivo* and *in vitro*. Experimental evidence from transgenic *Drosophila melanogaster* lines indicated that the DNA sequences upstream to both *Antryp1* and *Antryp2* can direct the constitutive expression of the reporter gene *ß*-galactosidase in the midgut of the transgenic flies (Skavdis *et al.*, 1996). Deletion analysis indicates that the putative regulatory DNA sequence of *Antryp1* extends from nucleotide −360 to −99 and that it contains the transcription control elements that confer tissue specificity in adult transgenic *Drosophila* (Skavdis *et al.*, 1996). These findings do not allow an inference to be made for the function of the conserved sequence found upstream to all trypsin genes as the transcriptional active element of *Antryp1* mapped upstream to this DNA motif. Constructs containing both the upstream and the downstream sequence of the early trypsin gene *Antryp4* failed to induce the expression of the reporter gene in the gut as well as in all the tissues of transgenic flies (C. Louis, personal communication). The results of the experiments indicate that in transgenic flies the *Anopheles* trypsin promoters show the same transcriptional activity pattern observed in mosquitoes after blood feeding: the transcription of early trypsin genes is suppressed whereas the late trypsin genes are induced. Transient transformation experiments indicate that the constructs containing the putative regulatory sequences of *Antryp1*, *Antryp2* and *Antryp4* induce the expression of a reporter gene in *Ae. aegypti* MOS20 cells. The levels of enzymatic activity (luciferase) induced in MOS20 cells by different constructs reflected the relative amounts of *Antryp1*, *Antryp2* and *Antryp4* transcripts that could be detected in unfed female mosquito guts (Crisanti and Müller, manuscript in preparation). Thus the level of luciferase activity induced by the upstream sequence of *Antryp4* was much higher than that induced by the corresponding sequences of *Antryp1*, *Antryp2*. Deletion analysis indicated that a DNA sequence extending from nucleotide −89 to −123 upstream to the transcription start site of *Antryp4* was involved in the transcription of the reporter gene (Crisanti and Müller, unpublished results). This region immediately precedes the conserved motif (−68 to −89) that is found upstream to all *Anopheles* trypsin genes thus indicating that it may also participate in the control of trypsin gene transcription.

The results of the *in vitro* and *in vivo* transformation experiments taken together allow the formulation of a transcription control model that may explain the expression of early and late trypsin genes in a tissue specific and time co-ordinated manner in *An. gambiae*. According to this model the conserved DNA motif found upstream to all the trypsin genes should function as a binding sequence for regulatory proteins. The presence of such proteins would exert opposite effects on the transcription of the early and the late trypsin genes. The presence of one copy of this DNA motif within the upstream sequences of the early trypsin genes would allow the binding of the regulatory proteins in a single orientation. This interaction may promote the transcription of the corresponding trypsin genes. The late trypsin promoters upstream of the conserved DNA motif contain an almost perfect palindromic sequence. The binding of the regulatory proteins in the opposite orientation may result in suppression of gene transcription. Early and late trypsin promoters should also contain, upstream to the shared DNA motif, a tissue responsive element that activates the transcription of the trypsin genes only in the mosquito gut cells. The experimental evidence so far gathered on the *in vivo* and *in vitro* activity of the trypsin promoters supports the proposed transcriptional model.

It can be anticipated that the tools of cell and molecular biology will allow rapid advances in our understanding of the processes in the control of digestive enzyme synthesis and secretion.

REFERENCES

Anderson, M. and Cochrane, D.G. (1978) Studies on the midgut of the locust *Schistocerca gregaria*. II Ultrastructure of the muscle coat and its innervation. *J. Morphol.*, **156**, 257–78.

Barillas-Mury, C., Graf, R., Hagedorn, H.H. and Wells, M.A. (1991) cDNA and deduced amino acid sequence of a blood meal-induced trypsin from the mosquito, *Aedes aegypti*. *Insect Biochem. Mol. Biol.*, **21**, 825–31.

Barillas-Mury, C., Noriega, F.G. and Wells, M.A. (1995) Early trypsin activity is part of the signal transduction system that activates transcription of the late trypsin gene in the midgut of the mosquito, *Aedes aegypti*. *Insect Biochem. Mol. Biol.*, **25**, 241–6.

Barillas-Mury, C. and Wells, M.A. (1993) Cloning and sequencing of the blood meal-induced late trypsin gene from the mosquito *Aedes aegypti* and characterization of the upstream regulatory region. *Insect Mol. Biol.*, **2**, 7–12.

Billingsley, P.F. and Hecker, H. (1991) Blood digestion in the mosquito, *Anopheles stephensi* Liston (Diptera: Culicidae), activity and distribution of trypsin, aminopeptidase and alpha-glucosidase in the midgut. *J. Med. Entomol.*, **28**, 865–71.

Blakemore, D., Lehane, M.J. and Williams, S. (1993) Cyclic AMP can promote the secretion of digestive enzymes in *Stomoxys calcitrans*. *Insect Biochem. Mol. Biol.*, **23**, 331–5.

Blakemore, D., Williams, S. and Lehane, M.J. (1995) Protein stimulation of trypsin secretion from the opaque zone midgut cells of *Stomoxys calcitrans*. *Comp. Biochem. Physiol.*, **110B**, 301–7.

Borovsky, D., Carlson, D.A., Griffin, P.R. *et al.* (1990) Mosquito oostatic factor: a novel decapeptide modulating trypsin-like enzyme biosynthesis in the midgut. *FASEB J.*, **4**, 3015–20.

Borovsky, D., Carlson, D.A., Griffin, P.R. *et al.* (1993) Mass-spectrometry and characterisation of *Aedes aegypti* trypsin modulating oostatic factor (TMOF) and its analogs. *Ins. Biochem. Mol. Biol.*, **23**, 703–12.

Briegel, H. and Lea, A.O. (1975) Relationship between protein and proteolytic activity in the midgut of mosquitoes. *J. Insect Physiol.*, **21**, 1597–604.

Briegel, H. and Lea, A.O. (1979) Influence of the endocrine system on tryptic activity in female *Aedes aegypti*. *J. Insect Physiol.*, **25**, 227–30.

Brown, M.R., Crim, J.W. and Lea, A.O. (1986) FMRFamide- and pancreatic polypeptide-like immunoreactivity of endocrine cells in the midgut of a mosquito. *Tissue Cell*, **18**, 419–25.

Champlain, R.A. and Fisk, F.W. (1956) The digestive enzymes of the stablefly *Stomoxys calcitrans* (L.). *Ohio J. Sci.*, **56**, 52–6.

Day, M.F. and Powning, R.F. (1949) A study of the processes of digestion in certain insects. *Aust. J. Sci. Res. B*, **2**, 175–215.

Detinova, T.S. (1962) Age-grouping methods in Diptera of medical importance. World Health Organization monograph **47**, Geneva, 216 pp.

Endo, Y., Iwanga, T. and Fumita, T. (1990) Gut endocrine cells of invertebrates, in *Progress in Comparative Endocrinology* (eds A. Epple, C.G. Scanes and M.H. Stetson), Wiley-Liss, New York, pp. 499–503.

Engelmann, F. (1969) Food-stimulated synthesis of intestinal proteolytic enzymes in the cockroach *Leucophaea maderae*. *J. Insect Physiol.*, **15**, 217–35.

Engelmann, F. and Wilkens, J.L. (1969) Synthesis of digestive enzyme in the fleshfly *Sarcophaga bullata* stimulated by food. *Nature*, **222**, 798.

Felix, C.R., Betschart, B., Billingsley, P.F. and Freyvogel, T.A. (1991) Post-feeding induction of trypsin in the midgut of *Aedes aegypti* L. (Diptera: Culicidae) is separable into two cellular phases. *Insect Biochem.*, **21**, 197–203.

Fujita, T., Kanno, T. and Kobayashi, S. (1988) *The Paraneuron*, Springer, Heidelberg.

Garcia, E. de S. and Garcia, M.L.M. (1977) Control of protease secretion in the intestine of fifth instar larvae of *Rhodnius prolixus*. *J. Insect Physiol.*, **23**, 247–51.

Gooding, R.H. (1977) Digestive processes of haematophagous insects. XII. Secretion of trypsin and carboxypeptidase B by *Glossina morsitans morsitans* Westwood (Diptera: Glossinidae).*Can. J. Zool.*, **55**, 215–22.

Graf, R. and Briegel, H. (1989) The synthetic pathway of trypsin in the mosquito *Aedes aegypti* L. (Diptera: Culicidae) and *in vitro* stimulation in isolated midguts. *Insect Biochem.*, **19**, 129–37.

Hori, K., Araki, S. and Kuramochi, K. (1982) Trypsin-like enzyme and aminopeptidase in the midguts of the adult horn fly, *Hematobia irritans* and of the stable fly, *Stomoxys calcitrans*: change of activity in relation to blood ingestion and age. *Entomol. Exp. App.*, **31**, 421–7.

Hörler, E. and Briegel, H. (1995) Proteolytic enzymes of female *Anopheles*: biphasic synthesis, regulation, and multiple feeding. *Arch. Insect Biochem. Physiol.*, **28**, 189–205.

Houseman, J.G., Aresta, F., Mark, W.A. and Morrison, P.E. (1988) Effect of fractionated blood components on trypsin activity in the stable fly *Stomoxys calcitrans* (L.) (Diptera: Muscidae). *Can J. Zool.*, **66**, 1188–90.

Houseman, J.G. and Downe, A.E.R. (1983) Activity cycles and the control of four digestive proteinases in the posterior midgut of *Rhodnius prolixus* Stal (Hemiptera: Reduviidae). *J. Insect Physiol.*, **29**, 141–8.

Houseman, J.G., Downe, A.E.R. and Morrison, P.E. (1985) Similarities in digestive proteinase production in *Rhodnius prolixus* (Hemiptera: Reduviidae) and *Stomoxys calcitrans* (Diptera: Muscidae). *Insect Biochem.*, **15**, 471–4.

Jenkins, A.C., Brown, M.R. and Crim, J.W. (1989) FMRF-amide immunoreactivity and the midgut of the corn earworm (*Heliothis zea*). *J. Exp. Zool.*, **252**, 71–8.

Jimenez, D.R. and Gilliam, M. (1989) Age-related changes in midgut ultrastructure and trypsin activity in the honey bee, *Apis mellifera*. *Apidologie*, **20**, 287–303.

Kalhok, S.E., Tabak, L.M., Prosser, D.E. *et al.* (1993) Isolation, sequencing and characterization of two cDNA clones coding for trypsin-like enzymes from the midgut of *Aedes aegypti*. *Insect Mol. Biol.*, **2**, 71–9.

Lehane, M.J. (1977) An hypothesis of the mechanism controlling proteolytic digestive enzyme production levels in *Stomoxys calcitrans*. *J. Insect Physiol.*, **23**, 713–15.

Lehane, M.J. (1996) The midgut, in *The Microscopic Anatomy of Invertebrates – Insect Structure* (ed. M. Locke), John Wiley, Chichester.

Lehane, M.J., Blakemore, D., Williams, S. and Moffatt, M.R. (1995) Minireview: Regulation of digestive enzyme secretion in insects. *Comp. Biochem. Physiol.*, **110B**, 285–9.

Müller, H.-M., Catteruccia, F., Vizioli, J. *et al.* (1995) Constitutive and blood meal-induced trypsin genes in *Anopheles gambiae*. *Exp. Parasitol.*, **81**, 371–85.

Müller, H.-M., Crampton, J.M., della Torre, A. *et al.* (1993) Members of a trypsin gene family in *Anopheles gambiae* are induced in the gut by blood meal. *EMBO J.*, **12**, 2891–900.

Muraleedharan, D. and Prabhu, V.K.K. (1979) Role of the median neurosecretory cells in secretion of protease and inverase in the red cotton bug, *Dysdercus cingulatus*. *J. Insect Physiol.*, **25**, 237–40.

Noriega, F.G., Pennington, J.E., Barillas-Mury, C. *et al.* (1996) *Aedes aegypti* midgut early trypsin is post-transcriptionally regulated by blood-feeding. *Insect Mol. Biol.*, **5**, 25–9.

Ramos, A., Mahowald, A. and Jacobs-Lorena, M. (1993) Gut-specific genes from the black fly *Simulium vittatum* encoding trypsin-like and carboxypeptidase-like proteins. *Insect Mol. Biol.*, **1**, 149–63.

Rounds, H.D. (1968) Diurnal variation in the effectiveness of extracts of cockroach midgut in the release of intestinal proteinase activity. *Comp. Biochem. Physiol.*, **25**, 1125–8.

Schneider, F., Houseman, J.G. and Morrison, P.E. (1987) Activity cycles and the regulation of digestive proteases in the posterior midgut of *Stomoxys calcitrans* (L.) (Diptera: Muscidae). *Insect Biochem.*, **17**, 859–62.

Sehnal, F. and Zitnan, D. (1990) Endocrines of insect gut, in *Progress in Comparative Endocrinology* (eds A. Epple, C.G. Scanes and M.H. Stetson), Wiley-Liss, New York, pp. 510–15.

Skavdis, G., Sidén-Kiamos, I., Müller, H.-M. *et al.* (1996) Conserved function of *Anopheles gambiae* midgut-specific promoters in the fruitfly. *EMBO J.*, **15**, 344–50.

Sreekumar, S. and Prabhu, V.K.K. (1988) Probable endocrine role of midgut tissue in stimulation of digestive enzyme secretion in *Oryctes rhinoceros* (Coleoptera: Scarabidae). *Proc. Indian Acad. Sci.*, **97**, 73–8.

Stiles, J.K., Wallbanks, K.R. and Molyneux, D.H. (1991) The use of casein substrate gels for determining trypsin-like activity in the midgut of *Glossina palpalis* spp. *J. Insect Physiol.*, **37**, 247.

Zitnan, D., Sauman, I. and Sehnal, F. (1993) Peptidergic innervation and endocrine cells of insect midgut. *Arch. Insect Biochem. Physiol.*, **22**, 113–32.

8

Compartmentalization of digestion

W.R. Terra, C. Ferreira and J.E. Baker

8.1 INTRODUCTION

Biochemical and physiological processes that result in digestion of nutrient polymers in the insect midgut are arranged both spatially and temporally. This spatial and temporal separation is a result of morphological features of the gut tract combined with fluid fluxes that occur within the midgut. These features effectively compartmentalize buffering mechanisms responsible for pH and redox potential, enzymatic activity, absorption of nutrients, as well as excretory functions of the gut to specific regions and allow an efficient, sequential breakdown of food polymers into utilizable nutrients. In this chapter we will review evidence for this spatial arrangement of digestion in the midgut. This review will also stress the importance of an evolutionary or phylogenetic approach to an understanding of the many convergent and divergent morphological and biochemical features responsible for digestion in insects.

The importance of compartmentalization of digestive processes in insects has been realized only recently. Qualitative and quantitative aspects of digestive enzymes have been studied for over a century (Wigglesworth, 1972). With the advent of advanced analytical techniques there has been excellent recent progress in the isolation and characterization of many of these proteins (Chapter 6). In addition, current emphasis on the insect midgut as the major interface between the insect

Biology of the Insect Midgut.
Edited by M.J. Lehane and P.F. Billingsley.
Published in 1996 by Chapman & Hall, London.
ISBN 0 412 61670 X.

and its trophic environment has stimulated an increase in overall research effort to understand physiological processes in the midgut and their importance in nutrient conservation. Nevertheless, despite decades of early research on digestion in insects, it was not until 1979 that the significance and importance of midgut compartments in regulating the initial, intermediate, and final stages of polymer digestion was demonstrated (Terra *et al.*, 1979). Since that time, however, significant progress has been made in our understanding of compartmentalization and its importance to the overall digestive process in insects. We have organized this chapter to review this progress from an evolutionary perspective as follows: (a) phylogenetic relationships of the major insect orders are presented and briefly reviewed; (b) the organization of digestion based on dietary habits is compared with an organization based on phylogenetic position; (c) spatial organization and the importance of the peritrophic matrix in the compartmentalization process is discussed; (d) apparent evolutionary trends of digestive systems are presented; and finally (e) compartmentalization of digestion in the major insect orders is compared.

8.2 PHYLOGENY AND EVOLUTION OF INSECTS

A brief review of the phylogenetic and evolutionary relationships and pathways among the major insect orders and families is presented below. For a more detailed and inclusive discussion of this aspect of insect biology see Kristensen (1981), Lawrence and Newton (1982), Schuch (1986), Terra (1988, 1990), and Thorne and Carpenter (1992).

All the winged insects except for Ephemeroptera (mayflies) and Odonata (dragonflies) are included in the Neoptera (Figure 8.1). Neoptera evolved along three main lines: (1) Polyneoptera (Orthoptera, Dictyoptera, Isoptera); (2) Paraneoptera (Hemiptera, Thysanoptera); and (3) Holometabola. Holometabola evolved along two main branches: one branch includes Coleoptera and the other branch includes Hymenoptera and the panorpoid orders (Trichoptera, Diptera, Lepidoptera).

8.2.1 Polyneoptera

Ancestors of Orthoptera gave rise to the suborder Ensifera (Gryllidae, crickets; Gryllotalpidae, mole crickets) and later to the suborder Caelifera (Acrididae, grasshoppers and locusts). Dictyoptera comprise two suborders: Blattaria (cockroaches) and Mantodea (mantids) and is considered to be a sister group of Isoptera. Isoptera (termites) are thought to have evolved from wood-feeding organisms similar to present-day *Cryptocercus punctulatus*. The lower termites include the drywood termites (Kalotermitidae), whereas the higher termites include

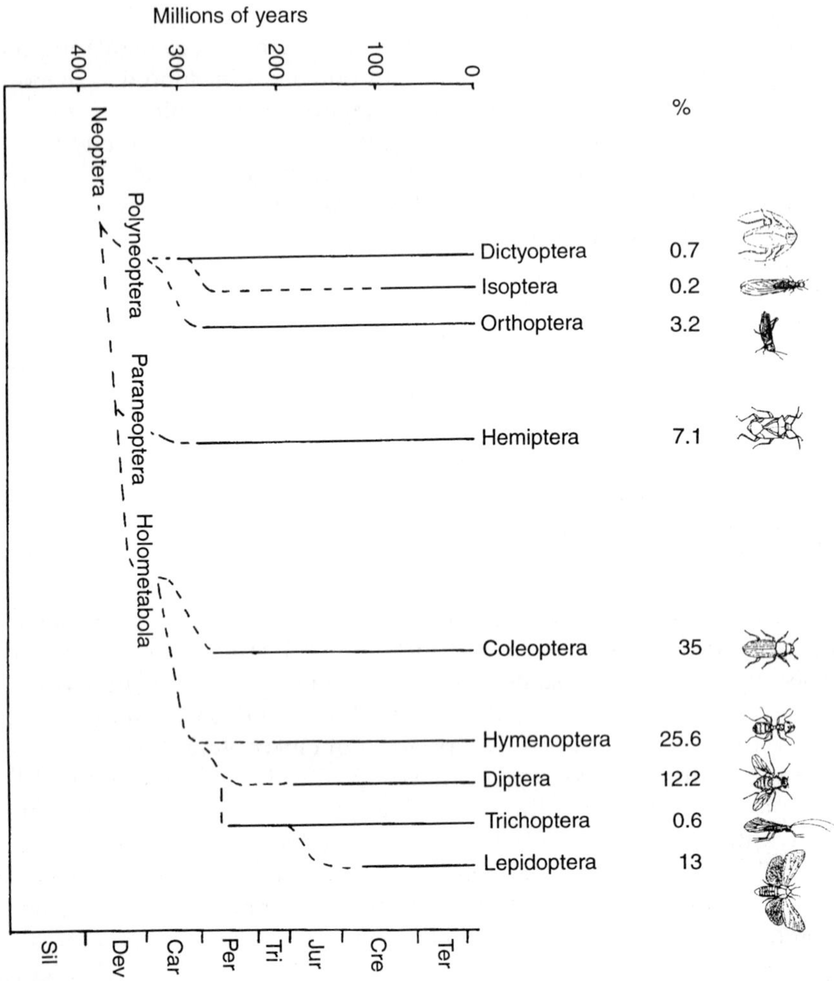

Figure 8.1 Phylogenetic relationships of the major insect orders and approximate numbers of living species (as percentage of total (860 000) described species). Solid lines show branches after they first appear in the fossil record; broken lines show probable geological ranges based on indirect evidence. Car, Carboniferous; Cre, Cretaceous; Dev, Devonian; Jur, Jurassic; Per, Permian; Sil, Silurian; Ter, Tertiary; Tri, Triassic. (Based on Terra, 1988.)

several subfamilies (e.g. Termitinae, Macrotermitinae) belonging to the family Termitidae.

8.2.2 Paraneoptera

Hemiptera comprise two suborders: Homoptera and Heteroptera. Ancestors of the entire order are thought to be sap-sucking insects similar to present-day Homoptera. Homoptera are divided among the minor, primitive series Colerrhyncha and two major series: Auchenorryncha (Fulgoridae, fulgorids; Cercopidae, spittle bugs; Cicadidae, cicadas; Cicadellidae, leafhoppers); and Sternorrhyncha (Coccidae, scale insects; Aphididae, aphids). Heteroptera includes the important infraorders Cimicomorpha (Cimicidae, bed bugs; Reduviidae, assassin bugs) and Pentatomomorpha (Lygaeidae, seed bugs; Pyrrhocoridae, cotton stainers; Coreidae, squash bugs; Pentatomidae, stink bugs).

8.2.3 Holometabola

The Coleoptera ancestor gave rise to the major suborder Adephaga (Carabidae, ground beetles), two minor suborders, and later to the major suborder Polyphaga. The ancestor of Polyphaga gave rise to two branches. The more primitive (series Staphyliniformia) contained the Staphylinidae (rove beetles) and Silphidae (carrion beetles). The second branch gave rise in turn to two subsequent branches: series Elateriformia (Scarabaeidae, scarab beetles; Buprestidae, flatheaded wood borers; Lampyridae, fireflies; Elateridae, click beetles) and the series Bostrychiformia–Cucujiformia. Members of the Bostrychiformia include Anobiidae (furniture beetles) and Dermestidae (dermestid beetles), whereas the Cucujiformia include Meloidae (blister beetles), Cucujidae (cucujids), Coccinellidae (ladybird beetles), Tenebrionidae (darkling beetles), and the principal groups of plant-feeding beetles (Cerambycidae, longhorned beetles; Bruchidae, seed beetles; Chrysomelidae, leaf beetles; Scolytidae, bark beetles; Curculionidae, weevils).

Hymenoptera comprise two suborders: Symphyta (sawflies and horntails) and Apocrita. A line presumably arising from a sawfly ancestor evolved into stem borers, comprising the group called horntails including Siricidae (wood wasps). Apocrita are thought to have arisen from the base of the horntail line. Larvae of Apocrita have a midgut which is closed as its posterior end, and which remains unconnected with the hindgut until pupation. From the ancestor of the primitive Aculeata, two branches originated: the Vespoidea and Sphecoidea. The Vespoidea gave rise to Formicidae and Vespidae, whereas the Sphecoidea–Apoidea branch gave rise first to Sphecoidea and then to Apoidea.

Diptera comprise three suborders: Nematocera, Brachycera and Cyclorrhapha. The Nematocera ancestor gave rise to two branches. The more primitive contained the Sciaridae (fungus gnats) and Cecidomyidae (gall midges). The other branch gave rise to Tipulidae (crane flies) and later to Culicidae (mosquitoes). The phylogenetic relationships of Simuliidae (blackflies) and Chironomidae (midges) are unknown. The Brachycera ancestor evolved along two branches: Asilidae (robber flies) and Tabanidae (horse flies). The Cyclorrhapha ancestor gave rise to the branches Aschiza (Syrphidae, flower flies) and Schizophora. The Schizophora gave rise to the lineage Acalyptratae (Tephritidae, fruit flies; Agromyzidae, leaf miner flies; Drosophilidae, vinegar flies) and the lineage Calyptratae (Muscidae, house flies; Gasterophilidae, horse bots; Calliphoridae, blowflies, Sarcophagidae, fleshflies; Tachinidae, tachina flies).

Zeugloptera is the most primitive suborder of Lepidoptera. From the base of the Zeugloptera stem, several branches evolved from which the suborder Ditrysia, containing the great majority of Lepidoptera species, gave rise to two lineages: one including the Tineidae (clothes moths) and the other leading to the tortricoid ancestor. This latter ancestral group gave rise to the Pyraloidea, Cossoidea, Tortricoidea, Castnioidea, and others. From Pyraloidea ancestors evolves the Pyralidae (pyralid moths), Geometridae (geometrid moths), Noctuidae (owlet moths) and the Sphingidae (sphinx moths). Cossoidea ancestors gave rise to Bombycidae (including the silkworm *Bombyx mori*) and Saturniidae (giant silkworm moths); Tortricoidea ancestors originated Tortricidae (leaf roller moths) and Castnioidea ancestors, the butterflies which include Pieridae (whites) and Papilionidae (swallowtail butterflies).

8.3 DIETARY HABITS VERSUS PHYLOGENY IN DIGESTIVE PHYSIOLOGY

By early convention, most studies of digestive processes in insects organized species by the types of diets they consume. Although this approach provides a useful framework in which to organize some current physiological data, it hinders characterization of groups with obvious similarities among their gut morphologies but with different feeding habits. For example, despite differences in feeding habits, cockroaches (generalists), grasshoppers (solid/plant feeders) and mantids (solid/animal feeders) generally have a large crop and relatively short midgut with diverticula (midgut caeca) at the anterior end. (Figure 8.2 gives a generalized diagram of the insect gut, also see Chapter 2.) As another example, the dietary approach to explaining gut morphology does not account for the presence or absence of midgut caeca in different insects, nor does it provide an explanation for the absence of a

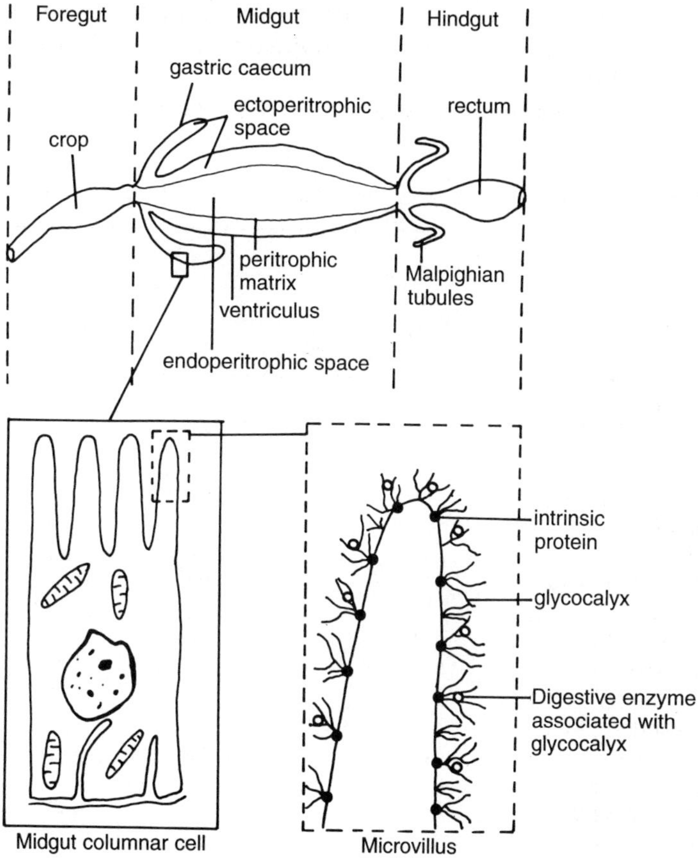

Figure 8.2 Diagrammatic representation of insect gut compartments. Glycocalyx: the carbohydrate moiety of intrinsic proteins and glycolipids occurring in the luminal face of microvillar membranes.

peritrophic matrix in all Hemiptera (see below), despite their contrasting dietary habits.

On physiological grounds, the dietary approach also does not explain a number of discrepancies between divergent insect groups that have a common food source. A comparison between the blood feeders (liquid/animal feeders) *Rhodnius prolixus* (Hemiptera: Reduviidae) and mosquitoes (Diptera: Culicidae) illustrates this point. Blood ingested by *R. prolixus* is stored in the anterior midgut and is digested in the posterior midgut by cathepsin-like enzymes in the absence of a peritrophic matrix, but with midgut cells showing perimicrovillar membranes (section 8.6.4). In mosquitoes, blood is stored and digested in the posterior midgut by the action of trypsin across a peritrophic matrix (section 8.6.7). These differences cannot be attributed to diet but to an adaptation from rather different ancestors.

Table 8.1 shows luminal gut pH values from insects in which

Table 8.1 pH of gut contents of representative families of major insect orders

			Ventriculus		
Orders *Family*	*Foregut*	*Caeca*	*Anterior*	*Middle*	*Posterior*
Dictyoptera					
Blattidae (A)	5.8	–	6.1	–	6.7
Orthoptera					
Acrididae (A)	5.8	6.2	6.2	–	7.3
Hemiptera					
Miridae (A)	–		5.2	5.4	4.8
Pyrrhocoridae (A)	–		6.4	5.7	5.4
Coleoptera					
Carabidae (A)	5.9		6.4	–	6.6
Scarabaeidae (L)	8.2		8.2	10.4	10.5
Passalidae (A)	8.3		8.6	9.5	9.3
Dermestidae (A)	5.8		7.2	–	7.2
Bostrichidae (A)	–		6.0	–	7.1
Cucujidae (A)	4.9		6.4	–	6.1
Coccinelidae (A)	5.2		5.5	5.4	5.5
Tenebrionidae (A)	5.1		7.1	–	6.3
Tenebrionidae (L)	6.0		5.6	–	7.9
Cerambycidae (L)	6.0		6.2	6.9	7.1
Chrysomelidae (L)	5.9		5.9	6.1	6.6
Curculionidae (A)	5.2		6.9	–	8.4
Hymenoptera					
Apidae (L)	–		6.0	5.7	5.6
Diptera					
Sciaridae (L)	–	7.2	9.5	9.4	8.7
Culicidae (L)	–	8.8	9.0	10.0	8.0
Simuliidae (L)	–	–	9.8	–	7.5
Muscidae (L)	–	–	6.1	3.1	6.8
Calliphoridae (L)	6.8	–	7.0	3.3	7.3
Tephritidae (L)	–	–	6.5	3.4	6.6
Lepidoptera					
Several families (L)	7.0		9.8	10.5	9.5

Data averaged within each family. A, adult stage; L, larval stage. A blank space indicates that the structure does not exist. Dash indicates values not determined. Lepidoptera families include: Sphingidae, Noctuidae and Pyralidae. (From Terra and Ferreira, 1994.)

determinations were performed in at least two different regions along their midguts. Although midgut pH is hypothesized to result from adaptation of an ancestral insect to a particular diet, its descendants may diverge, feeding on different diets, while still retaining the ancestral midgut pH condition. Thus it is not necessary that there is a relationship between midgut pH and diets. In fact, a relationship between midgut pH and insect phylogeny is more common. Nevertheless, there is generally a correlation between enzyme pH optima of α-amylases and luminal pH in insect guts. However, the pH optima of the other digestive enzymes (e.g. α-glucosidases, trehalases, aminopeptidases) is often different from that of the luminal contents (Chapter 6).

Redox conditions in the midgut are regulated and may be a result of phylogeny, although data are scarce. Reducing conditions are observed in clothes moths, sphinx moths, owlet moths and dermestid beetles (Appel and Martin, 1990) and in Hemiptera (Silva and Terra, 1994). Reducing conditions are important to open disulphide bonds in keratin ingested by some insects (clothes moths, dermestid beetles) (Appel and Martin, 1990), to maintain the activity of the major proteinase in Hemiptera (Chapter 6) and to reduce the impact of some plant allelochemicals, such as phenols, in some insect herbivores (Appel and Martin, 1990). Midgut detergency enhances protein solubility in the presence of tannins thus improving protein digestion (Felton and Duffey, 1991). Present knowledge is not sufficient to relate midgut detergency with diet or phylogeny or to both.

Types of digestive enzymes present in the alimentary canal of a given insect have been thought to depend on the chemical composition of their diet (Wigglesworth, 1972). A consequence of this view is the belief that adaptation to a feeding habit is more important than phylogenetic traits in determining the kinds of enzymes found in insect guts. Recent data have shown that the adaptation hypothesis is an oversimplification. The nectar-feeding adult blackfly *Cnephia docotensis* (Diptera: Simuliidae) displays high trypsin activity in the midgut (Yang and Davies, 1968), whereas many insects such as caterpillars and most grasshoppers that ingest leaves are unable to digest cellulose (Morgan, 1976). Furthermore, the inability to detect some enzyme activities in insect midguts (proteinases in adult moths and α-glucosidase in predatory beetles) has been shown to result from insensitive assays (Colepicolo-Neto *et al.*, 1986; Terra *et al.*, 1990). It is possible that most insect species have a full complement of digestive enzymes, and that only the relative amounts change in response to diet composition.

Data reviewed in this section challenge the view that insect digestion should be studied and organized only within dietary types, disregarding phylogenetic consideration. Furthermore, compartmentalization data on digestive enzymes suggest that the overall pattern of digestion, and the gross and ultrastructural morphology of the midgut, correlate well with phylogenetic position (Terra, 1988, 1990; Terra and Ferreira, 1994). Thus, adaptive features which are seen in insects within the same taxon but with different feeding habits seem to have evolved from the same basic digestive pattern. In contrast, insects with similar feeding habits, but within different taxa, show different spatial organizations of digestion, despite the existence of convergent features.

8.4 ORGANIZATION OF THE DIGESTIVE PROCESS

8.4.1 Overview

The organization of digestive processes depends on compartmentalization of digestive enzymes and on midgut fluid fluxes that are responsible for the translocation of enzymes and products of digestion. The absolute amount of each enzyme present in the midgut depends primarily on its secretory rate. Nevertheless, the relative distribution of digestive enzyme activities among different compartments is a property of each digestive system. Thus, despite secretory control of digestive enzymes, the spatial organization of enzymes is not affected by this process. Control of digestive enzyme secretion is discussed in Chapter 7.

Procedures used to study midgut fluid fluxes and compartmentalization of digestive enzymes were treated in detail in recent reviews (Dow, 1986; Terra and Ferreira, 1994). Properties of digestive enzymes and the nature and function of the peritrophic matrix have also been reviewed (Terra and Ferreira, 1994; Chapters 4 and 6). Thus, only a brief account will be presented here on the nature of midgut fluid fluxes and the roles played by the peritrophic matrix in digestion.

8.4.2 Spatial organization of digestion

Most food requiring digestion by insects consists of polymers such as proteins, starch, cellulose and hemicellulose. The digestive processes that degrade these polymers occur in three phases: initial, intermediate and final. The initial phase results in a decrease in molecular weight of polymeric food molecules through the action of polymer hydrolases, such as trypsin, α-amylase, cellulase and hemicellulase. The resulting oligomers then undergo further hydrolysis during the intermediate phase by polymer hydrolases exemplified by α-amylase acting on oligomaltodextrins or oligomer hydrolases exemplified by aminopeptidase acting on protein fragments. Products of this phase are dimers or small oligomers such as dipeptides, maltose and cellobiose derived from proteins, starch and cellulose, respectively. During the final phase of digestion dimers are split into monomers by dimer hydrolases, such as dipeptidase, maltase and cellobiase.

Describing the spatial organization of digestion in an insect consists of relating midgut compartments (Figure 8.2) to each of the three phases of digestion, and hence to the corresponding enzymes. To accomplish this, enzyme determinations must be performed in each midgut luminal compartment and in corresponding tissue. Cross-contamination of enzymes of one compartment with those of others should be evaluated. Assays in subcellular fractions of midgut cells should ideally include at least isolated microvilli and fractions enriched in glycocalyx-associated materials.

The first attempt to relate midgut compartments with each phase of digestion was accomplished with larvae of *Rhynchosciara americana* (Diptera: Nematocera) (Terra *et al.*, 1979). The results showed that trypsin and α-amylase are secreted into the ectoperitrophic space and diffused into the endoperitrophic space, whereas aminopeptidase and trehalase are secreted into the ectoperitrophic space but are unable to penetrate through the peritrophic matrix. Disaccharidases (except trehalase) occur in midgut caeca cell microvilli and are also found in minor amounts in posterior ventricular cell microvilli (Ferreira and Terra, 1980). Aminopeptidase was both bound to the microvillar membranes and present as a soluble enzyme trapped in the cell glycocalyx in caeca and the posterior ventriculus (Klinkowstrom *et al.*, 1994). These results led to the proposal that initial digestion occurs in the endoperitrophic space, whereas intermediate and final digestion occur in the ectoperitrophic space and midgut cells, respectively.

Although digestive enzymes were associated with the peritrophic matrix of *R. americana* and might have a physiological role in the initial phases of digestion in this species (Terra *et al.*, 1979), the enzyme activities were not quantified. The peritrophic matrix in *Spodoptera frugiperda* (Lepidoptera: Noctuidae) contains 13–18% of total midgut luminal activity of α-amylase and trypsin, respectively (Ferreira *et al.*, 1994).

Excretion rates of digestive enzymes by *R. americana* are surprisingly low (Terra and Ferreira, 1981), suggesting that at least 80% of the enzymes present in the endoperitrophic space are removed from the food bolus before it is expelled. A hypothetical endo-ectoperitrophic circulation of digestive enzymes may explain these results (Terra and Ferreira, 1981). As the food bolus moves in the endoperitrophic space from the anterior to the posterior region, water flows in the ectoperitrophic space from the posterior midgut to the anterior caeca (Figure 8.3c). Because of these fluxes, as soon as the polymeric molecules of food become sufficiently small to traverse the peritrophic matrix (with accompanying polymer hydrolases), they are displaced toward the

(a)

(b)

(c)

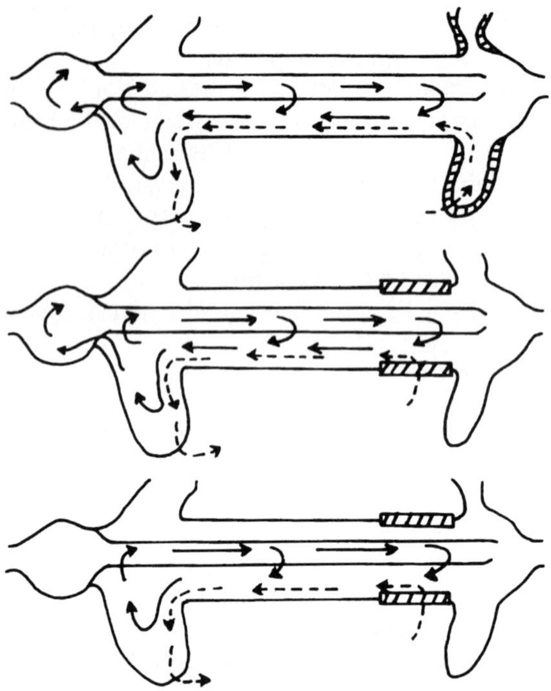

Figure 8.3 Diagrammatic representation of water fluxes (dotted arrows) and of the circulation of digestive enzymes (solid arrows) in putative insect ancestors. Cross-hatched areas correspond to fluid-secreting regions. See Figure 8.2 for gut compartments. In neopteran ancestors (a), midgut digestive enzymes pass into the crop. Countercurrent fluxes depend on the secretion of fluid by the Malpighian tubules and its absorption by the caeca. Holometabolan ancestors (b) are similar, except that secretion of fluid occurs in posterior ventriculus. Panorpoid ancestors (c) display countercurrent fluxes like holometabolan ancestors, midgut enzymes are not found in the crop, and only the polymer hydrolases pass through the peritrophic membrane. (From Terra, 1990.)

caeca, where intermediary and final digestion occur. From the caeca, polymer hydrolases may diffuse into the anterior endoperitrophic space, resulting in a new cycle of the endo-ectoperitrophic circulation of polymer hydrolases. Ultrastructural data on *R. americana* larval midgut (Ferreira *et al.*, 1981) support the above-mentioned model by describing adsorptive anterior midgut caecal cells and apparent secretory cells in the posterior midgut. This organization of digestion is typical of Diptera–Nematocera and modifications to it are observed in other taxa.

8.4.3 Peritrophic matrix and digestive physiology

The midgut lumen is divided into two compartments by the peritrophic matrix. This structure consists of a chitin network in a protein–carbohydrate matrix (Chapter 4). The capacity of the peritrophic matrix to compartmentalize the luminal digestive enzymes and ingested foods depends on its selective permeability, both to hydrolases and products of the stepwise degradation of nutrient polymers. Pore sizes of 7–9 nm diameter were observed in the peritrophic matrices of different insect species (Terra and Ferreira, 1994; Chapter 4).

The physiological roles of the peritrophic matrix are controversial. There is growing evidence that the peritrophic matrix provides mechanical protection for midgut cells, serves as a physical barrier for micro-organisms (Chapter 4), and has multiple functions in digestion (Terra, 1990; Terra and Ferreira, 1994).

Roles played by the peritrophic matrix in digestion include: (a) prevention of non-specific binding of undigested material on to membrane-bound hydrolases and/or transporting (active in absorption) proteins; (b) prevention of enzyme excretion by allowing the endo-ectoperitrophic circulation of digestive enzymes; and (c) permitting monomers to be produced from food oligomers in close proximity to the surface of midgut cells. Current data suggest that the peritrophic matrix of neopteran insects have functions (a) and (b), whereas function (c) is demonstrable only in the peritrophic matrix of panorpoid insects (see below).

Mosquitoes lacking peritrophic matrix are able to complete digestion of a blood meal (Billingsley and Rudin, 1992). In the absence of the peritrophic matrix, digestion is expected to be less efficient due to partial inhibition of microvillar hydrolases and/or transporting proteins, and also due to increased digestive enzyme excretion. Thus, unless the lack of peritrophic matrix is proved to have no effect on quantitative nutritional parameters (digestibility, efficiency of conversion of food into body mass and growth rate), there is no reason to suppose that functions of the peritrophic matrix in mosquitoes are different from those in other insects, as suggested by Billingsley and Rudin (1992).

8.5 EVOLUTIONARY TRENDS OF INSECT DIGESTIVE SYSTEMS

Research conducted primarily over the last decade has resulted in the recognition of several basic patterns of spatial organization of insect digestion (Terra *et al.*, 1985; Terra, 1988, 1990; Terra and Ferreira, 1994). These basic patterns are thought to correspond to putative ancestors from which specialized species evolved. The neopteran ancestors (Figure 8.3a) are thought to have the following properties in relation to their digestive physiology. First, digestive enzymes may pass forward from midgut to crop, which is the main site for digestion. Secondly, polymer, oligomer and dimer hydrolases are free and small (less than 7.5 nm in diameter) and are thus able to pass through the peritrophic matrix. Thirdly, the endo-ectoperitrophic circulation of digestive enzymes is driven by the secretion of fluid by Malpighian tubules and fluid absorption in the anteriorly placed midgut caeca. Fourthly, there is differentiation of an acidic anterior midgut (with a high carbohydrase activity) and an alkaline posterior midgut (with a high protease activity). This differentiation along the midgut may be an adaptation to instability of ancestral carbohydrases in the presence of proteinases.

Polyneopteran ancestors are thought to have retained characteristics of their neopteran ancestors. The same is true for ancestral Orthoptera, whereas the most remarkable derived character found in Dictyoptera is the enlargement of hindgut structures. These structures are thought to be associated with the utilization of refractory materials and are observed primarily in wood-feeding cockroaches. Isopteran ancestors have the following characteristics that are thought to be derived from their polyneopterous ancestors: (a) reduction of crop; (b) loss of caeca (retained in Mastotermitidae); and (c) differentiation of hindgut structures associated with the utilization of refractory materials.

Paraneopteran ancestors can not be discussed here because of a lack of sufficient data. Nevertheless, Condylognatha ancestors, a Paraneoptera taxon which includes Thysanoptera and Hemiptera, are thought to have the following characteristics which may be adaptations to phloem feeding: (a) loss of crop; (b) loss of polymer and oligomer hydrolases and of the peritrophic matrix associated with the lack of midgut luminal digestion; (c) loss of anterior caeca and of the endo-ectoperitrophic circulation of digestive enzymes; and (d) presence of perimicrovillar membranes that are responsible for the absorption of essential nutrients from dilute diet (section 8.6.4).

Digestive systems may change remarkably between larvae and adults of holometabolous insects. Despite these changes, adult digestive systems probably evolved in parallel to larval systems because, except for minor differences, the compartmentalization of digestion in larvae and adults seems to be similar (see below). Holometabolous ancestors (Figure 8.3b) are thought to differ from neopteran ancestors only in the site of fluid secretion in the gut. In holometabolous ancestors, the endo-ectoperitrophic circulation of digestive enzymes is presumed to be caused by the secretion of fluid in the posterior midgut and its absorption in midgut caeca. Since the posterior midgut fluid does not contain wastes, as is the case for Malpighian tubular fluid, the accumulation of wastes in caeca is decreased. There is an evolutionary trend leading to the loss of anterior midgut caeca in holometabolous insects and an increase in the use of anterior ventricular cells for water absorption. Caeca loss probably further decreases the accumulation of noxious substances in the caecal lumen, which would be more serious in insects that have high relative food consumption rates, such as is common among Holometabola.

Ancestors of the lower Homoletabola (Coleoptera and Hymenoptera) have not evolved dramatically from the holometabolan ancestor. Thus, coleopteran ancestors seem to have retained most characteristics of holometabolan ancestors, except for the anterior midgut caeca, which were lost and have been replaced in function by the anterior midgut. Nevertheless, evolved Coleoptera have several derived characters, some of which are similar to those found in panorpoid orders and probably confer greater efficiency to their digestive systems. They are: (a) a great reduction (or loss) of the crop; (b) enzymes involved in intermediate and final digestive phases are bound to the surface of midgut cells; (c) loss of differentiation of an acidic anterior midgut (with a high carbohydrase activity) and an alkaline posterior midgut (with a high protease activity). Another derived character is the presence of cathepsin-like proteinases primarily in species adapted to feed on seeds rich in trypsin inhibitors. Scarabaeidae larvae display the following characters which seem to be associated with cellulose degradation: (a) three rows of caeca along the midgut and (b) enlargement of hindgut structures.

The following characteristics of Hymenoptera ancestors are believed to have been derived from the putative holometabolan ancestors: (a) loss of anterior midgut caeca and (b) digestive enzymes that do not pass forward from midgut to crop. Derived characters found among evolved Hymenoptera include (a) loss of crop in the larval stage; and (b) the midgut is closed at its posterior end in larval Apocrita.

The evolution of panorpoid orders occurred through the occupation of ephemeral and mainly exposed habitat niches, which led to the appearance of several adaptations to assure insect survival. Among these adaptations, the most effective is probably the reduction in length of life cycle and resultant increases in growth rates and food consumption rates in panorpoid orders compared with other insects (Terra, 1988). In agreement with this, digestive systems in panorpoid insects are more sophisticated than those in other taxa.

Panorpoid ancestors (Figure 8.3c) are thought to differ from holometabolan ancestors in the following aspects: (a) digestive enzymes do not pass forward from midgut to crop; (b) loss of ancestral midgut differentiation in luminal pH and enzyme distribution; (c) oligomer and dimer hydrolases are unable to pass through pores of the peritrophic matrix. Consequently, initial digestion in these insects takes place in the endoperitrophic space and intermediate and final digestion occur in the ectoperitrophic space.

From the panorpoid ancestors, the ancestral Diptera and Lepidoptera diverged. Thus, dipteran ancestors are presumed to have possessed the following characteristics derived from their hypothetical panorpoid ancestors. Oligomer hydrolases are free in the ectoperitrophic fluid (mainly in the large caeca), whereas dimer hydrolases are plasma membrane proteins bound to the surface of midgut cell microvilli. The occurrence of dimer hydrolases in the plasma membrane of midgut cells may be seen as a device for releasing monomers close to the transport sites responsible for their absorption. Sciaridae flies, exemplified by *R. americana* (see previous sections), seem to have retained these characteristics. More evolved nematocerous larvae may show reduction in size of midgut caeca (Culicidae and Tipulidae) and enlargement of the hindgut associated with cellulose degradation (Terra, 1990).

Cyclorrhaphous ancestors are presumed to differ from dipteran ancestors in the following ways: (a) the presence in the middle midgut of specialized cells responsible for water absorption and for buffering the luminal contents in the acidic zone; (b) the presence of lysozyme and cathepsin D activity in middle midgut lumen; (c) endo-ectoperitrophic circulation of digestive enzymes in the posterior midgut caused by secretion of fluid in the posterior section of the posterior midgut and its absorption in the middle midgut; and (d) the oligomer hydrolases are microvillar enzymes. These derived characters displayed by cyclorrhaphous ancestors seem to be adaptations to a diet consisting mainly of bacteria.

Lepidopteran ancestors display the following characteristics derived from their hypothetical panorpoid ancestors: (a) anterior midgut cells that are involved in water absorption; (b) oligomer and dimer hydrolases that are trapped in the intermicrovillar glycocalyx space (except aminopeptidase, which is membrane bound); and (c) the presence of long-neck goblet cells and stalked goblet cells in the anterior and posterior larval midgut regions, respectively. Goblet cells excrete K^+ ions that are absorbed from leaves ingested by larvae (Dow, 1986). Goblet cells also seem to assist anterior columnar cells in water absorption and posterior columnar cells in water secretion (Terra, 1990).

8.6 DIGESTION IN THE MAJOR INSECT ORDERS

8.6.1 Orthoptera

Carbohydrate digestion in grasshoppers occurs primarily in the crop and utilizes enzymes passed forward from the midgut (Wigglesworth, 1972). The partially digested food bolus passes to the midgut and the dispersed material is probably directed to the anterior midgut caeca, where water absorption occurs (Dow, 1986). Protein digestion and final carbohydrate digestion should also occur there by the action of luminal and membrane-bound (at least part of the aminopeptidase) enzymes. Salivary enzymes play a minor role in digestion (Ferreira *et al.*, 1990a). Cellulase found in some grasshoppers is believed to facilitate the access of digestive enzymes to the plant cells ingested by the insects by degrading the cellulose framework of cell walls (Morgan, 1976). Midgut countercurrent flows are observed only in starving grasshoppers (Dow, 1986). This probably avoids excessive accumulation of noxious wastes in the caeca, and makes possible the high relative food consumption observed among locusts in their migratory phases.

Crickets and mole crickets are omnivorous or predatory insects. Their guts have an unusually large crop, a midgut divided into caeca, and an anterior (absent in mole crickets) and posterior ventriculus apparently covered with a cuticle. Despite several studies of digestive physiology with these insects (Terra, 1988), the only certainty is that most starch and protein digestion in crickets occurs in their crop (Teo and Woodring, 1988).

8.6.2 Dictyoptera

There are no data on the spatial organization of digestion in mantids, which are exclusively carnivorous. Digestion by cockroaches, which are generally omnivorous, occurs mainly in the crop with enzymes passed forward from the midgut (Bignell, 1981). Part of the final digestion may occur on the surface of midgut cells, since a fraction of the aminopeptidase in *Blabera gigantea* (Blaberidae) is found associated with midgut cell microvilli (Parenti *et al.*, 1986). *Panesthia cribrata* (Blaberidae) accumulates cellulase and other enzymes in the anterior midgut (Scrivener *et al.*, 1989). This and other indirect evidence (Terra, 1988) suggests that feeding cockroaches display an endo-ectoperitrophic circulation of digestive enzymes.

Cellulose and hemicellulose are digested in the colon by *Periplaneta americana* (Blattidae) by the action of bacteria associated with the luminal surface of this organ (Bignell, 1981). The products of bacterial fermentation are mainly short-chain fatty acids which pass through the hindgut wall. The wood-roach *Panesthia cribrata* feeds on dead wood and digests cellulose with its own cellulase (Scrivener *et al.*, 1989), whereas there is evidence that another wood-roach species, *Cryptocercus punctulatus* (Cryptocercidae), digests cellulose with a cellulase produced by a symbiotic protozoan (Cleveland *et al.*, 1934). Nevertheless, extrapolating from current data with Isoptera (Slaytor, 1992), it is very probable that cellulose digestion in *C. punctulatus* is accomplished through production of its own cellulase. Acetate and butyrate are transported to a significantly greater extent in hindguts of *P. cribrata* compared with those of *P. americana*, which agrees with the fact that *P. cribrata* depends primarily on wood for food (Hogan *et al.*, 1985).

8.6.3 Isoptera

Most evidence indicates that both lower and higher termites digest cellulose by using their own cellulase, despite the occurrence of cellulase-producing protozoa in the pounch, an enlarged region of the anterior hindgut in lower termites (Slaytor, 1992). Nitrogen is fixed in termites by the action of symbiotic bacteria present in the hindguts (Benemann, 1973), and is incorporated into the termite body mass after ingestion and digestion of faeces. This partly explains the ability of termites to develop successfully in diets very poor in protein.

Comprehensive data on the distribution of digestive enzymes in the gut system of lower termites are lacking. Termitidae hindgut is differentiated into five proctodeal segments (P_1 to P_5) (Bignell and Anderson, 1980). The mixed segment (a characteristic region between the midgut and hindgut) is responsible for an alkaline secretion (Bignell *et al.*, 1983). The high pH in this hindgut region could allow solubilization of materials present in undigested food (Terra, 1988), thus facilitating bacterial degradation and/or digestion of ingested faecal material. The fungus-growing higher termites (Macrotermitinae) feed on small amounts of a fungus they culture in their nests along with cellulose materials (Martin, 1987). Perhaps because Macrotermitinae are adapted to digest less refractory materials, they do not have the mixed segment in their gut. Protease activity in higher termites is restricted to the midgut, and most cellulase and α-amylase activities are also found in this region (Hogan *et al.*, 1988). These enzymes are stable and found predominantly in the midgut (Hogan *et al.*, 1988), so it is probable that feeding termites display an endo-ectoperitrophic circulation of digestive enzymes.

8.6.4 Hemiptera

Hemiptera lack a peritrophic matrix (Terra, 1988). Nevertheless, the microvillar membrane of most Heteroptera and Fulguroidea (Homoptera) midgut cells is ensheathed by an outer (perimicrovillar) lipid membrane which is almost free from intramembranous particles (Figure 8.4). The perimicrovillar membrane maintains a constant distance from the microvillar membrane, extends toward the luminal compartment with a dead end, and limits a closed compartment (the perimicrovillar space) (Reger, 1971; Lane and Harrison, 1979; Andries and Torpier, 1982; Baerwald and Delcarpio, 1983). The perimicrovillar membrane was not reported in Cicadoidea, Cercopoidea, Cicadelloidea and Aphididae

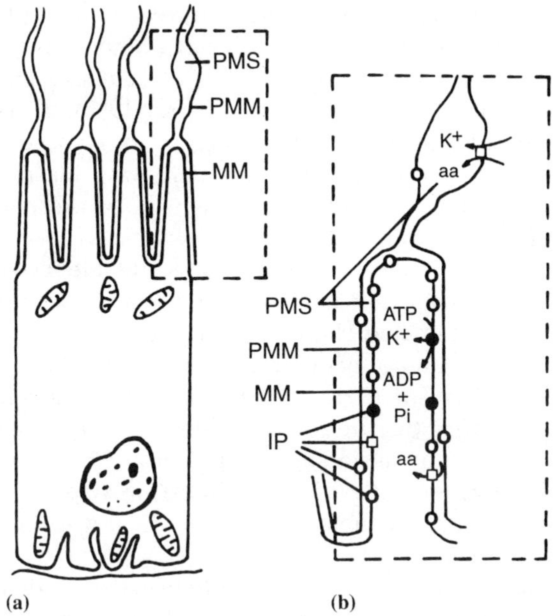

Figure 8.4 Model for the structure and physiological role of the microvillar border of midgut cells of Hemiptera. (a) Diagrammatic representation of a typical hemipteran midgut cell; (b) details of the apex. The microvillar membrane (MM) is ensheathed by the perimicrovillar membrane (PMM), which extends toward the luminal compartment with a dead end. The microvillar and perimicrovillar membranes delimit a closed compartment, i.e. perimicrovillar space (PMS). The microvillar membrane is rich and the perimicrovillar membrane is poor in integral proteins (IP). Microvillar membranes actively transport potassium ion (the most important ion in sap) from PMS into the midgut cells, generating a concentration gradient between the gut luminal sap and the PMS. This concentration gradient may be a driving force for the active absorption of organic compounds (amino acids (aa), for example) by appropriate carriers present in PMM. Organic compounds, once in the PMS, may diffuse up to specific carriers on the microvillar surface. This movement is probably enhanced by a transfer of water from midgut lumen to midgut cells, following (as solvation water) the trans-membrane transport of compounds and ions by the putative carriers. (From Terra and Ferreira, 1994.)

(O'Loughlin and Chambers, 1972; Marshall and Cheung, 1974; Lindsay and Marshall, 1980), but inspections of published photomicrographs and corresponding descriptions suggest that this membrane is also present in these groups.

Homoptera may suck phloem or xylem sap. Phloem fluid is rich in sucrose (0.15–0.73 M) and relatively poor in free amino acids (15–65 mM) and minerals, whereas xylem fluid is poor in amino acids (3–10 mM) and contains monosaccharides (about 1.5 mM), organic acids, potassium ions (about 6 mM) and other minerals (Terra, 1990). Thus, except for dimer (sucrose) hydrolysis, no food digestion is necessary in sap-suckers. The major problem facing a sap-sucking insect is to absorb nutrients, such as essential amino acids, which are present in very low concentrations. Insects other than Hemiptera never developed a sap-sucking habit, though many of them possess suitable mouthparts, so it is likely that the sap-sucking habit resulted from modifications of the alimentary canal that enable Hemipteran insects to absorb amino acids from dilute diets. Amino acids may be absorbed according to a hypothesized mechanism (Terra, 1988, 1990), that depends on perimicrovillar membranes (Figure 8.4). In phloem feeders such as aphids, this process may have an assimilation efficiency of 55% for amino acids but only 5% for sugars (Mittler, 1958), whereas in xylem feeders such as leafhoppers about 99% of dietary amino acids and carbohydrates are absorbed (Andersen *et al.*, 1989).

Organic compounds in xylem sap need to be concentrated before they are successfully absorbed by the perimicrovillar system. This occurs in the filter chamber (Chapter 1) of Cicadoidea and Cercopoidea which concentrates xylem sap about 10-fold (Cheung and Marshall, 1973). The filter chamber of Cicadelloidea (phloem feeders) seems to be secondarily simplified from a cercopoidean-type ancestral form and is able to concentrate dilute phloem about 2.5-fold (Lindsay and Marshall, 1981). Homopteran ancestors probably arose from insects adapted to scraping the surface of growing plants and sucking out phloem fluid, such as occurs in present-day Thysanoptera. The presence of perimicrovillar membranes in midgut cells of Thysanoptera supports this view (Del Bene *et al.*, 1991).

Digestion in Heteroptera Cimicomorpha is known mainly from studies with *R. prolixus* (Reduviidae). Blood ingested by *R. prolixus* is stored in the anterior midgut, where water is absorbed. Protein digestion begins in the posterior midgut lumen by the action of cathepsin-like proteinases (Houseman and Downe, 1980; Terra *et al.*, 1988). Oligopeptides resulting from haemoglobin digestion are transported into the perimicrovillar space to provide substrates for the aminopeptidase, whose products are further digested by dipeptidases and are subsequently absorbed (Billingsley and Downe, 1985; Ferreira *et al.*, 1988). Carbohydrate moieties of ingested glycoproteins are hydrolysed by the membrane-bound microvillar and perimicrovillar glycosidases (Ferreira *et al.*, 1988).

Among the infraorder Pentatomomorpha, studies have progressed furthest with the seed-sucking bug *Dysdercus peruvianus* (Pyrrhocoridae)

(Silva and Terra, 1994). The midgut of *D. peruvianus* is divided into three sections, V_1–V_3, which are linked through V_4 to the hindgut. The midgut contents are reducing with a pH of about 5.8. Most midgut carbohydrase activities are found in V_1, where water and glucose are absorbed independent of the presence of salts in the diet. In V_2, a cysteine proteinase predominates and amino acids are absorbed. This absorption is inhibited by sodium and putatively activated by potassium ions. Evidence indicates that digestion of proteins is completed in V_2.

8.6.5 Coleoptera

In predatory Carabidae, initial, intermediate and final digestion occur primarily in the crop by enzymes produced in the midgut (Crowson, 1981; Ferreira and Terra, 1989). In contrast, initial digestion occurs extra-orally in the predatory larvae of *Pyrearinus termitilluminans* (Elateridae) by action of enzymes present in a fluid that the larvae regurgitate onto their prey. In this species, intermediate and final digestion of pre-liquified material subsequently ingested by the larva takes place on the surface of midgut cells (Colepicolo-Neto *et al.*, 1986).

The entire digestive process seems to occur in the larval endo-peritrophic space of *Attagenus megatoma* (Dermestidae) (Baker, 1981). The same is true for *Tenebrio molitor* (Tenebrionidae) larvae (Terra *et al.*, 1985), except for the final digestion of proteins which takes place on the surface of midgut cells by the action of a microvillar aminopeptidase (Ferreira *et al.*, 1990b). Furthermore, some digestion occurs in the small crop of the mealworm through the action of digestive enzymes passed forward from the anterior midgut (Terra *et al.*, 1985). The distribution of enzymes in gut regions of adult *T. molitor* is similar to that of their larvae (Ferreira and Terra, unpublished). This suggests that the overall pattern of digestion in larvae and adults of Coleoptera is similar, despite the fact that in contrast to adults beetle larvae usually lack a crop.

Scarabaeidae and several related families are relatively isolated in the series Elateriformia. Scarabid larvae usually feed on cellulose materials undergoing degradation by a fungus-rich microflora. This degradation takes place in the midgut (Bayon, 1981), which has three rows of caeca, with a ventral groove between the middle and posterior row. The alkalinity of gut contents increases to almost pH 12 along the midgut ventral groove. This high pH probably enhances cellulose digestion which occurs mainly in the hindgut fermentation chamber (Bayon, 1981) through the probable action of bacterial cell-bound enzymes. The final product of cellulose degradation is mainly acetic acid which may pass through the hindgut wall (Bayon and Mathelin, 1980). There is evidence that contents of the fermentation chamber are periodically regurgitated forward into the midgut for further digestion and absorption (Crowson, 1981). This process makes the habit of feeding upon faecal material to obtain nitrogen, as described above for termites, unnecessary in scarab beetles.

Cellulose digestion in adult coccinellids seems to occur by the action of their own enzymes (Taylor, 1985). Cerambycidae larvae acquire the capacity to digest cellulose by ingesting fungal cellulases while feeding on fungus-ingested wood (Martin, 1987). Ingested fungal cellulase is maintained at high concentrations in Sciaridae larval midgut because its excretion is decreased by its endo-ectoperitrophic circulation (Espinoza-Fuentes *et al.*, 1984). Fungal enzymes are probably conserved in the cerambycid larval midgut by a similar mechanism.

Digestion in Curculionidae resembles digestion in Diptera and Lepidoptera. Thus, *Cylas formicarius* larvae have only a short foregut region with no distinct crop, and oligosaccharide digestion as well as oligopeptide digestion occurs mostly on the surface of midgut cells (Baker *et al.*, 1984). Insects of series Cucujiformia have cysteine proteinases in addition to (or in place of) serine proteinases as digestive enzymes (Chapter 6).

8.6.6 Hymenoptera

Woodwasp (Symphyta) larvae of the genus *Sirex* are able to digest and assimilate wood constituents by acquiring cellulase and xylanase, and possibly other enzymes, from fungi present in the wood on which they feed (Martin, 1987). The conservation of part of the ingested fungal enzymes in the *Sirex* midgut may be accomplished by their endo-ectoperitrophic circulation.

Adult *Apis mellifera* (Apidae, Apinae) ingest nectar and pollen. Sucrose from nectar is hydrolysed in the crop by the action of a sucrase from the hypopharyngeal glands. After ingestion, pollen grains extrude their protoplasm in the ventriculus, where digestion occurs (Klungness and Peng, 1984). Absorption sites for leucine and glucose have been found in the anterior two-thirds of *A. mellifera* adult ventriculus (Crailsheim, 1988). These data, together with a detailed ultrastructural and cytochemical study of adult *A. mellifera* midgut, led Jimenez and Gilliam (1990) to propose that enzymes and nutrients diffusing through the peritrophic matrix in these bees are translocated forward by a countercurrent flux. In *Scaptotrigona bipunctata* (Apidae, Meliponinae) larvae, except for a cysteine proteinase derived from ingested pollen, all digestive enzymes originate from the midgut tissue and are most active in the luminal contents. In addition, there is evidence suggesting the existence of an endo-ectoperitrophic circulation of digestive enzymes and nutrients, and that larval bees have lost the midgut luminal pH gradient hypothetically present in Hymenoptera ancestors (Schumaker *et al.*, 1993).

Worker ants feed on nectar, honeydew, plant sap, or on partly digested food regurgitated by their larvae (Terra, 1990). Thus, they seem to display only intermediate and (or) final digestion.

8.6.7 Diptera

The spatial organization of digestion in the majority of nematoceran larvae is probably similar to that of *R. americana* reviewed above. Larval culicids have midgut fluid fluxes (Wigglesworth, 1972). The high pH found along the midguts of larval tipulids would be expected to enhance the digestibility of proteins (Martin, 1987) and of hemicellulose and cellulose, which may in turn be digested by bacteria present in the hindgut caecum (Griffiths and Cheshire, 1987). Liquid food (nectar or decay products) ingested by *Trichosia pubescens* (Sciaridae) adults is stored in their crops. Digestion of starch by salivary α-amylase starts in the midgut with the resulting oligomaltodextrins and ingested α-glucosides being hydrolysed by a midgut membrane-bound α-glucosidase. Food proteins are hydrolysed initially by trypsin in the midgut lumen and the resulting oligopeptides by a midgut microvillar aminopeptidase (Espinoza-Fuentes *et al.*, 1984). This digestive pattern is probably also valid for other non-haematophagous nematoceran adult flies. Nectar ingested by mosquitoes (males and females) is stored in the crop, whereas blood, which is sucked only by females, passes to the posterior midgut. Based primarily on ultrastructural studies (Schneider *et al.*, 1987), evidence indicates that the anterior midgut of Culicidae mosquitoes is involved in the digestion and absorption of sugars, whereas the posterior midgut is responsible for the synthesis of precursors of the peritrophic matrix and digestive enzymes, as well as for the absorption of nutrients from blood.

The majority of cyclorrhaphous dipteran larvae are saprophagous. The spatial organization of digestion in these species is probably similar to that of *Musca domestica* (Muscidae) larvae. In this insect, the anterior midgut reduces the starch content of food ingested by larvae, making bacteria (the major food of houseflies) more susceptible to the combined action of low pH, cathepsin D and lysozyme that are found in the middle midgut. After bacteria are killed, and make free their contents, they pass to the posterior midgut. As the food polymeric molecules become sufficiently small to accompany the polymer hydrolases through the peritrophic matrix they are driven by a countercurrent of flux to the anterior region of the posterior midgut. In this region, terminal digestion and absorption of food occur at the surface of the cells (Espinoza-Fuentes and Terra, 1987; Lemos and Terra, 1991).

Cyclorrhaphous adults feed mainly on liquids associated with decaying material. The major food of these adults is probably bacteria. *M. domestica* adults salivate (or regurgitate their crop contents) onto their food. After the dispersed material is ingested, starch digestion is accomplished primarily in the crop by the action of salivary α-amylase. The resulting oligosaccharides are digested by a membrane-bound maltase on the surface of midgut cells. Bacteria present in the food are digested in a manner similar to that described for *M. domestica* larvae (Ferreira, Espinoza-Fuentes and Terra, unpublished). The stable fly (*Stomoxys calcitrans*, Muscidae) stores and concentrates the blood meal in the anterior midgut and gradually passes it to the posterior midgut. Blood proteins are at first digested by trypsin, chymotrypsin and carboxypeptidases A and B in the posterior midgut lumen. Final digestion presumably takes place at the surface of midgut cells by a cell-associated (perhaps microvillar) aminopeptidase activity (Houseman *et al.*, 1987). Proteinases are secreted by the middle midgut (Moffatt and Lehane, 1990). Blood carbohydrates are digested by an array of glycosidases, which include a midgut membrane-bound α-glucosidase and α-mannosidase (DeLoach and Spates, 1984). Tsetse flies (*Glossina* spp.) (Glossinidae) ingest only blood and store it in the crop; unloading of blood from the crop to the anterior midgut begins soon after ingestion (Moloo and Kutuza, 1970). Water and amino acids are absorbed primarily in the anterior and middle midgut regions, respectively (Brown, 1980). Water is presumably secreted in the posterior midgut, powering an endo-ectoperitrophic circulation of digestive enzymes and transporting the products of digestion to absorbing regions in the middle midgut or anterior midgut (Brown, 1980; Terra, 1988). In summary, the compartmentalization of digestion in adult Diptera, except for the occasional digestion of carbohydrate (usually starch) by salivary enzymes in the crop, is similar to that in corresponding larvae.

8.6.8 Lepidoptera

No digestion was detected in the foregut of larvae of *Erinnyis ello* (Sphingidae). In this species, initial digestion occurs in the endo-peritrophic space, whereas intermediate and final digestion is performed through enzymes associated with the midgut cell glycocalyx or bound to the microvillar membranes. Endo-ectoperitrophic circulation of digestive enzymes is driven by fluid fluxes produced by columnar cells with the probable interplay of goblet cells (Santos *et al.*, 1983, 1986). Similar results were obtained for *Spodoptera frugiperda* (Noctuidae) (Ferreira *et al.*, 1994). Most of the trypsin mRNA is found in the middle midgut, whereas activity is recovered mainly from anterior midgut lumen of

Manduca sexta (Sphingidae) larvae (Peterson *et al.*, 1994). This led to the proposition that trypsin is secreted from the middle midgut cells and carried forward in the ectoperitrophic space towards the anterior midgut. Thus, the spatial organization of the digestion process is probably comparable in most lepidopteran larvae. Amino acid absorption occurs in midgut columnar cells by a co-transport mechanism involving potassium ions (Chapter 10).

Although most lepidopteran larvae have a common pattern of digestion, species that feed on specific, unique diets generally display some adaptations. *Tineola bisselliella* (Tineidae) larvae that can feed on wool display a highly reducing midgut for cleaving the disulphide bonds in keratin in order to facilitate proteolytic hydrolysis of this otherwise insoluble protein (Waterhouse, 1952). Another adaptation has apparently occurred in lepidopteran adults which generally feed solely on nectar. Digestion of nectar only requires the action of an α-glucosidase (or a *β*-fructosidase) to hydrolyse sucrose, the major component present. Nevertheless, a study performed with nectar-feeding *E. ello* adults showed the presence of α-amylase in salivary glands and the occurrence of soluble *β*-fructosidase, trypsin and membrane-bound aminopeptidase and α-glucosidase in the midgut (Terra *et al.*, 1990). The presence of the complement of digestive enzymes in nectar-feeding moths may explain, at least on enzymological grounds, the adaptation of some adult Lepidoptera to new feeding habits such as blood and pollen (Terra, 1990). The compartmentalization of digestion in adult Lepidoptera seems to be similar to that in their larvae.

8.7 CONCLUSIONS AND FUTURE CONSIDERATIONS

Digestion of food polymers in the gut of insects takes place in three phases: initial, intermediate and final. These phases and the enzymes associated with each phase, are separated spatially and temporally by morphological features of the gut and fluid fluxes within the gut, respectively. The peritrophic matrix effectively compartmentalizes the midgut into an ectoperitrophic region and an endoperitrophic region. There is evidence that polymer hydrolases secreted into the lumen are conserved by an endo-ectoperitrophic circulation that depends on fluid fluxes from posterior to anterior midgut regions. This circulation is made possible by the permeability of the peritrophic matrix to hydrolases and nutrients. The fluid fluxes are driven by secretion of water from the Malpighian tubules or from the posterior regions of the midgut and absorption of water in the anterior or middle regions of the midgut. Comparative analyses of specific morphological features combined with studies of enzyme compartmentalization in a wide range of insect taxa indicate that phylogenic relationships, rather than ecological habitat, best describe the evolution of insect digestive processes and provide a sound framework or paradigm in which to advance our understanding of digestion in insects.

Studies designed to obtain detailed information on basic digestive processes, as outlined above, are also fundamental in advancing novel, applied methods for controlling agriculturally important insect pests. With the development of transformed plants containing engineered biochemical resistance factors or toxins targeted against phytophagous insects, a sophisticated understanding of midgut physiology, including compartmentalization of digestive enzymes in specific insect pests, will be necessary to evaluate fully the effectiveness of ingested toxins or enzyme inhibitors acting through the gut system. Toxins produced by *Bacillus thuringiensis* (Bt) (Bulla *et al.*, 1977) (Chapter 13), and chemical defence proteins (i.e. proteinase inhibitors, α-amylase inhibitors, vitamin-binding proteins, gut toxins, immunotoxins, lectins) (Kramer, 1995) (Chapter 14) ingested by insects interact with their targets in specific midgut regions, and in some cases, in a sequential manner. For example, interactions of ingested proteinaceous inhibitors in plants with polymer hydrolases present in insects could take place in either the endoperitrophic space or the ectoperitrophic space depending on the molecular size of the inhibitor and the degree to which the unbound or free inhibitor could penetrate the peritrophic matrix. After the inhibitor interacts with the enzyme, the larger molecular size of the complex itself may result in an effective compartmentalization in either the lumen or ectoperitrophic space because of the inability of the complex to subsequently pass through pores in the peritrophic matrix.

In the case of *Bacillus thuringiensis* toxin, the peritrophic matrix also plays a critical role in regulating or modulating the penetration of the toxin from the midgut lumen into the ectoperitrophic space where different enzymes as well as the midgut epithelium become accessible. Thus, following proteolytic activation of Bt protoxin in the gut lumen of lepidopterous larvae, the resulting lower molecular weight endotoxin can then pass through the peritrophic matrix and bind to brush border membranes of ventricular cells (Hofmann *et al.*, 1988; Hofte and Whiteley, 1989; Adang, 1991).

It is apparent that knowledge gained in analysing both morphological and biochemical aspects of compartmentalization of digestion in specific insects can be critical to understanding, evaluating or predicting the success or failure of biotechnological efforts to manage insect pests. This knowledge, based on unique or specific adaptive features in a given insect species, may even allow the modelling or design of more effective toxins or inhibitors in the future.

ACKNOWLEDGEMENTS

The work of Drs W.R. Terra and C. Ferreira was supported by Brazilian Research Agencies FAPESP, CNPq and FINEP. They are staff members of the Biochemistry Department and research fellows of CNPq and are grateful to Mrs Elisety de Andrade Silva for manuscript preparation. We thank K.J. Kramer and B. Oppert of the US Grain Marketing Research Laboratory, Manhattan, Kansas and T.L. Hopkings of the Department of Entomology, Kansas State University, Manhattan, Kansas, for comments on an earlier version of this manuscript.

REFERENCES

Adang, M.J. (1991) *Bacillus thuringiensis* insecticidal crystal proteins: gene structure, action, and utilization, in *Biotechnology for Biological Control of Pests and Vectors* (ed. K. Maramorosch), CRC Press, Boca Raton, pp. 3–24.

Andersen, P.C., Brodbeck, B.V. and Mizell, R.F. (1989) Metabolism of amino acids, organic acids and sugars extracted from the xylem fluid of four host plants by adult *Homalodisca coagulata*. *Entomol. Exp. Appl.*, **50**, 149–59.

Andries, J.C. and Torpier, G. (1982) An extracellular brush border coat of lipid membranes in the midgut of *Nepa cinerea* (Insecta: Heteroptera): ultrastructure and genesis. *Biol. Cell.*, **46**, 195–202.

Appel, H.M. and Martin, M.M. (1990) Gut redox conditions in herbivorous larvae. *J. Chem. Ecol.*, **16**, 3277–90.

Baerwald, R.J. and Delcarpio, J.B. (1983) Double membrane-bound intestinal microvilli in *Oncopeltus fasciatus*. *Cell Tissue Res.*, **232**, 593–600.

Baker, J.E. (1981) Localization of proteolytic enzymes in the midguts of larvae of the black carpet beetle. *J. Georgia Entomol. Soc.*, **16**, 495–500.

Baker, J.E., Woo, S.M. and Mullen, M.A. (1984) Distribution of proteinases and carbohydrases in the midgut of larvae of the sweetpotato weevil *Cylas formicarius elegantulus* and response of proteinases to inhibitors from sweet potato. *Entomol. Exp. Appl.*, **36**, 97–105.

Bayon, C. (1981) Modification ultrastructurales des parois vegetables dans le tube digestive d'une larve xylophage *Oryctes nasicornis* (Coleoptera, Scarabaeidae): rôle des bactéries. *Can. J. Zool.*, **59**, 2020–9.

Bayon, C. and Mathelin, J. (1980) Carbohydrate fermentation and by-product absorption studied with labelled cellulose in *Oryctes nasicornis* larvae (Coleoptera: Scarabaeidae). *J. Insect Physiol.*, **26**, 833–40.

Benemann, J.R. (1973) Nitrogen fixation in termites. *Science*, **181**, 164–5.

Bignell, D.E. (1981) Nutrition and digestion, in *The American Cockroach*, (eds W.J. Bell and K.G. Adiyodi), Chapman & Hall, London, pp. 57–86.

Bignell, D.E. and Anderson, J.M. (1980) Determination of pH and oxygen status in the guts of lower and higher termites. *J. Insect Physiol.*, **26**, 183–8.

Bignell, D.E., Oskarsson, H., Anderson, J.M. *et al.* (1983) Structure, microbial associations and function of the so-called 'mixed segment' of the gut in two soil-feeding termites, *Procubitermes aburiensis* and *Cubitermes severus* (Termitidae, Termitinae). *J. Zool. London*, **201**, 445–80.

Billingsley, P.F and Downe, A.E.R. (1985) Cellular localisation of aminopeptidase in the midgut of *Rhodnius prolixus* Stal (Hemiptera: Reduviidae) during blood digestion. *Cell Tissue Res.*, **241**, 421–8.

Billingsley, P.F. and Rudin, W. (1992) The role of the mosquito peritrophic membrane in bloodmeal digestion and infectivity of *Plasmodium* species. *J. Parasitol.*, **78**, 430–40.

Brown, R. (1980) Ultrastructure and function of midgut epithelium in the tsetse *Glossina morsitans*, Westw. (Diptera: Glossinidae). *J. Entomol. Soc. S. Afr.*, **43**, 195–214.

Bulla, L.A., Kramer, K.J. and Davidson, L.I. (1977) Characterization of the entomocidal parasporal crystal of *Bacillus thuringiensis*. *J. Bacteriol.*, **130**, 375–83.

Cheung, W.W.K. and Marshall, A.T. (1973) Water and ion regulation in cicadas in relation to xylem feeding. *J. Insect Physiol.*, **19**, 1801–16.

Cleveland, L.R., Hall, S.R., Sanders, E.P. and Collier, J. (1934) The wood-feeding roach, *Cryptocercus*, its protozoa, and the symbiosis between protozoa and the roach. *Mem. Am. Acad. Arts Sci.*, **17**, 185–343.

Colepicolo-Neto, P., Bechara, E.J.H., Ferreira, C. and Terra, W.R. (1986) Evolutionary considerations of the spatial organization of digestion in the luminescent predaceous larvae of *Pyrearinus termitilluminans* (Coleoptera: Elateridae). *Insect Biochem.*, **16**, 811–17.

Crailsheim, K. (1988) Transport of leucine in the alimentary canal of the honeybee (*Apis mellifera* L.) and its dependence on season. *J. Insect Physiol.*, **34**, 1093–100.

Crowson, R.A. (1981) *The Biology of Coleoptera*, Academic Press, London.

Del Bene, G., Dallai, R. and Marchini, D. (1991) Ultrastructure of the midgut and the adhering tubular salivary glands of *Frankliniella occidentalis* (Pergande) (Thysanoptera: Thripidae). *Int. J. Insect Morphol. Embryol.*, **20**, 15–24.

DeLoach, J.R. and Spates, G.E. (1984) Glycosidase activity from midgut region of *Stomoxys calcitrans* (Diptera: Muscidae). *Insect Biochem.*, **14**, 169–73.

Dow, J.A.T. (1986) Insect midgut function. *Adv. Insect Physiol.*, **19**, 187–328.

Espinoza-Fuentes, F.P., Ferreira, C. and Terra, W.R. (1984) Spatial organization of digestion in the larval and imaginal stages of the sciarid fly *Trichosia pubescens*. *Insect Biochem.*, **14**, 631–8.

Espinoza-Fuentes, F.P. and Terra, W.R. (1987) Physiological adaptations for digesting bacteria. Water fluxes and distribution of digestive enzymes in *Musca domestica* larval midgut. *Insect Biochem.*, **17**, 809–17.

Felton, G.W. and Duffey, S.S. (1991) Reassessment of the role of the gut alkalinity and detergency in insect herbivory. *J. Chem. Ecol.*, **17**, 1821–36.

Ferreira, C., Bellinello, G.L., Ribeiro, A.F. and Terra, W.R. (1990b) Digestive enzymes associated with the glycocalyx, microvillar membranes and secretory vesicles from midgut cells of *Tenebrio molitor* larvae. *Insect Biochem.*, **20**, 839–47.

Ferreira, C., Capella, A.N., Sitnik, R. and Terra, W.R. (1994) Digestive enzymes in midgut cells, endo- and ectoperitrophic contents and peritrophic membranes of *Spodoptera frugiperda* (Lepidoptera) larvae. *Arch. Insect Biochem. Physiol.*, **26**, 299–313.

Ferreira, C., Oliveira, M.C. and Terra, W.R. (1990a) Compartmentalization of the digestive process in *Abracris flavolineata* (Orthoptera: Acrididae). *Insect Biochem.*, **20**, 267–74.

Ferreira, C., Ribeiro, A.F., Garcia, E.S. and Terra, W.R. (1988) Digestive enzymes trapped between and associated with the double plasma membranes of *Rhodnius prolixus* posterior midgut cells. *Insect Biochem.*, **18**, 521–30.

Ferreira, C., Ribeiro, A.F. and Terra, W.R. (1981) Fine structure of the larval midgut of the fly *Rhynchosciara* and its physiological implications. *J. Insect Physiol.*, 27, 559–70.

Ferreira, C. and Terra, W.R. (1980) Intracellular distribution of hydrolases in midgut caeca cells from an insect with emphasis on plasma membrane-bound enzymes. *Comp. Biochem. Physiol.*, **66B**, 467–73.

Ferreira, C. and Terra, W.R. (1989) Spatial organization of digestion, secretory mechanisms and digestive enzyme properties in *Pheropsophus aequinoctialis* (Coleoptera: Carabidae). *Insect Biochem.*, **19**, 383–91.

Griffiths, B.S. and Cheshire, M.V. (1987) Digestion and excretion of nitrogen and carbohydrate by the cranefly larva *Tipula paludosa* (Diptera: Tipulidae). *Insect Biochem.*, **17**, 277–82.

Hogan, M.E., Slaytor, M. and O'Brien, R.W. (1985) Transport of volatile fatty acids across the hindgut of the cockroach *Panesthia cribrata* Sausurre and the termite, *Mastotermes darwiniensis* Frogatt. *J. Insect Physiol.*, **31**, 587–91.

Hogan, M., Veivers, P.C., Slaytor, M. and Czolij, R.T. (1988) The site of cellulose breakdown in higher termites. *J. Insect Physiol.*, **34**, 891–9.

Hofmann, C.H., Vanderbruggen, H., Höfte, J. *et al.* (1988) Specificity of *Bacillus thuringiensis*-endotoxins is correlated with the presence of high-affinity binding sites in the brush border membrane of target insect midguts. *Proc. Natl Acad. Sci. USA*, **85**, 7844–8.

Hofte, H. and Whiteley, H.R. (1989) Insecticidal crystal proteins of *Bacillus thuringiensis*. *Microbiol. Rev.*, **53**, 242–55.

Houseman, J.G., Campbell, F.C. and Morrison, P.E. (1987) A preliminary characterization of digestive proteases in the posterior midgut of the stable fly *Stomoxys calcitrans* (L.) (Diptera: Muscidae). *Insect Biochem.*, **17**, 213–18.

Houseman, J.G. and Downe, A.E.R. (1980) Endoproteinase activity in the posterior midgut of *Rhodnius prolixus* Stal (Hemiptera: Reduviidae). *Insect Biochem.*, **10**, 363–6.

Jimenez, D.R. and Gilliam, M. (1990) Ultrastructure of ventriculus of the honey bee *Apis mellifera* (L.): cytochemical localization of the acid phosphatase, alkaline phosphatase, and non-specific esterase. *Cell Tissue Res.*, **261**, 431–43.

Klinkowstrom, A.M., Terra, W.R. and Ferreira, C. (1994) Aminopeptidase A from *Rhynchosciara americana* (Diptera) larval midguts: properties and midgut distribution. *Arch. Insect Biochem. Physiol.*, **27**, 301–5.

Klungness, L.M. and Peng, Y.S. (1984) A histochemical study of pollen digestion in the alimentary canal of the honeybee (*Apis mellifera* L.). *J. Insect Physiol.*, **30**, 511–21.

Kramer, K.J. (1995) Protein resources for insect pest management, in *Proc. of the 33rd US–Japan Natural Resources Protein Panel* (ed. A. Pavlath), Berkeley Press, Berkeley, CA (in press).

Kristensen, N.P. (1981) Phylogeny of insect orders. *Annu. Rev. Entomol.*, **26**, 135–57.

Lane, N.J. and Harrison, J.B. (1979) An unusual cell surface modification: a double plasma membrane. *J. Cell Sci.*, **39**, 355–72.

Lawrence, J.F. and Newton, A.F. Jr (1982) Evolution and classification of beetles. *Annu. Rev. Ecol. Syst.*, **13**, 261–90.

Lemos, F.J.A. and Terra, W.R. (1991) Properties and intracellular distribution of a cathepsin D-like proteinase active at the acid region of *Musca domestica* midgut. *Insect Biochem.*, **21**, 457–65.

Lindsay, K.L. and Marshall, A.T. (1980) Ultrastructure of the filter chamber complex in the alimentary canal of *Eurymela distincta* Signoret (Homoptera, Eurymelidae). *Int. J. Insect Morphol. Embryol.*, **9**, 179–98.

Lindsay, K.L. and Marshall, A.T. (1981) The osmoregulatory role of the filter-chamber in relation to phloem-feeding in *Eurymela distincta* (Cicadelloidea, Homoptera). *Physiol. Entomol.*, **6**, 413–19.

Marshall, A.T. and Cheung, W.W.K. (1974) Studies on water and ion transport in homopteran insects: ultrastructure and cytochemistry of the cicadoid and cercopoid Malpighian tubules and filter chamber. *Tissue Cell.*, **6**, 153–71.

Martin, M.M. (1987) *Invertebrate–Microbial Interactions: Ingested Fungal Enzymes in Arthropod Biology*, Ithaca, Cornell.

Mittler, T.E. (1958) Studies on the feeding and nutrition of *Tuberolachnus salignus* (Gmelin) (Homoptera, Aphididae). II. The nitrogen and sugar composition of ingested phloem sap and excreted honeydew. *J. Exp. Biol.*, **35**, 74–84.

Moffatt, M.R. and Lehane, M.J. (1990) Trypsin is stored as an inactive zymogen in the midgut of *Stomoxys calcitrans*. *Insect Biochem.*, **20**, 719–23.

Moloo, S.K. and Kutuza, S.B. (1970) Feeding and crop emptying in *Glossina brevipalpis* Newstead. *Acta Trop.*, **27**, 356–77.

Morgan, M.R.J. (1976) Gut carbohydrases in locusts and grasshoppers. *Acrida*, **5**, 45–58.

O'Loughlin, G.T. and Chambers, T.C. (1972) Extracellular microtubules in the aphid gut. *J. Cell Biol.*, **53**, 575–8.

Parenti, P., Sacchi, F.V., Hanozet, G.M. and Giordana, B. (1986) Na-dependent uptake of phenylalanine in the midgut of a cockroach (*Blabera gigantea*). *J. Comp. Physiol. B*, **156**, 549–56.

Peterson, A.M., Barillas-Mury, C.V. and Wells, M.A. (1994) Sequence of three cDNAs encoding an alkaline midgut trypsin from *Manduca sexta*. *Insect Biochem. Mol. Biol.*, **24**, 463–71.

Reger, J.F. (1971) Fine structure of the surface coat of the midgut epithelial cells in the homopteran *Phylloscelis atra* (Fulgorid). *J. Submicrosc. Cytol.*, **3**, 353–8.

Santos, C.D., Ferreira, C. and Terra, W.R. (1983) Consumption of food and spatial organization of digestion in the cassava hornworm, *Erinnyis ello*. *J. Insect Physiol.*, **29**, 707–14.

Santos, C.D., Ribeiro, A.F. and Terra, W.R. (1986) Differential centrifugation, calcium precipitation and ultrasonic disruption of midgut cells of *Erinnyis ello* caterpillars. Purification of cell microvilli and inferences concerning secretory mechanisms. *Can. J. Zool.*, **64**, 490–500.

Schneider, M., Rudin, W. and Hecker, H. (1987) Absorption and transport of radioactive tracers in the midgut of the malaria mosquito, *Anopheles stephensi*. *J. Ultrastruc. Mol. Struct. Res.*, **97**, 50–63.

Schuch, R.T. (1986) The influence of cladistics on heteropteran classification. *Annu. Rev. Entomol.*, **31**, 67–93.

Schumaker, T.T.S., Cristofoletti, P.T. and Terra, W.R. (1993) Properties and compartmentalization of digestive carbohydrases and proteases in *Scaptotrigona bipunctata* (Apidae: Meliponinae) larvae. *Apidologie*, **24**, 3–17.

Scrivener, A.M., Slaytor, M. and Rose, H.A. (1989) Symbiont-independent digestion of cellulose and starch in *Panesthia cribrata* Saussure, an Australian wood-eating cockroach. *J. Insect Physiol.*, **35**, 935–41.

Silva, C.P. and Terra, W.R. (1994) Digestive and absorptive sites along the midgut of the cotton seed sucker bug *Dysdercus peruvianus* (Hemiptera: Pyrrhocoridae). *Insect Biochem. Mol. Biol.*, **24**, 493–505.

Slaytor, M. (1992) Cellulose digestion in termites and cockroaches: what role do symbionts play? *Comp. Biochem. Physiol.*, **103B**, 775–84.

Taylor, E.C. (1985) Cellulose digestion in a leaf eating insect, the Mexican beetle, *Epilachna varivestis*. *Insect Biochem.*, **15**, 315–20.

Teo, L.H. and Woodring, J.P. (1988) The digestive protease and lipase in the house cricket *Acheta domesticus*. *Insect Biochem.*, **18**, 363–7.

Terra, W.R. (1988) Physiology and biochemistry of insect digestion: an evolutionary perspective. *Braz. J. Med. Biol. Res.*, **21**, 675–734.

Terra, W.R. (1990) Evolution of digestive systems of insects. *Annu. Rev. Entomol.*, **35**, 181–200.

Terra, W.R. and Ferreira, C. (1981) The physiological role of the peritrophic membrane and trehalase: digestive enzymes in the midgut and excreta of starved larvae of *Rhynchosciara*. *J. Insect Physiol.*, **27**, 325–31.

Terra, W.R. and Ferreira, C. (1994) Insect digestive enzymes: properties, compartmentalization and function. *Comp. Biochem. Physiol.*, **109B**, 1–62.

Terra, W.R., Ferreira, C. and Bastos, F. (1985) Phylogenetic considerations of insect digestion. Disaccharidases and the spatial organization of digestion in the *Tenebrio molitor* larvae. *Insect Biochem.*, **15**, 443–9.

Terra, W.R., Ferreira, C. and De Bianchi, A.G. (1979) Distribution of digestive enzymes among the endo- and ectoperitrophic spaces and midgut cells of *Rhynchosciara* and its physiological significance. *J. Insect Physiol.*, **25**, 487–94.

Terra, W.R., Ferreira, C. and Garcia, E.S. (1988) Origin, distribution, properties and functions of the major *Rhodnius prolixus* midgut hydrolases. *Insect Biochem.*, **18**, 423–34.

Terra, W.R, Santos, C.D. and Ribeiro, A.F. (1990) Ultrastructural and biochemical basis of the digestion of nectar and other nutrients by the moth *Erinnyis ello*. *Entomol. Exp. Appl.*, **56**, 277–86.

Thorne, B.L. and Carpenter, J.M. (1992) Phylogeny of the Dictyoptera. *Syst. Entomol.*, **17**, 253–68.

Waterhouse, D.F. (1952) Studies on the digestion of wood by insects. VI. The pH and oxidation potential of the alimentary canal of the clothes moth larvae (*Tineola bisselliella* (Humm.)). *Aust. J. Biol. Sci.*, **5B**, 178–88.

Wigglesworth, V.B. (1972) *The Principles of Insect Physiology*, 7th edn, Methuen, London.

Yang, Y.J. and Davies, D.M. (1968) Digestion, emphasizing trypsin activity, in adult simuliids (Diptera) fed blood, blood–sucrose mixtures, and sucrose. *J. Insect Physiol.*, **14**, 205–22.

9

Ion transport in Lepidoptera

U. Klein, A. Koch and D.F. Moffett

9.1 INTRODUCTION

9.1.1 Overview

The lepidopteran midgut has received intensive study, and from a biophysical point of view is undoubtedly the best understood of insect guts. This chapter will concentrate on movement of ions between midgut lumen and haemolymph. This movement includes absorption of ingested ions from the midgut and, in addition, secretion of ions from haemolymph and the production of very high luminal pH values. An excellent and comprehensive review of insect midgut function was published by Dow (1986). Accordingly, the present chapter will concentrate on the more recent findings and interpretations.

In the guts of metazoans, uptake of molecular sources of energy and nitrogen and essential inorganic solutes is all ultimately driven by active ionic transport. In vertebrates, the primary motor for these transports is active extrusion of Na^+ across the basolateral membrane by Na^+/K^+-ATPase. The activity of this Na^+ pump lowers cellular Na^+ concentration, producing a driving force for Na^+ to move from gut lumen into the cytoplasm of the epithelial cell. Na^+ crosses the apical membrane via a number of secondary processes. These include co-transport of Na^+ and glucose, co-transport of Na^+ and amino acids and countertransport of Na^+ with H^+. Cl^- follows, driven by the transgut voltage derived from active extrusion of Na^+. In addition, absorption of Na^+ and Cl^- leads to absorption of solute and water *pari passu*. Thus the variety of different absorptive processes carried on by vertebrate intestine is accomplished

Biology of the Insect Midgut.
Edited by M.J. Lehane and P.F. Billingsley.
Published in 1996 by Chapman & Hall, London.
ISBN 0 412 61670 X.

by coupling secondary processes onto the primary Na^+/K^+-ATPase that resides in the basolateral membrane.

From the few examples that have been studied, insects follow the same general model of coupling one primary ion transport ATPase to a variety of secondary processes. However, the primary motor as well as the secondary processes differ between insect groups; the differences in part reflect ancestral adaptation to particular diets (Chapter 7). Carnivorous and parasitic insects receive considerable Na^+ in their diet. These animals have high- $[Na^+]$ haemolymph and use intestinal absorptive processes similar to those used in vertebrates. The diet of herbivorous insects is typically low in Na^+ and high in K^+. Most of them nevertheless ingest enough Na^+ to be able to regulate haemolymph $[Na^+]$ above 50 mM (Sutcliffe, 1963). From the few representative species studied, most herbivorous insects use the Na^+/K^+-ATPase to drive solute and fluid absorption. However, the larvae of lepidopterans are unable to retain sufficient Na^+ from the diet to maintain a concentration of Na^+ above 10 mM in their haemolymph. The low $[Na^+]$ in the haemolymph precludes use of the sodium pump (Jungreis and Vaughn, 1977). The primary motor they use appears to be a vacuolar-type proton ATPase, often coupled to a H^+/K^+ antiport to produce what was taken for several decades of study to be a primary active K^+ pump (Harvey *et al.*, 1981). The site of the primary motor is moved from the basal membrane, the location of the Na^+/K^+ pump in the gut epithelia of most species, to the apical membrane, the typical location for externally facing proton pumps in those epithelia studied thus far (Ehrenfeld and Garcia-Romeu, 1977; Ehrenfeld *et al.*, 1985; Wieczorek *et al.*, 1991; Gluck and Nelson, 1992; Klein, 1992; Maddrell and O'Donnell, 1992; Wessing *et al.*, 1993). This strategy shift was surely fraught with implications for ionic and volume homeostasis of the cells of species that adopted it, but these problems have as yet received little study.

9.1.2 General features and evidence for regional specialization

The larval midgut epithelium consists of a single layer of cells containing two major cell types: columnar and goblet cells (Anderson and Harvey, 1966; Cioffi, 1979), as well as smaller numbers of regenerative or stem cells (Cioffi, 1979; Baldwin and Hakim, 1991), and endocrine cells (Endo and Nishiitsutsuji-Uwo, 1982). The goblet cells contain a central cavity filled with a glycoprotein matrix (Schultz *et al.*, 1981). Numerous projections of apical membrane (GCAM) line the cavity; others converge to form a valve-like structure (Anderson and Harvey, 1966) at its apical end. The goblet cells of anterior and middle midgut are characterized by large, basally located cavities with apical necks. Each of their GCAM projections contains a mitochondrion. In contrast, goblet cells of the posterior midgut have cavities that occupy only the apical half of the cell and their apical projections do not contain mitochondria. The conformation of the apical membrane of columnar cells changes gradually from branched folds in the anterior part to true microvilli in the posterior part (Cioffi, 1979). In addition to the ultrastructural differences, the regions can be distinguished on the basis of the density and pattern of their epithelial folds (Cioffi, 1979; Cioffi and Harvey, 1981).

9.2 FLOWS AND FORCES FOR K^+, H^+, Cl^- AND WATER

9.2.1 *In vivo*; open-circuit *in vitro*

The open circuit condition *in vitro* probably corresponds best to the situation *in vivo*. In the open circuit condition (OC), some of the ionic distributions and transepithelial movements observed are secondary to the OC voltage across the epithelium (V_t), which is normally lumen side positive. This potential will, in itself, drive ions through the cells if they can cross both basal and apical membranes. In addition, it will drive ions through any paracellular shunt that is permeable to that ion. In the live *Manduca*, transepithelial voltage is generally at least 100 mV, midgut lumen positive (Moffett and Cummings, 1994).

With the normal high K^+ diet, there is a small net secretion of K^+ into the midgut *in vivo*. When animals are placed on a low K^+, high Na^+ diet, K^+ secretion is much increased (Ramsay, 1976; Chamberlin, 1990a). The easiest explanation of these findings is that K^+ is secreted and then subsequently used to drive the symport absorption of amino acids. Na^+ can also be used in some of these symports so that during consumption of the experimental high Na^+ diet, which causes $[Na^+]$ to be high and $[K^+]$ to be low in the midgut contents, much of the later co-transport uses Na^+ and most of the secreted K^+ stays in the midgut lumen (Giordana *et al.*, 1994; Sacchi *et al.*, 1994). On high K^+ diets, midgut $[Na^+]$ is low and much of the K^+ that is secreted is used in the subsequent organic co-transport. A considerable amount of K^+ is reabsorbed in this way. Details of the co-transport are given in Chapter 10. This result suggests that K^+ secretion by the midgut is mainly directed toward the digestive process; regulation of body K^+ balance seems to take place in the Malpighian tubules and the more distal portions of the gut.

Most base secretion appears to take place in the anterior midgut; the pH gradient is maintained in the middle midgut and then dissipates somewhat in the posterior portion (Dow, 1984; Dow and Harvey, 1988; Moffett, 1994). The alkalinization is strongly O_2 dependent (Dow and O'Donnell, 1990), presumably via the O_2 dependence of K^+ transport since alkalinization is inhibited when haemolymph $[K^+]$ is low (Chamberlin, 1990b). Either transepithelial reacidification takes place, or acidic material is refluxed from the ileum into the posterior midgut region. Secretion of Cl^- is seen in all portions of the *Manduca* midgut (Chamberlin, 1990a). Water enters the midgut in the anterior portion, but is then partially reabsorbed in the middle and posterior region (Ramsay, 1976). Moffett (1994) calculated a net secretion of water of about 50 µl/h from guts of 6 g *Manduca*, but Nedergaard (1972) found only 9 µl/h secreted in *Hyalophora cecropia* larvae.

The magnitude of the passive pathway available to small ions can be estimated from current voltage curves when the active process has been inhibited by anoxia (Moffett, 1980), by inhibitors (Schirmanns and Zeiske, 1994a) or from the back flux of isotopic K^+ (Moffett and Koch, 1985). Such estimates are in the range 3–10 mS/cm^2. Most of this conductance is probably paracellular and cation-selective. It has been estimated that about 90% of this conductance is due to K^+ and only about 10% to Cl^- (Chao *et al.*, 1989). Nevertheless it is probable that Cl^- secretion observed (Chamberlin, 1990a) is a passive consequence of V_t.

Finally, lepidopteran midgut is also reported to absorb Ca^{2+} (Wood and Harvey, 1976) and Mg^{2+} (Wood *et al.*, 1975) and to secret dyes such as phenol red (a surrogate for hippurate or urate) (Nijout, 1975). The mechanisms of these processes and any relationships with H^+ or K^+ transport are essentially unknown.

9.2.2 Short circuit

The preparation that is bathed in identical solutions on both sides and maintained at zero electrical potential across the epithelium avoids external driving forces and allows the definition of what transport requires active energy input. In short-circuited (SC) preparations, K^+ is secreted by all portions of the midgut at similar rates (Cioffi and Harvey, 1981). This active K^+ secretion, of course, was the process that initially drew attention to the lepidopteran midgut (Harvey and Nedergaard, 1964). Rates of transport ranging up to 3200 $\mu A/cm^2$, that is up to 2 µmol of $K^+/min/cm^2$, have been reported in *Manduca*, *Hyalophora* and *Bombyx*. These rates are exaggerated by the extensive folding of the tissue, which causes the actual surface area to be about 20 times that of the apparent area. Correction for folding yields rates of about 160 $\mu A/cm^2$ of flat epithelium, comparable to the rates of alkali metal ion transport of other well-studied tissues such as frog skin. K^+ secretion accounts for at least 95% of the SC current (I_{sc}) which is normally used to measure the active process (Cioffi and Harvey, 1981).

Potassium secretion is closely coupled to oxygenation, and falls within seconds of a reduction in O_2. There is an approximately three-fold drop in I_{sc} between 100% and 21% of O_2 and a further six-fold drop when O_2 is reduced to 5% (Moffett and Koch, 1991). Interestingly, all blocking agents show increased potency at lower O_2 tensions. This is true of agents that block basal K^+ entry such as Ba^{2+} and lidocaine and quinidine (Moffett and Koch, 1988a, 1991) as well as agents such as nitrate that block at the pump (Chao *et al.*, 1989).

When the tissue is bathed in a solution containing KCl, at least 1 mM Ca^{2+} and Mg^{2+}, at least 100 mM sucrose and buffered to pH 8.0 (minimal saline), I_{sc} shows apparent Michaelis–Menten kinetics over the range 5–90 mM K^+ (Harvey and Nedergaard, 1964; Moffett and Koch, 1982; Zerahn, 1985; Schirmanns and Zeiste, 1994). However, when external $[K^+]$ was elevated to 128 mM, there was a decline in I_{sc} (Zerahn, 1985). When bathed in solutions that contain no Ca^{2+}, this curve is compressed so that I_{sc} is higher than in minimal saline at low $[K^+]$, but the roll off occurs at 90 mM (Moffett and Koch, 1985). Clearly K^+ secretion depends on extracellular $[K^+]$, but the relationship is not simple and probably does not directly reflect changes in intracellular $[K^+]$ (see below). When K^+ is balanced by the halides Cl^-, Br^- and I^-, I_{sc} is higher than when it is balanced by gluconate (Zeiske and Marin, 1992; Zeiske *et al.*, 1992). However, isothionate acts in the same way as Cl^- (Chao *et al.*, 1989) and replacement of Cl^- by PO_4^{3-} and SO_4^{2-} does not affect I_{sc} (Chamberlin, 1990a). It seems possible that gluconate is a mild inhibitor of K^+ transport in itself.

In addition to its active secretion of K^+, the most spectacular activity of the midgut *in vivo* is its alkalinizing power; pH values approaching 12 have been found in *Manduca* (Dow, 1984). Alkalinization has been remarkably difficult to show in the isolated SC preparation. Chamberlin (1990a) obtained such data, but was not certain that the results did not come from a transient leakage of material from the cell.

Chloride is absorbed from the posterior midgut (Chao *et al.*, 1989) and also secreted from all portions of the midgut (Chamberlin, 1990a). The values are low, less than 10% of the simultaneous K^+ movement, and not apparently coupled to it. Zerahn (1985) measured water movement in SC *Hyalophora cecropia* using minimal salines with varying $[K^+]$. The water movement depended on I_{sc}, but was far from enough to produce an isosmotic secretion.

9.3 DIFFERENTIATION OF BASAL AND APICAL PROCESSES

9.3.1 Cell to cell coupling

The midgut epithelium has been studied with standard single-barrelled microelectrodes, and with electrodes loaded with Lucifer yellow dye for iontophoretic cell marking with simultaneous voltage recording (Moffett *et al.*, 1982; Thomas and May, 1984; Moffett and Koch, 1988; Dow and Peacock, 1989). It is now accepted that midgut epithelial cells in isolated,

SC tissues in standard saline have a transbasal potential (V_b) of approximately 35–55 mV, and that columnar and goblet cells are electrically indistinguishable to microelectrodes advanced across the basal membrane (Moffett *et al.*, 1982; Moffett and Koch, 1988a). Similarly, the cytoplasmic ionic activity of K^+, $(K^+)_i$, averages 80–100 mM in such tissues (Moffett and Koch, 1988a) and the distribution of values is unimodal, both within individual tissues and for a database of more than 1000 penetrations (Moffett and Koch, unpublished). These findings are strong circumstantial evidence for effective coupling between columnar and goblet cells for small ions, although this coupling might be intermittent (Moffett and Koch, 1982; Thomas and May, 1984; Chapter 2). Assuming that all K^+ secretion proceeds through goblet cells, such coupling would allow the basal membranes of both cell types to contribute to K^+ uptake.

The distribution of cell types is such that each goblet cell is encircled by a meshwork of columnar cells (Schultz *et al.*, 1981; Hakim *et al.*, 1988). Each goblet cell is probably surrounded by three or more columnar cells to form an electrically coupled functional unit (Moffett and Koch, 1988b). Because of the coupling of columnar to goblet cells, the tissue can be considered to be composed of two fluid compartments, cellular and goblet cavity. There are four barriers for ionic transfer, basolateral membrane, columnar cell apical membrane (CCAM), goblet cavity or apical membrane (GCAM) and the valve between the goblet cavity and the midgut lumen.

9.3.2 Voltages

The voltage across the basal membrane (V_b) has a value of about −35 mV in the SC preparation and of about −30 mV in the OC preparation when both sides are bathed in a minimal saline containing 32 mM K^+ (Moffett *et al.*, 1982; Moffett and Koch, 1988a). Similar values have been found in the middle and anterior portions (Dow and Peacock, 1989; Moffett *et al.*, unpublished). The basal potentials are more negative at lower $[K^+]$ and depolarize at higher $[K^+]$.

In the SC preparation, the goblet cavity averages 45 mV positive to the external solution. It is much less positive during OC, averaging 7 mV positive to the apical side (V_g) (Moffett and Koch, 1988b). Again, similar values have been found in the middle and anterior regions. The value of V_g increases with increasing value of I_{sc} in the SC preparation (Chao *et al.*, 1989; Dow and Peacock, 1989). V_g is higher in middle midgut, averaging 20 mV in OC preparations (Moffett and Koch, 1988).

The voltage step between the cell and the goblet cavity (V_{cg}) is the sum of these two voltages, plus any transepithelial voltage (V_t) present. Thus, in the SC preparation, V_{cg} has an average value of about 84 mV,

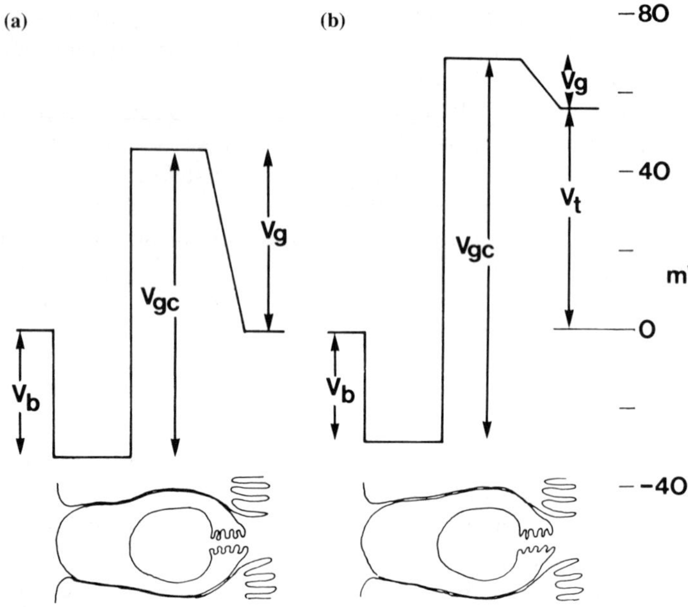

Figure 9.1 Idealized electrical potential profiles of short-circuited (a) and open-circuited (b) midgut. Each trace shows mean voltage recorded by advancing microelectrodes from the reference solution on the haemolymphal (left) side to the luminal solution on the right side. In each case the horizontal axis reflects the position of the electrode in the epithelium as indicated by the goblet cell diagram under each trace. V_b, transbasal voltage step; V_{cg}, voltage step between cytoplasm and goblet cavity; V_g voltage step between goblet cavity and luminal solution; V_t, transepithelial potential. Under short-circuit, V_t is zero. (From Chao *et al.*, 1991, with permission of the Company of Biologists, Ltd).

goblet cavity positive to cell. An OC preparation with a V_t of 75 mV would have a value of $V_{cg} = 30 + 7 + 75 = 112$ mV. The V_t *in vivo* is probably higher than 75 mV, so the preceding calculation gives a minimum value. Figure 9.1 shows typical values for open and short circuited preparations in minimal saline with 32 mM K^+.

9.3.3 Concentrations

The first measures of ionic concentrations were made in *Manduca* posterior midgut using X-ray microprobe techniques (Dow *et al.*, 1984). Subsequently, measurements of K^+, H^+ and Cl^- activity have been made in the same region with ion selective microelectrodes. In the SC posterior midgut using bilaterally symmetric minimal saline with K^+ activity = 30, the cellular K^+ activity $(K^+)_i$ averages 94 mM within cells

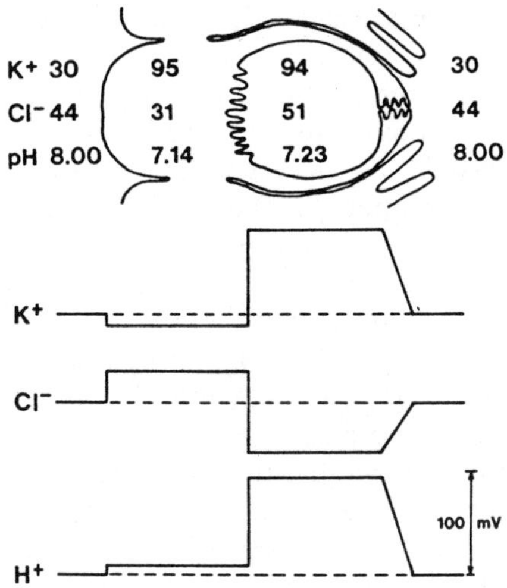

Figure 9.2 Ionic activities and driving forces for K^+, H^+ and Cl^- under short-circuit. Upper part shows mean extracellular, cytoplasmic and goblet cavity pH and activities of K^+ and Cl^- (in mM). Lower part shows electrochemical gradients of the three ions across the basal membrane, GCAM and goblet valve. (Data from Chao *et al.*, 1989, 1991 and Moffett and Koch, 1988a,b (from Chao *et al.*, 1991, with permission of the Company of Biologists, Ltd).)

and 95 mM in goblet cavity (Moffett and Koch, 1988a,b). The value does not change appreciably in the OC preparation. Chloride activity is 31 mM within cells and 51 mM in goblet cavity. The pH of the cells is 7.14, whereas that of the goblet cavity is 7.23 (Chao *et al.*, 1989, 1991). Activities of K^+, H^+ and Cl^- in the SC posterior midgut are shown in Figure 9.2.

9.3.4 Driving forces

The concentrations and voltages expressed above can be combined to an electrochemical potential ϕ. ϕ is defined as $58.Z^{-1}.\log (C_1/C_2)$, where Z is the valence of the ion and C_1 and C_2 are its concentrations in two different compartments. The net driving force then is the difference between ϕ and the membrane potential

$$\text{Driving force} = Z.\ (\phi - \Delta V)$$

where $\Delta V = V_1 - V_2$.

This expression is not quite the same as the expression for electrochemical potential because the signs have been adjusted so that a positive driving force means there can be passive movement from compartment 1 to compartment 2, regardless of whether we are discussing a cation or an anion.

Figure 9.2 also shows the driving forces computed for K^+, Cl^- and H^+ across the goblet cell of the posterior midgut under SC conditions when external [K^+] is 32 mM. In Figure 9.2 when the line goes down, passive movement can occur in that direction; when it goes up, the passive movement will be in the opposite direction. Thus, K^+ can cross the basal membrane passively, but requires coupling to external energy to move from goblet cell to goblet cavity. It can move passively from goblet cavity to midgut lumen. Both Cl^- and H^+ will move passively from goblet cell to haemolymph. Cl^- will move passively into the goblet cavity from either direction and H^+ will move passively out of the goblet cavity in either direction. The driving forces for either membrane of columnar cells are the same as for the basal membrane of goblet cells.

Comparable data are available for K^+ in the posterior midgut as a function of external [K^+] (Moffett and Koch, 1988a,b). Two significant conclusions stand out. First, at low external [K^+], the driving force for K^+ across the basal membrane becomes unfavourable for K^+ entry. Yet K^+ secretion continues, albeit at a lower rate. Second, the main factor that correlates with the reduction in I_{sc} is the voltage step across the GCAM. Neither the cellular pH nor the cellular K^+ activity change much immediately after a change in external [K^+], although the change in I_{sc} is immediate. Figure 9.3 shows the immediate effect of changes in the

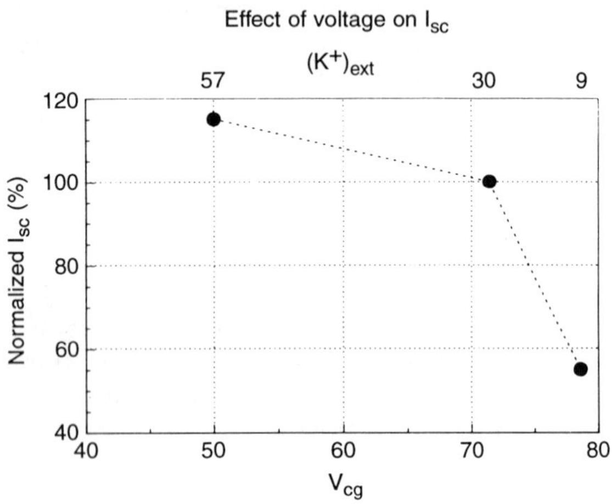

Figure 9.3 Current–voltage curve of K^+ transport in *Manduca sexta* posterior midgut. Three different concentrations of K^+ were used. At each concentration the short-circuit current and the voltage step between cell and goblet cavity were determined. (Data from Moffett and Koch, 1988a,b.)

voltage across GCAM on I_{sc}. These results are typical of the voltage effect seen on electrogenic transport systems. At low $[K^+]_{ext}$, V_b becomes more negative and hence V_{cg} is increased. This bucking voltage inhibits the electrogenic pump.

The results obtained from posterior midgut are probably not very relevant to the processes involved in luminal alkalinization since most if not all alkalinization takes place in the anterior and middle midgut. There is a paucity of electrochemical data from those regions. Only Dow and Peacock (1989) have reported voltages from both cells and goblet cavities of middle midgut. The intracellular voltages are similar to those seen in the posterior region, whereas goblet cavity voltage is much higher, averaging nearly 50 mV positive to the lumen. There are no published values for intracellular or intragoblet Cl^- or pH in middle midgut and there are no published data on any of ions in anterior midgut.

9.4 CELLULAR ANALYSIS

9.4.1 Basal K^+ uptake

The first study of K^+ uptake channels in the midgut (Zeiske *et al.*, 1986) utilized current–noise analysis. In these studies, the SC midgut displayed a Lorentzian component suggestive of single-channel currents of approximately 0.23 pA that was abolished with substitution of Na^+ for K^+ and by addition of millimolar concentrations of Ba^{2+}. Assuming a driving force of 10 mV for basal K^+ entry, these results suggest unit channels with 23 pS conductance. Subsequent patch-clamp studies of the basal membrane of goblet cells showed spontaneously active channels with current–voltage relationships and reversal potential consistent with selectivity for K^+ over anion. Channel activity in excised patches was similar to that in cell-attached patches. Unitary conductances of the channels observed were remarkably diverse, ranging from 20 to 220 pS (Moffet and Lewis, 1990).

Barium, a known blocker of K^+ channels, inhibits I_{sc}. This inhibition occurs at the basal membrane. Blockade of basal channels was inferred from the noise analysis data of Zeiske *et al.* (1986). Microelectrode studies showed that Ba^{2+} causes a large and immediate hyperpolarization of the basolateral membrane (Moffett and Koch, 1988a). As Ba^{2+} blocks some of the basolateral K^+ channels, the resistance of the basal membrane increases and the flow of K^+ across this elevated resistance produces an increased voltage drop. The noise analysis and the microelectrode studies clearly show the presence of K^+ channels in the basal membrane.

The basal channels are not, however, either simple or uniform. At high levels of $[K^+]_{ext}$, where an increase in $[K^+]_{ext}$ leads to a reduction of I_{sc}, addition of Ba^{2+} transiently stimulates I_{sc} (Moffett and Koch, 1985). This stimulation is paralleled by changes in basal channel activity. Moffett and Lewis (1990) observed a transient opening of very large channels (conductances > 100 pS) immediately after addition of Ba^{2+}. Schirmanns and Zeiske (1994b) found that Ba^{2+} blockade is voltage-dependent with greatest effectiveness at V_b between −10 and −70 mV. Furthermore, a Ba^{2+}-stimulated increase in conductance was noted when the membrane was clamped at V_b outside the range −70 to +70. These results suggest that Ba^{2+} exerts two different actions on basal K^+ channels. As in other tissues, it blocks in a voltage-dependent manner. In addition, it seems to promote the production of super-large channels. When the membrane voltage is beyond the range of effective blockade, the dominant result becomes the formation of super channels and a concurrent stimulation of I_{sc}.

Lidocaine blocks basal K^+ permeation in the posterior midgut, leading to basal hyperpolarization and inhibition of I_{sc} (Moffett and Koch, 1991). Unlike Ba^{2+}, lidocaine must dissolve in the cell membrane, since quaternary analogues of lidocaine are ineffective in the midgut. The population of basal channels blocked by lidocaine shows some overlap and some difference from the population of channels blocked by Ba^{2+}.

Evidence for a second uptake process for K^+ is provided by the observation that K^+ secretion continues when $[K^+]_{ext}$ is sufficiently low that the passive driving force for K^+ is out of the cell into the haemolymph (Moffett and Koch, 1988a; Chao *et al.*, 1990). The (K_i) does not fall, so K^+ must still be entering the cells across the basal membrane. Application of Ba^{2+} under these conditions leads to a depolarization of the basolateral membrane, showing that K^+ is indeed leaking out of the cell through the basal channels. Hence the basolateral membrane contains a K^+ uptake mechanism that works despite an unfavourable driving force. A possible mechanism is a K^+/H^+ antiport, since H^+ is sufficiently above its equilibrium concentration to drive K^+ uptake.

9.4.2 K^+ pump across GCAM

Detailed study of the nature of the apical K^+ pump became possible with the development of a procedure for isolation of plasma membrane fractions from the posterior midgut (Cioffi and Wolfersberger, 1983). These included GCAM, columnar cell apical membrane (CCAM), and basal and lateral membrane fractions. Of these, only the GCAM contained considerable K^+-stimulated ATPase activity (Wieczorek *et al.*, 1986, 1989). The ATPase was solubilized by non-ionic detergent and purified by sucrose density gradient centrifugation; its functional and molecular properties identified it as a V-ATPase (Schweikl *et al.*, 1989).

Such ATPases are ubiquitous in endomembranes of acidic organelles, but are increasingly found in plasma membranes of animal cells (Gluck and Nelson, 1992).

Further studies of GCAM vesicles suggested the K^+ pump consists of two molecular components: a primary H^+ pump (the V-ATPase) and a secondary K^+/H^+ antiport Wieczorek *et al.*, 1991; Lepier *et al.*, 1994; Azuma *et al.*, 1995). This is the first instance in which a V-ATPase was found to energize a secondary active transport in animal plasma membranes. With ATP present and K^+ absent, GCAM vesicles developed a pH gradient of 2–3 units and an electrical potential

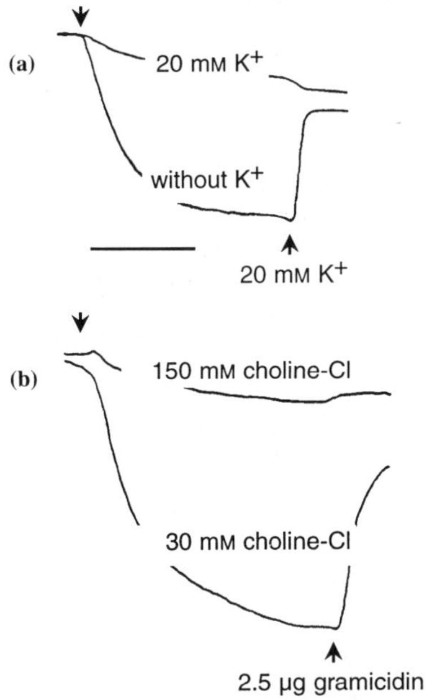

Figure 9.4 ATP-dependent ion transport in GCAM vesicles – evidence for an electrogenic proton pump. Highly purified GCAM vesicles used in (a) or vesicles from the enriched fraction B2 used in (b) were prepared according to Cioffi and Wolfersberger (1983). Experiments were begun by addition of GCAM vesicles to a Tris–HCl buffered solution (pH 8.1) containing 1 mM ATP (standard conditions). Horizontal time bar = 30 s. (a) Proton transport was measured by the fluorescence quench of acridine orange (downward). Acidification was observed under standard conditions, but could be prevented by K^+ in the incubation mixture or by subsequent addition (arrow). (b) Voltage generation was measured by the fluorescence quench of Oxonol V (downward). The voltage generated could be dissipated by gramicidin. It did not occur when external Cl^- was high, thus suggesting the presence of Cl^- channels. (Data from Wieczorek *et al.*, 1989, 1991.)

difference (Figure 9.4), suggesting that the primary transporter is an electrogenic proton pump. Both gradients are sensitive to micromolar concentrations of bafilomycin A_1, a specific inhibitor of V-ATPase (Bowman *et al.*, 1988; Dröse *et al.*, 1993; Crider *et al.*, 1994; Zhang *et al.*, 1994). The presence of K^+ dissipated both the electrical and chemical gradients, but stimulated ATP hydrolysis (Wieczorek *et al.*, 1991). The dissipation of the gradients is consistent with K^+/H^+ exchange; the stimulation of ATP hydrolysis would reflect sensitivity of the electrogenic H^+ pump to the H^+ gradient. When a K^+ gradient across the vesicle membrane was present, both acidification and charge transfer took place in the absence of ATP (Figure 9.5) (Wieczorek *et al.*, 1991), consistent with K^+/H^+ exchange driven by the K^+ gradient. These findings indicate

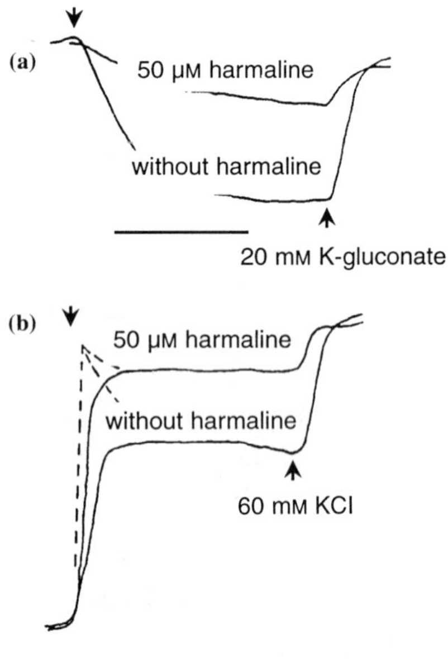

Figure 9.5 ATP-independent H^+ transport in GCAM vesicles – evidence for electrogenic K^+/H^+ antiport. Highly purified GCAM vesicles were preloaded with K^+ by incubation in standard buffer with 20 mM K^+. Fluorescence quench measurements of pH and voltage were made as in Figure 9.4, with the K^+-loaded vesicles bathed initially in K^+-free standard solution without ATP. (Data from Wieczorek *et al.*, 1991.) (a) Generation of a pH gradient secondary to K^+ efflux was measured with acridine orange. The pH gradient was dissipated by the addition of K^+ to the outer solution. (b) Voltage generation secondary to K^+ efflux was measured with Oxonol V. The addition of vesicles led to a fast increase in fluorescence superimposed on the intitial fluorescence decrease as indicated by the hatched lines. The fluorescence decrease was reduced in the presence of the antiport inhibitor, harmaline.

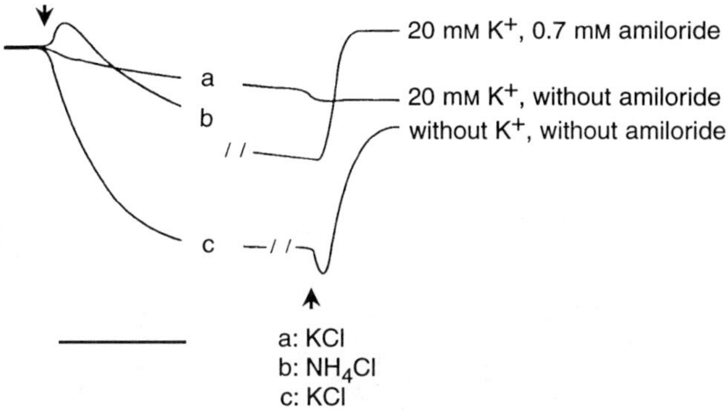

Figure 9.6 ATP-dependent H^+ transport in the presence of K^+ – evidence that the V-ATPase is an exclusive H^+ pump. Highly purified GCAM vesicles were used under standard conditions (Figure 9.4). Traces a and c are comparable to Figure 9.4; ATP induces acidification which can be dissipated by external K^+. Trace b shows that acidification takes place in the presence of K^+ when the antiport is blocked with amiloride. NH_4^+ dissipated this H^+ gradient. (Data from Wieczorek *et al.*, 1991.)

the presence of a K^+/H^+ antiport which both modifies chemical composition and transfers charge. Subsequent experiments using static head measurements in vesicles have shown the stoichiometry to be $2H^+$ per K^+ (Azuma *et al.*, 1995). The antiport is sensitive to 1 mM amiloride and to hamaline (Wieczorek *et al.*, 1991).

The separate nature of the ATPase and the antiport is reinforced by the finding that antibodies to the V-ATPase block ATP-dependent acidification but not K^+/H^+ exchange (Wieczorek *et al.*, 1991). An alternative hypothesis, that the V-ATPase itself can accept K^+ as well as H^+, was disproved by experiments in which vesicle acidification still occurred in the presence of K^+ if the antiport was simultaneously blocked by amiloride (Figure 9.6).

Immunohistochemistry using antibodies raised against the V-ATPase confirmed that it is confined to the GCAM and showed that it is present in the anterior and middle regions as well as in the posterior midgut (Figure 9.7) (Klein *et al.*, 1991; Russell *et al.*, 1992; Jäger *et al.*, 1996). The immunoreactivity co-localizes with the 10 nm 'portasomes' that can be detected in ultrathin sections of the midgut. Such portasomes are typical features of epithelial cells containing a V-ATPase and are now known to be identical with the V_1-sector of the V-ATPase molecule (Cioffi, 1979; Harvey *et al.*, 1981; Klein *et al.*, 1991; Klein, 1992). The head structures of the large multisubunit enzyme can be visualized by electron microscopy in negatively stained purified GCAM vesicles (Figure 9.8). Even though the structure of the goblet cells and of the goblet cavity itself differs

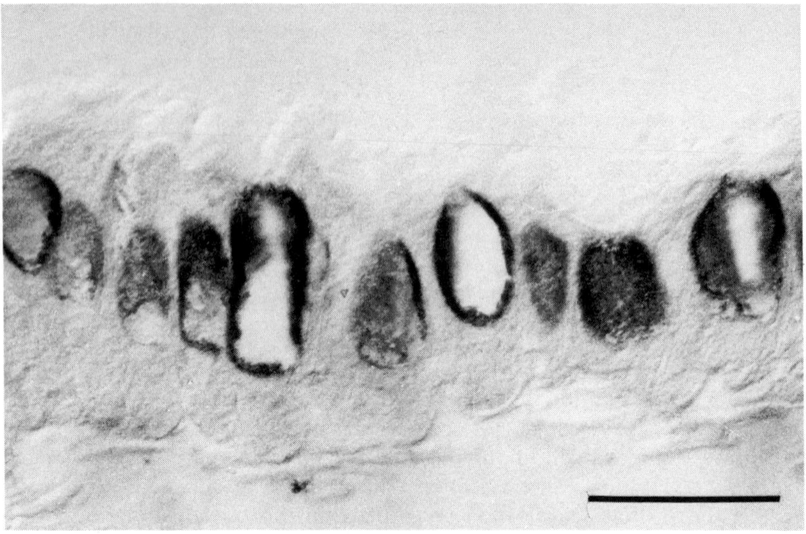

Figure 9.7 Localization of V-ATPase in the GCAM of *Manduca sexta* posterior midgut. Cryosections were labelled with mouse monoclonal antibody to subunit A of the V-ATPase. The antibody binding was visualized by gold-conjugated secondary antimouse antibody with silver enhancement. Scale bar = 50 μm. (For method see Jäger *et al.*, 1996.)

considerably between different parts of the midgut, no regional differences in particle density were detected. This finding correlates with the parallel failure to find regional differences in K^+ transport activity (Cioffi, 1979; Cioffi and Harvey, 1981). In addition, recent studies confirmed that GCAM vesicles from anterior midgut showed V-ATPase content and activity and K^+/H^+ antiport properties similar to those of posterior midgut GCAM (M. Huss and H. Wieczorek, personal communication).

9.4.3 Alkalinization

The three regions of the midgut display differences in the cellular distribution of carbonic anhydrase (Ridgeway and Moffett, 1986) and alkaline phosphatase (Azuma *et al.*, 1991). These differences might relate to the finding that alkalinization occurs mainly in the anterior and middle regions. Carbonic anhydrase is characteristically found in cells that mediate secretion or absorption of protons and/or bicarbonate (Maren, 1967). Abundant carbonic anhydrase is associated with the apical microvilli of goblet cells of anterior and middle midgut. In contrast, the bulk of carbonic anhydrase in posterior midgut was

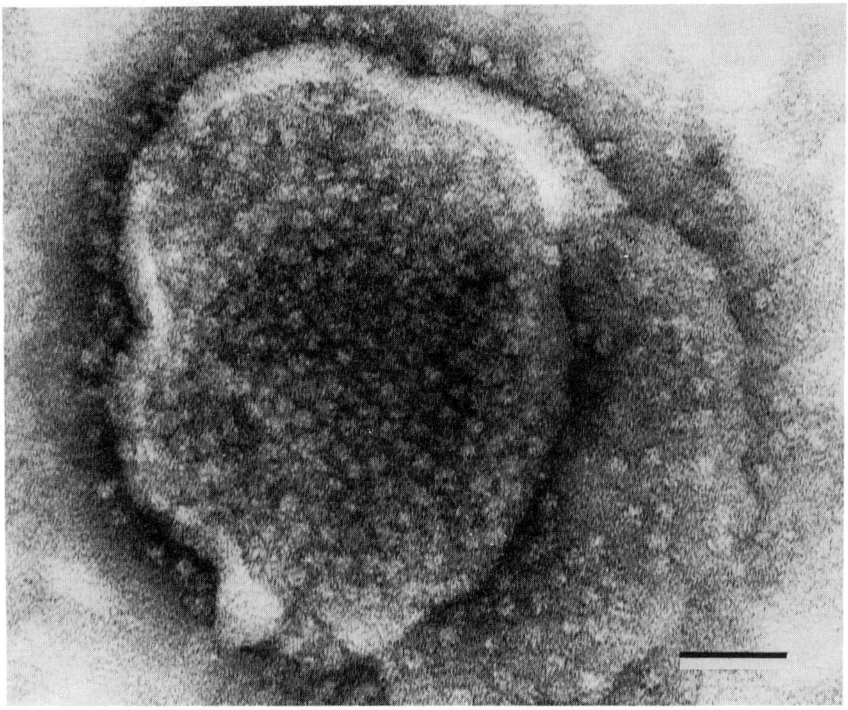

Figure 9.8 V-ATPase molecules in a GCAM vesicle of *Manduca sexta* midgut. A highly purified GCAM vesicle (Cioffi and Wolfersberger, 1983) negatively stained with 2% ammonium molybdate. V_1 sectors of individual V-ATPase molecules can be seen *en face* on the vesicle surface and in lateral view projecting from the edge of the flattened vesicle. Scale bar = 50 μm.

associated with the microvilli of columnar cells (Ridgeway and Moffett, 1986). These results correlate with the studies of regional alkalinization in that the anterior and middle regions appear to have the greatest alkalinizing power (Dow, 1984; Dow and Harvey, 1988; Chamberlin, 1990b; Moffett, 1994). The midgut possesses two forms of alkaline phosphatase. A membrane-bound form is covalently bound to the apical membranes of columnar cells, whereas a 'soluble' form (presumably an extrinsic membrane protein) is associated with the apical microvilli of goblet cells (Akai, 1969; Azuma and Eguchi, 1989; Azuma *et al.*, 1991) and is also present in the goblet matrix. The soluble form was shown to be an HCO_3^--sensitive Mg^{2+} ATPase (Azuma *et al.*, 1991), possibly identical with the HCO_3^--sensitive ATPase earlier reported from midgut by Deaton (1984). It is tempting to associate the alkalinizing activity of the midgut with this enzyme, but two factors suggest that caution is appropriate. First, the soluble alkaline phosphatase is found throughout the midgut, but is most abundant in the posterior midgut, the most weakly alkalinizing region. Second, no bicarbonate-based alkalinization can account for the very high pH values attained by the midgut (Moffett and Cummings, 1994).

9.5 MOLECULAR CHARACTERIZATION

9.5.1 V-ATPase

The insect plasma membrane V-ATPase consists of at least nine subunits (Wieczorek, 1992). Six of these with relative molecular masses (M_r) of 67 kDa, 56 kDa, 43 kDa, 28 kDa, 17 kDa and 14 kDa are common to all V-ATPases. Three others of 40 kDa, 20 kDa (both glycoproteins) and 16 kDa so far have no correspondents in other V-ATPases (Figure 9.9) (Russell *et al.*, 1992). Six subunits have been sequenced. Subunit A (67 kDa) is 83% identical to the corresponding subunit isoform of bovine V-ATPases (Gräf *et al.*, 1992). Subunit B (56 kDa) shows a high degree of identity to all eukaryotic V-ATPases (Novak *et al.*, 1992). Subunit E (28 kDa) shows regions of high identity with mammalian and yeast V-ATPases in addition to a number of highly differing regions (Gräf *et al.*, 1994a). The 17 kDa subunit is a proteolipid highly homologous to that found in other V-ATPases (Dow *et al.*, 1992; Pietrantonio and Gill, 1993).

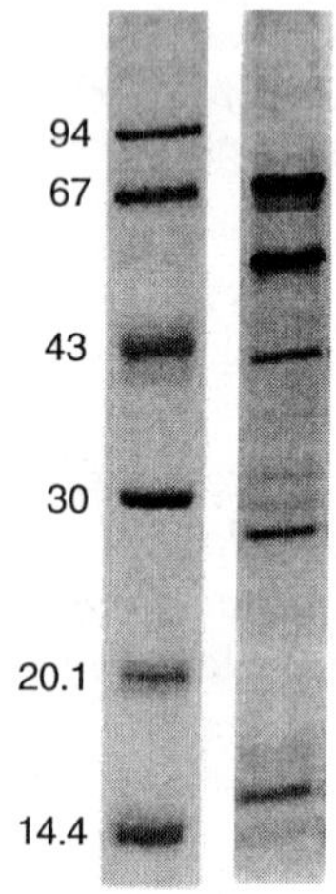

Figure 9.9 Subunit pattern of insect plasma membrane V-ATPase. SDS-polyacrylamide gel electrophoresis. Left lane: protein standards with relative molecular masses × 10^{-3}. Right lane: V-ATPase purified from the GCAM of *Manduca sexta* midgut (methods described in Schweikl *et al.*, 1989).

The 14 kDa subunit was first found in the insect V-ATPase. This subunit may be involved in the energy transmission from ATP hydrolysis to H^+ extrusion since antibodies to this subunit recognized only the ATP-loaded conformation of the enzyme (Gräf *et al.*, 1994b). Finally, there is a subunit with 13 kDa based on its sequence. This subunit shows anomalous 16 kDa in sodium dodecyl sulphate electrophoresis, probably because of its high content of acidic amino acids (Lepier *et al.*, 1996).

Subunit stoichiometry is believed to be similar in all V-ATPases (Forgac, 1992). The catalytic subunit A and the regulatory subunit B appear to bind ATP and both exist in three copies each. Six copies of the 17 kDa subunit compose the proton channel; they bind *N,N'*-dicyclohexylcarbodiimide (DCCD) and are believed to be the target of the specific V-ATPase inhibitors, bafilomycin A_1 and folimycin (Bowman *et al.*, 1988; Dröse *et al.*, 1993; Crider *et al.*, 1994; Zhang *et al.*, 1994). The other subunits are probably present as single copies. As in all V-ATPases, the peripheral subunits 67 kDa (A), 56 kDa (B), 28 kDa (E), 13 kDa and 14 kDa belong to the V_1 sector that is stripped from the membrane-integrated V_0 sector by chaotropic anions (Gräf *et al.*, 1994b). Thus far, no K^+/H^+ antiport has been chemically isolated nor has an immunochemical identification been made, despite recent attempts (Lepier *et al.*, 1994).

9.6 INTEGRATION OF THE SYSTEM

Three different manifestations of the V-ATPase must be addressed. Transepithelial secretion of K^+ is the main reflection of the V-ATPase during SC and this is seen in all portions of the midgut. In OC conditions, the main reflections are the production of a transepithelial voltage and luminal alkalinization. The minimum requirement for integration is an explanation of the events across the GCAM, of how these events are translated to secretion of K^+ during SC and of how they are translated to alkalinization of anterior or middle midgut during OC.

9.6.1 How do GCAM processes interact *in vivo*?

In the intact goblet cavity, the results of the combination of V-ATPase and $K^+/2H^+$ antiport depend on the relative turnover rates of the pump and the antiport. If the ATPase pumps 1 H^+ for each K^+ extruded, then the antiport must get the additional H^+ from the goblet cavity. The result of this would be that the goblet cavity would become alkalinized and K^+ would be extruded from the cell. Goblet cavity pH is 0.15–0.25 pH units higher than cytoplasmic pH and $[K^+]$ is incrementally higher in goblet cavity than in cells (Chao *et al.*, 1991). The problem here is that the pump–antiport combination is electrically neutral. One H^+ is pumped out and one K^+ is exchanged for 2 H^+ coming back into the cell. This system could not produce a voltage across GCAM and could only sustain one if the membrane were totally impermeable to other ions. In vesicles, at least, there is a proton leak. At sufficiently high [Cl^-], vesicles also show a Cl^- leak (Wieczorek, 1992). Such chloride gating of a Cl^- channel has previously been noted in frogs, toads and crayfish (Larsen, 1991; Zetino and Kirschner, 1993). The measured Cl^- activities in cell and goblet cavity may be high enough to open this leak; at any rate, Cl^- leaves the intact goblet cavity when the tissue is bathed in Cl^--free solution (Chao *et al.*, 1989).

It thus seems necessary for the ATPase to pump 2 H^+ for each K^+ exchanged. This coupling results in a continuous transfer of charge and will produce and maintain a positive V_{cg}. It will produce the slight elevation of (K^+) observed in goblet cavity and can account for the equivalence of I_{sc} and net K^+ transport.

9.6.2 How does K^+ pass from goblet cavity to gut lumen?

The presumption is that K^+ flows from goblet cavity to lumen across the valve. There is certainly a strong driving force for such movement (cf. Figure 9.2). In SC posterior midgut, this electrochemical driving force is about 80 mV. It is somewhat lower in OC preparations, but still favourable for movement of K^+ from goblet cavity to midgut lumen. In contrast, the driving force across the columnar cell apical membrane is always unfavourable to K^+ secretion. Further, the voltage between the goblet cavity and the midgut lumen during SC varies linearly with I_{sc} as would be expected for a current passing through the valve resistance (Moffett and Koch, 1988a; Chao *et al.*, 1989).

Several kinds of data pose problems for this explanation. Dow and Peacock (1989) obtained an average of 34 mΩ for the electrical resistance across single valve structures, more than five times the resistance they found from cell to lumen. Chao *et al.* (1989) found that when the tissue is maintained in Cl^--free solution, Cl^- leaches out of the goblet cavity with a half time of about 35 min, at least four times slower than from cells. The short circuit current delivered by one goblet cell is in the range of 1 nA. The diameter of any pore that would allow that much current to flow with a driving force of 50 mV (roughly V_g) would have to be at least 0.5 μm. A pore so large would be unselective to ions. In the SC posterior midgut, H^+ has a driving force that is comparable to that of K^+, but has a mobility about five times higher. With a non-selective pore, one would expect to see H^+ rather than K^+ secretion. In the middle midgut, the driving force for H^+ is much higher than it is for K^+ (the pH difference of up to 4.5 units is the equivalent of 260 mV). If the valve passage were an open pore, the middle midgut would be expected to be a strongly acidifying tissue. Finally, it has been found that tetramethylammonium did not enter or leave the goblet cavity via a direct path to the midgut lumen (Koch and Moffett, 1995; Moffett and Koch, 1992; Moffett *et al.*, 1995).

Because the K^+ pumped into the goblet cavity is balanced by K^+ entering the lumen and because the positive goblet cavity to lumen voltage varies with I_{sc}, we tentatively conclude that K^+ enters the midgut lumen through the goblet valve during SC. It probably does not go through a wide aqueous passage, but rather through a K^+ selective channel or carrier. Any carrier must carry a net of one positive charge from goblet cavity to lumen, precluding either a neutral K^+/H^+ exchange or the $K^+/2H^+$ antiport.

9.6.3 How does luminal alkalinization occur?

One of the prime barriers to understanding alkalinization is the paucity of published data on voltages and ionic activities in the anterior or middle midgut under OC conditions. Only Dow and Peacock (1989) have published data for voltages of cells and goblet cavities. There are no published data of cellular or goblet cavity pH. Chao, Moffett, Woods and Koch have a small amount of unpublished data on voltages, K^+ activities and pH from middle midgut under both SC and OC (pH = 6.8 on basal and 8.0 on apical side) conditions. Their voltage data are similar to the higher readings obtained by Dow and Peacock (1989). Although the data are both preliminary and incomplete, they are included here because nothing else is available (Table 9.1).

The first suggestion for the cause of alkalinization was that the transepithelial voltage in itself could cause the alkalinization by pulling H^+ across the epithelium (Dow, 1984; Dow and Harvey, 1988; Dow and

Table 9.1 Voltage, K^+ activity and pH data from the middle midgut under SC and OC conditions

I_{sc} or V_t ($\mu A/cm^2$ or mV)	V_b (mV)	V_g1 (mV)	$(K^+)_l$ (mM)	$(K^+)_a$ (mM)	pH_l	pH_a
Short circuit						
453±54.3	−35.3±2.88	61.2±4.35	83.3±4.72	87.6±10.7	7.17±.0.45	
n^a 5	45/5	12/3	16/4	12/3	1/1	
Open circuit. 6.8–8.0						
54.8±1.83	−30.1±0.69	38			7.14±.046	7.20
n 1	8/1	1/1			8/1	1/1

[a] n is the number of animals when referring to I_{sc} or V_{oc}. In the second to last columns, n depicts the number of impalements/the number of animals.

O'Donnel, 1990). In this theory, the role of the K^+ pump was simply to set up V_t. However, as mentioned above, when Moffett and Cummings (1994) simultaneously measured both transepithelial pH and voltage differences in middle midgut of moderately intact animals, the value of V_t they obtained was less than half that required to produce the observed pH difference. Although this voltage difference would cause some H^+ to move out of the midgut lumen, the maximum alkalinization that could be produced from their values of V_t is about 8.2 (assuming electrochemical equilibrium is reached), whereas their measured midgut pH was 10.05. Further, recent studies of isolated midgut revealed no passive transepithelial H^+ movement in response to values of V_t ranging from −100 to +100 mV (Bower, Koch and Moffett, unpublished results).

It has been suggested that $K^+/2H^+$ antiport could account for the *in vivo* alkalinization if it faced the lumen (Azuma *et al.*, 1995). The question then is: where could this antiport be located? The only lumen-facing cell membranes are the goblet valve and the CCAM. The question is whether a second antiport molecule in one of these locations could accomplish alkalinization. The energetic requirement that a $K^+/2H^+$ antiport extrude K^+ from a cell compartment into the lumen in exchange for 2 H^+ can be expressed in the following voltage equation:

$$58 \cdot \log \frac{(H_1)^2\,(K_x)}{(H_x)^2\,(K_1)} - V = \Delta\,\mu$$

where H_1 and K_1 are the hydrogen and potassium activities in midgut lumen, H_x and K_x in the tissue compartment and V is the voltage in the tissue compartment with reference to the lumen. If $\Delta\mu$ is positive, luminal alkalinization can occur, if it is negative, the antiport would acidify the lumen. Figure 9.10 shows net driving potentials, in mV, for posterior and middle midgut under SC and OC (pH = 6.8 basal and 8.0 apical). The driving potentials are computed for three different locations: across GCAM, goblet valve and CCAM. The data for posterior midgut were obtained in our laboratory (Moffett *et al.*, 1982; Moffett and Koch, 1988a,b; Chao *et al.*, 1989). The data for middle midgut, which should be regarded as tentative, were taken from Table 9.1 with the extrapolation that the values used for K^+ activities in cell and goblet cavity under OC were those determined from the SC conditions. No goblet punctures were obtained for middle midgut during SC, so no estimate can be made of driving forces across GCAM or valve of the SC middle midgut.

In all conditions for which there are data, the value of $\Delta\mu$ across GCAM is strongly positive, indicating that the energy is available for the V-ATPase – $K^+/2H^+$ antiport combination to act as a K^+ pump and

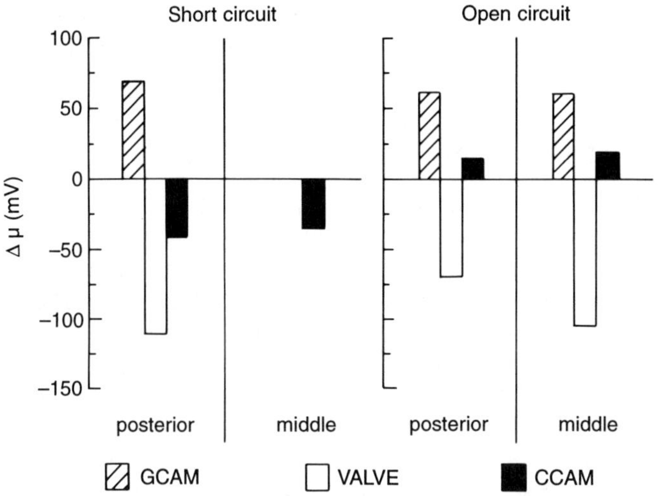

Figure 9.10 Driving force of $K^+/2H^+$ antiport for alkalinization. Δ μ across possible barriers in posterior and middlegut of *Manduca sexta*. The positive values for GCAM indicate that the $K^+/2H^+$ antiport will lead to K^+ secretion and alkalinization of goblet cavity. The negative valves for goblet valve indicate that the antiport located in the valve can never alkalinize gut lumen. An antiport located in the CCAM can alkalinize the lumen somewhat under open circuit conditions, but not during short circuit. (Data from Moffett *et al.*, 1982; Moffett and Koch, 1988a,b and Table 9.1.)

goblet cavity alkalinizer. However, if goblet cavity alkalinization causes luminal alkalinization, protons from the lumen must exchange with goblet cavity K^+ across the goblet valve. The value of Δμ computed across that valve was at least 70 mV negative in all cases.

The possibility of an antiport in CCAM is provocative. During SC, the antiport would work backwards. In other words, cell fluid is sufficiently acid to the luminal pH of 8 to force movement of H^+ into the lumen and consequent absorption of K^+. In OC, however, a $K^+/2H^+$ antiport would continue to alkalinize the midgut at the pH of 8, the value of luminal pH in these experiments. This is true in both posterior and middle midgut preparations. It thus appears that the CCAM is the only location where the antiport could alkalinize the midgut. On the basis of the values in Table 9.1, the highest pH that could be attained would be about 8.3. In living animals, where V_t is normally higher, the highest pH would be near 9.0. All measurements of the pH of middle midgut contents are above 10, so cell fluid must be above 8.3 if this mechanism is to explain alkalinization *in vivo*. Although either V_t or a $K^+/2H^+$ antiport on CCAM would alkalinize, the limiting pH for either case is in the range of 8.3. These processes cannot be used to assist any other mechanism. Indeed, they would oppose any other mechanism that was capable of elevating lumen pH above 8.3.

9.7 CONCLUSIONS AND FUTURE DIRECTIONS

A V-ATPase $K^+/2H^+$ antiport combination is located in GCAM and accounts adequately for all the data obtained concerning this membrane. This system is the ultimate energy source required both for K^+ secretion during SC and generation of V_t. In SC conditions, the ATPase probably operates two cycles for each K^+ exchanged. Although the K^+ pumped into goblet cavity is probably the K^+ secreted during SC, there is another element present in the goblet valve to explain the apparent selectivity of this route to K^+ over H^+. There is as yet no satisfactory explanation of luminal alkalinization during OC. Considerable advances have been made in the characterization of the chemistry and molecular biology of the V-ATPase.

There is still a good deal of work to be done at each level of integration. At the level of the intact cell, data are needed on ionic (including H^+) activities and voltages in each of the three portions of the midgut under open circuit conditions. Although the SC preparation was very useful in defining the locus of energy input, the short circuit condition can uncouple processes that are normally tied together.

In addition, more information is needed on current–voltage properties of the intact GCAM, including a complete and accurate description of the pump system. Extension of the I–E curve across GCAM shown in Figure 9.3, suggests that the pump might reach its limit when opposed by 100–120 mV. This is a lower voltage limit than is seen in other V-ATPases. It may well be a reflection of the proton leak that has been observed in vesicles, but it does not change the observation that there may be no net H^+ pumping during OC.

Perhaps the greatest need at the level of individual membranes is to isolate and localize any or all K^+/H^+ antiporters. Although the evidence for a $K^+/2H^+$ exchange in GCAM is completely convincing, the responsible molecule has not been identified either chemically or immunologically. There might be a K^+/H^+ antiport on the basal or the lateral membrane as well as on CCAM. Very little chemical work has been done on basal or lateral membranes and most of the work on CCAM has been directed toward the amino acid symports.

The increasing understanding of the V-ATPase molecule opens the way to its use as as marker for intracellular protein sorting to different target membranes. In development the goblet starts as an intracellular vacuole (Hakim *et al.*, 1988). However, V-ATPase sorting to this membrane continues even after the vacuole is inserted into the apical membrane domain (Sumner *et al.*, 1994). So far, the signals that target the V-ATPase to the GCAM are not known.

An important question relates to the endocrine control of midgut function in the developmental programme. V-ATPase activity and K^+ transport capacity are maintained at a high level throughout the feeding phase of each of the five larval instars, but are abruptly decreased after the cessation of feeding at the end of each instar, remain turned off during the period of moulting, and are restored within a period of two hours at the beginning of the next instar (Sumner *et al.*, 1994). Larval–larval moults are triggered by ecdysone secretion against a background of high juvenile hormone levels (Riddiford, 1980). Inactivation of the V-ATPase at the cessation of feeding for a larval–larval moult is apparently exerted by stripping the enzyme of peripheral V_1 sectors (Sumner *et al.*, 1994). In contrast, a separate mechanism involving decline in mitochondrial energy production has been proposed for the general loss of gut function that occurs at the beginning of prepupation wandering (Chamberlin, 1994). The hormones of insect development integrate gut function into the developmental programme; the cellular mechanisms of this control are just beginning to be investigated.

REFERENCES

Akai, H. (1969) Ultrastructural localization of phosphatases in the midgut of the silkworm *Bombyx mori*. *J. Insect Physiol.*, **15**, 1623–8.

Anderson, E. and Harvey, W.R. (1966) Active transport in the *Cecropia* midgut II. Fine structure of the midgut epithelium. *J. Cell Biol.*, **31**, 107–34.

Azuma, M. and Eguchi, M. (1989) Discrete localization of distinct alkaline phosphatase isozymes in the cell surface of silkworm midgut epithelium. *J. Exp. Zool.*, **251**, 108–12.

Azuma, M., Harvey, W.R. and Wieczorek, H. (1995) Stoichiometry of K^+/H^+ antiport helps to explain extracellular pH 11 in a model epithelium. *FEBS Lett.*, **361**, 153–6.

Azuma, M., Takeda, S., Yamamoto, H. *et al.* (1991) Goblet cell alkaline phosphatase isozymes in the cell surface of silkworm midgut epithelium. *J. Exp. Zool.*, **258**, 294–302.

Baldwin, K.M. and Hakim, R.S. (1991) Growth and differentiation of the larval midgut epithelium during molting in the moth, *Manduca sexta*. *Tissue Cell*, **32**, 411–22.

Bowman, E.J., Siebers, A. and Altendorf, K. (1988) Bafilomycins: a class of inhibitors of membrane ATPases from microorganisms, animal cells, and plant cells. *Proc. Natl Acad. Sci. USA*, **85**, 7972–6.

Chamberlin, M.E. (1990a). Ion transport across the midgut of the tobacco hornworm (*Manduca sexta*). *J. Exp. Biol.*, **150**, 425–42.

Chamberlin, M.E. (1990b) Luminal alkalinization by the isolated midgut of the tobacco hornworm (*Manduca sexta*). *J. Exp. Biol.*, **150**, 467–71.

Chamberlin, M.E. (1994) Developmental changes in midgut ion transport and metabolism in the tobacco hornworm (*Manduca sexta*). *Physiol. Zool.*, **67**, 82–94.

Chao, A.C., Koch, A.R. and Moffett, D.F. (1989) Active chloride transport in isolated posterior midgut of tobacco hornworm (*Manduca sexta*). *Am. J. Physiol.*, **257** R752–R761.

Chao, A.C., Koch, A.R. and Moffett, D.F. (1990) Basal membrane uptake in potassium-secreting cells of midgut of tobacco hornworm (*Manduca sexta*). *Am. J. Physiol.*, **258**, R112–R119.

Chao, A.C., Moffett, D.F. and Koch, A. (1991) Cytoplasmic pH and goblet cavity pH in the posterior midgut of the tobacco hornworm *Manduca sexta*. *J. Exp. Biol.*, **155**, 403–14.

Cioffi, M. (1979) The morphology and fine structure of the larval midgut of a moth (*Manduca sexta*) in relation to active ion transport. *Tissue Cell*, **11**, 467–79.

Cioffi, M and Harvey, W.R. (1981) Comparison of potassium transport in three structurally distinct regions of the insect midgut. *J. Exp. Biol.*, **91**, 103–16.

Cioffi, M. and Wolfersberger, M.G. (1983) Isolation of separate apical, lateral and basal plasma membrane from cells of insect epithelium. A procedure based on tissue organization and ultrastructure. *Tissue Cell*, **15**, 781–803.

Crider, B.P., Xie, X.S. and Stone, D.K. (1994) Bafilomycin inhibits proton flow through the H^+ channel of vacuolar proton pumps. *J. Biol. Chem.*, **269**, 17379–81.

Deaton, L.E. (1984) Tissue K^+ stimulated ATPase and HCO_3-stimulated ATPase in the tobacco hornworm, *Manduca sexta*. *Insect Biochem.*, **14**, 109–114.

Dow, J.A.T. (1984) Extremely high pH in biological systems: a model for carbonate transport. *Am. J. Physiol.*, **246**, R633–R635.

Dow, J.A.T. (1986) Insect midgut function. *Adv. Insect Physiol.*, **19**, 187–238

Dow, J.A.T., Goodwin, S.F. and Kaiser, K. (1992) Analysis of the gene encoding a 16 K proteolipid subunit of the vacuolar H^+-ATPase from *Manduca sexta* midgut and tubules. *Gene*, **122**, 355–60.

Dow, J.A.T., Gupta, B.L., Hall, T.A. and Harvey, W.R. (1984) X-ray microanalysis of elements in frozen-hydrated sections of an electrogenic K^+ transport system: the posterior midgut of tobacco hornworm (*Manduca sexta*) *in vivo* and *in vitro*. *J. Membr. Biol.*, **77**, 223–41.

Dow, J.A.T., and Harvey, W.R. (1988) Role of midgut electrogenic K^+ pump potential difference in regulating lumen K^+ and pH in larval Lepidoptera. *J. Exp. Biol.*, **140**, 455–63.

Dow, J.A.T. and O'Donnell, M.J. (1990) Reversible alkalinization by *Manduca sexta* midgut. *J. Exp. Biol.*, **150**, 247–56.

Dow, J.A.T. and Peacock, J.M. (1989) Microelectrode evidence for the electrical isolation of goblet cell cavities in *Manduca sexta* middle midgut. *J. Exp. Biol.*, **143**, 101–14.

Dröse, S., Bindseil, K.U., Bowman, E.J. *et al.* (1993) Inhibitory effect of modified bafilomycins and concanamycins on P-type and V-type adenosine triphosphatases. *Biochemistry*, **32**, 3902–6.

Ehrenfeld, J. and Garcia-Romeu, F. (1977) Active hydrogen excretion and sodium absorption through isolated frog skin. *Am. J. Physiol.*, **233**, F46–F54.

Ehrenfeld, J., Garcia-Romeu, F. and Harvey, B.J. (1985) Electrogenic active proton pump in *Rana esculenta* skin and its role in sodium ion transport. *J. Physiol.*, **359**, 331–55.

Endo, Y. and Nishiitsutsuji-Uwo, J. (1982) Exocytotic release of secretory granules from endocrine cells in the midgut of insects. *Cell Tissue Res.*, **222**, 515–22.

Forgac, M. (1992) Structure, function and regulation of the coated vesicle V-ATPase. *J. Exp. Biol.*, **172**, 155–69.

Giordana, B., Leonardi, M.G., Tasca, M. *et al.* (1994) The amino acid/K^+ symporters for neutral amino acids along the midgut of lepidopteran larvae: functional differentiations. *J. Insect Physiol.*, **40**, 1059–68.

Gluck, S. and Nelson, R. (1992) The role of V-ATPase in renal epithelial H^+ transport. *J. Exp. Biol.*, **172**, 205–18.

Gräf, R., Harvey, W.R. and Wieczorek, H. (1994a) Cloning, sequencing and expression of cDNA encoding an insect V-ATPase subunit E. *Biochim. Biophys. Acta.*, **1190**, 193–6.

Gräf, R., Lepier, A., Harvey, W.R. and Wieczorek, H. (1994b) A novel 14-K V-ATPase subunit in tobacco hornworm midgut. *J. Biol. Chem.*, **269**, 3737–74.

Gräf, R., Novak, F.J.S., Harvey, W.R. and Wieczorek, H. (1992) Cloning and sequencing of cDNA encoding the putative insect plasma membrane V-ATPase subunit-A. *FEBS Lett.*, **300**, 119–22.

Hakim, R.S., Baldwin, K.M. and Bayer, P.E. (1988) Cell differentiation in the embryonic midgut of the tobacco hornworm, *Manduca sexta*. *Tissue Cell*, **20**, 51–62.

Harvey, W.R., Cioffi, M. and Wolfersberger, M.G. (1981) Portasomes as coupling factors in active ion transport and oxidative phosphorylation. *Am. Zool.*, **21**, 777–91.

Harvey, W.R. and Nedergaard, S. (1964) Sodium-independent active transport of potassium in the isolated midgut of *Cecropia* silkworm. *Proc. Natl Acad. Sci. USA*, **51**, 757–65.

Jägar, D., Novak, F.J.S., Harvey, W.R. *et al.* (1996) Temporal and spatial distribution of V-ATPase and its mRNA in the midgut of moulting *Manduca sexta* larvae. *J. Exp. Biol.*, **199**, 1019–1027.

Jungreis, A.M. and Vaughn, G.L. (1977) Insensitivity of lepidopteran tissues to ouabain: absence of ouabain binding of Na^+, K^+-ATPases in larval and adult midgut. *J. Insect Physiol.*, **23**, 503–9.

Klein, U. (1992) The insect V-ATPase, a plasma membrane proton pump energizing secondary active transport: immunological evidence for the occurrence of a V-ATPase in insect ion transporting epithelia. *J. Exp. Biol.*, **172**, 345–54.

Klein, U., Löffelmann, G. and Wieczorek, H. (1991) The midgut as a model system for insect K^+-transporting epithelia – Immunocytochemical localization of a vacuolar-type H^+ pump. *J. Exp. Biol.*, **161**, 61–75.

Koch, A. and Moffett, D.F. (1995) Electrophysiology of K^+ transport by midgut epithelium of lepidopteran insect larvae. IV. A multicompartmental model accounts for tetramethylammonium entry into goblet cavities. *J. Exp. Biol.*, **198**, 2115–2125.

Larsen, E.H. (1991) Chloride transport by high-resistance heterocellular epithelia. *Physiol. Rev.*, **71**, 235–83.

Lepier, A., Azuma, M., Harvey, W.R. and Wieczorek, H. (1994) K^+/H^+ antiport in the tobacco hornworm midgut. The K^+ transporting component of the K^+ pump. *J. Exp. Biol.*, **196**, 361–73.

Lepier, A., Gräf, R., Azuma, M. *et al.* (1996) The peripheral complex of the tobacco horn worm V-ATPase contains a novel 13 kDa subunit G. *J. Biol. Chem.* (in press).

Maddrell, S.H.P. and O'Donnell, M.J. (1992) Insect Malpighian tubules: V-ATPase action in ion and fluid transport. *J. Exp. Biol.*, **172**, 417–29.

Maren, T.H. (1967) Carbonic anhydrase: chemistry, physiology, and inhibition. *Physiol. Rev.*, **47**, 595–781.

Moffett, D.F. (1980) Voltage–current relation and K^+ transport in tobacco hornworm (*Manduca sexta*) midgut. *J. Memb. Biol.*, **54**, 213–19.

Moffett, D.F. (1994) Recycling of K^+ acid–base equivalents, and fluid between gut and hemolymph in lepidopteran larvae. *Physiol. Zool.*, **67**, 68–81.

Moffett, D.F. and Cummings, S.A. (1994) Transepithelial potential and alkalinization in an *in situ* preparation of tobacco hornworm (*Manduca sexta*) midgut. *J. Exp. Biol.*, **194**, 341–5.

Moffett, D.F., Hudson, R.L., Moffett, S.B. and Ridgway, R.L. (1982) Intracellular K^+ activities and cell membrane potentials in a K^+-transporting epithelium, the midgut of tobacco hornworm (*Manduca sexta*). *J. Membr. Biol.*, **70**, 59–68.

Moffett, D.F. and Koch, A.R. (1982) The kinetics of active K transport by the midgut of lepidopteran larvae: effects of divalent ions. *J. Exp. Biol.*, **105**, 403–5.

Moffett, D.F. and Koch, A.R. (1985) Barium modifies the concentration dependence of active potassium transport by insect midgut. *J. Membr. Biol.*, **86**, 89–97.

Moffett, D.F. and Koch, A.R. (1988a) Electrophysiology of K^+ transport by midgut epithelium of lepidopteran insect larvae. I. The transbasal electrochemical gradient. *J. Exp. Biol.*, **135**, 25–38.

Moffett, D.F. and Koch, A.R. (1988b) Electrophysiology of K^+ transport by midgut epithelium of lepidopteran insect larvae. II. The transapical electrochemical gradients. *J. Exp. Biol.*, **135**, 39–49.

Moffett, D.F. and Koch, A. (1991) Lidocaine and barium distinguish separate routes of transbasal K^+ uptake in the posterior midgut of the tobacco hornworm (*Manduca sexta*). *J. Exp. Biol.*, **157**, 243–56.

Moffett, D.F. and Koch, A. (1992) Driving forces and pathways for H^+ and K^+ transport in insect midgut goblet cells. *J. Exp. Biol.*, **172**, 403–15.

Moffett, D.F., Koch, A. and Woods, R. (1995) Electrophysiology of K^+ transport by midgut epithelium of lepidopteran insect larvae III. Goblet valve patency. *J. Exp. Biol.*, **198**, 2013–2113.

Moffett, D.F. and Lewis, S.A. (1990) Cation channels of insect midgut goblet cells: conductance diversity and Ba^{++} activation. *Biophys. J.*, **57**, 85a.

Nedergaard, S. (1972) Active transport of α-aminoisobutyric acid by the isolated midgut of *Hyalophora cecropia*. *J. Exp. Biol.*, **56**, 167–72.

Nijout, H.F. (1975) Excretory role of the midgut in larvae of the tobacco hornworm. *J. Exp. Biol.*, **62**, 221–30.

Novak, F.J.S., Gräf, R., Waring, R.B. *et al.* (1992) Primary structure of V-ATPase subunit-B from *Manduca sexta* midgut. *Biochim. Biophys. Acta*, **1132**, 67–71.

Pietrantonio, P.V. and Gill, S.S. (1993) Sequence of a 17K vacuolar H^+ ATPase proteolipid subunit from insect midgut and Malpighian tubules. *Insect Biochem. Mol. Biol.*, **23**, 675–80.

Ramsay, J.A. (1976) The rectal complex in the larvae of lepidoptera. *Proc. R. Soc. B.*, **274**, 205–26.

Riddiford, L.M. (1980) Interaction of ecdysteroids and juvenile hormone in the regulation of larval growth and metamorphosis of the tobacco hornworm, in *Progress in Ecdysone Research* (ed. J.A. Hoffman), Elsevier, North Holland, pp. 409–430.

Ridgway, R. and Moffett, D.F. (1986) Regional differences in the histochemical localization of carbonic anhydrase in the midgut of tobacco hornworm (*Manduca sexta*). *J. Exp. Zool.*, **237**, 407–12.

Russell, V.E.W., Klein, U., Reuveni, M. *et al.* (1992) Antibodies to mammalian and plant V-ATPases cross-react with vacuolar-type ATPase from insect cation-transporting plasma membranes. *J. Exp. Biol.*, **166**, 131–43.

Sacchi, V.F., Parenti, P., Perego, C. and Giordana, B. (1994) Interaction between Na^+ and K^+-dependent amino acid transport in midgut BBMV from *Philosamia cynthia* larvae. *J. Insect. Physiol.*, **40**, 68–74.

Schirmanns, K. and Zeiske, W. (1994a) An investigation of the midgut K^+ pump of the tobacco hornworm (*Manduca sexta*) using specific inhibitors and amphotericin B. *J. Exp. Biol.*, **188**, 191–204.

Schirmanns, K. and Zeiske, W. (1994b) K^+ channel permeation and block in the midgut epithelium of the tobacco hornworm *Manduca sexta*. *J. Exp. Biol.*, **197**, 179–204.

Schultz, T.W., Lozano, G. and Cajina-Quezada, M. (1981) Histochemical analysis of the goblet cell matrix in the larval midgut of *Manduca sexta*. *Trans. Am. Microsc. Soc.*, **100**, 204–9.

Schweikl, H., Klein, U., Schindlbeck, M. and Wieczorek, H. (1989) A vacuolar-type ATPase, partially purified from potassium transporting plasma membranes of tobacco hornworm midgut. *J. Biol. Chem.*, **264**, 11136–42.

Sumner, J.P., Dow, J.A.T., Early, F.G.P. *et al.* (1994) Regulation of plasma membrane V-ATPase activity by dissociation of peripheral subunits. *J. Biol. Chem.*, **270**, 5649–53.

Sutcliffe, D.W. (1963) The chemical composition of hemolymph in insects and other arthropods in relation to their phylogeny. *Comp. Biochem. Physiol.*, **9**, 121–35.

Thomas, M.V. and May, T.E. (1984) Active potassium ion transport across the caterpillar midgut. II. Intracellular microelectrode studies. *J. Exp. Biol.*, **108**, 293–304.

Wessing, A., Bertram, G. and Zierold, K. (1993) Effects of Bafilomycin A_1 and amiloride on the apical potassium and proton gradients in *Drosophila* Malpighian tubules studied by X-ray microanalysis and microelectrode measurements. *J. Comp. Physiol. B*, **163**, 452–62.

Wieczorek, H. (1992) The insect V-ATPase, a plasma membrane proton pump energizing secondary active transport. Molecular analysis of electrogenic potassium transport in the tobacco hornworm midgut. *J. Exp. Biol.*, **172**, 335–43.

Wieczorek, H., Putzenlechner, M., Zeiske, W. and Klein, U. (1991) A vacuolar-type proton pump energizes K^+/H^+ antiport in animal plasma membrane. *J. Biol. Chem.*, **266**, 15340–7.

Wieczorek, H., Weerth, S., Schindlbeck, S. and Klein, U. (1989) A vacuolar-type proton pump in a vesicle fraction enriched with potassium transporting plasma membranes from tobacco hornworm midgut. *J. Biol. Chem.*, **264**, 11143–8.

Wieczorek, H., Wolfersberger, M.G., Cioffi, M. and Harvey, W.R. (1986) Cation-stimulated ATPase activity in purified plasma membranes from tobacco hornworm midgut. *Biochem. Biophys. Acta*, **857**, 271–81.

Wood, J.L. and Harvey, W.R. (1976) Active transport of calcium across the isolated midgut of *Hyalophora cecropia*. *J. Exp. Biol.*, **65**, 313–20.

Wood, J.L., Jungreis, A.M. and Harvey, W.R. (1975) Active transport of magnesium across the isolated midgut of *Hyalophora cecropia*. *J. Exp. Biol.*, **63**, 347–60.

Zeiske, W. and Marin, H. (1992) K^+ current simulation by Cl^- in the midgut epithelium of tobaco hornworm (*Manduca sexta*). II. Analysis of Ba^{2+}-induced K^+ channel noise. *J. Comp. Physiol.*, **162**, 340–4.

Zeiske, W., Schröder, H. and Alpert, G. (1992) K^+ current stimulation by Cl^- in

the midgut epithelium of tobacco hornworm (*Manduca sexta*). I. Kinetics and effect of Cl^--site-specific agents. *J. Comp. Physiol.*, **162**, 331–9.
Zeiske, W., Van Driessche, W. and Ziegler, R. (1986) Current–noise analysis of the basal route for K^+ ions across a K^+-secreting insect midgut epithelium (*Manduca sexta*). *Pflügers Arch.*, **407**, 657–63.
Zerahn, K. (1985) Water transport across the short-circuited midgut of the American silkworm. *J. Exp. Biol.*, **116**, 481–5.
Zetino, A.M. and Kirschner, L.B. (1993) On the existence of epithelial chloride channels in intact frogs and crayfish. *J. Exp. Zool.*, **265**, 366–72.
Zhang, J.M., Feng, Y. and Forgac, M. (1994) Proton conduction and bafilomycin binding by the V_0 domain of the coated vesicle V-ATPase. *J. Biol. Chem.*, **269**, 23518–23.

10

Amino acid absorption

V.F. Sacchi and M.G. Wolfersberger

10.1 INTRODUCTION

Amino acid absorption in insect midgut has been discussed previously in reviews of insect midgut function (Dow, 1986) and amino acid transport in invertebrates (Nedergaard, 1977). About 40% of the latter review was devoted to a discussion of amino acid absorption in insect midgut. More than half of that discussion concerned studies of amino acid absorption in midguts of lepidopteran larvae. The emphasis on larval lepidopteran midgut was reasonably representative of the literature at the time of the review. Since then the literature on amino acid absorption by insect midguts has become even more unbalanced in favour of lepidopteran larvae. Consequently, nearly 90% of the information discussed in this chapter is derived from studies conducted on midguts of lepidopteran larvae. Therefore, although we have attempted to discuss all studies of amino acid absorption in insect midguts published since those covered in the review by Nedergaard (1977), this chapter unavoidably focuses on amino acid absorption in midguts of lepidopteran larvae.

After a brief discussion of absorption mechanisms in general and a short review of the structure and steady-state physiological conditions in larval lepidopteran midgut, there is a comprehensive discussion of neutral amino acid (leucine) absorption in the midgut of *Philosamia cynthia* larvae. This is easily the most extensively studied paradigm for amino acid absorption in insect midgut. The remainder of the chapter discusses what is known about absorption of other amino acids in *P. cynthia* as well as amino acid absorption in midguts of several other

Biology of the Insect Midgut.
Edited by M.J. Lehane and P.F. Billingsley.
Published in 1996 by Chapman & Hall, London.
ISBN 0 412 61670 X.

lepidopterans plus one coleopteran and one orthopteran. In these sections we have attempted to highlight similarities and differences between amino acid uptake systems in *P. cynthia* and other insects as well as interesting questions to be addressed in future research.

10.2 ABSORPTION MECHANISMS

Solutes can cross cell membranes by simple diffusion, simple facilitated diffusion, coupled facilitated diffusion (co-transport) or primary transport. The latter mechanism, with its high ratio of energy consumed to solute translocated, is generally not utilized for nutrient absorption by metazoan gut cells (Stevens *et al.*, 1984). Since the ability of a solute to cross a cell membrane by simple diffusion is related to its lipophilicity, this mechanism is of limited usefulness for the absorption of hydrophilic solutes like most amino acids and sugars. Hydrophilic nutrients that are abundant in the diet may be absorbed by facilitated diffusion (uniport). Essential nutrients or ones that are less abundant in the diet are more frequently absorbed by co-transport (symport or antiport). Co-transport processes are driven by energy stored in ionic electrochemical potential differences between intracellular and extracellular compartments. Nutrient uptake occurs through transport proteins (symporters) which couple the flux of the nutrient moving against its electrochemical potential gradient with the flux of an ion moving down its electrochemical potential gradient. Symport and antiport mechanisms are ubiquitous in absorptive and secreting epithelia which carry out net transepithelial fluxes of ions and metabolites. Usually the driving cations are H^+ or Na^+ and their gradients between compartments are maintained by primary active transport. In epithelia these primary transporters are biochemically identified with ATPases.

10.3 TRANSPORT PHYSIOLOGY OF LARVAL LEPIDOPTERAN MIDGUT

Since the study by Tobias (1948) and the subsequent systematic analysis by Florkin and Jeuniaux (1974), it has been apparent that the ionic composition of the haemolymph in different orders of insects has a great variability. Lepidoptera represent an extreme case in which the $[Na^+]/[K^+]$ ratio is <1 and the Mg^{2+} and Ca^{2+} concentrations are very high. As a consequence of these ionic features, the physiology of their plasma membranes has been considered an intriguing subject which has received increasing interest in recent years.

The midgut of lepidopteran larvae is a K^+-secreting epithelium composed mainly of two types of enterocytes: goblet and columnar cells. The former are responsible for net potassium secretion and determine the high transepithelial electrical potential difference ($\Delta\Psi$) as well as the large pH difference (ΔpH) between lumen (pH $\approx$ 10) and cells (pH $\approx$ 7) (Harvey and Nedergaard, 1964; Zerahn, 1977; Harvey *et al.*, 1983; Dow, 1984, 1992; Chao *et al.*, 1990; Wieczorek *et al.*, 1991; Chapter 9), whereas the latter are absorptive cells with a well-developed brush border.

The absorption of amino acids, measured in isolated midguts of lepidopteran larvae, is sodium independent, sensitive to the trans-epithelial $\Delta\Psi$ and drastically reduced by anoxia as well as the metabolism inhibitor dinitrophenol (DNP) (Nedergaard, 1972, 1973, 1977; Sacchi and Giordana, 1980; Sacchi *et al.*, 1981, 1984; Giordana *et al.*, 1982). Absorption includes transport across the mucosal border, mixing in the cytoplasmic pool and exit from the cell across the basal membrane into the blood. This entire process can be fractionated through the use of purified plasma membranes which, under appropriate conditions, form sealed vesicles so that an intravesicular aqueous space with a known composition is separated from the external medium. These preparations, which have the advantage of avoiding the interference of cell metabolism, have been used to demonstrate the existence of a unique K^+-dependent amino acid symport localized on the brush border membrane of columnar cells (Hanozet *et al.*, 1980).

10.4 NEUTRAL AMINO ACID ABSORPTION IN THE MIDGUT OF *PHILOSAMIA CYNTHIA* LARVAE

Hanozet *et al.* (1980) reported that phenylalanine uptake was energized either by sodium or potassium gradients in brush border membrane vesicles (BBMVs) from *Philosamia cynthia* midgut. In experiments with mammalian BBMVs a potassium gradient is often used as a control to measure a merely equilibrative uptake of a nutrient molecule into membrane vesicles, but in BBMVs from *Philosamia cynthia* midgut a K^+ gradient was able to drive a four-fold accumulation of phenylalanine over the equilibrium value. To evaluate if this symport could actually occur in the midgut *in vivo*, the electrochemical potential gradient for potassium was estimated. On the basis of the values of transmucosal electrical potential difference (up to 150 mV, lumen positive) and of potassium concentrations, large negative electrochemical potential differences between lumen and cells were calculated to be present under both *in vivo* or *in vitro* conditions. It was apparent that the negative value of the electrochemical potential differences, which means a favourable entry from the lumen into the cell, is almost completely dependent on the electrical component. This conclusion was in agreement with experimental results previously reported in *Hyalophora cecropia* where a 70% drop in the net aminoisobutyric acid (AIB) transport was observed when the transepithelial electrical potential was eliminated by short circuiting the tissue (Nedergaard, 1973). On the basis of these results, a general model for amino acid absorption in the midgut of lepidopteran larvae was proposed (Giordana *et al.*, 1982). Figure 10.1 shows a schematic model for leucine uptake in a columnar cell of *Philosamia cynthia* midgut. K^+ is the physiological driver of the symport responsible for amino acid absorption in the midgut of lepidopteran larvae. Both goblet and columnar cells co-operate in ionic homeostasis and metabolite absorption. A vacuolar-type proton ATPase coupled with a K^+/nH^+ antiporter localized on the apical membrane of goblet cells energizes a number of symports on the brush border of columnar cells. In vertebrates and in some insects two different portions of the plasma membrane of the same cell, namely the basolateral membrane where a Na^+/K^+-ATPase activity is located, and the brush border membrane of the enterocyte containing several different transport proteins, are involved in transepithelial absorption.

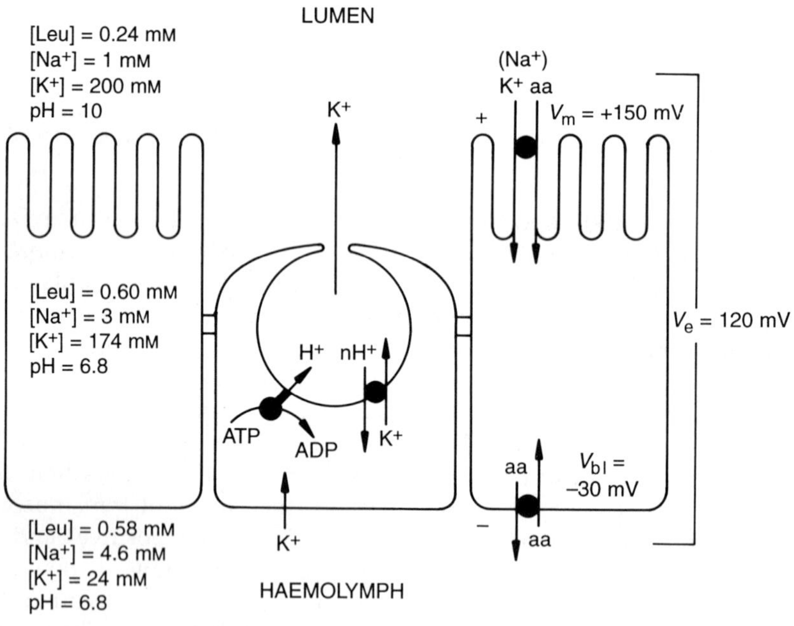

Figure 10.1 Schematic model of a goblet cell between two columnar absorptive cells in the midgut of *Philosamia cynthia* larva. On the left, leucine, sodium and potassium concentrations and pH values in the lumen, midgut tissue and haemolymph. In the middle, mechanisms involved in potassium transport. On the right, mechanisms involved in amino acid (aa) absorption and transmucosal (V_m), transbasolateral (V_{bl}) and transepithelial (V_e) electrical potential differences.

Little is known about the exit of amino acid from the cell through the basolateral membrane. One study reports the presence of an amino acid exchange mechanism in the basolateral membrane of the midgut epithelium from the larva of *Hyalophora cecropia* (Nedergaard, 1981). The existence of this exchange mechanism could explain the accumulation of AIB in the enterocytes when the labelled amino acid is present at the haemolymph side of the midgut epithelium of *Philosamia cynthia* larva (Sacchi *et al.*, 1984).

All the amino acids so far tested are symported with potassium in BBMVs from *Philosamia cynthia* midgut, therefore, amino acid symporters on the apical border of columnar cells seem to belong to a new group of transporters which couple intracellularly directed K^+ and amino acid fluxes. However, studies with brush-border membrane vesicles showed that this symport is not strictly K^+-dependent since sodium and in some case lithium can activate the transport. The net absorption of amino acids in the isolated midgut of *Philosamia cynthia* was only slightly diminished by replacing luminal K^+ with Na^+ which also had a large negative electrochemical potential difference between lumen and absorptive cell (Giordana *et al.*, 1982). Table 10.1 shows the effects of gradients (100 mM, out; 0 mM in) of different alkali cations on amino acid uptake in BBMVs. The experiments were performed at pH 7.5.

Several transport systems with different amino acid and cation specificities have been identified in *Philosamia cynthia* midgut. Among these the neutral amino acid transport system is the best characterized and among the amino acids that can be handled by this system, L-leucine is the most carefully studied. The accumulation ratio, i.e. the ratio between the concentration of leucine inside the vesicles at the peak of the uptake curve, which occurs after ~3 min of incubation, and the equilibrium value, at 60 min, is about 4. This K^+-dependent accumulation of leucine is abolished by addition of the K^+ ionophore valinomycin, which dissipates the potassium gradient, and by phenylglyoxal (PGO) a specific reagent for arginine residues. Kinetic analysis of the interaction of PGO with K^+-dependent leucine transport in *Philosamia cynthia* midgut revealed that inhibition was related to a decrease in the V_{max} value, and that neither leucine nor K^+ were able to prevent inhibition. These results suggest that the PGO-sensitive arginine residues are not involved in either the K^+ or the leucine binding sites (Parenti *et al.*, 1994).

10.4.1 Effects of $\Delta\Psi$ and ΔpH

Most leucine uptake takes place through a potassium-dependent symport mechanism which is strongly dependent on the transmembrane potential as indicated by three different lines of experimental evidence: (1) the uptake value depends on the anions of the potassium salts used

Table 10.1 Effect of monovalent cations on amino acid uptake in BBMV

	L-*Phe* (1 mM)	L-*His* (0.5 mM)	L-*Lys* (0.5 mM)	L-*Leu* (0.5 mM)	L-*Ala* (0.5 mM)	D-*Ala* (0.5 mM)
	(pmol/mg protein/3 min)		*(pmol/mg protein/min)*			*(pmol/mg/3 min)*
LiCl	3940 ± 833	2663 ± 143	2004 ± 41	1316 ± 52	3496 ± 25	850 ± 63
NaCl	5893 ± 940	2178 ± 25	1644 ± 53	2309 ± 144	8367 ± 151	6828 ± 457
KCl	7186 ± 600	2129 ± 222	2447 ± 73	2275 ± 58	8532 ± 61	3406 ± 277
RbCl	2890 ± 890	644 ± 40	1054 ± 61	407 ± 70	1377 ± 16	818 ± 37
CsCl	1660 ± 73	529 ± 17	682 ± 15	–	1144 ± 50	1529 ± 357
CholineCl	1515 ± 465	–	–	–	–	–
TMACl	–	400 ± 13	427 ± 8	–	–	–

in a way presumably related to anion permeabilities and therefore to $\Delta\Psi$ modulations; (2) experiments performed with the electrical potential difference short-circuited by the highly permeant anion NO_3^- showed a reduction of the uptake; and (3) the protonophore carbonyl cyanide *p*-(trifluoro-methoxy) phenylhydrazone (FCCP), which induces a proton diffusion potential in the presence of a pH gradient, caused a large increase in leucine uptake. A ΔpH (7.4 in/8.8 out) similar to that occurring *in vivo* improved the kinetic properties of the transporter causing a relevant increase in V_{max} (Sacchi *et al.*, 1990). In a similar way, a $\Delta\Psi$ generated by proton diffusion in BBMVs (5.5 pH in/7.2 pH out) reduced by 10 times the histidine K_m value. However, in this case the ΔpH was not physiological, and a reduction of the accumulation ratio compared with that measured at neutral pH was observed (Giordana *et al.*, 1985).

About 20% of 0.5 mM leucine uptake measured at saturating potassium concentration was potassium-independent. This component is also carrier-mediated since leucine uptake as a function of leucine concentration displayed saturation. A modest positive effect of $\Delta\Psi$ and ΔpH was observed on leucine kinetic parameters even in the absence of potassium. The similar inhibition patterns exerted by a number of amino acids on leucine uptake in the presence and in the absence of potassium suggested the existence of a single transporter that can cross the membrane as a binary (carrier and leucine) or ternary (carrier, leucine and potassium) complex. The fully loaded form of the transporter, in the presence of a potassium gradient, a ΔpH and $\Delta\Psi$ (inside negative), seemed to have the highest efficiency in translocating substrates across the membrane (Sacchi *et al.*, 1990; Parenti *et al.*, 1992). A single transporter for both leucine uniport and leucine-K^+ symport seems to be in contrast with the current opinion that amino acid transport systems can be divided into at least two categories: those that catalyse uniport and those that catalyse symport (McGivan and Pastor-Anglada, 1994). Only the isolation and reconstitution of amino acid transport proteins can shed light on whether symport and uniport are carried out by the same protein.

10.4.2 Affinity of system for substrates

In *P. cynthia* midgut the transport of most neutral amino acids and histidine seems to occur through a system which excludes methylaminoisobutyric acid (MeAIB) and proline but can transport with a relatively low affinity AIB (K_m = 2.36 mM) and D-alanine (K_m = 0.95 mM). This system seems to have much in common, apart from cation specificity, with system B, a vertebrate epithelial amino acid transport system of broad specificity which excludes MeAIB, an amino acid analogue efficiently translocated by system A in non-polarized cells. Usually the initial uptake of amino acid as a function of amino acid concentration is a curve described by the sum of a linear component and a rectangular hyperbola. The linear component, interpreted as a diffusion pathway, was not observed for leucine. The K_m values for neutral amino acids range between 0.1 and 0.7 mM at saturating K^+ concentration (Giordana *et al.*, 1989).

Since Na^+ can efficiently substitute for K^+ in leucine symport (Figure 10.2) the ability of Na^+ and K^+ to activate leucine uptake was measured. Figure 10.3 shows sodium and potassium activation curves of 0.2 mM leucine uptake in BBMVs. Both curves are hyperbolic and the calculated $K_{½}$ and V_{max} values are 37 ± 5 mM and 567 ± 31 pmol/s/mg protein for potassium and 2.1 ± 0.2 mM and 229 ± 9 pmol/s/mg protein for sodium. These results demonstrate that the transporter can discriminate between the two cations and that the affinity of the transporter for sodium is

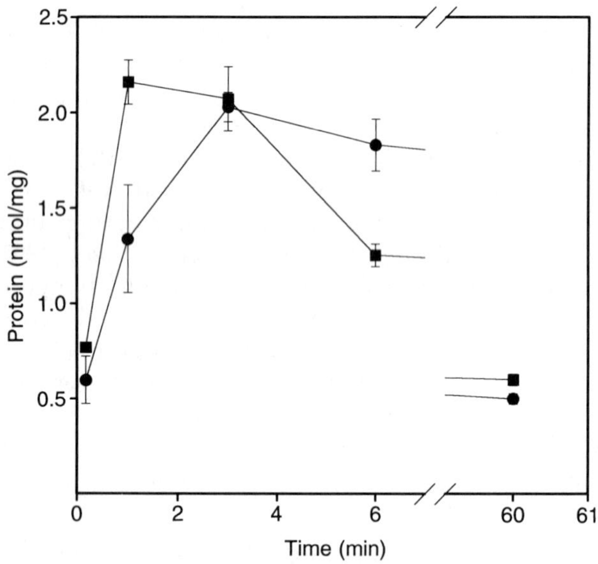

Figure 10.2 Time courses of leucine uptake in BBMVs driven by sodium (■) or potassium (●). Vesicle internal solution was: 160 mM mannitol, 90 mM Hepes, 45 mMm tris (pH 7.4). The vesicles were diluted 1:5 in solutions having the following final composition: 118 mM mannitol, 18 mM Hepes, 57 mM tris (pH 8.9), 0.2 mM [^{3}H]L-leucine (10 μCi/ml), 0.1 mM FCCP and K_2SO_4 or Na_2SO_4. Each point is the mean ± SE of a typical experiment performed in quadruplicate. (From Sacchi *et al.*, 1994, with permission from Elsevier Science Ltd.)

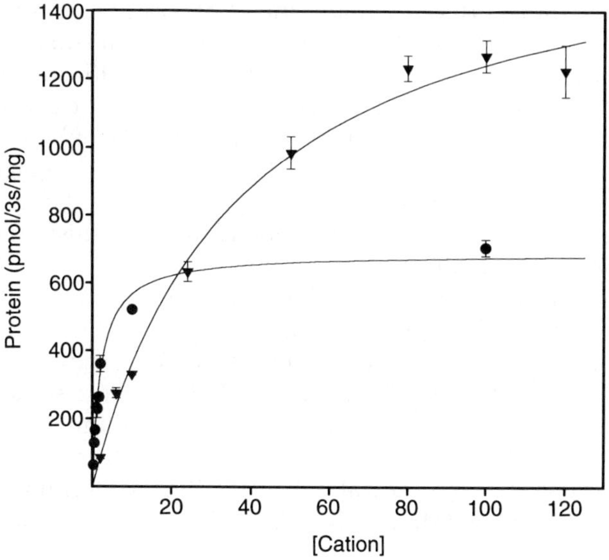

Figure 10.3 Leucine uptake in BBMVs as a function of sodium or potassium concentrations (mM). Vesicle internal solution was: 225 mM mannitol, 90 mM Hepes, 45 mM tris, pH 7.4. The vesicles were diluted 1:5 in a solution of the following final composition: 85 mM mannitol, 18 mM Hepes, 57 mM tris (pH 8.9) and 0–100 mM NaCl plus 200–0 mM mannitol (●) or 0–100 mM KCl plus 200–0 mM mannitol (▼), 0.1 mM FCCP and 0.2 mM [^{3}H]leucine (80 μCi/ml). Incubations lasted 3 s. Each point is the mean ± SE of a typical experiment performed in quadruplicate. When not present, error bars were smaller than the symbols. (From Sacchi *et al.*, 1994, with permission from Elsevier Science Ltd.)

about 18 times that for potassium, whereas leucine V_{max} was 2.5 times higher with potassium (Sacchi *et al.*, 1994).

Studies were undertaken to shed light on whether sodium and potassium ions were handled by the same mechanism and to compare the effects of these cations on amino acid uptake. The existence of two independent transporters, one Na^+- and the other K^+-dependent, is disproved by the lack of additivity when both cations are present (Hanozet *et al.*, 1992; Sacchi *et al.*, 1994). Instead, Na^+ reduces K^+-dependent leucine uptake which supports the hypothesis that these cations interact with the same transporter. Na^+ inhibits K^+-dependent leucine uptake even at low concentrations because of its high affinity for the transporter. However, the translocation step is favoured greatly by K^+ binding and the change in conformation that leads to the internal release of ligands is more rapid when potassium rather than sodium is bound to the transporter. Therefore the transport process can occur by the formation of three complexes, one with potassium and amino acid, a second with amino acid and sodium, and a third with the amino acid alone as depicted in the model shown in Figure 10.4. The concentrations of leucine, K^+ and Na^+ in the lumen contents of *Philosamia cynthia* larvae are 0.24 (Parenti *et al.*, 1985) 200 and 1 mM, respectively (Figure 10.1). From these values it was estimated that 95% of the amino acid absorption should occur via the K^+-dependent mechanism (Hanozet *et al.*, 1992; Sacchi *et al.*, 1994).

It is difficult to understand how the broadening of the cation specificity is compatible with an increased sodium affinity which, even at low sodium concentration, hampers amino acid uptake. This apparent disadvantage may have the function to ensure sodium uptake in an epithelium which contains no conventional sodium pump (Harvey and Nedergaard, 1964; Jungreis and Vaughan, 1977) but has to absorb sodium because it is present at relatively high concentration in the central nervous system (Monticelli *et al.*, 1985) where a ouabain sensitive

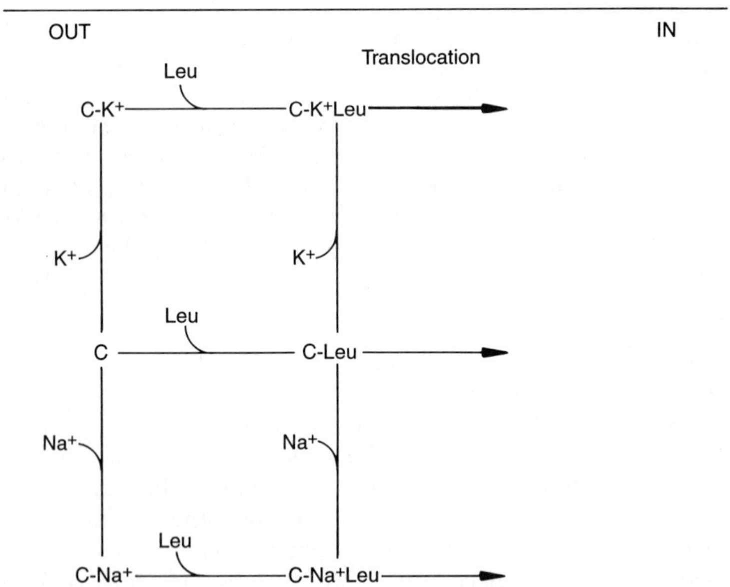

Figure 10.4 Kinetic model for a random mechanism, with two alternative drivers, occurring on the external side of brush-border membranes. C, transporter; Leu, leucine; C–K^+, complex with potassium, C–Na^+, complex with sodium; C–Leu, complex with leucine; C–N^+Leu, complex with potassium and leucine; C–Na^+Leu, complex with sodium and leucine. (From Sacchi *et al.*, 1994, with permission from Elsevier Science Ltd.)

Na^+/K^+-ATpase was found (Abbot and Treherne, 1977; Jungreis and Vaughan, 1977). Although the amino acid symporter can explain the uptake pathway for sodium into columar cells, it is not yet known how sodium ions leave the cells, and yet a net sodium absorption was measured in the midgut of *Manduca sexta* larvae (Chamberlin, 1990).

10.4.3 Mechanism deduced from kinetic studies

A complete kinetic analysis of K^+-dependent leucine uptake in BBMVs in the presence of $\Delta\Psi$ and ΔpH was performed by Parenti *et al.* (1992). Changing the external K^+ concentration from 10 to 150 mM caused a two-fold increase in V_{max} and a six-fold decrease in the K_m for leucine. Increasing the concentration of leucine from 0.1 to 5.0 mM caused a 12-fold decrease of the apparent K_m for K^+ and a 2.5-fold increase in V_{max}. Furthermore, either potassium or leucine preloaded inside the vesicles caused a mixed-type inhibition of leucine uptake. The data presented were compatible with a system in which K^+ and leucine bind randomly to the carrier, with a coupling ratio of 1 K^+ to 1 leucine, and both the fully loaded complex and the leucine-only form are able to isomerize and release their substrates at the cytoplasmatic side of the membrane. In the terminology of enzyme kinetics, the system can be described as an iso-random Bi Bi reaction. Table 10.2 reports a summary of the effect of the external K^+ and L-leucine on the kinetics of L-leucine uptake.

Table 10.2 Effects of external K^+ and leucine on the kinetics of leucine uptake in BBMVs. (From Parenti *et al.*, 1992, with permission from The American Society for Biochemistry and Molecular Biology, Inc.)

[Leu] (*mM*)	*[K]* (*mM*)	K_m (*mM*)	V_{max} (*pmol/3 s/ mg/protein)*
0.1	0–100	108.0 ± 8.0	1572 ± 70
0.14	0–100	57.7 ± 6.1	2163 ± 100
0.2	0–100	35.0 ± 5.3	2494 ± 169
0.8	0–100	27.2 ± 0.9	3420 ± 144
1.0	0–100	15.7 ± 6.2	3620 ± 121
5.0	0–100	8.3 ± 1.5	4073 ± 212
0.05–5	10	0.78 ± 0.22	2452 ± 124
0.05–5	20	0.65 ± 0.05	3250 ± 130
0.05–5	40	0.36 ± 0.05	3887 ± 202
0.05–5	150	0.13 ± 0.02	4949 ± 331

10.4.4 Progress toward isolation of a leucine-K^+ symporter

The knowledge of the physiological and kinetic properties of a transport system is a necessary prerequisite for using the expression cloning technique. This method can lead to the identification, cloning and sequencing of the cDNA encoding a transporter from which the primary structure of the protein can be deduced (Sigel, 1990). The first step of this experimental approach is the expression of a new protein after injection of exogenous mRNA into the cytoplasm of *Xenopus laevis* oocytes. The expression can cause the appearance of a new function or modify a function already present in the oocytes. Recently mRNA purified from the midgut of *Philosamia cynthia* larvae was injected into *Xenopus laevis* oocytes and the increased leucine transport was considered the result of expression of a new transport system. This new system had properties, different from the endogenous ones, which resemble closely those of a system in the tissue from which the mRNA was purified (Sacchi *et al.*, 1995). The data in agreement with this hypothesis are:

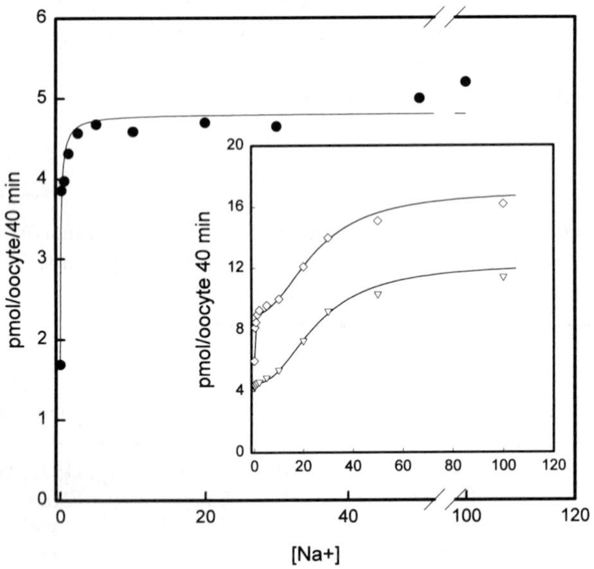

Figure 10.5 Net mRNA-induced uptake by *Xenopus* oocytes of 0.1 mM L-leucine as a function of Na^+ concentration. Uptake values are the differences between the curves shown in the inset. Inset: 0.1 mM L-leucine uptake as a function of Na^+ concentration in water-injected (▽) and in mRNA-injected oocytes (◊). NaCl was replaced by cholineCl. Data are means ± SE for 10 oocytes in a representative experiment. (From Sacchi *et al.*, 1995, with permission from The Company of Biologists Ltd.)

(1) the lack of stereospecificity of the induced leucine transport; (2) alkaline pH in the bathing solution increased the induced leucine uptake in mRNA-injected oocytes; (3) phenylglyoxal inhibited the induced leucine transport as it did on leucine uptake in BBMVs; and (4) the sodium activation curves in control and mRNA injected oocytes were different (Figure 10.5 inset). The presence of an endogenous Na^+-dependent L-leucine uptake in oocytes makes it more difficult to analyse the induced L-leucine transport. However, the kinetic features of the endogenous and induced transports are so different that it is possible to distinguish between them. The relationship between leucine uptake and sodium concentration was sigmoidal in control oocytes and non-sigmoidal in mRNA-injected oocytes. The difference between the two curves should give the hyperbolic sodium activation of the newly expressed leucine transport (Figure 10.5). The mean $K_{1/2}$ value for sodium was 0.4 ± 0.24 mM (mean ± SEM calculated from three independent kinetics). This value is similar to that found in BBMVs from *Philosamia cynthia* larvae (Figure 10.3) and an order of magnitude lower than the mean $K_{1/2}$ value measured in control oocytes ($K_{1/2}$ = 19.70 ± 4.69 mM; mean ± SEM, calculated from five independent kinetics). Unfortunately, in *Xenopus* oocytes, no potassium activation could be demonstrated in uptake experiments. However, it should be noted that in *Xenopus* oocytes K^+ cannot drive L-leucine uptake because, even at a high external K^+ concentration, there is little or no inwardly directed electrochemical potential difference of the driver.

10.5 OTHER SYSTEMS FOR AMINO ACID UPTAKE IN LARVAL *PHILOSAMIA CYNTHIA* MIDGUT

The presence of different K^+-dependent amino acid transport systems in the luminal membrane of the larval *Philosamia cynthia* midgut was investigated by means of countertransport and inhibition experiments performed with BBMVs (Giordana *et al.*, 1989; Hanozet *et al.*, 1989). In counterflow experiments the uptake of a labelled amino acid into the BBMVs was measured in the presence of unlabelled amino acids inside the vesicles. Potassium concentration was the same on both sides of the membrane, and any electrical coupling via transmembrane potential was avoided by the addition of the potassium ionophore valinomycin. Under these experimental conditions, a counterflow accumulation of a labelled amino acid, in the presence of an imposed oppositely directed (*trans*) gradient of a different amino acid, is obtained only if they share the same transporter and the accumulation is the result of the inhibition exerted by the intravesicular unlabelled amino acid on the outflux of the labelled one once it has entered the vesicles. It should be noted that the presence of a substrate on the *trans* side of the membrane can also modify the kinetic features of the transporter. *Trans*-acceleration occurs if the amino acid–carrier complex crosses the membrane faster than the free carrier; on the contrary, *trans*-inhibition results if the mobility of the complex is slower than that of the free carrier. In any event these kinetic modifications should not account for an intravesicular accumulation of the labelled amino acid. Therefore, counterflow accumulation can be used in principle to discriminate the transporters present on the plasma membrane. The agreement between counterflow accumulation experiments and more conventional inhibition experiments has been tested. The rationale was that those amino acids that caused an intravesicular accumulation of a labelled amino acid should also inhibit in a competitive way the uptake of that amino acid from the *cis* side of the membrane. Anionic and cationic amino acids (with the exception of lysine that elicited a counterflow accumulation of leucine) failed to elicit counterflow accumulation of neutral amino acids and histidine whereas alanine, leucine, phenylalanine, serine, aspargine and glutamine but not proline and glycine were effective. The same amino acids were able to inhibit histidine uptake when present in the external solution, therefore the authors concluded that this is a suitable test for an initial screening of the interactions of amino acids with different transport systems (Hanozet *et al.*, 1989). The main criticism of this experimental approach is that *cis* inhibition by the substance preloaded into the vesicles can always be present on the external side of the membrane vesicles. Therefore, counterflow accumulation, if present, is a clear indication that the two molecules tested share the same transporter, but the reverse does not prove that they do not. As a matter of fact glycine, which in this first report did not elicit transaccumulation of neutral amino acids, was later considered an amino acid transported by the neutral system on the basis of inhibition experiments (Hanozet *et al.*, 1992). Despite the fact that this amino acid is probably transported by the neutral amino acid transport system, its transport presents some particular features. Sodium is more effective than potassium in driving glycine uptake into BBMVs at pH 7.4, whereas potassium rather than sodium caused the highest value of glycine uptake and accumulation in BBMVs in the presence of ΔpH (7.4 in/8.9 out) and $\Delta\Psi$ (inside negative) (Hanozet *et al.*, 1992). This observation confirms that the pH affects the behaviour of the transport protein and suggests some caution in comparing experiments performed at different pH values.

Proline is a poor inhibitor of neutral amino acid uptake in larval *P. cynthia* midgut BBMVs and does not elicit counterflow accumulation of alanine, leucine, phenylalanine or serine (Giordana *et al.*, 1989), therefore experiments should be performed to verify the presence of a transporter for amino acids.

The model proposed for the absorption of neutral amino acids in the

K^+-transporting midgut epithelium of *Philosamia cynthia* larva also applies to the transport of the basic amino acids. Histidine and lysine are transported by different transport agencies as indicated by counterflow and inhibition experiments (Giordana *et al.*, 1985; Hanozet *et al.*, 1989). Furthermore, the neutral amino acid transport system is not the only one sensitive to $\Delta\Psi$. The K_m for lysine is reduced by an intravesicular negative $\Delta\Psi$ in the presence of potassium (Hanozet *et al.*, 1985). Arginine as well as neutral amino acids appear unable to interact with the lysine transporter (Giordana, 1989).

L-Glutamate uptake is activated by potassium in larval *P. cynthia* midgut BBMVs and its transient accumulation into the vesicles is prevented by valinomycin. However, unlike those involved in the uptake of neutral, basic and D-amino acids, this transport system does not seem to be influenced by membrane potential. Countertransport experiments indicate that aspartic acid does not interact with the glutamate transporter (Giordana *et al.*, 1989).

Since the maximal percentage inhibitions of histidine are quite high for all the amino acids tested (aspartic acid excluded), a broad range of overlapping specificities for the diverse amino acid transport systems was suggested (Hanozet *et al.*, 1989). Furthermore, the transport system for neutral amino acids is not stereospecific. The D-isomer of each amino acid inhibits the uptake of the L-form. However, inhibition of L-alanine uptake by D-alanine is competitive, whereas inhibition of D-alanine uptake by L-alanine is non-competitive. These kinetic data together with counterflow accumulation experiments and the different pattern of specificity to cations (Table 10.1) suggest the existence of two transport systems in *P. cynthia* midgut that can handle D-alanine (Hanozet *et al.*, 1984).

10.6 AMINO ACID UPTAKE IN LARVAL *BOMBYX MORI* MIDGUT

All the species of phytophagous larvae of lepidoptera so far studied share several physiological features. As in *P. cynthia* the sodium concentration in the haemolymph, lumen content and intestinal tissue of *Bombyx mori* larvae is below 5 mM and the midgut tissue isolated and perfused *in vitro* exhibits a high lumen-positive transepithelial electrical potential difference (69.1 ± 2.3 mV, mean ± SEM from 35 experiments, measurements performed 15 min after isolation) (Giordana and Sacchi, 1977a,b, 1978a). The transepithelial potential decay with time is faster in *B. mori* than in *P. cynthia* midgut but can be counteracted by the presence of alanine in the solution bathing the haemolymphatic side of the tissue. Since D-alanine had no effect on the transepithelial electrical potential difference decay whereas pyruvic acid increased the potential difference, it was concluded that L-alanine is utilized by the midgut as a metabolic substrate (Giordana and Sacchi, 1978b, 1979).

An analysis of the metabolic activity related to the potassium pump in the midgut of *B. mori* larvae showed that amino acid metabolism directly related to the tricarboxylic acid cycle seems to be the primary source of energy for the potassium pump. Measurements of the free amino acid concentration in the lumen content, intestinal cell and haemolymph showed that glutamic acid, asparagine and glutamine are accumulated in the cell, whereas the haemolymph is enriched with basic amino acids plus glycine, alanine, serine and tyrosine, the major components of silk fibroin (Parenti *et al.*, 1985).

L-Alanine, glycine and phenylalanine are absorbed in the midgut of *B. mori* larvae *in vitro* but only L-alanine is metabolized by the tissue (Sacchi and Giordana, 1980). L-Phenylalanine and α-aminoisobutyric acid are actively transported from the lumen to the haemolymph in the isolated midgut of *B. mori* larva and the net flux shows saturation with increasing amino acid concentration, is sodium independent and inhibited by DNP and anoxia (Sacchi *et al.*, 1981). Transient accumulation of L-phenylalanine, L-alanine and α-aminoisobutyric acid was observed in larval *B. mori* midgut BBMVs in the presence of a potassium gradient (Giordana *et al.*, 1982; Sacchi *et al.*, 1984). Therefore, the general model for amino acid absorption appears to be valid in this species.

Marked differences in the functional properties of the symporter for neutral amino acids were observed between BBMVs prepared from the posterior and anterior-middle regions of the midgut of *B. mori* larvae (Giordana *et al.*, 1994). Leucine V_{max} in the posterior midgut was more than 11-fold higher than that of the anterior-middle tract, and only in this region the transport was remarkably influenced by extravesicular pH and $\Delta\Psi$. The affinity for leucine was similar along the midgut (0.15 mM) and only in the posterior region was reduced at neutral pH.

A direct demonstration of the $\Delta\Psi$ dependence of leucine uptake in BBMVs prepared from the posterior midgut used the fluorescent potential-sensitive dye 3,3′-diethylthiacarbocyanine iodide (Dis-C2(5)) to measure the depolarization induced by the cation-dependent amino acid symporter in BBMVs with an imposed, inside negative, $\Delta\Psi$ (Giordana and Parenti, 1994).

K^+ is by far the most effective cation in promoting leucine uptake in both the anterior-middle and the posterior regions of larval *B. mori* midgut. Uptake of 1 mM leucine as a function of external potassium concentration was sigmoidal in BBMVs prepared from the anterior midgut and hyperbolic in BBMVs prepared from the posterior midgut. In both preparations the potassium concentration giving half-maximal leucine uptake was 49 mM. The presence in both midgut regions of a broad specificity system for neutral amino acids was supported by inhibition experiments. However, the functional differences between leucine uptake observed in BBMVs prepared from the anterior-middle and the posterior regions of the midgut suggested that two isoforms of the neutral amino acid–K^+ symporter are expressed in the two regions.

The activated toxin of *Bacillus thuringiensis* seems to act as a non-competitive inhibitor of amino acid–K^+ symport in *B. mori* midgut in the absence of a K^+ gradient (Giordana *et al.*, 1993) as well as in the complete absence of K^+ (Parenti *et al.*, 1995). Therefore, the authors concluded that the inhibition of amino acid transport is not secondary to the well-documented pore formation (Carroll and Ellar, 1993) but it might involve the binding of the toxin to the amino acid symporter or a strictly associated membrane protein.

10.7 AMINO ACID ABSORPTION IN LARVAL *MANDUCA SEXTA* MIDGUT

10.7.1 Uptake of neutral (zwitterionic) amino acids

Tobacco hornworm (*Manduca sexta*) eggs, larvae and pupae are available commercially. The larvae grow rapidly on commercially available synthetic diet to fifth instars weighing 5–7 g. Because of these and other favourable characteristics *M. sexta* has become a favourite model for all lepidopterans and ranks among the most widely studied of all insects (Law and Wells, 1989). None the less, amino acid uptake in larval *M. sexta* midgut remained uninvestigated until the late 1980s. These first experiments showed that, as in *P. cynthia*, phenylalanine accumulation was driven by either sodium or potassium gradients (Hennigan and Wolfersberger, 1989). Further investigation of amino acid uptake by BBMVs prepared from larval *M. sexta* midguts (Eisen *et al.*, 1989) revealed the presence of a B-like system, in many respects quite similar to that found in *P. cynthia* (Hennigan *et al.*, 1993a,b). This system seemed to be the primary agency for absorption, by amino acid–K^+ symport, of most neutral (zwitterionic) amino acids. On the basis of both countertransport and *cis* inhibition studies, this system was implicated in the absorption of all common amino acids except arginine, aspartate, glutamate, cysteine and proline. As in *P. cynthia*, AIB was symported by this system but MeAIB was not (Hennigan *et al.*, 1993b). In most tests, histidine ranked close to leucine as the favoured substrate for this symport system. Lysine proved to be among the most potent inhibitors of leucine uptake but only an average elicitor of leucine accumulation by countertransport at pH 10. Arginine, aspartate, glutamate, cystine and proline were found to be poor *cis* inhibitors of KCl gradient-driven leucine uptake by frozen and thawed larval *M. sexta* midgut BBMVs at pH 7.4 (Reuveni and Dunn, 1993). However, in these experiments lysine was a poor inhibitor of leucine uptake. That lysine was a potent inhibitor of leucine uptake at pH 10 and a weak inhibitor at pH 7.4 may be due in large part to lysine being mainly zwitterionic at pH 10 and predominantly cationic at pH 7.4. One striking difference between the B-like systems in *M. sexta* and *P. cynthia* is that the *M. sexta* system seems to be insensitive to phenylglyoxal (Parthasarthy and Harvey, 1994a).

A systematic investigation of the organic substrate selectivity of zwitterionic amino acid–K^+ symport in larval *M. sexta* midgut BBMVs revealed relatively broad tolerances for size and shape but rather narrow tolerances for changes in charge distribution (Parthasarthy *et al.*, 1994). D-Leucine was as effective as L-leucine in eliciting L-leucine accumulation by countertransport. This lack of chiral selectivity is again reminiscent of neutral amino acid–K^+ symport in *P. cynthia*. However, at least in *M. sexta*, it seems to pertain only to aliphatic amino acids. D-Phenylalanine was a very weak inhibitor of L-phenylalanine uptake and did not elicit accumulation of L-phenylalanine by countertransport. Within the range occupied by the amino acids found in common proteins neither the size nor the shape of a neutral side chain had much influence on the suitability of an amino acid as a substrate. L-Isoleucine is as effective as L-leucine in eliciting L-leucine accumulation and more effective than L-phenylalanine in eliciting L-phenylalanine accumulation. This is not the case when the side chain contains a charged group. The effectiveness of diaminoalkanoic acids in inhibiting lysine uptake increased almost linearly as the position of the second amino group was moved from the third to the sixth carbon. The most stringent criterion for substrate selectivity seems to be the position of the first (or only) amino group. Only α-amino acids are accepted as substrates.

10.7.2 Effects of pH, ΔpH and ΔΨ on zwitterionic amino acid–K^+ symport

The initial rate of KSCN gradient driven uptake of alanine, leucine and phenylalanine by larval *M. sexta* midgut BBMVs was found to be greatest at pH 9.5 to 10. The initial rate of phenylalanine symport at pH 10 was nearly double that at pH 8 whereas the initial rate of leucine symport increased by less than 20% over the same pH range (Hennigan *et al.*, 1993a). pH also affected the cation selectivity of phenylalanine symport. At pH 8 Na^+ and K^+ were equally effective in driving phenylalanine uptake (Hennigan and Wolfersberger, 1989) whereas at pH 10 K^+ was significantly more effective than Na^+ (Hennigan *et al.*, 1993a). Since the sodium concentration in larval *M. sexta* midgut lumen is less than 2% of the potassium concentration (Dow *et al.*, 1984) and the vast majority of neutral amino acid uptake occurs in the posterior third of the midgut where the median pH is ~8.3 (Dow, 1984), the physiological implications of these pH effects seem to be limited. However, it is clear from these studies that neutral amino acid–K^+ symport can function effectively over at least most of the pH range found in larval *M. sexta* midgut.

The affect of $\Delta\Psi$, the only significant driving force for amino acid–K^+ symport *in vivo*, on neutral amino acid uptake by larval *M. sexta* midgut BBMVs was first deduced from the affects of anions on K^+-driven phenylalanine and leucine accumulations. Maximal phenylalanine accumulation with a KSCN gradient was more than twice that with a K_2SO_4 gradient (Hennigan and Wolfersberger, 1989) and maximal leucine accumulation decreased with potassium salts in the order KSCN $> KNO_3 >> KCl > K_2SO_4 \approx K_2HPO_4$ (Hennigan *et al.*, 1993a). Larval *M. sexta* midgut BBMVs have been shown to be much more permeable to SCN^- than to Cl^- (Carroll and Ellar, 1993). More quantitative studies of the effects of $\Delta\Psi$ on neutral amino acid–Na^+ symport by larval *M. sexta* midgut BBMVs revealed that amino acid affinity increased as membrane potential was varied between 0 and −70 mV (inside of vesicles negative relative to outside). More negative potentials had no further affect on $K_{1/2}$ but resulted in substantial increases in V_{max} (Parthasarthy and Harvey, 1994b).

In vivo there is a pH difference of 1–4 units as well as a large electrical potential difference across the apical brush-border membrane of larval *M. sexta* midgut columnar cells (Chapter 9). In an attempt to investigate the possible affects of ΔpH on neutral amino acid–K^+ symport BBMVs were prepared with internal pH buffered at either 7 (physiological), 8 or 9. They were then used in leucine uptake experiments where the final pH of the external solution was either 7, 8 or 9 (Wolfersberger, unpublished observations). With a KSCN gradient providing the driving force, the initial rate of leucine uptake with pH 9 solutions both inside and outside was more than twice that with pH 7 solutions both inside and outside the BBMVs. With pH 7 inside and pH 9 outside, the initial rate of leucine uptake was less than that with pH 9 solutions both inside and outside but not significantly different from the rate with pH 8 solutions both inside and outside the BBMVs. Even in the presence of a pH gradient, FCCP had no effect on the rate of KSCN gradient-driven leucine uptake. A rather different set of results was obtained when KCl was substituted for KSCN. As expected, FCCP more than doubled the initial rate of KCl gradient-driven leucine uptake by BBMVs with pH 7 solution inside and pH 9 solution outside. These results confirm previous work reporting acceleration of leucine–K^+ symport by alkaline pH and membrane potential while showing that at least a two unit pH gradient alone is without effect on leucine–K^+ symport in larval *M. sexta* midgut BBMVs.

Glycine and proline were among the weakest inhibitors of leucine uptake and the poorest elicitors of leucine accumulation by countertransport into larval *M. sexta* midgut BBMVs. However, since glycine was able to elicit a statistically significant accumulation of both alanine and phenylalanine whereas proline was not, glycine but not proline was considered to be a substrate for the B-type amino acid–K^+ symport system in larval *M. sexta* midgut (Hennigan *et al.*, 1993b). More recent studies (Dixon, 1995) have shown that KSCN gradients drive transient accumulation of both glycine and proline to several times their equilibrium concentration in larval *M. sexta* midgut BBMVs. Furthermore, glycine drives countertransport accumulation of proline to more than twice its equilibrium concentration and vice versa. These observations are consistent with there being a second neutral amino acid–K^+ symport system in larval *M. sexta* midgut that prefers these amino acids to leucine as organic substrate.

10.7.3 Acidic (anionic) amino acid uptake in larval *M. sexta* midgut

The acidic amino acids aspartate and glutamate are among the amino acids that seemed to be, at best, very poor substrates for the neutral amino acid–K^+ symport system in larval *M. sexta* midgut BBMVs (Hennigan *et al.*, 1993a,b). Reuveni and Dunn (1993) reported that aspartate uptake by frozen and thawed larval *M. sexta* midgut BBMVs at pH 7.4 appeared to be the sum of two components, neither of which seemed to be involved in leucine uptake. One component was non-saturable and K^+ independent. The other was saturable and its rate was increased by a KCl gradient. The KCl gradient effect on aspartate uptake was reduced three-fold by valinomycin. KCl gradient driven aspartate uptake was inhibited significantly by cysteine, glycine, leucine, methionine, phenylalanine, tryptophan and valine. Since all these amino acids except cysteine are substrates for the B-like system in larval *M. sexta* midgut, their inhibition of KCl gradient driven aspartate uptake can be ascribed to their ability to compete favourably for the K^+ gradient. Since cysteine is at best a marginal substrate for the B-like system, its inhibition of aspartate uptake is more likely due to its dissipating the K^+ gradient through a third symport system or competing with aspartate for the aspartate transporting agency. Data to distinguish between these alternatives are unavailable. Glutamate was a rather poor inhibitor of KCl gradient-driven aspartate uptake.

Xie *et al.* (1994) reported that KSCN as well as KCl gradients increased the initial rate of glutamate uptake and drove a transient accumulation of glutamate to more than twice its equilibrium concentration in larval *M. sexta* midgut BBMVs at pH 9. The transient K^+ gradient-driven glutamate accumulation was eliminated completely by valinomycin. NaCl but not LiCl, RbCl or CsCl gradients also drove glutamate accumulation by the vesicles at pH 8.9 and 9.7. However, the initial rate of K^+ gradient-driven glutamate uptake by larval *M. sexta* midgut BBMVs was low (<15%) compared to that for leucine and insensitive to $\Delta\Psi$. Furthermore, in the presence of K^+ but in the absence of a K^+ gradient, no common amino acid, although present in 20-fold excess, was able to inhibit glutamate uptake by more than 50%. From these observations, plus the inability of either L-aspartate, D-glutamate or L-glutamate to drive glutamate accumulation by countertransport, the authors concluded that glutamate–K^+ symport was, at best, a marginal mechanism for glutamate uptake by larval *M. sexta* midgut columnar cells.

10.7.4 Basic (cationic) amino acid absorption in larval *M. sexta* midgut

At pH 10 the three basic amino acids histidine, lysine and arginine are approximately 13, 77 and 99% in their zwitterionic form. Nevertheless, at pH 10 both histidine and lysine but not arginine seem to be readily accepted substrates for the B-like zwitterionic amino acid–K^+ symport system in larval *M. sexta* midgut BBMVs (Hennigan *et al.*, 1993b). At pH 7.4 lysine is more than 97% in its cationic form and a very poor substrate for the B-like neutral amino acid–K^+ symport system in larval *M. sexta* midgut BBMVs (Reuveni and Dunn, 1993). Lysine–K^+ symport by larval *M. sexta* midgut BBMVs is inhibited strongly by leucine and essentially unaffected by arginine at pH 10 whereas essentially the opposite results are obtained at pH 7.4. These observations suggested that lysine–K^+ symport was mediated by at least two distinct systems. Closer examination of arginine and lysine uptake by larval *M. sexta* midgut BBMVs at pH 7.4 revealed the presence of a unique amino acid–alkali metal cation symport system that functioned equally well with K^+ or Na^+ as the inorganic substrate but accepted only the net positively charged forms of arginine, lysine, homoarginine or ornithine as organic substrates (Liu, 1994). The initial rate of arginine–K^+ or lysine–K^+ symport by this system was greatest at pH $\approx$ 8 and decreased as the pH was raised or lowered. The decrease in symport rate with increasing pH coincided with the decrease in the proportion of amino acid substrate in its cationic ionization state. However, since both arginine and lysine are at least 90% in the net +1 ionization state at all pH values between 8.0 and 3.2, the decrease in rate with decreasing pH was more likely due to an affect on the symporter.

Lysine elicited arginine accumulation by countertransport and behaved as a competitive inhibitor of arginine–K^+ symport, suggesting that these two amino acids shared a common transporter. However, their interactions with the transporter were not identical. Although $K_{1/2}$ for arginine (0.35 mM) was not significantly different from $K_{1/2}$ for lysine

(0.55 mM), arginine acted as a non-competitive inhibitor of lysine–K^+ symport with an apparent K_i of 0.15 mM (Liu, 1994). Considering its relative affinities for arginine and lysine, it seems likely that *in vivo* the cationic amino acid–K^+ symport system functions almost exclusively as an arginine absorption system with histidine and lysine uptake being mediated by the systems involved in the absorption of leucine and most other neutral amino acids.

As mentioned above, one of the most striking differences between the leucine–K^+ symport systems in *P. cynthia* and *M. sexta* is that whereas the former is quite sensitive to phenylglyoxal the latter is essentially unaffected. However, although PGO treatment had no significant effect on the initial rate of K^+ gradient-driven leucine uptake by larval *M. sexta* midgut BBMVs at pH 7.4, it reduced the initial rate of lysine uptake by ~70% (Parthasarthy and Harvey, 1994a). The virtual absence of any effect of PGO treatment on leucine–K^+ symport plus the decreasing effect of PGO treatment on lysine–K^+ symport at more alkaline pH values indicated that PGO was mainly affecting the agency involved in cationic amino acid–K^+ symport. Unlike in the case of *P. cynthia* leucine–K^+ symport, PGO seemed to react at two sites and affected both the K_m and V_{max} of lysine–K^+ symport in *M. sexta*.

10.8 AMINO ACID ABSORPTION IN *PIERIS BRASSICAE* AND *LYMANTRIA DISPAR* MIDGUTS

KSCN gradients (100 mM_{out}, 0 mM_{in}) drove transient accumulations of alanine, histidine and phenylalanine to approximately seven times their equilibrium concentration in BBMVs prepared from larval *P. brassicae* midguts (Wolfersberger *et al.*, 1987). These transient accumulations were abolished by treatment of the vesicles with valinomycin. In confirmation of the relatively broad cation selectivity of larval lepidopteran midgut zwitterionic amino acid symporters, NaCl gradients were as effective as KCl gradients in increasing the initial rates of alanine, histidine and phenylalanine uptake by larval *P. brassicae* midgut BBMVs. RbCl gradients failed to increase the initial rate of alanine or histidine uptake but increased the initial rate of phenylalanine uptake by approximately 30%. LiCl gradients were more effective than equivalent KCl or NaCl gradients in increasing the initial rate of histidine uptake but significantly less effective in increasing the initial rates of alanine and phenylalanine uptake.

At pH 7.4 KSCN gradients were approximately 60% as effective in driving lysine accumulation as in driving alanine accumulation by larval *P. brassicae* midgut BBMVs. A KCl gradient was somewhat more effective than equivalent NaCl or LiCl gradients in increasing the initial rate of lysine uptake by these vesicles (Wolfersberger *et al.*, 1987).

Unfortunately, there are no inhibition, countertransport, or pH dependence studies of lysine–K^+ symport by *P. brassicae* midgut BBMVs that might provide further insight regarding the agencies involved.

A KSCN gradient drove glutamate accumulation by larval *P. brassicae* midgut BBMVs to a maximum concentration of only 1.6 times its equilibrium concentration (Wolfersberger *et al.*, 1987). Equivalent KCl, LiCl or RbCl gradients increased the initial rate of glutamate uptake by approximately 80%, 67% and 30%, respectively, whereas a NaCl gradient gave rise to a 460% increase in the initial rate of glutamate uptake by larval *P. brassicae* midgut BBMVs. Since a NaCl gradient of the same magnitude increased the initial rate of alanine uptake by these vesicles by only approximately half this amount, this unusual and seemingly very specific Na^+ effect seems worthy of further investigation.

Potassium salt gradients (75 mM_{out}, 0 mM_{in}) drove transient accumulations of phenylalanine and leucine to three or four times their equilibrium concentrations in BBMVs prepared from larval *Lymantria dispar* midguts (Wolfersberger, 1991, 1993). Both the initial rate of leucine uptake and the maximal leucine accumulation by the vesicles at pH 8 increased in parallel with the presumed membrane permeability of the anions (KSCN > KNO_3 > KCl > K_2SO_4) suggesting involvement of $\Delta\Psi$ in driving symport. A KCl gradient was much more effective than an equivalent NaCl or LiCl gradient in driving leucine accumulation. Neither RbCl nor CsCl gradients drove any significant leucine accumulation by larval *L. dispar* BBMVs. These results are consistent with the relatively broad cation selectivity of larval lepidopteran midgut zwitterionic amino acid symporters but indicate a somewhat greater potassium specificity in this forest pest than in other tested species. A KCl gradient also drove transient accumulation of lysine to more than twice its equilibrium concentration in larval *L. dispar* midgut BBMVs at pH 8 (Wolfersberger, 1993). RbCl, CsCl and NaCl but not LiCl gradients resulted in modest transient lysine accumulations by these vesicles. The caesium gradient-driven lysine accumulation seems to be unique to larval *L. dispar* midgut BBMVs. Neither LiCl, NaCl, KCl, RbCl nor CsCl gradients drove accumulation of glutamic acid by larval *L. dispar* midgut BBMVs at pH 8 (Wolfersberger, 1993). Furthermore, equilibrative glutamate uptake by these vesicles was unusually slow. Whereas the leucine concentration in these vesicles reached 94% of its equilibrium value during 6 min of incubation in the absence of potassium, the glutamate concentration in the vesicles was only 87% of its equilibrium value after 30 min of incubation in the presence of potassium. The failure of salt gradients to increase the initial rate of glutamate uptake or drive glutamate accumulation by larval *L. dispar* midgut BBMVs argues against the presence of glutamate–alkali cation symporters in these vesicles and the extremely slow rate of glutamate uptake indicates, at best, a very low concentration of glutamate transporters of any variety (Wolfersberger, 1994) in the apical membrane of larval *L. dispar* midgut columnar cells.

Studies of midgut amino acid uptake by these two pest lepidopteran larvae contribute to one's confidence in the general applicability of the model depicted in Figure 10.1 for absorption of zwitterionic amino acids. They also provide additional support for the suggestion that symport with potassium may not be a particularly important mechanism for acidic amino acid absorption by lepidopteran larvae. Finally, they provide leads on some interesting questions to be addressed by future research.

10.9 AMINO ACID ABSORPTION IN *BLABERA GIGANTEA* AND *LEPTINOTARSA DECEMLINEATA* MIDGUTS

The only recent studies of midgut amino acid absorption by insects other than lepidopteran larvae are on the adult stage of the orthopteran *Blabera gigantea* and the larval stage of the coleopteran *Leptinotarsa decemlineata*. When applied to frozen midguts dissected from Colorado potato beetle larvae the widely used magnesium precipitation and differential centrifugation method for isolating BBMVs (Bieber *et al.*, 1981) yielded a subcellular fraction enriched tenfold relative to the homogenate in leucine aminopeptidase activity. Since putative leucine uptake by this midgut fraction was not stimulated by either KCl or NaCl, Reuveni *et al.* (1993) concluded that leucine absorption in larval *L. decemlineata* midgut was unlikely to be mediated by amino acid–alkali ion symport.

Net absorption of phenylalanine by isolated cockroach midgut was accompanied by a net absorption of sodium ions. There was no net flux of phenylalanine in the absence of sodium and both sodium and phenylalanine absorption were abolished by ouabain (Parenti *et al.*, 1986). These observations suggested an overall absorption mechanism similar to that found in mammalian intestine: amino acid–Na^+ symport driven by a sodium gradient established by Na/K-ATPase activity. Evidence for the presence of these elements in cockroach midgut were obtained by establishing the presence of Na/K-ATPase activity in midgut homogenates and showing that sodium salt gradients were able to drive phenylalanine accumulation by *B. gigantea* midgut BBMVs. In contrast to the now familiar case in lepidopterans, potassium as well as rubidium and lithium salt gradients were ineffective in driving phenylalanine accumulation by *B. gigantea* midgut BBMVs. However, since the rate of sodium gradient-driven phenylalanine uptake decreased with salts in the order NaSCN > NaCl > Na_2SO_4, it appeared that the sodium specific phenylalanine symporter of cockroach midgut was influenced by membrane potential in the same manner as the less cation specific amino acid symporters in the midguts of lepidopteran larvae. Indirect evidence suggests that similar systems for coupling amino acid absorption to sodium transport may be present in the midguts of other orthopterans (Dow, 1986).

The studies with *B. gigantea* show that amino acid absorption by amino acid–alkali ion symport occurs not only in lepidopteran but also orthopteran midguts. However, the work with *L. decemlineata* suggests the need for caution in generalizing from our very limited database to amino acid absorption in the midguts of other orders of insects. Perhaps the broadest conclusion is that much more work needs to be done before we will be able to make informed general statements about amino acid absorption in insect midguts.

REFERENCES

Abbot, N.J. and Treherne, J.E. (1977) Homeostasis in the brain microenviroment: a comparative account, in *Transport of Ions and Water in Animals* (eds B.L. Gupta, R.B. Moreton, J.L. Oschman and B.J. Wall), Academic Press, New York, pp. 481–509.

Bieber, J., Stieger, B., Haase, W. and Murer, H. (1981) A high yield preparation for rat kidney brush border membranes. Different behaviour of lysosomal markers. *Biochim. Biophys. Acta*, **647**, 169–76.

Carroll, J. and Ellar, D.J. (1993) An analysis of *Bacillus thuringiensis* δ-endotoxin action on insect midgut membrane permeability using a light-scattering assay. *Eur. J. Biochem.*, **214**, 771–8.

Chamberlin, M.E. (1990) Ion transport across the midgut of the tobacco hornworm, (*Manduca sexta*). *J. Exp. Biol.*, **150**, 425–42.

Chao, A.C., Koch, A.R. and Moffet, D.F. (1990) Basal membrane uptake in potassium-secreting cells of midgut of tobacco hornworm (*Manduca sexta*). *Am. J. Physiol.*, **258**, R112–R119.

Dixon, A.L. (1995) MA Thesis, Temple University.

Dow, J.A.T. (1984) Extremely high pH in biological systems: a model for carbonate transport. *Am. J. Physiol.*, **246**, R633–R635.

Dow, J.A.T. (1986) Insect midgut function. *Adv. Insect Physiol.*, **19**, 187–328.

Dow, J.A.T. (1992) pH gradients in lepidopteran midgut. *J. exp. Biol.*, **172**, 355–375.

Dow, J.A.T, Gupta, B.L., Hall, T.A. and Harvey, W.R. (1984) X-ray microanalysis of elements in frozen-hydrated sections of an electrogenic K^+ transport system: the posterior midgut of tobacco hornworm (*Manduca sexta*) *in vivo* and *in vitro*. *J. Membr. Biol.*, **77**, 223–41.

Eisen, N.S., Fernandes, V.F., Harvey, W.R. *et al.* (1989) Comparison of brush border membrane vesicles prepared by three methods from larval *Manduca sexta* midgut. *Insect Biochem.*, **19**, 337–42.

Florkin, M. and Jeuniaux, C. (1974) Haemolymph composition, in *The Physiology of Insecta* (ed. Rockstein), Academic Press, New York, pp. 255–307.

Giordana, B., Leonardi, M.G., Tasca, M. *et al.* (1994) The amino acid/K^+ symporters for neutral amino acids along the midgut of lepidopteran larvae. *J. Insect Physiol.*, **40**, 1059–68.

Giordana, B. and Parenti, P. (1994) Determinants for the activity of the neutral amino acid/K^+ symport in lepidopteran larval midgut. *J. Exp. Biol.*, **196**, 145–55.

Giordana, B., Parenti, P., Hanozet, G.M. and Sacchi, V.F. (1985) Electrogenic K^+-basic amino-acid cotransport in the midgut of lepidopteran larvae. *J. Membr. Biol.*, **88**, 45–53.

Giordana, B. and Sacchi, V.F. (1977a) Some ionic and electrical parameters of the intestinal epithelium in three mature larvae of lepidoptera. *Comp. Biochem. Physiol.*, **56A**, 95–9.

Giordana, B. and Sacchi, V.F. (1977b) Extracellular space values and intracellular ionic concentration in isolated midgut of *Philosamia cynthia* and *Bombyx mori*. *Experientia*, **33**, 1065.

Giordana, B. and Sacchi, V.F. (1978a) Cellular ionic concentration in the midgut of two larvae of lepidoptera *in vivo* and *in vitro*. *Comp. Biochem. Physiol.*, **59A**, 17–20.

Giordana, B. and Sacchi, V.F. (1978b) Glycine and L-alanine influence on transepithelial electrical potential difference in the midgut of *Bombyx mori* larva *in vitro*. *Comp. Biochem. Physiol.*, **61A**, 605–9.

Giordana, B. and Sacchi, V.F. (1979) The transepithelial electrical potential decay across the isolated midguts of two larvae of lepidoptera (*Bombyx mori* and *Philosamia cynthia*). *Comp. Biochem. Physiol.*, **66A**, 533–6.

Giordana, B., Sacchi, V.F. and Hanozet, G.M. (1982) Intestinal amino acid absorption in lepidopteran larvae. *Biochim. Biophys. Acta*, **692**, 81–8.

Giordana, B., Sacchi, V.F., Parenti, P. and Hanozet, G.M. (1989) Amino acid transport systems in intestinal brush-border membranes from lepidopteran larvae. *Am. J. Physiol.*, **257**, R494–R500.

Giordana, B., Tasca, M., Villa, M. *et al.* (1993) *Bacillus thuringiensis* subsp. *aizawai* δ-endotoxin inhibits the K^+/amino acid cotransporters of lepidopteran larval midgut. *Comp. Biochem. Physiol.*, **106C**, 403–7.

Hanozet, G.M., Giordana, B., Parenti, P. and Guerritore, A. (1984) L- and D-alanine transport in brush border membrane vesicles from lepidopteran midgut: evidence for two transport systems. *J. Membr. Biol.*, **81**, 233–40.

Hanozet, G.M., Giordana, B. and Sacchi, V.F. (1980) K^+-dependent phenylalanine uptake in membrane vesicles isolated from the midgut of *Philosamia cynthia* larvae. *Biochim. Biophys. Acta*, **596**, 481–6.

Hanozet, G.M., Giordana, B., Sacchi, V.F. and Parenti, P. (1989) Amino acid transport systems in brush border membranes vesicles from lepidopteran enterocytes. *J. Exp. Biol.*, **143**, 87–100.

Hanozet, G.M., Sacchi, V.F., Nedergaard, S. *et al.* (1992) The K^+-driven amino acid cotransporter of the larval midgut of lepidoptera: is Na^+ an alternative substrate? *J. Exp. Biol.*, **162**, 281–94.

Harvey, R.W. and Nedergaard, S. (1964) Sodium independent active transport of potassium in the isolated midgut of *Hyalophora cecropia* silkworm. *Proc. Natl Acad. Sci. USA*, **51**, 757–62.

Harvey, W.R., Cioffi, M. and Wolfersberger, M.G. (1983) Chemiosmotic potassium ion pump of insect epithelia. *Am. J. Physiol.*, **244**, 91–118.

Hennigan, B.B. and Wolfersberger, M.G. (1989) Intestinal amino acid absorption in tobacco hornworm larvae is stimulated by potassium and sodium but not rubudium or lithium. *Arch. Insect Biochem. Physiol.*, **11**, 21–8.

Hennigan, B.B., Wolfersberger, M.G. and Harvey, W.R. (1993b) Neutral amino acid symport in larval *Manduca sexta* midgut brush-border membrane vesicles deduced from cation-dependent uptake of leucine, alanine, and phenylalanine. *Biochim. Biophys. Acta*, **1148**, 216–22.

Hennigan, B.B., Wolfersberger, M.G., Parthasarthy, R. and Harvey, W.R. (1993a) Cation-dependent leucine, alanine, and phenylalanine uptake at pH

10 in brush-border membrane vesicles from larval *Manduca sexta* midgut. *Biochim. Biophys. Acta*, **1148**, 209–15.

Jungreis, A.M. and Vaughan, G.L. (1977) Insensitivity of lepidopteran tissues to ouabain: absence of ouabain binding and Na^+–K^+ ATPase in larval and adult midgut. *J. Insect Physiol.*, **23**, 503–9.

Law, J.H. and Wells, M.A. (1989) Insects as biochemical models. *J. Biol. Chem.*, **264**, 16335–8.

Liu, Z. (1994) PhD Thesis, Temple University.

McGivan, J.D. and Pastor-Anglada, M. (1994) Regulatory and molecular aspects of mammalian amino acid transport. *Biochem. J.*, **999**, 321–34.

Monticelli, G., Giordana, B., Sacchi, V.F. and Simonetta, M.P. (1985) An analysis of potassium distribution in the central nervous system of *Bombyx mori*. *Comp. Biochem. Physiol.*, **80A**, 425–31.

Nedergaard, S. (1972) Active transport of α-aminoisobutyric acid by the isolated midgut of *Hyalophora cecropia*. *J. Exp. Biol.*, **58**, 175–9.

Nedergaard, S. (1973) Transport of amino acids in *Hyalophora cecropia* midgut, in *Transport Mechanism in Epithelia* (eds H.H. Ussing and M.A. Thorm), Munksgard, Copenhagen, pp. 372–81.

Nedergaard, S. (1977) Amino acid transport, in *Transport of Ions and Water in Animals* (eds B.L. Gupta, R.B. Moreton, J.L. Oschman and B.J. Wall), Academic Press, London, pp. 381–401.

Nedergaard, S. (1981) Amino acid exchange mechanism in the basolateral membrane of the midgut epithelium of the larva of *Hyalophora cecropia*. *J. Membr. Biol.*, **58**, 175–9.

Parenti, P., Cidaria, D., Hanozet, G.M. and Giordana, B. (1985) Free amino acid composition of the intestinal contents of intestinal cells and haemolymph of *Philosamia cynthia* larvae. *Experientia*, **41**, 1158–9.

Parenti, P., Giordana, B., Sacchi, V.F. *et al.* (1985) Metabolic activity related to the potassium pump in the midgut of *Bombyx mori* larvae. *J. Exp. Biol.*, **116**, 69–78.

Parenti, P., Hanozet, G.M., Villa, M. and Giordana, B. (1994) Effect of arginine modification on K^+-dependent leucine uptake in brush-border membrane vesicles from the midgut of *Philosamia cynthia* larvae. *Biochim. Biophys. Acta*, **1191**, 27–32.

Parenti, P., Sacchi, V.F., Hanozet, G.M. and Giordana, B. (1986) Na-dependent uptake of phenylalanine in the midgut of a cockroach (*Blabera gigantea*). *J. Comp. Physiol.*, **156**, 549–56.

Parenti, P., Villa, M. and Hanozet, G.M. (1992) Kinetics of leucine transport in brush-border membrane vesicles from lepidopteran larvae midgut. *J. Biol. Chem.*, **267**, 15391–7.

Parenti, P., Villa, M., Hanozet, G.M. *et al.* (1995) Interaction of the insecticidal crystal protein CryIA from *Bacillus thuringiensis* with amino acid transport into brush border membranes from *Bombyx mori* larval midgut. *J. Invertebr. Pathol.*, **65**, 35–42.

Parthasarthy, R. and Harvey, W.R. (1994a) Phenylglyoxal suppresses cationic lysine/K^+ symport under alkaline conditions in brush-border membrane vesicles from larval *Manduca sexta* midgut. *Arch. Insect Biochem. Physiol.*, **28**, 237–45.

Parthasarthy, R. and Harvey, W.R. (1994b) Potential differences influence amino acid/Na^+ symport rates in larval *Manduca sexta* midgut brush border membrane vesicles. *J. Exp. Biol.*, **189**, 55–67.

Parthasarthy, R., Xie, T., Wolfersberger, M.G. and Harvey, W.R. (1994) Substrate structure and amino acid/K^+ symport in brush-border membrane vesicles from larval *Manduca sexta* midgut. *J. Exp. Biol.*, **197**, 237–50.

Reuveni, M. and Dunn, P.E. (1993) Absorption pathways of amino acids in the midgut of *Manduca sexta* larvae. *Insect Biochem. Mol. Biol.*, **23**, 959–66.

Reuveni, M., Hong, Y.S., Dunn, P.E. and Neal, J.J. (1993) Leucine transport into brush border membrane vesicles from guts of *Leptinotarsa decemlineata* and *Manduca sexta*. *Comp. Biochem. Physiol.*, **104A**, 267–72.

Sacchi, V.F., Cattaneo, G., Carpentieri, M. and Giordana, B. (1981) L-Phenylalanine active transport in the midgut of *Bombyx mori* larva. *J. Insect Physiol.*, **27**, 211–14.

Sacchi, V.F. and Giordana, B. (1980) Absorption of glycine, L-alanine, L-phenylalanine in the midgut of the larvae of *Bombyx mori*. *Experientia*, **36**, 659–60.

Sacchi, V.F., Giordana, B., Campanini, F. *et al.* (1990) Leucine uptake in brush-border membrane vesicles from the midgut of a lepidopteran larva, *Philosamia cynthia*. *J. Exp. Biol.*, **149**, 207–21.

Sacchi, V.F., Hanozet, G.M. and Giordana, B. (1984) α-Aminoisobutyric acid transport in the midgut of two lepidopteran larvae. *J. Exp. Biol.*, **108**, 329–39.

Sacchi, V.F., Parenti, P., Perego, C. and Giordana, B. (1994) Interaction between Na^+ and the K^+-dependent amino acid transport in midgut brush-border membrane vesicles from *Philosamia cynthia* larvae. *J. Insect Physiol.*, **40**, 69–74.

Sacchi, V.F., Perego, C. and Magagnin, S. (1995) Functional characterization of leucine transport induced in *Xenopus laevis* oocytes injected with mRNA isolated from midguts of lepidopteran larvae (*Philosamia cynthia*). *J. Exp. Biol.*, **198**, 961–6.

Sigel, E. (1990) News of *Xenopus* oocytes for the functional expression of plasma membrane proteins. *J. Membr. Biol.*, **117**, 201–21.

Stevens, B.R., Kaunitz, J.D. and Wright, E.M. (1984) Intestinal transport of amino acids and sugars: advances using membrane vesicles. *Annu. Rev. Physiol.*, **46**, 417–33.

Tobias, J.M. (1948) The high potassium and low sodium in the body fluid of a phytophagous insect, the silkworm *Bombyx mori* and the change before pupation. *J. Cell. Comp. Physiol.*, **31**, 143–8.

Wieczorek, H., Putzenlechners, M., Zeiske, W. and Klein, U. (1991) A vacuolar-type proton pump energizes K^+/H^+ antiport in an animal plasma membrane. *J. Biol. Chem.*, **266**, 15340–7.

Wolfersberger, M.G. (1991) Inhibition of potassium gradient driven phenylalanine uptake in *Lymantria dispar* midgut by two *Bacillus thuringiensis* delta-endotoxins correlates with the activity of the toxins as gypsy moth larvicides. *J. Exp. Biol.*, **161**, 519–25.

Wolfersberger, M.G. (1993) Preparation and partial characterization of amino acid transporting brush border membrane vesicles from the larval midgut of the gypsy moth (*Lymantria dipar*). *Arch. Insect Biochem. Physiol.*, **24**, 139–47.

Wolfersberger, M.G. (1994) Uniporters, symporters and antiporters. *J. Exp. Biol.*, **196**, 5–6.

Wolfersberger, M.G., Luethy, P., Maurer, A. *et al.* (1987) Preparation and partial characterization of amino acid transporting brush border membrane vesicles from the larval midgut of the cabbage butterfly (*Pieris brassicae*). *Comp. Biochem. Physiol.*, **68A**, 301–8.

Xie, T., Parthasarthy, R., Wolfersberger, M.G. and Harvey, W.R. (1994) Anomalous glutamate/alkali cation symport in larval *Manduca sexta* midgut. *J. Exp. Biol.*, **194**, 181–94.

Zerahn, K. (1977) Potassium transport in insect midgut, in *Transport of Ions and Water in Animals* (eds B.L. Gupta, R.B. Moreton, J.L. Oschman and B.J. Wall), Academic Press, New York, pp. 381–400.

11

Lipid and sugar absorption

S. Turunen and K. Crailsheim

Dietary lipids and carbohydrates ingested by insects vary widely depending on the foods consumed. Approximately half the known insect species are phytophagous, consuming photosynthetic or other plant tissues. Insect and angiosperm coevolution has also led to a more specialized type of herbivory, namely the use of nectar and pollen as sources of sugars and amino acids. Insect carnivores feed on animal material ranging from insect tissues to mammalian blood.

Lipids and sugars serve as energy sources in all animals. In addition, lipids contain essential polyunsaturated fatty acids and provide sterols, carotenoids, lipid-soluble vitamins, choline, *myo*-inositol and, for example, the sugar galactose (Dadd, 1985; Turunen, 1985). Because substances released from the gut into the haemolymph are available to all tissues, the insect gut may function as a gateway organ similar to the vertebrate liver in modifying dietary substrates. Thus, the midgut is known to shorten the chain length of absorbed fatty acids, synthesize acylglycerols from potentially toxic free fatty acids, convert lysophospholipids into corresponding phospholipids and dietary galactose into glucose and other metabolites (Turunen, 1993).

11.1 TYPES OF DIETARY LIPIDS

11.1.1 Plant leaf lipids

Insect herbivores feeding on green leaves consume lipids which differ markedly from those of animal tissues. Major photosynthetic tissue

Biology of the Insect Midgut.
Edited by M.J. Lehane and P.F. Billingsley.
Published in 1996 by Chapman & Hall, London.
ISBN 0 412 61670 X.

lipids are the glycosyl glycerides monogalactosyldiglyceride (MGDG) and digalactosyldiglyceride (DGDG), which have been suggested to be the world's commonest lipids (Harwood, 1980). MGDG is present mainly in chloroplasts, is rich in polyunsaturated fatty acids and has galactose attached glycosidically to the 3-(*sn*)-position of glycerol (Figure 11.1). DGDG is less abundant and usually found with MGDG, and also has a high proportion of polyunsaturated fatty acids, especially linolenic acid ($C_{18:3n-3}$). In both lipids glycerol has the same configuration as in phospholipids (Douce *et al.*, 1990). Another type of glycosyl glyceride found in all green plants are sulpholipids (sulphoquinovosyldiglyceride), containing the carbohydrate *S*-sulphoquinovose (Figure 11.1). Because of the ubiquity of galactosyldiglycerides, galactose is the most abundant sugar in higher plants and algae. These lipids, including their polyunsaturated fatty acids, are formed only in the light and are therefore much reduced in etiolated plant tissues.

In addition to glycolipids plant leaves contain glycerophospholipids, of which phosphatidylcholine, phosphatidylethanolamine and phosphatidylglycerol are quantitatively prominent (Harwood, 1980; Turunen and Chippendale, 1989) (Figure 11.1). Major fatty acids in plant leaf phospholipids are linoleic ($C_{18:2-n6}$), linolenic, oleic ($C_{18:1-n9}$) and palmitic ($C_{18:0}$) acids, although phospholipids accumulate less linolenate than do glycolipids (Turunen, 1990a).

Waxes formed from fatty acids and higher-molecular-weight alcohols are present in all photosynthetic tissues. Insects have a dietary requirement for a sterol, which may be met with phytosterols such as beta-sitosterol, stigmasterol, campesterol or brassicasterol. Lepidoptera and several other phytophagous species convert these C_{29}- and C_{28}-sterols into the C_{27} sterol, cholesterol (Svoboda and Thompson, 1983).

Triacylglycerols are minor lipids in plant leaves, but predominate in plant storage tissues such as seeds, and are used in insect artificial diets as sources of essential and other fatty acids. Lipid-soluble plant compounds of dietary significance also include carotenoids, tocopherols (vitamin E) and phylloquinone (vitamin K) (Turunen, 1985).

11.1.2 Animal lipids

Triacylglycerol is the main lipid type in whole body extracts of animals. Other lipids of dietary significance to carnivores are phospholipids, mainly phosphatidylcholine and phosphatidylethanolamine, and smaller proportions of other phospholipids, cholesterol and waxes.

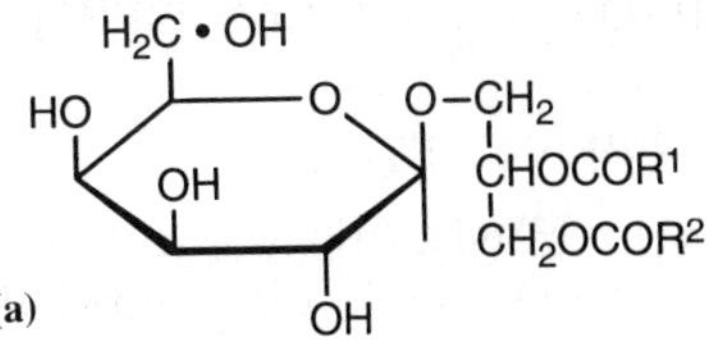

(a)

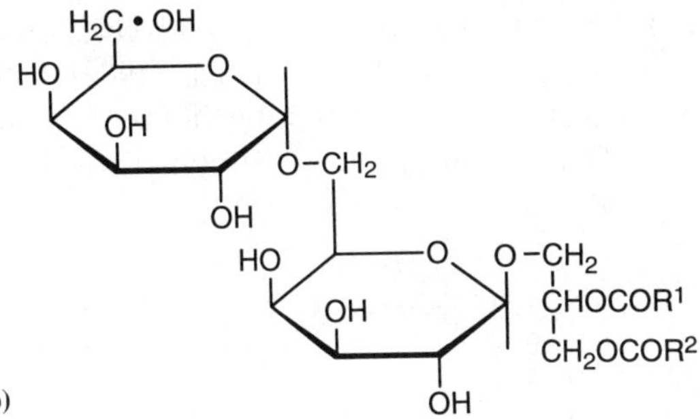

(b)

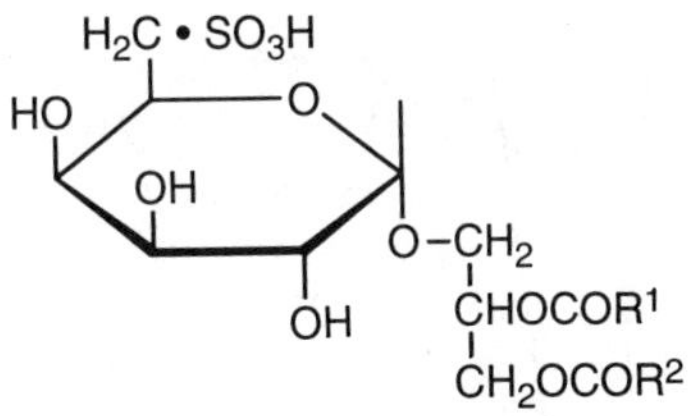

(c)

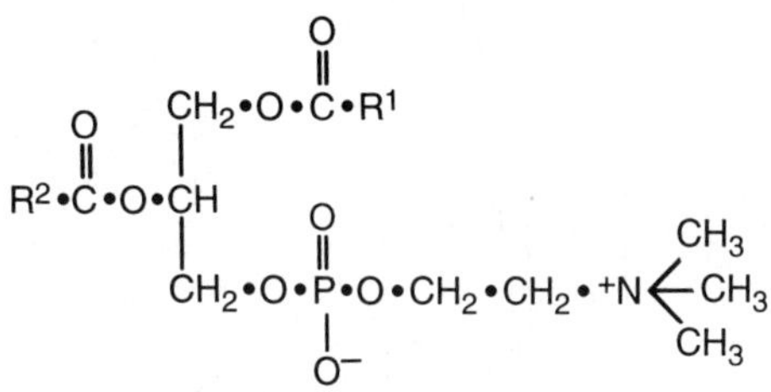

(d)

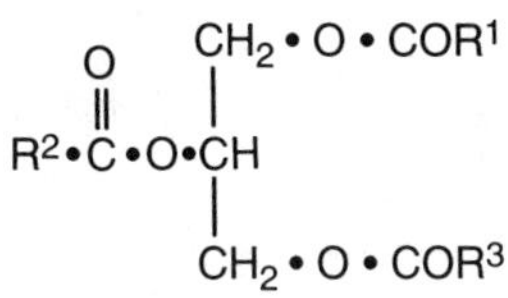

(e)

(f) $CH_3(CH_2)_4CH{=}CH \cdot CH_2 \cdot CH{=}CH(CH_2)_7 \cdot COOH$

Figure 11.1 Lipids of dietary importance to insects. (a) Monogalactosyl diglyceride; (b) digalactosyl diglyceride; (c) plant sulpholipid (sulphoquinovosyl diglyceride); (d) phosphatidylcholine; (e) triacylglycerol; (f) linoleic acid.

11.2 PROCESSING OF LIPIDS PRIOR TO ABSORPTION

11.2.1 Digestion of neutral lipids

Insect digestive enzymes hydrolyse triacylglycerols, phospholipids, glycolipids and sulpholipids, although our knowledge is confined mainly to lipases cleaving the ester bonds of triacylglycerols and partial glycerides (Chapter 6). Digestion of triacylglycerols has been examined in representatives of Odonata (Komnick *et al.*, 1984), Blattodea (Treherne, 1958c; Hoffman and Downer, 1979a), Orthoptera (Weintraub and Tietz, 1973; Thomas, 1984), Lepidoptera (Turunen, 1975; Tsuchida and Wells,

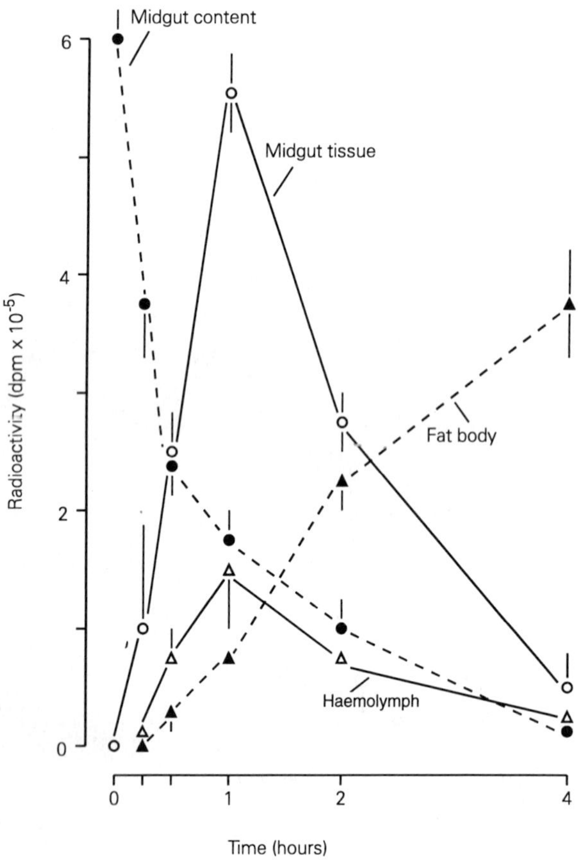

Figure 11.2 Fate of dietary triolein in the lepidopteran *Manduca sexta* after feeding the larvae on tritiated triolein. Total radioactivity recovered from midgut content, midgut tissue, haemolymph and fat body is shown as a function of time after feeding. (From Tsuchida and Wells, 1988, with permission.)

1988), Diptera (Lehane, 1977; Langley *et al.*, 1987) and Hymenoptera (Barlow and Jones, 1981). The digestion of triacylglycerol is followed by a transient increase in midgut tissue and haemolymph lipid, and subsequent storage of lipid in fat body (Figure 11.2).

Both complete and partial lipolysis of triacylglycerols have been described in insects (Figure 11.3). Midgut lipases of *Locusta migratoria* appear to have high activity toward monoacylglycerol, resulting in complete hydrolysis of trioleoylglycerol. Similar results were obtained in larvae of *Bombyx mori* (Weintraub and Tietz, 1973, 1978). In both examples the labelled lipid under study was applied on plant leaves. Complete hydrolysis of trioleoylglycerol was also suggested in *Manduca sexta* (Tsuchida and Wells, 1988). In contrast, midgut lipase activity in

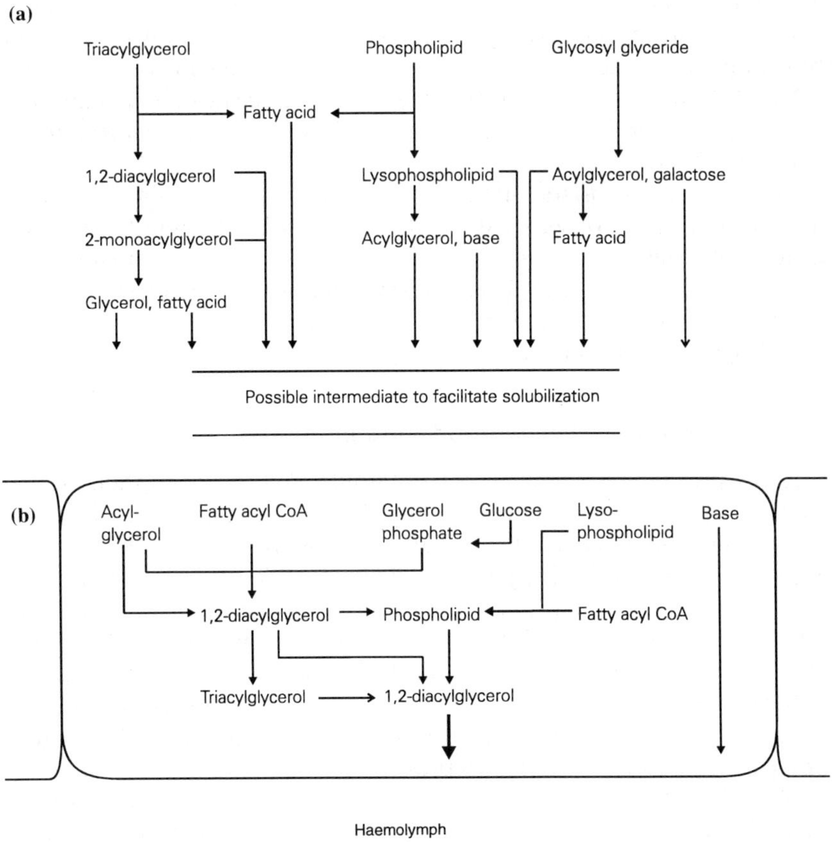

Figure 11.3 Overview of pathways observed in lipid digestion and absorption in insects. (a) Digestion of lipids in the midgut lumen; (b) lipid resynthesis in midgut cells. Based on data from several insect species.

Periplaneta americana yielded primarily 2-monoacylglycerol and free fatty acids (Hoffman and Downer, 1979a; Male and Storey, 1981), a situation analogous to that in vertebrates. Hydrolysis of trioleoylglycerol in the midgut of *Pieris brassicae* and *Diatraea grandiosella* fed a semi-artificial diet yielded free fatty acids and only a small amount of diacylglycerol. Unhydrolysed trioleoylglycerol was excreted in faeces, showing inefficient hydrolysis probably as a result of lipase saturation (Turunen, 1975; Turunen and Chippendale, 1977b). More recent data in *P. brassicae* suggest the possibility that released fatty acids and partial glycerides are temporarily incorporated into luminal polar lipids before absorption (section 11.2.3).

It may be significant for the study of lipid absorption that in the mammalian duodenum the rate of hydrolysis of partial glycerides diminishes *in vivo*, explaining the intraluminal accumulation of partial glycerides. Under *in vitro* conditions, resynthesis of di- and triacylglycerols by intestinal lipase is less likely to affect the lipolysis, which is thus more likely to proceed to completion (Thomson and Dietschy, 1981). Midgut lipase of *P. americana*, which produces 2-monoacylglycerol *in vivo*, promotes further hydrolysis to glycerol *in vitro* (Hoffman and Downer, 1979a).

Work on larvae of the dragonfly *Aeshna cyanea* has shown midgut luminal hydrolysis of wax esters (Komnick and Bauerfeind, 1991) and some hydrolysis of cholesterol oleate (Komnick and Giesa, 1994). Sterol ester hydrolysis was slow and a substantial portion of cholesterol oleate was excreted. In *Eurycotis floridana* some absorption of intact sterol ester was observed (Clayton *et al.*, 1964).

11.2.2 Digestion of phospholipids and glycolipids

In contrast to the digestion of neutral lipids, data on phospholipids are available on only a few species of Lepidoptera, Coleoptera and Diptera (Somerville and Pockett, 1976; Turunen and Kastari, 1979; Turunen, 1988a; Turunen and Chippendale, 1989; Uschian *et al.*, 1995). In *P. brassicae* hydrolysis of phosphatidylcholine *in vivo* yields unesterified fatty acids, lysophosphatidylcholine, choline or its water-soluble derivative and acylglycerols (Turunen and Kastari, 1979; Turunen, 1988b) (Figure 11.3). The release of lysophospholipid indicates phospholipase A activity, and the presence of the base moiety and acylglycerol suggests the activity of midgut digestive phospholipase C (and/or D) (Turunen, 1990b). Phospholipase C activity, possibly cytosolic, was implicated in the hydrolysis of phosphatidylcholine by midgut tissue homogenates in the dipteran *Stomoxys calcitrans* (Spates *et al.*, 1990).

The use of dietary glycosyl glycerides appears to have been studied only in lepidopterous larvae. *P. brassicae* larvae reared on leaves of

Brassica oleracea used their dietary phospho- and glycolipids more completely than neutral lipids (Kastari and Turunen, 1977). Analysis of lipids in the diet, midgut contents and faeces of the larvae of the peacock, *Inachis io*, feeding on leaves of *Urtica dioica* showed digestion of dietary monogalactosyl diglycerides, digalactosyldiglycerides and sulpholipids. Data on larvae of *Anticarsia gemmatilis* feeding on leaves of *Zea mays* indicated digestion of dietary digalactosyldiglycerides, phosphatidylcholines, waxes and triacylglycerols (Turunen, 1990a). Similar results were also obtained on *P. brassicae* feeding on *Brassica oleracea* and *Heliothis zea* feeding on *Zea mays* (Turunen and Chippendale, 1989).

Very little is known of the enzymes involved in glycolipid hydrolysis. A combination of alpha- and beta-galactosidases can degrade these compounds to free galactose and 1,2-diacylglycerol. Digestive galactosidase activity has been found in several species of insects, for example *L. migratoria* (Morgan, 1975), the Coleopteran *Acanthoscelides obtectus* (Leroy *et al.*, 1984), and the dipterans *Calliphora erythrocephala* (House and Ginsborg, 1985) and *Rhynchosciara americana* (Terra *et al.*, 1979; Terra, 1990), but their possible role in glycolipid digestion remains to be examined. Plant leaves contain galactosidases and lipase and the possible function of these enzymes in plant lipid digestion in the neutral and alkaline contents of the insect midgut would be worth examining.

11.2.3 Lipid solubilization

Lipids are relatively insoluble in water, but in order to be absorbed they must diffuse through the layer of water adjacent to absorptive cells. In vertebrates this problem is solved by the secretion of bile, which contains bile acids, phosphatidylcholine and cholesterol. Bile salts are soluble, polar substances with strongly hydrophilic and hydrophobic sides and perform a detergent function in the emulsification of lipids.

Observations in *P. brassicae* reared on triacylglycerol-supplemented diets or on cabbage leaves have indicated transient incorporation of dietary lipids into three major polar fractions in the midgut lumen (designated GLa, GLb, GLc) containing lipids and carbohydrates but no amino groups (Turunen, 1988a,b; Turunen and Chippendale, 1989). The fractions differ in lipid composition, suggested by feeding the insects [1-^{14}C]linolenic acid or di[1-^{14}C]oleoylphosphatidylcholine. Radioactivity from linolenic acid was detected mainly in GLb with none in GLa. However, radioactivity from phosphatidylcholine was equally divided between GLa and GLb with some present in GLc (Table 11.1). All fractions can be labelled by feeding the larvae D-[U-^{14}C]galactose (Turunen, 1992). The polar molecules are absent from midgut cells and

Table 11.1 Incorporation of radioactivity from dietary substrates into three glycolipid-like compounds proposed to facilitate lipid solubilization in the midgut lumen of larval *Pieris brassicae*[a]

Substrate	*GLa*	*GLb*	*GLc*
Di[1-^{14}C]oleoylphosphatidylcholine			
4 h	+	++	++
48 h	+++	+++	++
Glycerol tri[1-^{14}C]oleate			
48 h	+++	++++	++
[1-^{14}C]Linolenic acid			
24 h	–	++++	+
D-[U-^{14}C]Galactose			
24 h	++	++++	+

[a]Labelled lipids and galactose were incorporated into the semi-artificial diet of larvae (Turunen, 1988a,b, 1992).

only a small portion is excreted, suggesting that they aid in lipid solubilization.

Possible other means used by insects to increase lipid solubilization include the formation of fatty acyl-amino acid complexes (Collatz and Mommsen, 1974), saponification under alkaline midgut conditions (Lehane, 1977) and the formation of the strongly detergent lysophospholipids (Turunen and Kastari, 1979; Turunen, 1988a). Lysophospholipids react with and may be precipitated by tannins, however, lessening their detergent action in some diets. In the gypsy moth this effect of tannins has been suggested to reduce lipid absorption (DeVean and Schultz, 1992).

11.3 ABSORPTION OF LIPIDS INTO THE MIDGUT EPITHELIUM

11.3.1 Evidence of functional differentiation of the midgut in lipid absorption

Although the midgut is the main site of lipid absorption, lipids may diffuse through the cuticular lining of the foregut. Data from *P. americana* have shown absorption of free fatty acid and resynthesis of glyceride in the crop (Hoffman and Downer, 1976). Functional differentiation has been implied in lipid absorption also in the midgut of Diptera and Lepidoptera on the basis of sectional tissue analyses, tracer studies and enzyme assays (Lehane, 1977; Turunen and Chippendale, 1977a,b). Studies of lipid turnover in midgut sections of *P. brassicae* have shown shorter half-lives of triacylglycerol and phosphatidylcholine in the anterior than in the posterior midgut (Turunen, 1988a). The data were interpreted to indicate more rapid lipid absorption across the anterior than across the posterior midgut (Turunen, 1990b).

The distribution of fatty acid binding proteins (FABP) in the midgut of *M. sexta* also suggests functional differentiation (Smith *et al.*, 1992). Two cytosolic FABPs were identified, both of which bind fatty acids in a 1:1 molar ratio. One (MFB1) is more concentrated in the anterior two thirds of the midgut and exchanges fatty acids more readily than MFB2, which is more posterior in location. Little is known of the function of FABPs, which are known to occur in both vertebrates and invertebrates. They have been suggested to facilitate fatty acid uptake by cells and to target fatty acids to organelles and specific pathways (Sweetser *et al.*, 1987). Similar ubiquitous proteins facilitate cytosolic transfer of phospholipids (Wirtz, 1991).

11.3.2 Acylglycerols: absorption and resynthesis in midgut cells

Insects resemble vertebrates in the overall resynthesis of lipids following uptake into the enterocytes (Figure 11.3). In all species studied, absorbed fatty acids or partial glycerides are incorporated into intestinal tissue 1,2-diacylglycerols, triacylglycerols and phospholipids (Weintraub and Tietz, 1973; Turunen, 1975; Turunen and Chippendale, 1977a; Chino and Downer, 1979; Komnick *et al.*, 1984; Thomas, 1984; Tsuchida and Wells, 1988). The synthesis could proceed by acylation of 2-monoacylglycerol or 1,2-diacylglycerol following absorption from the lumen, or via the glycerol phosphate pathway from absorbed fatty acids and glycerol. The relative contributions of the acylglycerol and the glycerol phosphate pathways to glyceride synthesis have not been established in any of the insects studied. It is also possible that both pathways are present and are used simultaneously, as in mammals.

In *P. americana*, midgut and crop homogenates contain monoacylglycerol acyltransferase activity and are able to catalyse synthesis of diacylglycerol from 2-mono-oleoylglycerol and palmitic acid (Hoffman and Downer, 1979b). *In vitro*, the reaction produced mainly diacylglycerol, with a small amount of triacylglycerol. Enzyme activity exhibited linear dependence on the titre of 2-monoacylglycerol. These data suggest that the monoacylglycerol pathway is involved in glyceride absorption in *P. americana*. The data do not necessarily reflect the *in vivo* situation as regards the proportion of diacylglycerol/triacylglycerol formed, for in mammals the pathway is known to yield a higher proportion of triacylglycerol *in vivo* than *in vitro* (Breckenridge and Kuksis, 1975). Data in *P. brassicae* leave open the possibility of 1,2-diacylglycerol uptake from diet, since little or no monoacylglycerol has been found in luminal or mucosal lipids (Turunen, 1985). Absorption of diacylglycerol from the lumen was suggested in *Glossina morsitans* (Langley *et al.*, 1987).

The glycerol phosphate pathway, implied by the complete hydrolysis of triacylglycerol into glycerol and fatty acids in *L. migratoria* and *B. mori*, entails formation of glycerol-3-phosphate and its acylation into 1,2-diacyl-3-glycerophosphate (phosphatidic acid). In general, the major portion of the required glycerophosphate is thought to originate from carbohydrate. The main products of the glycerol phosphate pathway in *P. brassicae* midgut *in vivo* are triacylglycerol and phosphatidylcholine (Figure 11.3). Absorbed [1(3)-^{3}H]glycerol was also recovered from midgut tissue diacylglycerol, phosphatidylethanolamine and amino acids, including glycine, proline and aspartic acid (Turunen, 1993). In the haemolymph, the only labelled lipid following absorption was diacylglycerol. The bulk of haemolymph radioactivity from glycerol was recovered from amino acids and peptides, and some from glyceraldehyde and trehalose. Based on these data, the half-life of midgut tissue 1,2-diacylglycerol in feeding, last instar larvae of *P. brassicae* was estimated to be about 30–60 min.

The role of carbohydrate as a source of glycerophosphate was examined with dietary D-[U-^{14}C]galactose, a constituent of plant lipids (Turunen, 1992, 1993). Absorption was followed by rapid increase in haemolymph trehalose. Because little or no labelled galactose was recovered from haemolymph, galactose appeared to be completely converted in midgut cells into glucose, glyceraldehyde, amino acids and a significant portion into midgut tissue lipids (Figure 11.3).

In the hymenopterous parasites *Exeristes roborator* and *Itoplectis conquisitor* the glycerol phosphate pathway may be absent (Thompson and Barlow, 1983). Microsomal preparations of these insects failed to incorporate glycerol phosphate into glycerides, and larvae seemed to absorb partial glycerides from their hosts.

Larvae and nymphs of the dragonfly *A. cyanea* were found to absorb several fatty alcohols, including lauryl, myristyl, cetyl and stearyl alcohol, from the diet (Komnick and Bauerfeind, 1991). In midgut cells, palmityl alcohol was oxidized to the corresponding fatty acid and incorporated into phospholipids and acyl glycerols (Komnick and Wachtmann, 1994). A small portion of the alcohol was esterified to a wax ester and released into the haemolymph. Digestive enzymes of *A. cyanea* also synthesized wax esters in the midgut lumen, although the esters were probably not absorbed before hydrolysis to free alcohol.

Rapid labelling of phosphatidylcholine in midgut cells following the absorption of free fatty acids (Turunen, 1988b; Wachtman and Komnick, 1993) or glycerol (Turunen, 1993) has been interpreted as evidence of phospholipid involvement in lipid transfer into the haemolymph. In *A. cyanea* the pattern of lipid synthesis in midgut tissue depended on the amount of fatty acid absorbed. At low doses little or no synthesis of midgut tissue triacylglycerol was observed, and haemolymph diacylglycerol was thought to originate from midgut tissue phospholipids, but at higher doses both phospholipids and triacylglycerols were apparently sources of haemolymph diacylglycerol (Wachtman and Komnick, 1993). In *P. brassicae* the turnover of midgut tissue phospholipids is much slower than that of triacylglycerols, suggesting that acylglycerol translocation via triacylglycerol is a major pathway of fatty acid absorption.

11.3.3 Phospholipid and glycolipids

Phospholipid absorption has been examined in only a few species but it is apparent that phospholipids contribute to the nutrition of *P. brassicae* and some other phytophagous lepidopterans (Kastari and Turunen, 1977; Turunen, 1988a; Turunen and Chippendale, 1989). After hydrolysis of phosphatidylcholine in *P. brassicae* larvae, lysophosphatidylcholine and fatty acid are absorbed into the mucosa, where resynthesis of phosphatidylcholine occurs (Figure 11.3). Phospholipase activity also releases soluble choline, most of which is passed into the haemolymph without resynthesis into mucosal phospholipid (Turunen, 1993). Some midgut synthesis of phosphatidylcholine and sphingomyelin also occurs from absorbed choline (Table 11.2).

Galactosyl diglyceride hydrolysis has evidently not been examined in much detail in animals. Lipolysis may yield free galactose and diacylglycerol, both of which are known to be utilized by larvae of *P. brassicae* (Turunen, 1992, 1993). Galactose is probably the main dietary sugar ingested by leaf-feeding animals. In *P. brassicae* larvae it was found to be assimilated at an efficiency of ca. 90%, and metabolized in midgut cells to the extent that little or no free galactose or galactitol was present in haemolymph.

11.3.4 Sterols

Insects have a dietary requirement for sterol (Dadd, 1985). In many species, including representatives of Hemiptera, Lepidoptera, Diptera, Hymenoptera and Coleoptera, the midgut is the main site of cholesterol absorption, but significant absorption was found in the crop, e.g. in *Eurycotis floridana* (Clayton *et al.*, 1964), *Gryllodes sigilatus* and *Camponotus compressus* (Joshi and Agarwall, 1976, 1977). Both free sterol and intact cholesterol stearate were absorbed in *E. floridana*, but hydrolysis of cholesterol oleate preceded absorption in *G. morsitans* (Langley *et al.*, 1987). In larvae of *D. grandiosella* absorption of free cholesterol was followed by some esterification, so that after 48 h of feeding on a diet containing 4-[^{14}C]cholesterol, about 7% of the radioactivity in midgut

Table 11.2 Use of absorbed substrates in midgut cells of last instar *Pieris brassicae* larvae during nutrient absorption

Substrate absorbed from the lumen	*Use in midgut cells*	*Released into haemolymph*	*Reference*
Free fatty acids	1,2-Diacylglycerol	+	Turunen (1975, 1993)
	Triacylglycerol	–	
	Phospholipids	?	
Glycerol	Glycerol	+	Turunen (1993)
	Glyceraldehyde	?	
	1,2-Diacylglycerol	+	
	Triacylglycerol	–	
	Phospholipids	?	
	Glucose	+	
	Amino acids	+	
Choline chloride	Choline	+[a]	Turunen (1993)
	Lysophosphatidylcholine	–	
	Phosphatidylcholine	?	
	Sphingomyelin	?	
Lysophosphatidylcholine	Phosphatidylcholine	?	Turunen and Kastari (1979); Turunen (1988a)
Myo-inositol	*Myo*-inositol	+[b]	Turunen (1987)
	Phosphatidylinositol	?	
Galactose	Galactose	–	Turunen (1992, 1993)
	Glucose	+	
	1,2-Diacylglycerol	+	
	Triacylglycerol	–	
	Amino acids	+	
	Glycogen	–	

[a]Choline is present mainly as a soluble metabolite in haemolymph.
[b]Released in water-soluble form of unidentified structure.

tissue was present as sterol esters (Turunen and Chippendale, 1977a). Most of the esterification occurred in the posterior midgut. Cholesterol was released into the haemolymph mainly as free sterol. Cholesterol oleate was hydrolysed in the midgut lumen of larval *A. cyanea* but absorption of sterol was slow. A small proportion of sterol was esterified in midgut cells but release into the haemolymph appeared to be in the form of free sterol (Komnick and Giesa, 1994). Intracellular esterification of cholesterol was also described in *E. floridana*, and sterol esters were a prominent lipid fraction in midgut tissues in the lepidopterans *Spodoptera frugiperda*, *Heliothis zea* and *H. subflexa*, but not in *P. brassicae* (Clayton *et al.*, 1964; Turunen and Chippendale, 1989). Although some esterification occurs in the vertebrate intestinal mucosa, most of the sterol remains unesterified until incorporation into lipoproteins and transport into the lymph.

Phytophagous insects use plant sterols, usually present as fatty acid esters in their diet. Cholesterol may also be found in plant tissues at trace concentrations and in the khapra beetle, *Trogoderma granarium*, is selectively absorbed, together with campesterol (Svoboda and Thompson, 1983). Utilization of sterol esters may depend on the rate of luminal hydrolysis, as suggested in the khapra beetle (Nair and Agarwall, 1977). As further evidence for selective uptake, ergosterol and stigmasterol were not utilized in the beetle *Dermestes maculatus*, although beta-sitosterol and campesterol were assimilated (Budowski *et al.*, 1967). In accordance with the situation in vertebrates, phytosterols may compete for absorption with cholesterol. In *D. maculatus*, for example, tissue cholesterol was reduced by 50% or more with the addition of beta-sitosterol, campesterol or stigmasterol into the diet (Katz *et al.*, 1971).

Conversion of phytosterols into cholesterol may occur in midgut tissue but it is not necessary for sterol release into the haemolymph. Absorbed phytosterols may be retained unchanged even in predacious species, as shown in a comparison of the predacious ladybird *Coccinella septempunctata* and the phytophagous Mexican bean beetle *Epilachna varivestis*. Among the sterols of *C. septempunctata* were beta-sitosterol (29.5%), campesterol (12%) and stigmasterol (5.5%), which were absorbed from the phytophagous prey (Svoboda and Robbins, 1979).

11.3.5 Other lipid-derived components

Only plants and micro-organisms appear to synthesize carotenoids *de novo*. Internal micro-organisms may contribute to the synthesis of insect carotenoids, for example in the seven-spot ladybird (Britton *et al.*, 1977). Insects absorb a variety of plant carotenoids and show selectivity in their uptake from the diet. Of the six carotenoids found in leaves of *Lilium hansonii*, only beta-carotene and 5,6-monoepoxy-beta-carotene were present in the lily beetle *Lilioceris lilii* (Mummery and Valadon, 1974). In larvae of *P. brassicae* the use of dietary carotenoids varied considerably: over 30% of the ingested 5,6-monoepoxy-beta-carotene was excreted, compared with 11.5% of beta-carotene and 6.7% of lutein (Feltwell and Valadon, 1972).

Microbial fermentation of cellulose in the alimentary canal of both vertebrates and insects yields volatile, short chain fatty acids, primarily acetate, propionate and butyrate. Fermentation may be restricted mainly to the hindgut, as in *P. americana* (Bracke and Markovetz, 1980), but in the scarabeid beetle *Oryctes nasicornis*, acetate and propionate, probably the main products of polysaccharide digestion, were absorbed in the proctodeal dilation and midgut *in vitro* (Bayon and Mathelin, 1980). Absorption of the unionized form has been shown to occur by passive diffusion in the mammalian alimentary canal (Bugaut, 1987). Acetate and butyrate are ketogenic or lipogenic, whereas propionate is glucogenic. The significance of volatile fatty acids in the nutrition of xylophagous insects presents many fascinating problems of intermediary metabolism.

A dietary requirement for choline is probably common to all insects and many require *myo*-inositol (Dadd, 1985). In larvae of *P. brassicae* most of the radioactivity from dietary [methyl-^{3}H]choline chloride was transferred in water-soluble form across the midgut. Here it was converted into a less polar metabolite, which accumulated in the haemolymph, thus facilitating overall absorption (Turunen, 1993). Results with tritiated *myo*-inositol showed similar rapid transfer and accumulation of water-soluble inositol radioactivity in the haemolymph, but the exact mode of uptake remains to be studied (Turunen, 1989, 1990). It is noteworthy that the uptake of dietary choline can be inhibited by ethanol in *Drosophila*, suggesting an effect on choline receptors in the gut (Miller *et al.*, 1993).

11.4 NUTRIENT METABOLISM IN MIDGUT CELLS AND OTHER FACTORS INFLUENCING ABSORPTION

A summary of some pathways involved in dietary lipid metabolism in the lepidopteran midgut is shown in Table 11.2. The significance of triacylglycerol and phosphatidylcholine as intermediates in lipid transfer across the midgut is suggested by studies of lipid turnover in the midgut tissue. After prelabelling the midgut of *P. brassicae* for 24 h with dietary [1-^{14}C]oleic acid, larvae were allowed to continue feeding on an unlabelled diet, and the decline of midgut tissue radioactivity was monitored from different lipids. The turnover of oleate radioactivity was most rapid in triacylglycerols (Table 11.3). Quantitatively, triacylglycerols and phosphatidylcholines contributed most to the oleate turnover.

Potentially harmful dietary fatty acids may be metabolized in midgut cells before release into the haemolymph. Midguts of *A. cyanea* were found to oxidize dietary erucic ($C_{22:1}$) and nervonic acid ($C_{24:1}$). Chain-shortening of erucic to oleic acid was attributed to peroxisomal beta-oxidation (Fischer and Komnick, 1992). A previous finding in *P. brassicae* had also suggested restricted uptake of dietary erucic acid (Turunen, 1973). Although some erucate was released unchanged from the midgut into haemolymph, tissue erucate was restricted mainly to neutral

Table 11.3 Decline of radioactivity in midgut tissue of *Pieris brassicae* larvae prelabelled with [1-^{14}C]oleic acid[a]. (From Turunen and Chippendale, 1989 with permission.)

Time (h)	*Total ^{14}C in midgut tissue (d.p.m.)*	*Radioactivity (d.p.m./mg midgut tissue[b])*				
		PC	*PI*	*PE*	*DG*	*TG*
0	6237 ± 1192	71 ± 5	6.5 ± 1	53 ± 2	9 ± 1	45 ± 5
6	5448 ± 772	52 ± 3	5.5 ± 1	49 ± 2	4 ± 1	16 ± 3
24	3090 ± 246	21 ± 2	3 ± 0.5	33 ± 2	2 ± 1	4 ± 0.5

[a]Last instar larvae were fed a diet containing labelled oleate for 24 h and then transferred on to a non-radioactive diet (zero time). Data are presented as mean ± SD. PC, phosphatidyl choline; PI, phosphatidylinositol; PE, phosphatidylethanolamine; DG, diacylglycerol; TG, triacylglycerol.
[b]The fresh weight of midgut tissue was: 28.9 ± 4.4 mg/larva (0 h), 34.2 ± 2.3 mg (6 h), and 36.7 ± 2.3 mg (24 h).

glycerides, with little incorporation into phospholipids (Turunen, 1990a). In *O. nasicornis* absorbed acetate and propionate were metabolized in midgut cells, because none was found in the haemolymph (Bayon and Mathelin, 1980).

At the luminal level uptake may be influenced by selective lipolysis or absorption (Small, 1991). There is evidence that polyunsaturated fatty acids could be preferentially absorbed in *Heliothis virescens* (Dikeman *et al.*, 1981) and that midgut lipase of *L. migratoria* shows specificity toward unsaturated fatty acids (Weintraub and Tietz, 1973). Uptake of linolenic and linoleic acids is diminished in *P. brassicae* larvae fed a triacylglycerol-supplemented diet, if compared to the uptake of the same fatty acids from plant leaf galactosyl diglycerides, an effect which may in part be explained by differences in lipolysis (Turunen, 1973, 1985, 1990a). Low tissue polyunsaturated fatty acid titres have been observed in many species reared on triacylglycerol-supplemented diets. They are to some extent explained by tissue synthesis of palmitoleic ($C_{16:1}$) and oleic ($C_{18:1}$) acids, but such synthesis is primarily brought about by diminished absorption of the essential linoleic and linolenic acids (Thompson and Barlow, 1983; Turunen, 1983, 1985).

11.5 RELEASE INTO THE HAEMOLYMPH AND TRANSPORT OF DIETARY LIPIDS

Release of midgut tissue lipid into the haemolymph has been studied *in vivo* and *in vitro* and results from several insect orders (Odonata, Blottodea, Orthoptera, Lepidoptera) suggest diacylglycerols are the major lipids appearing in haemolymph as a result of glyceride digestion and absorption (Weintraub and Tietz, 1973; Turunen, 1975; Chino and Downer, 1979; Wachtmann and Komnick, 1993). The central role of diacylglycerol is an aspect of lipid absorption clearly distinguishing insects from mammals, in which triacylglycerols are the main lipid incorporated into mucosal lipoproteins. Released diacylglycerol is carried in the haemolymph bound to lipophorin (Shapiro *et al.*, 1988; Blacklock and Ryan, 1994). In addition to diacylglycerol and phospholipids, lipophorin transports cholesterol (Chino, 1985; Komnick and Giesa, 1994), hydrocarbons (Chino and Kitazawa, 1981) and carotenoids (Bergman and Chippendale, 1992).

In *M. sexta* the apoproteins of lipophorin are synthesized in the fat body, which secretes a lipophorin particle containing apoproteins and phospholipids. The secreted particle travels to the midgut to acquire diacylglycerol, being thereby converted into a high density lipoprotein and apparently serving as a reusable shuttle in the transport of lipids (Prasad *et al.*, 1986; Tsuchida and Wells, 1988) as originally suggested by Chino and Kitazawa (1981). No midgut synthesis of the apoprotein components of lipophorin or internalization of existing lipophorin by the midgut seemed to occur in *M. sexta*. In *A. cyanea* lipophorin was detected in the basolateral intercellular clefts of the midgut epithelium but again was not endocytosed (Bauerfeind and Komnick, 1992). In mammals, in contrast, chylomicrons and VLDL are synthesized in the cells of the small intestine.

11.6 TYPES AND SOURCES OF DIETARY CARBOHYDRATES

Carbohydrates are necessary in the diets of most insects, at least for optimum growth; only some specialists, for instance some that feed on vertebrate flesh, blood, carrion or beeswax can be reared on carbohydrate-free diets (Singh, 1977; Dadd, 1985; Reinecke, 1985). On the other hand, animals specializing on non-carbohydrate food, e.g. larvae of the waxmoth *Galleria mellonella* and hide beetle *Dermestes maculatus*, develop on fat-free carbohydrate diets (Dadd, 1964, 1966; Appelbaum *et al.*, 1971).

Carbohydrates come from any organic sources, i.e. animals, plants and micro-organisms. They are consumed in a relatively pure form as mono-, di-, or trisaccharides by nectar feeders and phloem sap suckers, as starch by insects feeding on seeds or stored products, and as glycogen by predators. Species feeding on cellulose and lignin produce the required enzymes or co-operate with micro-organisms (Chapters 6 and 15). In most cases, however, carbohydrates are consumed as mixtures of different types of carbohydrates or with large proportions of other food constituents. Some insects can feed on a single diet throughout life, whereas others depend on different foods at different developmental stages (Reinecke, 1985). Thus, almost all larvae of Lepidoptera are phytophagous, whereas adults feed on nectar, fruit juice or survive without food (Dow, 1986).

Most carbohydrates consumed are polymerized and must be broken down to monomers. The best-utilized monomers are glucose and fructose, constituting starch, glycogen and sucrose, and galactose and mannose. Other hexoses, pentoses and sugar alcohols are of varying and minor importance (Dadd, 1985). Recently, hemicellulose has been reported as an important component in the diet of *Diatraea grandiosella* (Hedin *et al.*, 1994).

11.7 DIGESTION

Enzymatic breakdown may begin in the foregut or midgut, and can be supported by micro-organisms (Chapters 2, 6 and 8; Dadd, 1985). The ultimate products are usually monosaccharides or acetic acid. Membrane-associated and intracellular disaccharidases have been described in the midguts of some species (Terra *et al.*, 1985; Ferreira *et al.*, 1988; Terra, 1990), indicating that disaccharides enter the cells. Data suggest the transport of sucrose through the midgut wall in *Aulacophora foveicollis* (Srivastava and Krishna, 1977) and *Anthonomus grandis* (Nettles *et al.*, 1971).

11.8 ABSORPTION OF MONOSACCHARIDES

Glucose transporters have been studied extensively in mammals in the last 10 years. Three types have been described: Na^+-independent, insulin sensitive and Na^+-dependent, and the amino acid sequences of some have been deduced from cloned cDNA (Silverman, 1991). A transporter similar to those in mammals was recently described in *Drosophila* Kc cell lines (Wang and Wang, 1993). It is stereospecific for D-glucose but also transports 2-deoxy-D-glucose, D-fructose and D-galactose.

Although carbohydrates are used by almost all insects, data on absorption in the midgut are scarce. In most of the publications cited in Table 11.4 transport is thought to be of the diffusion type. Treherne (1958a,b), Gelperin (1966) and Droste and Zebe (1974a) used *in vitro* preparations with KCN to inhibit energy availability for active transport and could demonstrate unaffected sugar transport. No transport against a concentration gradient could be shown in *Melanoplus differentialis* (Randall and Derr, 1965). Crailsheim (1988a,b) tested transport of glucose at different concentrations (33–1000 mM) in *Apis mellifera in vivo* and showed linear correlation between concentration and amount of sugar transported through the midgut wall. Mannitol, fructose and 3-*O*-methylglucose, which is not metabolized by the bee, did not affect glucose transport and were transported at the same rate as glucose. 3-*O*-Methylglucose injected into the haemolymph was found within 30 min to be equally distributed in the haemolymph and the liquid phase of the midgut. Linear dependence of transport on concentration over a wide range, lack of any competitive inhibition between the sugars tested and equal distribution of 3-*O*-methylglucose all indicate physical diffusion at least at the concentrations tested. It should be noted that these concentrations were relevant for the honeybee. In *Dysdercus peruvianus*, glucose transport could be demonstrated without the presence of any dietary salt that might be necessary for a co-transport mechanism (Silva and Terra, 1994). That result is, nevertheless, no proof for the absence of such a mechanism, since sufficient amounts of ions might be present in the gut.

Tentative data of a glucose transporter were found in the larval midgut of *Ostrinia nubilalis*. The transporter disappears after starvation, reappears on feeding, and is suppressed during diapause (Trabelsi and Lavenseau, 1994; Lavenseau, personal communication).

Availability and absorption of food in the midgut depend on the rate of feeding and can be additionally regulated by release from the foregut (Treherne, 1957; Chapman, 1985). The latter mechanism is most pronounced in *Apis mellifera* (Crailsheim, 1988c), where the crop also functions as a container to transport food to other members of the colony. The crop is very small when empty, but may carry up to 70 mg of nectar when filled (Fukuda *et al.*, 1969) and occupies almost the whole abdomen of the bee (which has an 'empty' weight of ca. 90 mg).

The data on absorption sites in Table 11.4 might be influenced by the insects' mode of feeding and by the consistency of the food used in the experiments. For example, in the honeybee not only the sugar but also most of the solvent water is absorbed before reaching the posterior region of the midgut, making it difficult to estimate the occurrence of glucose absorption in that region (Schneider *et al.*, 1987; Crailsheim, 1988a,b). Because apparently only physical diffusion is likely in the guts of many species, absorption probably occurs in all parts of the midgut.

11.9 MECHANISMS SUPPORTING SUGAR TRANSPORT

11.9.1 A high concentration of sugars in the gut lumen

Gaining molecules by physical diffusion, or by facilitated diffusion, from the midgut lumen requires a concentration gradient. Insects feeding on carbohydrate-rich foods would easily obtain an adequate monosaccharide titre in the haemolymph. Nevertheless, insects need special

Table 11.4 Studies of sugar absorption in the insect midgut[a]

Species (Reference)	*Substrate*	*Method*	*Part of the midgut*	*Transport*
Periplaneta americana (Treherne, 1957)	Glucose	Dye/[^{14}C]sugar *in vivo*	Caeca > midgut	–
Schistocerca gregaria (Treherne, 1958a)	Glucose	Dye/[^{14}C]sugar *in vivo*, *in vitro*	Caeca > midgut	Diffusion
Schistocerca gregaria (Treherne, 1958b)	Glucose, mannose, fructose	Dye/[^{14}C]sugar *in vivo*, *in vitro*	Caeca > midgut	Diffusion
Melanoplus differentialis (Randall and Derr, 1965)	Glucose	Chemical *in vitro*	Midgut	Diffusion
Phormia regina (Gelperin, 1966)	Glucose, fructose	[^{14}C]sugar *in vitro*	Midgut	Diffusion
Bombyx mori (Shayamala and Bhat, 1965)	Glucose	*in vitro*	Midgut	Diffusion
Locusta migratoria (Droste and Zebe, 1974a)	Glucose	[^{14}C]sugar *in vivo*, *in vitro*	Midgut	Diffusion
Gryllus assimilis *Gryllus rubens* *Scapteriscus acletus* (Thomas and Nation, 1984)	Glucose	[^{14}C]sugar *in vitro*	Midgut	–
Anopheles stephensi (Schneider *et al.*, 1987)	Glucose	[^{14}C]sugar autoradiography	Posterior midgut > anterior midgut	–
Apis mellifera (Crailsheim, 1988a)	Glucose, mannitol, 3-*O*-methylglucose	[^{14}C]sugar *in vivo*	Anterior midgut	Diffusion
Apis mellifera (Crailsheim, 1988b)	Glucose	[^{14}C]sugar *in vivo*, flight	Anterior midgut	Diffusion
Pieris brassicae (Turunen, 1992)	Galactose	[^{14}C]sugar *in vivo*	Gastrointestine	No transport to haemolymph
Pieris brassicae (Turunen, 1993)	Galactose	[^{14}C]sugar *in vivo*	Midgut	Uptake into enterocytes
Disdercus peruvianus (Silva and Terra, 1994)	Glucose	Dye/sugar *in vivo*	Anterior midgut	No dietary salt needed

[a]The type of transport is as reported by the respective authors. A carrier-mediated system tentatively reported in *Ostrinia nubilalis* is discussed in the text.

mechanisms to avoid a loss of monosaccharides in the faeces. In fact, many phloem-sucking Aphidae, which may even have specialized guts to shorten the route of the food (Munk, 1968a,b; Kunkel and Kloft, 1985), excrete large amounts of sugars in the faeces (Mittler, 1958). The loss is fully compensated for, however, by the large amount of sugar present in the diet.

Trehalose (1-[α-D-glucosido]-α-D-glucose), the most common haemolymph sugar in insects, can diffuse into the midgut. Trehalase occurs in gastrointestinal cells in many insects. Hydrolysis of trehalose in the gut can establish or intensify a glucose gradient (Wyatt, 1967).

11.9.2 Removal of transported sugars from the haemolymph

Mammals with high and relatively constant levels of glucose in their blood can live on diets deficient in or lacking carbohydrates, or starve for long periods. They must avoid equilibration of glucose between the intestinal lumen and blood. Facilitated and energy-consuming active transport mechanisms have evolved for that purpose.

In the absence of such transporters a concentration gradient from the midgut to the haemolymph is necessary. In addition to or instead of a high sugar concentration in food, such a gradient is obtained by rapid removal from blood of the transported molecule. In most insects the disaccharide trehalose is the predominant blood sugar (Florkin and Jeuniaux, 1974; Mullins, 1985). Treherne (1958a,b) had suggested that synthesis of trehalose from glucose in the fat body is the mechanism establishing a glucose gradient necessary for effective transport into the haemolymph (Friedman, 1985; Keeley, 1985). Furthermore, energy-consuming tissues such as active muscles, or tissues storing glycogen and fat, remove large amounts of glucose from the haemolymph.

11.9.3 Water absorption

Some insects have little or no trehalose, including the larvae of *Phormia regina* (Wimer, 1969), or *Agria affinis*, in which only 1–2% of haemolymph sugar is trehalose (Barlow and House, 1969). In *Apis mellifera*, glucose and trehalose are present at about equal concentrations (Alumot *et al.*, 1969; Woodring *et al.*, 1993), and the haemolymph sugar concentration (glucose, fructose and trehalose) may reach 4%. Nevertheless, excreta of bees fed radiolabelled glucose and allowed to fly until exhaustion, contained no radiolabelled sugars (Gmeinbauer and Crailsheim, 1993). This indicated complete absorption of the fed glucose in spite of the high haemolymph sugar concentration. Rapid water absorption occurs in such experiments, generating a gradient for the sugar (Crailsheim,

1988b). Such transepithelial flux of water is common in the midgut (Chapter 9).

11.10 CARBOHYDRATE METABOLISM AND CONVERSION IN THE ENTEROCYTES

From the first to the last larval instar insects have tremendous growth rates. For example, the body mass of *Manduca sexta* increases nearly 10 000-fold in about 16 days (Goodman *et al.*, 1985). Because of this extreme growth, the gut must transport and convert large amounts of various substances and sustains an intense rate of metabolism (Mandel *et al.*, 1980). The pattern of metabolism may vary during larval growth (Gibellato and Chamberlin, 1994) and can be fuelled by carbohydrates, lipids and amino acids (Chamberlin and Phillips, 1983; Chamberlin, 1987).

Sugar uptake itself does not need energy, but the processes described

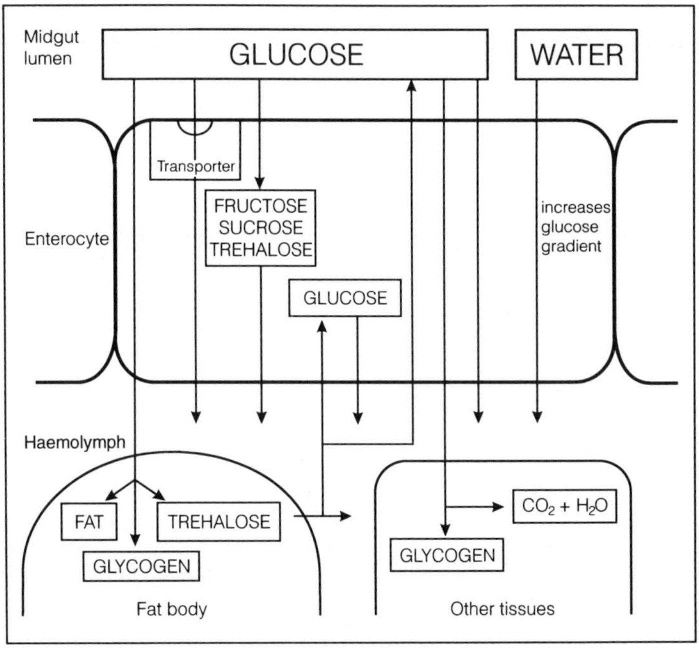

Figure 11.4 Overview of pathways observed in glucose absorption and metabolism in insects. Some mechanisms have been described for single species only. For simplicity all arrows are unidirectional. The whole midgut wall is represented by an enterocyte. A facilitated glucose transport system has been tentatively described in *Ostrinia nubilalis* (Lavenseau, personal communication).

above do, namely the establishment of concentration gradients. Sugars may be hydrolysed by membrane-bound enzymes and converted within the enterocytes. Thus after uptake of [^{14}C]galactose into midgut cells of *Pieris brassicae*, ^{14}C can be found in glucose, glycogen, glycolytic intermediates, amino acids and lipids (Table 11.2). Glycogen or glycogen-like material may also be found in gut tissue in other insects with different feeding behaviour, for example *Aedes aegypti* (Rudin and Hecker, 1979), *Anopheles gambiae* (Hecker, 1977), or *A. mellifera* (Crailsheim *et al.*, 1994). Arabinose and fructose must be converted before use into glucose in the midgut in *Aulacophora foveicollis* (Srivastava and Krishna, 1977). Conversion of glucose into trehalose in gut tissue *in vitro* has been suggested in *Bombyx mori* (Shayamala and Bhat, 1965) and in crickets (Thomas and Nation, 1984). A summary of possible glucose pathways during absorption is shown in Figure 11.4.

A most elaborate work on sugar conversion in *Locusta migratoria* was done by Droste and Zebe (1974), who demonstrated ready conversion of glucose into fructose and incorporation of both into sucrose. Both processes also clearly depended on the concentrations of the different sugars. Trehalase was distributed rather evenly in the enterocytes of all gut sections. In this insect, at least, the midgut is not only important in sugar absorption but is also a centre of carbohydrate metabolism.

REFERENCES

Alumot, E., Lensky, Y. and Holstein, P. (1969) Sugars and trehalase in the reproductive organs and hemolymph of the queen and drone honeybees (*Apis mellifera* L. var. *ligustica* Spin.). *Comp. Biochem. Physiol.*, **28**, 1419–25.

Appelbaum, S.W., Konijn, A.M. and Menco, B. (1971) Growth and biochemical adaptation of larvae of the beetle *Dermestes maculatus* to carbohydrate free diets. *Insect Biochem.*, **1**, 1–13.

Barlow, J.S. and House, H.J. (1960) Carbohydrates in the haemolymph of *Agria* (Dipt.). *J. Insect Physiol.*, **5**, 181–9.

Barlow, J.S. and Jones, D. (1981) A comparative study of transacylation in three species of insects. *Can. J. Zool.*, **59**, 1141–7.

Bauerfeind, R. and Komnick, H. (1992) Lipid-loading and unloading of lipophorin in the midgut of dragonfly larvae (*Aeshna cyanea*). A biochemical and immunocytochemical study. *J. Insect Physiol.*, **38**, 147–60.

Bayon, C. and Mathelin, J. (1980) Carbohydrate fermentation and by-product absorption studied with labelled cellulose in *Oryctes nasicornis* larvae (Coleoptera: Scarabaeidae). *J. Insect Physiol.*, **26**, 833–40.

Bergman, D.K. and Chippendale, G.M. (1992) Carotenoid transport by the larval lipophorin of the southwestern corn borer, *Diatraea grandiosella*. *Entomol. Exp. Appl.*, **62**, 81–5.

Blacklock, B.J. and Ryan, R.O. (1994) Haemolymph lipid transport. *Insect Biochem. Mol. Biol.*, **24**, 855–73.

Bracke, J.W. and Markovetz, A.J. (1980) Transport of bacterial end products from the colon of *Periplaneta americana*. *J. Insect Physiol.*, **26**, 85–9.

Breckenridge, W.C. and Kuksis, A. (1975) Triacylglycerol biosynthesis in everted sacs of rat intestinal mucosa. *Can. J. Biochem.*, **53**, 1184–95.

Britton, G., Lockley, W.J.S., Harriman, G.A. and Goodwin, T.W. (1977) Pigmentation of the ladybird beetle *Coccinella septempunctata* by carotenoids not of plant origin. *Nature*, **266**, 49–50.

Budowski, P., Ishaaya, I. and Katz, M. (1967) Growth inhibition of *Dermestes maculatus* by phytosterols. *J. Nutr.*, **91**, 201–7.

Bugaut, M. (1987) Occurrence, absorption and metabolism of short chain fatty acids in the digestive tract of mammals. *Comp. Biochem. Physiol.*, **86B**, 439–72.

Chamberlin, M.E. (1987) Enzyme activities and mitochondrial substrate oxidation in tobacco hornworm midgut. *J. Comp. Physiol. B*, **157**, 643–9.

Chamberlin, M.E. and Phillips, J.E. (1983) Oxidative metabolism in the locust rectum. *J. Comp. Physiol. B*, **151**, 191–8.

Chapman, R.F. (1985) Coordination of digestion, in *Comprehensive Insect Physiology, Biochemistry and Pharmacology* Vol. 4 (eds G.A. Kerkut and L.I. Gilbert), Pergamon Press, Oxford, pp. 213–40.

Chino, H. (1985) Lipid transport: biochemistry of haemolymph lipophorin, in *Comprehensive Insect Physiology, Biochemistry and Pharmacology* Vol. 10 (eds G.A. Kerkut and L.I. Gilbert), Pergamon Press, Oxford, pp. 115–35.

Chino, H. and Downer, R.G.H. (1979) The role of diacylglycerol in absorption of dietary glyceride in the American cockroach, *Periplaneta americana* L. *Insect Biochem.*, **9**, 379–82.

Chino, H. and Kitazawa, K. (1981) Diacylglycerol-carrying lipoprotein of hemolymph of the locust and some other insects. *J. Lipid Res.*, **22**, 1042–52.

Clayton, R.B., Hinkle, P.C., Smith, D.A. and Edwards, A.M. (1964) The intestinal absorption of cholesterol, its esters and some related sterols and analogues in the roach, *Eurycotis floridana*. *Comp. Biochem. Physiol.*, **11**, 333–50.

Collatz, K.-G. and Mommsen, T. (1974) Die Struktur der emulgieranden Substanzen verschiedener Invertebraten. *J. Comp. Physiol.*, **94**, 339–52.

Crailsheim, K. (1988a) Intestinal transport of sugars in the honeybee (*Apis mellifera* L.). *J. Insect Physiol.*, **34**, 839–45.

Crailsheim, K. (1988b) Intestinal transport of glucose solution during honeybee flight. *Biona Report* **6** (ed. W. Nachtigall), Gustav Fischer, Stuttgart, pp. 119–28.

Crailsheim, K. (1988c) Regulation of food passage in the intestine of the honeybee (*Apis mellifera* L.). *J. Insect Physiol.*, **34**, 85–90.

Crailsheim, K., Panzenböck, U., Gmeinbauer, R. and Leonhard, B. (1994) Glycogen metabolism of honeybee workers and drones during flight. *Apidologie*, **25**, 467–8.

Dadd, R.H. (1964) A study of carbohydrate and lipid nutrition in the wax moth, *Galleria mellonella* (L.), using partially synthetic diets. *J. Insect Physiol.*, **10**, 161–78.

Dadd, R.H. (1966) Beeswax in the nutrition of the wax moth, *Galleria mellonella* (L.). *J. Insect Physiol.*, **12**, 1479–92.

Dadd, R.H. (1985) Nutrition: organisms, in *Comprehensive Insect Physiology, Biochemistry and Pharmacology* Vol. 14 (eds G.A. Kerkut and L.I. Gilbert), Pergamon Press, Oxford, pp. 313–90.

DeVean, E.J.I. and Schultz, J.C. (1992) Reassessment of interaction between gut detergents and tannins in Lepidoptera and significance for gypsy moth larvae. *J. Chem. Ecol.*, **18**, 1437–54.

Dikeman, R.N., Lambremont, E.N. and Allen, R.S. (1981) Evidence for selective absorption of polyunsaturated fatty acids during digestion in the tobacco budworm, *Heliothis virescens* F. *J. Insect Physiol.*, **27**, 31–3.

Douce, R., Joyard, J., Block, M.A. and Dornes, A.-J. (1990) Glycolipid analysis and synthesis in plastids, in *Methods in Plant Biochemistry* Vol. 4. *Lipids, Membranes and Aspects of Photobiology* (eds J.L. Harwood and J.R. Bowyer), Academic Press, London, pp. 71–103.

Dow, J.A.T. (1986) Insect midgut function. *Adv. Insect Physiol.*, **19**, 187–328.

Droste, H.J. and Zebe, E. (1974) Resorption und Stoffwechsel von Glucose in Darmtrakt der Wanderheuschrecke *Locusta migratoria*. *J. Insect Physiol.*, **20**, 2385–96.

Feltwell, J.S.E. and Valadon, L.R.G. (1972) Carotenoids of *Pieris brassicae* and of its food plants. *J. Insect Physiol.*, **18**, 2203–15.

Ferreira, C., Ribeiro, A.F., Garcia, E.S. and Terra, W.R. (1988) Digestive enzymes trapped between and associated with the double plasma membranes of *Rhodnius prolixus* posterior midgut cells. *Insect Biochem.*, **18**, 521–30.

Fischer, R. and Komnick, H. (1992) Peroxisomal acyl-CoA oxidase and chain-length shortening of dietary fatty acids in the midgut of dragonfly *Aeshna cyanea*. *Insect Biochem. Mol. Biol.*, **22**, 793–801.

Florkin, M. and Jeuniaux, C. (1974) Haemolymph: composition, in *The Physiology of Insecta* Vol. V (ed. M. Rockstein), Academic Press, New York, pp. 255–307.

Friedman, S. (1985) Carbohydrate metabolism, in *Comprehensive Insect Physiology, Biochemistry and Pharmacology* Vol. 10 (eds G.A. Kerkut and L.I. Gilbert), Pergamon Press, Oxford, pp. 43–76.

Fukuda, H., Moriya, K. and Skiguchi, K. (1969) The weight of crop contents in foraging honeybee workers. *Annot. Zool. Japonenses*, **42**, 80–90.

Gelperin, A. (1966) Control of crop emptying in the blowfly. *J. Insect Physiol.*, **12**, 331–45.

Gibellato, C.M. and Chamberlin, M.E. (1994) Midgut metabolism in different instars of the tobacco hornworm (*Manduca sexta*). *J. Exp. Zool.*, **270**, 405–9.

Gmeinbauer, R. and Crailsheim, K. (1993) Glucose utilization during flight of honeybee (*Apis mellifera*) workers, drones and queens. *J. Insect Physiol.*, **39**, 959–67.

Goodman, W.G., Carlson, R.O. and Nelson, K.L. (1985) Analysis of larval and pupal development in the tobacco hornworm (Lepidoptera: Sphingidae), *Manduca sexta*. *Ann. Entomol. Soc. Am.*, **78**, 70–80.

Harwood, J.L. (1980) Plant acyl lipids. Structure, distribution, and analysis, in *The Biochemistry of Plants. A Comprehensive Treatise* Vol. 4. *Lipids: Structure and Function* (ed. P.K. Stumpf), Academic Press, New York, pp. 1–56.

Hecker, H. (1977) Structure and function of midgut epithelial cells in culicidae mosquitoes (Insecta, Diptera). *Cell Tissue Res.*, **184**, 321–41.

Hedin, P.A., Davis, F.M., Callahan, F.E. and Dollar, D.A. (1994) Wheat germ hemicellulose is an absolute requirement for growth and development of the southwestern corn borer. *J. Nutr.*, **124**, 2458–65.

Hoffman, A.G.D. and Downer, R.G.H. (1976) The crop as an organ of glyceride absorption in the American cockroach, *Periplaneta americana* L. *Can. J. Zool.*, **54**, 1165–71.

Hoffman, A.G.D. and Downer, R.G.H. (1979a) End product specificity of triacylglycerol lipases from intestine, fat body, muscle and haemolymph of the American cockroach, *Periplaneta americana* L. *Lipids*, **14**, 893–9.

Hoffman, A.G.D. and Downer, R.G.H. (1979b) Synthesis of diacylglycerols by monoacylglycerol acyltransferase from crop, midgut and fat body tissues of the American cockroach, *Periplaneta americana* L. *Insect Biochem.*, **9**, 129–34.

House, C.R. and Ginsborg, B.L. (1985) Salivary glands, in *Comprehensive Insect Physiology, Biochemistry and Pharmacology* Vol. 11 (eds G.A. Kerkut and L.I. Gilbert), Pergamon Press, Oxford, pp. 195–224.

Joshi, M. and Agarwall, H.C. (1976) Cholesterol absorption in the roach, *Periplaneta americana*. *Entomon*, **1**, 93–100.

Joshi, M. and Agarwall, H.C. (1977) Site of cholesterol absorption in some insects. *J. Insect Physiol.*, **23**, 403–4.

Kastari, T. and Turunen, S. (1977) Lipid utilization in *Pieris brassicae* reared on meridic and natural diets: implications for dietary improvement. *Entomol. Exp. Appl.*, **22**, 71–80.

Katz, M., Budowski, P. and Bondi, A. (1971) The effect of phytosterols on the growth and sterol composition of *Dermestes maculatus*. *J. Insect Physiol.*, **17**, 1295–303.

Keeley, L.L. (1985) Physiology and biochemistry of the fat body, in *Comprehensive Insect Physiology, Biochemistry and Pharmacology* Vol. 3 (eds G.A. Kerkut and L.I. Gilbert), Pergamon Press, Oxford, pp. 211–48.

Komnick, H. and Bauerfeind, R. (1991) Intestinal absorption of defined lipids by the larval dragonfly *Aeshna cyanea* (Insecta: Odonata): wax esters and fatty alcohols. *J. Insect Physiol.*, **37**, 179–91.

Komnick, H. and Giesa, U. (1994) Intestinal absorption of cholesterol, transport in the haemolymph, and incorporation into the fat body and Malpighian tubules of the larval dragonfly *Aeshna cyanea*. *Comp. Biochem. Physiol.*, **107A**, 553–7.

Komnick, H., Kukulies, J., Bongers, J. and Fischer, W. (1984) Absorption of dietary triacylglycerol by lipolysis and lipid resynthesis in the mesenteron of larval *Aeshna cyanea* (Insecta, Odonata). *Protoplasma*, **123**, 57–69.

Komnick, H. and Wachtmann, D. (1994) Assimilation and esterification of dietary fatty alcohol by the nymphal dragonfly, *Aeshna cyanea*. *Insect Biochem. Mol. Biol.*, **24**, 319–28.

Kunkel, H. and Kloft, W.J. (1985) Die Honigtau-Erzeuger des Waldes, in *Waldtracht und Waldhonig in der Imkerei* (eds W.J. Kloft and H. Kunkel), Ehrenwirt, Munich, pp. 48–265.

Langley, P.A., Ogwal, L.M., Felton, T. and Stafford, K. (1987) Lipid digestion in the tsetse fly, *Glossina morsitans*. *J. Insect Physiol.*, **33**, 981–6.

Lehane, M.J. (1977) Transcellular absorption of lipids in the midgut of the stablefly, *Stomoxys calcitrans*. *J. Insect Physiol.*, **23**, 945–54.

Leroy, B., Chararas, C. and Chipoulet, J.M. (1984) Etude des activites osidasiques du tube digestif des adultes et des larves de la bruche du haricot, *Acanthoscelides obtectus* (Coleoptera: Brucidae). *Entomol. Exp. Appl.*, **35**, 269–73.

Male, K.B. and Storey, K.B. (1981) Enzyme activities and isozyme composition of triglyceride, diglyceride and monoglyceride lipases in *Periplaneta americana*, *Locusta migratoria* and *Polia adjuncta*. *Insect Biochem.*, **11**, 423–7.

Mandel, L.J., Moffett, D.F., Riddle, T.G. and Grafton, M.M. (1980) Coupling between oxidative metabolism and active transport in the midgut of tobacco hornworm. *Am. J. Physiol.*, **238**, C1–C9.

Miller, R.R., Jr, Yates, J.W. and Green, B.W. (1993) Dietary ethanol reduces phosphatidylcholine levels and inhibits the uptake of dietary choline in *Drosophila melanogaster* larvae. *Comp. Biochem. Physiol.*, **104A**, 837–44.

Mittler, T.E. (1958) Studies on the feeding and nutrition of *Tuberolachnus salignus* (Gmelin) (Homoptera, Aphidae) 2. The nitrogen and sugar composition of ingested phloem sap and excreted honeydew. *J. Exp. Biol.*, **35**, 74–84.

Morgan, M.R.J. (1975) A qualitative survey of the carbohydrates of the alimentary tract of the migratory locust, *Locusta migratoria migratorioides*. *J. Insect Physiol.*, **21**, 1045–53.

Mullins, D.E. (1985) Chemistry and physiology of the hemolymph, in

Comprehensive Insect Physiology, Biochemistry and Pharmacology Vol. 3 (eds G.A. Kerkut and L.I. Gilbert), Pergamon Press, Oxford, pp. 355–400.

Mummery, R.S. and Valadon, L.R.G. (1974) Carotenoids of the lily beetle (*Lilioceris lilii*) and of its food plant (*Lilium hansonii*). *J. Insect Physiol.*, **20**, 429–33.

Munk, R. (1968a) Uber den Feinbau der Filterkammer der Kleinzikade *Euscelidus variegatus* Kbm. (Jassidae). *Z. Zellforsch. mikrosk. Anat.*, **85**, 210–24.

Munk, R. (1968b) Die Richtung des Nahrungsflusses im Darmtrakt der Kleinzikade *Euscelidus variegatus* Kbm. (Jassidae). *Z. vergleich. Physiol.*, **58**, 423–8.

Nair, M.G.A. and Agarwall, H.C. (1977) Sterols and sterol esters in nutrition of the beetle *Trogoderma granarium* Everts. *Indian J. Exp. Biol.*, **15**, 576–8.

Nettles, W.C., Parro, B., Sharbaugh, C. and Magnum, C.L. (1971) Trehalose and other carbohydrates in *Anthonomus grandis*, *Heliothis zea*, and *Heliothis virescens* during growth and development. *J. Insect Physiol.*, **17**, 657–75.

Prasad, S.V., Fernando-Warnakulasuriya, G.J.P., Sumida, M. *et al.* (1986) Lipoprotein biosynthesis in the larvae of the tobacco hornworm, *Manduca sexta*. *J. Biol. Chem.*, **261**, 17174–6.

Randall, D.D. and Derr, R.F. (1965) Trehalose: occurrence and relation to egg diapause and active transport in the differential grasshopper, *Melanoplus differentialis*. *J. Insect Physiol.*, **11**, 329–35.

Reinecke, J.P. (1985) Nutrition: artificial diets, in *Comprehensive Insect Physiology, Biochemistry and Pharmacology* Vol. 4 (eds G.A. Kerkut and L.I. Gilbert), Pergamon Press, Oxford, pp. 391–419.

Rudin, W. and Hecker, H. (1979) Functional morphology of the midgut of *Aedes aegypti* L. (Insecta, Diptera) during blood digestion. *Cell Tissue Res.*, **200**, 193–203.

Schneider, M., Rudin, W. and Hecker, H. (1987) Absorption and transport of radioactive tracers in the midgut of the malaria mosquito, *Anopheles stephensi*. *J. Ultrastruct. Mol. Struct. Res.*, **97**, 50–63.

Shapiro, J.P., Wells, M.A. and Law, J.H. (1988) Lipid transport in insects. *Annu. Rev. Entomol.*, **33**, 297–318.

Shayamala, M.B. and Bhat, J.V. (1965) Intestinal transport of glucose in the silkworm *Bombyx mori* L. *Indian J. Biochem.*, **2**, 101–4.

Silva, C.P. and Terra, W.R. (1994) Digestive and absorptive sites along the midgut of the cotton seed sucker bug *Dysdercus peruvianus* (Hemiptera: Pyrrhocoridae). *Insect Biochem. Mol. Biol.*, **24**, 493–505.

Silverman, M. (1991) Structure and functions of hexose transporters. *Annu. Rev. Biochem.*, **60**, 757–94.

Singh, P. (1977) *Artificial Diets for Insects, Mites and Spiders*, Plenum Press, New York.

Small, D.M. (1991) The effects of glyceride structure on absorption and metabolism. *Annu. Rev. Nutr.*, **11**, 413–34.

Smith, A.F., Tsuchida, K., Hannemann, E. *et al.* (1992) Isolation, characterization, and cDNA sequence of two fatty acid-binding proteins from the midgut of *Manduca sexta* larva. *J. Biol. Chem.*, **267**, 380–4.

Somerville, H.J. and Pockett, H.V. (1976) Phospholipase activity in gut juice of Lepidopterous larvae. *Insect Biochem.*, **6**, 351–4.

Spates, G.E., Bull, D.L. and Chen, A.C. (1990) Hydrolysis of sphingomyelin and phosphatidylcholine by midgut homogenates of the stable fly. *Arch. Insect Biochem. Physiol.*, **14**, 1–12.

Srivastava, A.K. and Krishna, S.S. (1977) Mode of utilization of certain carbohydrates in the red pumpkin beetle, *Aulacophora foveicollis* Lucas (Chrysomelidae-Coleoptera). *Z. Angew. Zool.*, **64**, 93–9.

Svoboda, J.A. and Robbins, W.E. (1979) Comparison of sterols from a phytophagous and predacious species of the family Coccinellidae. *Experientia*, **35**, 186–7.

Svoboda, J.A. and Thompson, M.J. (1983) Comparative sterol metabolism in insects, in *Metabolic Aspects of Lipid Nutrition in Insects* (eds T.E. Mittler and R.H. Dadd), Westview Press, Boulder, Colorado, pp. 1–16.

Sweetser, D.A., Heuckeroth, R.O. and Goedon, J.I. (1987) The metabolic significance of mammalian fatty-acid-binding proteins: abundant proteins in search of a function. *Annu. Rev. Nutr.*, **7**, 337–59.

Terra, W.R. (1990) Evolution of digestive systems of insects. *Annu. Rev. Entomol.*, **35**, 181–200.

Terra, W.R., Ferreira, C. and Bastos, F. (1985) Phylogenetic considerations of insect digestion. Disaccharidases and spatial organization of digestion in the *Tenebrio molitor* larvae. *Insect Biochem.*, **15**, 445–9.

Terra, W.-R., Ferreira, C. and DeBianchi, A.G. (1979) Distribution of digestive enzymes among the endo- and ectoperitrophic spaces and midgut cells of *Rhynchosciara* and its physiological significance. *J. Insect Physiol.*, **25**, 487–94.

Thomas, K.K. (1984) Studies on the absorption of lipid from the gut of desert locust, *Schistocerca gregaria*. *Comp. Biochem. Physiol.*, **77A**, 707–12.

Thomas, K.K. and Nation, J.L. (1984) Absorption of glucose, glycine and palmitic acid by isolated midgut and hindgut from crickets. *Comp. Biochem. Physiol.*, **79A**, 289–95.

Thompson, S.N. and Barlow, J.S. (1983) Metabolic determination and regulation of fatty acid composition in parasitic Hymenoptera and other animals, in *Metabolic Aspects of Lipid Nutrition in Insects* (eds T.E. Mittler and R.H. Dadd), Westview Press, Boulder, Colorado, pp. 73–106.

Thomson, A.B.R. and Dietschy, J.M. (1981) Intestinal lipid absorption: major extracellular and intracellular events, in *Physiology of the Gastrointestinal Tract* Vol. 2 (ed. L.R. Johnson), Raven Press, New York, pp. 1147–220.

Trabelsi, M.M. and Lavanseau, L. (1994) Modifications fonctionelles de l'intestin moyen au cours de la diapause de la pyrale du mais. *Bull. Soc. Zool. France*, **119**, 391.

Treherne, J.E. (1957) Glucose absorption in the cockroach. *J. Exp. Biol.*, **34**, 478–85.

Treherne, J.E. (1958a) The absorption of glucose from the alimentary canal of the locust *Schistocerca gregaria* (Forsk.). *J. Exp. Biol.*, **35**, 297–306.

Treherne, J.E. (1958b) The absorption and metabolism of some sugars in the locust, *Schistocerca gregaria* (Forsk.). *J. Exp. Biol.*, **35**, 611–25.

Treherne, J.E. (1958c) The digestion and absorption of tripalmitin in the cockroach, *Periplaneta americana* L. *J. Exp. Biol.*, **35**, 862–70.

Tsuchida, K. and Wells, M.A. (1988) Digestion, absorption, transport and storage of fat during the last larval stadium of *Manduca sexta*. Changes in the role of lipophorin in the delivery of dietary lipid to the fat body. *Insect Bicohem.*, **18**, 263–8.

Turunen, S. (1973) Utilization of fatty acids by *Pieris brassicae* reared on artificial and natural diets. *J. Insect Physiol.*, **19**, 1999–2009.

Turunen, S. (1975) Absorption and transport of dietary lipids in *Pieris brassicae*. *J. Insect Physiol.*, **21**, 1521–9.

Turunen, S. (1983) Absorption and utilization of essential fatty acids in lepidopterous larvae: metabolic implications, in *Metabolic Aspects of Lipid*

Nutrition in Insects (eds T.E. Mittler and R.H. Dadd), Westview Press, Boulder, Colorado, pp. 57–71.

Turunen, S. (1985) Absorption, in *Comprehensive Insect Physiology, Biochemistry and Pharmacology* Vol. 4 (eds G.S. Kerkut and L.I. Gilbert), Pergamon Press, Oxford, pp. 241–77.

Turunen, S. (1988a) Digestion and absorption of glycerophospholipid in *Pieris brassicae*. *Comp. Biochem. Physiol.*, **89A**, 19–24.

Turunen, S. (1988b) Uptake of dietary lipids: a novel pathway in *Pieris brassicae*. *Insect Biochem.*, **18**, 499–505.

Turunen, S. (1990a) Plant leaf lipids as fatty acid sources in two species of Lepidoptera. *J. Insect Physiol.*, **36**, 665–72.

Turunen, S. (1990b) Absorption of choline, myo-inositol, and oleic acid in the midgut of *Pieris brassicae*: sectional differentiation and uptake into the haemolymph. *J. Insect Physiol.*, **36**, 737–41.

Turunen, S. (1992) Efficient use of dietary galactose in *Pieris brassicae*. *J. Insect Physiol.*, **38**, 503–9.

Turunen, S. (1993) Metabolic pathways in the midgut epithelium of *Pieris brassicae* during carbohydrate and lipid assimilation. *Insect Biochem. Mol. Biol.*, **23**, 681–9.

Turunen, S. and Chippendale, G.M. (1977a) Lipid absorption and transport: sectional analysis of the larval midgut of the corn borer, *Diatraea grandiosella*. *Insect Biochem.*, **7**, 203–8.

Turunen, S. and Chippendale, G.M. (1977b) Ventricular esterases: comparison of their distribution within the larval midgut of four species of Lepidoptera. *Ann. Entomol. Soc. Am.*, **70**, 146–9.

Turunen, S. and Chippendale, G.M. (1989) Relationship between dietary lipids, midgut lipids, and lipid absorption in eight species of Lepidoptera reared on artificial and natural diets. *J. Insect Physiol.*, **35**, 627–33.

Turunen, S. and Kastari, T. (1979) Digestion and absorption of lecithin in larvae of the cabbage butterfly, *Pieris brassicae*. *Comp. Biochem. Physiol.*, **62A**, 933–7.

Uschian, J.M., Miller, J.S., Sarath, G. and Stanley-Samuelson, D.W. (1995) A digestive phospholipase A_2 in the tiger beetle *Cicindella circumpicta*. *J. Insect Physiol.*, **41**, 135–41.

Wachtmann, D. and Komnick, H. (1993) Dose-dependent incorporation of orally infused (1-^{14}C)oleic acid into the lipid classes of midgut, haemolymph and fat body of dragonfly larvae (*Aeshna cyanea*). *Comp. Biochem. Physiol.*, **106A**, 397–402.

Wang, M. and Wang, C. (1993) Characterization of glucose transport system in *Drosophila* Kc cells. *FEBS Lett.*, **317**, 241–4.

Weintraub, H. and Tietz, A. (1973) Triglyceride digestion and absorption in the locust, *Locusta migratoria*. *Biochim. Biophys. Acta*, **306**, 31–41.

Weintraub, H. and Tietz, A. (1978) Lipid absorption by isolated intestinal preparations. *Insect Biochem.*, **8**, 267–74.

Wimer, L. (1969) A comparison of the carbohydrate composition of the haemolymph and fat body of *Phormia regina* during larval development. *Comp. Biochem. Physiol.*, **29**, 1055–62.

Wirtz, K.W.A. (1991) Phospholipid transfer proteins. *Annu. Rev. Biochem.*, **60**, 73–99.

Woodring, J., Boulden, M., Das, S. and Gäde, G. (1993) Studies on blood sugar homeostasis in the honeybee (*Apis mellifera*, L.). *J. Insect Physiol.*, **39**, 89–97.

Wyatt, G.R. (1967) The biochemistry of sugars and polysaccharides in insects. *Adv. Insect Physiol.*, **4**, 287–360.

Part Three

The Midgut as a Target for Control Strategies

12

Immune intervention against blood-feeding insects

P. Willadsen and P.F. Billingsley

Parasites, including haematophagous arthropods, which feed on blood or tissue fluid will ingest such components of the host's immune system as antibody, complement and a limited repertoire of the cellular effectors of the immune system. From the many studies of the interactions between hosts and haematophagous arthropods, it is obvious that the host is exposed normally to a limited repertoire of arthropod-derived antigens. These are usually molecules which the parasite secretes with saliva into the host as it feeds, those which are located in such structures as the mouthparts which make direct contact with the host, or excretory products deposited at the lesion. Reactions against such molecules become manifested in such host reactions as resistance to ectoparasite infection, common in ticks (Trager, 1939), as various allgergenic reactions at the bite lesion (Wikel, 1982), and as measurable, exposure-related antibody titres to salivary gland proteins (Brummer-Korvenkontio *et al.*, 1994). However, it is unlikely that the antibody and other immune components taken up by the feeding parasite would have specificity for the parasite organs such as the midgut, simply because there had been no prior exposure of the host to them. The idea that the midgut of a blood-feeding ectoparasite could be the target of a protective immunological response is relatively recent. This idea has culminated in the marketing of a vaccine against the tick *Boophilus microplus*, thus demonstrating that the idea was not just a piece of interesting academic theory but something that could be turned to practical advantage against

Biology of the Insect Midgut.
Edited by M.J. Lehane and P.F. Billingsley.
Published in 1996 by Chapman & Hall, London.
ISBN 0 412 61670 X.

a range of pests of economic and social importance. Indeed, the breadth of application of this idea is largely unexplored. There is potential to apply the principles that led to the development of the anti-tick vaccine to novel forms of human, animal and plant protection.

The development of such vaccines is based on two simple propositions. First, that the gut and other internal organs are normally 'concealed' from the host, such that natural exposure would not normally lead to the development of immunity directed against these organs. This is the proposition just described. Secondly, that artificial vaccination of the host with appropriate ectoparasite structures would lead not only to significant immunological damage to the ectoparasite subsequently ingesting blood or tissue fluid, but also to interference with the vector capacity of the arthropod. The midgut is probably the best potential source of protective antigens for a range of species as it receives the blood, usually directly, and is thus ideally suited to targeted immunological attack. There is a further possibility, namely that immunological attack on the species being transmitted may occur in the vector midgut (Chapter 16), and the success of these transmission-blocking studies is a clear indication that host immune responses can be effective in the vector midgut. In this chapter we examine data concerning several insect and tick species, and discuss the difficulties of interpreting data obtained with crude midgut extracts. More established examples are reviewed in some detail, focusing on the immunological responses, effector mechanisms and the critical antigens of anti-midgut vaccines.

12.1 THE FEASIBILITY OF ANTI-MIDGUT VACCINES

The usual and most significant assays for the efficacy of anti-arthropod vaccine candidates have been to examine the effects on the arthropod of feeding on blood of an immunized host. This usually means examining such factors as mortality rates, feeding success and effects on fecundity of the females. In addition, for true vectors, feasibility can be measured in terms of a reduced ability to transmit disease organisms (Billingsley, 1994a). In several studies, each or all of these parameters have been monitored to varying degrees, and the 'feasibility' or efficacy described in terms of statistically measurable effects. Such demonstrations of feasibility are usually very difficult to interpret from the data presented. Reproducibility has often been very poor, compounded often by inadequate sample sizes and limited experimental and statistical detail (for review see Jacobs-Lorena and Lemos, 1995). Even within a single experimental system, the innate variability can be frustratingly high (Figure 12.1), making what appear to be trends in mortality and fecundity very difficult to validate.

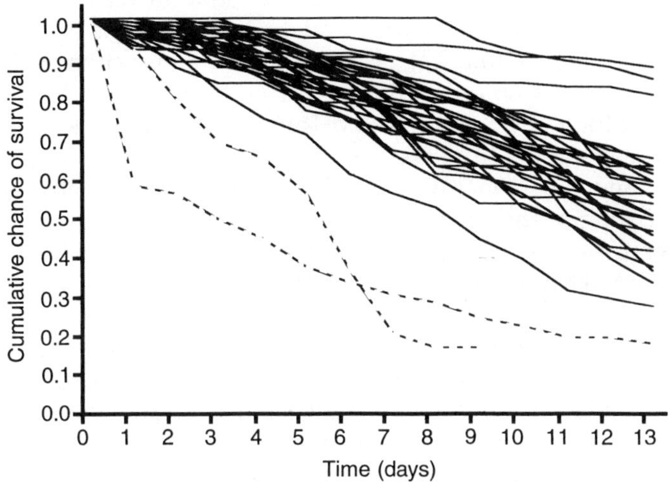

Figure 12.1 Variation in the death rates of *Anopheles stephensi* mosquitoes fed on mice. Each line represents the death rate (calculated as the cumulative chance of survival) of a group of mosquitoes fed on naive mice on days 0, 4 and 8 within a series of antibody feeding experiments (Almeida, 1994). Despite all feeds being made under identical conditions, variation between groups is extremely high and some groups (– – –) can be defined statistically as outliers. (Adapted from Figure 4.1 in Almeida, 1994.)

Among the earliest demonstrations of feasibility were those of Alger and Cabrera (1972), Sutherland and Ewen (1974) and Schlein and Lewis (1976). These workers vaccinated rabbits or guinea pigs with internal organs of, respectively, mosquitoes (*Anopheles stephensi*, *Aedes aegypti* and *Culex tarsalis*) and the stable fly (*Stomoxys calcitrans*), and on subsequent feeding the insects exhibited a variety of pathological symptoms. These observations have been confirmed by other workers and extended to a variety of other species. Those for which gut preparations are effective sources of protective antigens include the stable fly *Stomoxys calcitrans* (Webster *et al.*, 1992), the cat flea *Ctenocephalides felis felis* (Heath *et al.*, 1994), the human body louse (*Pediculus humanus humanus* (Ben-Yakir *et al.*, 1994), and the sheep blowfly *Lucilia cuprina* (East *et al.*, 1993). Similar observations have been made with a number of tick species including *Rhipicephalus appendiculatus* (Essuman *et al.*, 1991), and repeatedly for the tick *Boophilus microplus* (for example, Johnston *et al.*, 1986; Opdebeeck *et al.*, 1988; Wong and Opdebeeck, 1988) (for reviews see Wikel, 1988; Tellam *et al.*, 1992; Opdebeeck, 1994). In most of these cases the effects of host immunity on the feeding insect or tick are similar and include increased mortality, impairment of feeding performance and a reduction in the ability of the engorged adult female to lay eggs. However, demonstration that gut or midgut material is a source of protective antigens is not the same as showing that the midgut is the site of immunological attack, nor is the fact that antigens come from an internal organ sufficient evidence on its own that they are 'concealed' in the sense described above. This problem of antigen specificity will be addressed below.

Few attempts have been made to predict the efficacy of an anti-midgut vaccine required for a beneficial reduction in pest impact. The involvement of biomathematical modellers in predicting efficacy is long overdue. On this larger scale, the feasibility can be examined at the population level, and used to provide potential cost–benefit analyses (Floyd *et al.*, 1995). The feasibility studies must consider the biology of the vector and, in particular, its feeding and reproductive strategies. Thus the anti-tick vaccines may have pronounced effects on tick populations (Floyd *et al.*, 1995), whereas in a very simple model, an anti-mosquito vaccine of low efficacy could theoretically reduce malaria transmission without significantly affecting mosquito populations (Billingsley, 1994b).

12.2 THE MIDGUT AS THE SITE OF IMMUNE ATTACK: DIFFICULTIES IN INTERPRETATION OF THE EVIDENCE

In most of the above studies, midgut material was used to immunize the host directly. This still leaves open the question of whether immunological reactions in the midgut play a part in naturally acquired immunity, that is, whether antigens are truly 'concealed'. There is limited evidence that this can occur. Antisera from rabbits and guinea pigs resistant to the tick *Amblyomma americanum* reacted on Western blots with a range of polypeptides present in midgut extracts (Brown, 1988). As these polypeptides were not present in salivary glands, this suggested that the ticks regurgitated gut material during feeding and this led to the development of an anti-gut immune response. There is also direct evidence to an effect of acquired immunity on tick guts. Walker and Fletcher (1987) described damage to stem cells, moribund nuclei, poorly differentiated cytoplasm, a decrease in the number of digestive and secretory cells, and a lack of gut expansion to accommodate the blood meal in guts of *Rhipicephalis appendiculatus* ticks feeding on resistant rabbits and cattle. Similar pathological changes occurred in the tick salivery glands (Walker and Fletcher, 1990). In both cases, and in a way that is not understood, there is a feedback from the host resistance status to the pathological condition of these two tick organs.

Interpretation of such experiments remains difficult unless it is clear how immunologically unique gut antigens are. Many antigens are shared between a variety of tissues. Mice immunized with various tissue extracts from the mosquito, *Anopheles stephensi*, demonstrate extensive cross-reactivities between sera tested against each extract (Almeida and Billingsley, unpublished observation), even when the sera were raised against specific membrane fractions (Almeida and Billingsley, unpublished observation). Some of the observed cross-reactivity was clearly against conserved epitopes, but the nature of the epitopes is unknown. Moreover, immunological cross-reactivity between a range of proteins within a single tissue has been well documented. One reason for this, glycosylation, will be discussed in more detail below. Occasionally the immunological cross-reactivities are sufficiently unexpected that their significance is uncertain. To give two examples: sera from sheep resistant to the tick *Amblyomma americanum* react with up to 44 antigens, up to half of which are also recognized by pre-infestation sera (Barriga *et al.*, 1991), whereas sera from rabbits infested with the mite *Psoroptes cuniculi* cross-react with antigens from the tick *Dermacentor variabilis* (den Hollander and Allen, 1986). Whether such cross-reactions can lead to significant cross-protection or involve midgut antigens is unknown. However, such observations do show that the reaction of an antibody with insect or tick midgut material is insufficient proof that the midgut is a primary site of immunological attack or that the antibody has been raised by exposure to midgut material. Although further evidence is needed, this may all be regarded as a moot point if immunization with midgut material is effective.

12.3 EXAMPLES OF THE MIDGUT AS AN IMMUNE TARGET

Specific examples that can unequivocally illustrate the role of the midgut as an immune target are actually quite limited. Two for which detailed knowledge is available are the cattle tick, *Boophilus microplus*, and the sheep blowfly, *Lucilia cuprina*. Conversely, there have been numerous attempts to immunize with mosquito midgut material (Jacobs-Lorena and Lemos, 1995), but the state of knowledge concerning antigen identification and the immune response is very poor. These are also good examples because they represent different organizations of the midgut and digestive functions. The tick has a midgut dominated by large digest cells which carry out active endocytosis, most digestion occurring within the cells rather than in the lumen. There is no peritrophic matrix. In *Lucilia cuprina* a well-defined, continuously-produced peritrophic matrix separates the midgut cells from the lumen.

Active proteolytic digestion of the meal of blood and tissue fluid occurs within a defined region of the midgut, and digestion products then pass through the peritrophic matrix into the ectoperitrophic space. Synthesis and secretion of proteases and peritrophic matrix in mosquitoes are initiated by ingestion of the blood meal (Billingsley and Rudin, 1992), and active proteolytic digestion and absorption occur in the same, relatively undifferentiated part of the midgut. Thus, as targets for immunological attack, these organisms represent very different situations. In the tick, one would expect ingested antibody and host cells to remain in the gut in a relatively benign environment, as would be the case for insects ingesting blood into an enzyme-free crop or stomach (for example *Rhodnius prolixus* or *Glossina* spp.), and to have free access to the digest cells of the gut. In the blowfly ingested antibody and cells will reside in a hostile environment as their access to the underlying midgut cells is restricted by the peritrophic matrix. In the mosquito, the antibodies and cells are taken into a seemingly non-hostile environment that changes rapidly as digestion proceeds.

12.3.1 Vaccination against *Boophilus microplus*

The research leading to the development of a commercial vaccine against *Boophilus microplus* has been summarized a number of times (Cobon and Willadsen, 1990; Tellam *et al.*, 1992; Willadsen *et al.*, 1995). The starting point was the demonstration that crude extracts of semi-engorged adult female ticks could be used to vaccinate cattle. Subsequent infestations of the vaccinated cattle were very variable, but effective immunity could be produced. Many of the surviving ticks showed a striking red coloration due to leakage of host erythrocytes through a gut suffering gross immunological damage, including lysis of digest cells and permeabilization of the basal lamina. This was a clear demonstration that such vaccination was able to attack the tick midgut. Observed effects were a reduction in the number of ticks engorging, a reduction in the weight of engorged female ticks, a reduction of their ability to lay eggs, and a slight effect on the viability of the eggs laid. The antigens which were targets for the protective immunological response were subsequently identified by a laborious and time-consuming procedure of protein fractionation, vaccination and parasite challenge. Identification of the first protective antigen took 4 years. This antigen, named Bm86 after the year of its first identification, has subsequently been fully sequenced, expressed as a recombinant protein in both *Escherichia coli* and insect cells using a baculovirus vector, and been taken through extensive efficacy trials under both control and field conditions. The commercial development of this single antigen vaccine proceeded until in 1994 the registered vaccine became commercially available under the name, TickGARD. This was simultaneously both the first vaccine against any ectoparasite and the first recombinant antiparasite vaccine to be marketed.

The Bm86 antigen, though particularly effective on its own, is not the only protective antigen present in ticks. Details of a second protective antigen, Bm91, have just been published (Riding *et al.*, 1994; Jarmey *et al.*, 1995). A third protective antigen of high molecular weight has also been purified using antibody affinity chromatography and has been partially characterized (Lee and Opdebeeck, 1991). Still others have been identified and patent applications filed but details of these characterizations are, as yet, unpublished.

12.3.2 Vaccination against *Lucilia cuprina*

The sheep blowfly, *Lucilia cuprina*, is responsible for the overwhelming majority of primary blowfly strikes in sheep in Australia, the cost of control, losses of production and sheep death exceeding US$150 million per annum. Larvae feeding on sheep vaccinated with crude midgut extract and peritrophic matrix material from larval *Lucilia cuprina* are affected in a number of ways. Growth is retarded during *in vivo* and *in vitro* feeding in which larvae are fed on blood or serum of vaccinated sheep. In extreme cases, the retardation of larval growth is accompanied by an increased larval mortality (East and Eisemann, 1993; East *et al.*, 1993). Using a pragmatic protocol of protein fractionation, vaccination and parasite challenge analogous to that used with the cattle tick, a number of effective antigens have been purified and characterized from the blowfly. These include two proteins which have been isolated from peritrophic matrix material, peritrophin 44 and peritrophin 95 (Chapter 4), as well as several other antigens which have yet to be described in the literature. Reduction in growth is also observed with larvae feeding on serum from sheep previously infested with *Lucilia* (Eisemann *et al.*, 1990) but whether this effect has the same cause as vaccination is not known. It is clear, however, that the growth retardation resulting from vaccination is much stronger than that seen after even extensive natural exposure of sheep to the parasite. It is also clear that the effector mechanism is an immunological reaction against the peritrophic matrix of the larval midgut.

12.3.3 Vaccination against mosquitoes

In contrast to the two previous examples, vaccination against mosquito midguts has been tried repeatedly but the state of knowledge remains very poor. Few attempts have been made to characterize prospective antigens (Ramasamy *et al.*, 1992; Almeida and Billingsley, unpublished).

Rabbit anti-sera raised against the midguts of sugar-fed *Anopheles tessellatus* reduced the fecundity of the homologous mosquitoes (Ramasamy *et al.*, 1992). Anti-midgut antiserum recognized a 47–50 kDa protein specific to the midgut, and some antigens specific to the mosquito species itself. However, there was considerable cross-reactivity against a distantly related mosquito, *Culex quinquefasciatus*, but without any concurrent reduction in fecundity. Thus it is possible that a midgut-specific, effective antigen is present in sugar-fed *Anopheles tessellatus*, but this has not been investigated further.

Antisera raised in mice against crude midgut extracts, midgut microvilli and midgut basolateral plasma membranes of *Anopheles stephensi* and at least six midgut-specific proteins have been observed in Western blots (Almeida and Billingsley, unpublished). It has not been possible to correlate the presence of antibodies against any of these proteins with increased mosquito mortality or decreased fecundity or transmission (Almeida and Billingsley, unpublished observations). Furthermore, gut specific proteins have been identified using monoclonal antibodies (mAb) raised against the midgut, two of which (mAb G8 and mAb 35.2) are effective; both show increased mortality and one a decrease in the prevalence of *Plasmodium berghei* infections after feeding mAb and ookinetes (Billingsley, unpublished; Almeida, 1994). mAb 35.2 binds to *Anopheles stephensi* midgut microvilli and recognizes bands at 87 kDa and 25 kDa, whereas mAb G8 localizes to the apical part of the midgut epithelial cells and recognizes a set of bands of 25–28 kDa. The antigens have not been examined further, but results demonstrate that midgut-specific antibodies may affect mosquito longevity, albeit in a very limited way.

12.4 MECHANISMS OF IMMUNITY: THE EFFECTOR ARM OF THE RESPONSE

The effector components of the host immune system that are active in an insect or tick midgut represent a potentially small proportion of the immunological repertoire, and include antibody, complement and some cells of the immune system. Thus the immunological effector mechanisms are probably restricted compared to the complete repertoire present in the host. The limited available experimental evidence supports this.

Most of the information on immune mechanisms against *Boophilus microplus* and *Lucilia cuprina* has been obtained from *in vitro* feeding systems. Freshly moulted or semi-engorged adult *Boophilus microplus* can be fed on blood, serum or antibody from vacinated cattle. The result of the immunity is increased permeability of the gut and measurable leakage into the haemolymph of erythrocytes or dye coupled to bovine serum albumin. Damage to the gut can be produced by antibody alone, as addition of granulocytes and phagocytic cells to the feeding medium did not increase the observable damage. The role of complement is uncertain. Sera from cattle protected by vaccination with crude, complex antigen mixtures were equally effective whether or not they had been heat inactivated prior to being fed to ticks, suggesting that complement was not critical for damage. With antisera to purified Bm86 antigen, however, the amount of damage is normally reduced by heat inactivation of the serum. As a working hypothesis it seems that damage can be caused by antibody alone but, if the serum is slightly less effective, then complement may enhance the damage. It must be remembered that such *in vitro* feeding systems are only a poor model for what could happen to a tick feeding on a vaccinated host. *In vitro* feeding can only be conveniently carried out on ticks for a small proportion of the life cycle, and the observed effect, namely damage to the tick's gut, is only one measure of vaccine efficacy and the least meaningful. Increased mortality, reduction in engorgement weight and inhibition of egg laying relate directly to the efficacy of vaccination and none of these is measured in the *in vitro* tick feeding system. Therefore, it cannot be concluded from work *in vitro* that antibody alone is important *in vivo* (Kemp *et al.*, 1986, 1989).

From *in vivo* experiments and field trials it appears that the efficacy of vaccination with Bm86 correlates broadly with antibody titre against Bm86 (Willadsen, 1994). The antibody titre is thus the major determinant of the vaccine effect and the influence of other factors appears secondary. Similarly, for *Lucilia cuprina* the growth of larvae feeding *in vitro* is inhibited by antibody alone.

In the studies with mosquitoes, antibody titre correlates poorly with protective effect. Titres have only been measured in animals injected with crude mosquito extracts and against crude extracts, rather than against any specific midgut antigen. It is therefore probable that most of the antibodies raised are irrelevant in terms of effect. However, dose-dependent anti-mosquito effects can be demonstrated with mAbs (Billingsley and Clare, unpublished), showing that cellular and complement immune factors may not be obligatory for immunological damage. Thus antibody-dependent effects seem to be the norm so far for the few species where characterization of anti-midgut immunization has been attempted. The importance of other components of the immune response remains largely unexplored, even though cellular (Sinden and Smalley, 1976) and complement (Grotendorst *et al.*, 1986; Grotendorst and Carter, 1987) factors remain active and effective in the mosquito midgut long enough to influence malaria transmission.

12.5 NATURE OF THE PROTECTIVE ANTIGENS

It appears that the effector arm of the immune response is relatively easy to determine mainly because the immune system is so well characterized. However, determining the nature of the protective antigens is much more challenging. Immunity may be directed against the midgut as a bulk attack against an abundant antigen or as a more specific attack on a limited repertoire of molecules essential to the insect's survival. The evidence suggests that the immune mechanism operating with the current antigens from *Lucilia cuprina* is an example of the first of these possibilities whereas the mechanism for *Boophillus microplus* antigens is an example of the latter.

In the peritrophic matrix of *Lucilia cuprina*, peritrophin-44 and peritrophin-95 appear by immunofluorescence localization to be uniformly distributed. The effects of feeding on antiperitrophin antibody can be examined by including colloidal gold particles in the diet. Larvae feeding on blood from non-vaccinated sheep allow 6 nm colloidal gold particles to pass readily through the peritrophic matrix into the ectoperitrophic space and reach the midgut cells. Once antibody to peritrophin-44 or -95 is included in the meal, a thick, amorphous layer forms on the luminal side of the peritrophic matrix and separates the 6 nm gold particles aggregating in the lumen from the peritrophic matrix. The layer is thicker than expected for an antibody monolayer and the mechanism of its formation is unclear (Willadsen *et al.*, 1993). A tentative model for the mode of action of antibodies against the peritrophic matrix suggests that normal digestion is prevented by formation of a barrier on the peritrophic matrix which is relatively impenetrable to small molecules including digested food and digestive enzymes. This effectively leads to starvation. The peritrophins are glycoproteins which are heavily disulphide bonded using 6-cysteine motifs (Chapter 4). Such disulphide bonding is typical of proteins whose structural stability makes them resistant to proteolytic degradation, a useful attribute for any protein existing in the lumen of an insect midgut. It is important that both proteins are quite abundant in the peritrophic matrix, as the immunity may be mostly a bulk action of antibody coating the peritrophic matrix surface as it forms.

The countervailing idea, that the most efficacious antigens are likely to be low abundance molecules of critical function also has many attractions. Such targets should be more susceptible in that a smaller quantity of ingested, active antibody should be required for effective immune attack. This appears to be the case with the Bm86 antigen. In *Boophilus microplus* indirect immunofluorescence has localized Bm86 to the surface of the midgut digest cells and immunogold labelling has shown that the antigen is predominantly located on the microvilli

(Gough and Kemp, 1993). The low abundance is reflected in the low yields of Bm86 during purification. Nevertheless, location alone is not sufficient for a protective antigen. For example, during the isolation of the Bm86 antigen, a large number of partially purified fractions of membrane glycoproteins were used to vaccinate cattle and were ineffective in producing effective immunity. Antisera from some of these vaccinated animals also reacted strongly in immunofluorescence with the digest cells of the tick midgut. In vaccinations with such crude fractions, as in the mosquitoes (see above) there was little correlation between the efficacy of vaccination and the titre of antibodies reacting with the digest cell surface. Furthermore, in experiments to be described below, antibodies to carbohydrate determinants of Bm86 were not protective. Despite the detailed studies, the function of Bm86 is currently unknown. The structure of the protein has been described (Tellam *et al.*, 1992); it is a typical cell surface glycoprotein with a leader peptide for export, a number of glycosylation sites and a single C-terminal transmembrane segment. This is replaced in the mature protein by a glycosyl phosphatidylinositol membrane anchor. Bm86 also has a series of EGF repeats, a 6-cysteine structural motif that is found in a range of receptors and proteins controlling cellular function.

It is easier to deduce the function of the recently identified Bm91 antigen. A number of peptide sequences obtained from the native protein by gas phase sequencing identified significant sequence similarity between Bm91 and mammalian angiotensin converting enzymes (ACE) (Riding *et al.*, 1994). This has been confirmed by the sequencing of the Bm91 gene (Whitfield *et al.*, unpublished) where the overall sequence identity between Bm91 and the testicular form of human ACE is 41%. The isolated antigen is active in an ACE assay, and its biochemical characteristics are strikingly similar to those of the mammalian enzymes (Jarmey *et al.*, 1995). Bm91 cleaves carboxyterminal dipeptides from a range of peptides, and is inhibited by a number of mammalian ACE inhibitors at concentrations comparable to those for the mammalian enzymes. The natural substrate(s) for Bm91 in the tick is unknown, but the enzyme could control peptide hormone action in the tick, as it does in mammals, or take part in non-specific peptide processing. It therefore represents a good example of an effective antigen with an identifiable and presumably critical biochemical function.

Given the subject of this review, namely immunity in and against midguts, it is unfortunate that the site of effective immunological attack against Bm91 is unclear, so the question of whether Bm91 is a true midgut antigen cannot be answered. Bm91 is present in the midgut (Riding *et al.*, 1994) where its cellular site has yet to be determined, but it is also present in larger amounts and at higher concentrations in the salivary glands. Anti-Bm91 antibody presented in the blood meal can bind in detectable quantities to the protein in the salivary gland. This is not surprising as host immunoglobulin, albeit at low concentrations, will cross into the haemolymph of ticks (Ben-Yakir *et al.*, 1987), the buffalo fly *Haematobia irritans exigua* (Allingham *et al.*, 1992), the tsetse fly *Glossina morsitans* (Nogge and Giannetti, 1980) and mosquitoes (Vaughan *et al.*, 1990). In the latter case, antibodies remain active in the haemolymph for up to 18 h after feeding and can affect malarial infections in both the midgut (Ranawaka *et al.*, 1993) and the salivary glands (Hollingdale and do Rosario, 1987).

The best example of a protective midgut antigen of identifiable biochemical function in fact comes not from an arthropod but from the nematode, *Haemonchus contortus* (Smith *et al.*, 1993). Here the protective activity is against an intestinal microvillar membrane aminopeptidase, termed the H11 antigen (Munn *et al.*, 1993; Newton, 1995). H11 thus shares one similarity with the Bm91 antigen from *Boophilus microplus* in that it is involved in peptide processing. Again, the substrate peptides may be of digestive origin or perhaps have specific pharmacological activities.

There has been speculation that other proteolytic enzymes, particularly the digestive proteases, could be effective protective antigens against insects and ticks, and there is some evidence that protease inhibition is detrimental to blood meal processing. Soya bean trypsin inhibitor, fed with whole blood at a high concentration of 3 mg/ml, inhibited egg production in *Stomoxys calcitrans* by 71% (Spates, 1979). Similar observations have been made on *Lucilia cuprina* (Casu *et al.*, 1994a). Conversely, whereas aminopeptidase activity can be inhibited using a range of inhibitors in the midgut of the mosquito, *Anopheles stephensi*, the blood meal is processed to completion and longevity is unaffected (Docherty and Billingsley, unpublished observations). Such studies may be useful in identifying or discarding targets for immune attack (Billingsley, 1994a). If proven to be suitable targets, proteases would have the advantage of being almost ubiquitous among the insects (Purcell *et al.*, 1992; Chapter 6). Similarly, proteases in hemipterans and ticks would be common although of a different nature.

The feasibility of vaccinating against proteases has also been speculated on (Elvin and Kemp, 1994; Lehane, 1994) but experimental evidence is not encouraging. Sheep antibodies against purified secretory proteases, LCT25a and LCT25b, from the first instar *Lucilia cuprina* have no effect on growing and feeding larvae (Tellam *et al.*, 1994). It might also be argued that cell-associated rather than secretory proteases would be better targets but, despite the success of the H11 vaccine, there are reasons for assuming that such an approach is unlikely to be generally successful (Tellam *et al.*, 1994). It is argued that protease activity may not be inhibited by antibodies; that proteases may be stage specific; that the quantity of active antibody required to inhibit proteases may be difficult to achieve (particularly if proteases are simultaneously degrading the antibody); that there may be feedback control on protease expression such that reducing protease activity may simply induce greater gene or mRNA expression; or that there may be a larger repertoire of proteases than previously thought, inhibition of any one being unlikely to be particularly deleterious. It is interesting that Elvin *et al.* (1994) have used a combination of sequence data and statistical methods to estimate that approximately 200 serine proteinases are present in the buffalo fly *Haematobia irritans*, an unknown number of which will be present in the midgut. Such arguments may seem daunting with regard to raising effective antibodies, but careful studies on digestive biology may reveal more suitable targets among the proteases. In the mosquito midgut there is differential temporal expression of trypsin genes, which operate in a cascade fashion to digest the blood meal (Felix *et al.*, 1991; Müller *et al.*, 1993; Barrilas-Mury *et al.*, 1995). The inhibition with antibody of the first trypsin in the cascade inhibits all subsequent trypsin activity, though mosquitoes are not killed (Crisanti, personal communication).

12.6 THE IMPORTANCE OF GLYCOSYLATION

It is particularly interesting and important to consider the role of glycosylation in the antigenicity and protective immunity of an immunological response directed at the midgut. The well-characterized antigens. Bm86, Bm91, peritrophin-44 and peritrophin-95, are all glycosylated. This is not surprising as glycosylation, like heavy disulphide cross-linking, can protect against proteolytic degradation, and these antigens exist in the hostile environment of the digestive system. It is still an open question, however, what role carbohydrate epitopes play in protective immunity. Currently, there are two lines of evidence that they are important.

A number of lectins are highly toxic to larvae of *Lucilia cuprina* when fed *in vitro*, suggesting that lectin binding to carbohydrate-containing molecules is detrimental to the insect. Unfortunately, lectins react with a variety of different glycoconjugates depending on the sugars they express, so it may be impossible to identify any particular target molecule. Far more convincing are experiments where *Boophilus microplus* midgut membrane proteins and a series of immunoprecipitated proteins were treated with periodate and the protective activity of the antigen mixtures was lost (Lee *et al.*, 1991). Although periodate, like many other chemical reagents, can have some non-specific effects, particularly in attacking residues like methionine and tryptophan, the experiments certainly suggest that destruction of carbohydrate structures can lead to loss of protective antigenicity.

There is also evidence to the contrary. The availability of native and recombinant Bm86 antigens, the latter produced in *E. coli* and SF9 cells using a baculovirus expression system, has allowed the dissection of anti-protein from anti-carbohydrate immune reactions (Willadsen and McKenna, 1991). Antibody to the carbohydrate of Bm86 was cross-reactive with a range of other tick glycoproteins which were not protective. Conversely, antibody to the protein component of Bm86 produced in recombinant *E. coli* was immunologically specific and protective. Thus the protective epitopes of Bm86 reside on the polypeptide and not the carbohydrate. Evidence is still being collected on the other antigens described in this review. Information from other parasites is also worth considering in the context of vaccine development. Carbohydrate epitopes appear to play no role in immunity to *Trichinella spiralis* (Jarvis and Pritchard, 1992), where native and periodate-treated excretory and secretory proteins, in which presumably at least some of the carbohydrate was degraded, were just as effective as the native antigens, suggesting that the protective epitopes were peptide or periodate-sensitive carbohydrate.

The immunological cross-reaction, however, reported for the carbohydrate epitopes of the Bm86 antigen is not unique and may be common. For example, there is considerable cross-reactivity between proteins of *Trypanosoma cruzi* (Xu and Powell, 1991), and periodate treatment again demonstrated that most of the cross-reactive antibodies were directed towards carbohydrate. Glycosylation may also explain much of the observed immunological cross-reactivity in other cases. Numerous glycoproteins can be visualized by lectin blotting of microvillar fractions from the midgut of the mosquito, *Anopheles stephensi*, yet surprisingly the variety of oligosaccharide side chains is limited. *N*-acetyl glucosamine and *N*-acetyl galactosamine predominate as terminal sugars and the variety of linkages of these sugars, as demonstrated by specific glycosidase treatments, is restricted to eight types (Wilkins and Billingsley, 1996) (Figure 12.2). The immunological significance of mosquito midgut glycosylation has yet to be examined in detail, but preliminary experiments have shown that crude midgut antigen is equally protective when injected into BAlb/C mice, which have a full immune response, or into CAB mice which respond poorly to carbohydrate (Billingsley, Rudston and Almeida, unpublished).

The immunological cross-reactivity of carbohydrate epitopes, whether protective or not, is important as it is likely to confuse the interpretation of immunological data. After extensive natural exposure to *Boophilus microplus* some cattle develop quite a good antibody response to the

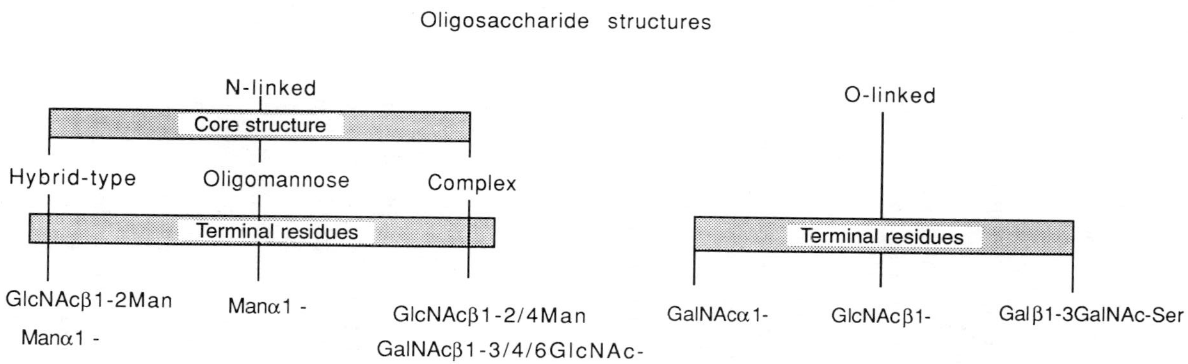

Figure 12.2 Oligosaccharide structures on the midgut microvilli of the mosquito, *Anopheles stephensi*. The variation in oligosaccharide structure is largely restricted to eight definable types, but these are expressed on a larger number of glycoproteins, and could therefore account to a large degree for immunological cross-reactivity and antigen masking. (Summarized from Wilkins and Billingsley, 1996.)

native Bm86 antigen. However, they show no detectable antibody to the polypeptide component of the antigen and the immunological reaction is partially blocked by low-molecular-weight tick carbohydrate (Willadsen and McKenna, 1991). The simple interpretation of a positive Western blot reaction by sera from naturally exposed cattle and against native Bm86 antigen, namely that cattle are naturally exposed to the antigen, is therefore at worst erroneous, at best confusing. It does not indicate exposure to this particular antigen but rather to another carbohydrate structure which is immunologically cross-reactive with Bm86 carbohydrate. As a second example of complexity, Lee *et al.* (1991) have shown a monoclonal antibody which recognizes carbohydrate epitopes and immunoprecipitates at least six *Boophilus microplus* glycoproteins. This glycoprotein mixture was protective after repeated vaccinations.

Currently, the technology is not available for large-scale economic production of specific patterns of glycosylation. Therefore the identification of a specific protective carbohydrate epitope, although of scientific interest, is unlikely to be turned into a practical vaccine, particularly as part of a recombinant protein.

12.7 DIRECTIONS FOR THE FUTURE

Developing midgut antigens for anti-ectoparasite vaccines based on midgut antigens has been hampered by a number of factors. Discouraging results have come from the many attempts to use crude material in feasibility studies and, where results have been successful, identifying the appropriate antigens has been particularly difficult. These factors may be closely linked. Poor results from immunizations with crude homogenate may result from the complex biology of the system. The nature of the peritrophic matrix, the presence and type of proteolytic systems, the longevity of antibody in the midgut, even the size of the blood meal all contribute to variability and emphasize the need for a better understanding of the basic biology of each ectoparasite examined. Furthermore, crude extracts contain a plethora of antigenic molecules, only a small proportion of which will activate the appropriate immune response. Antigenic competition between these molecules in vaccination experiments may result in the protective response being absent or at best variable in its expression. Thus the probability of provoking the desired response using crude antigens may be extremely small. The limitation in identifying and characterizing the correct antigen(s) then producing it in suitable quantity and quality is as much a limitation of time and resources as of technical ability. The successes with the cattle tick, with *Haemonchus contortus*, and with the sheep blowfly have been obtained by tedious application of protein purification methodology to the identification of antigens. Such methods can always be applied to each new pest species, but the disadvantages would remain. We see this as the central problem which must be solved if this novel form of immunological pest control is to progress more rapidly and effectively. What is needed is more ingenious, more efficient and more effective ways of identifying target antigens. To achieve this, two generic approaches can be imagined.

First, the nature and characteristics of some of the small number of known protective antigens have been described above. Bm91 from *Boophilus microplus* is a carboxydipeptidase analogous to angiotensin converting enzyme, H11 from *Haemonchus contortus* is a membrane-bound aminopeptidase, and the peritrophins from *Lucilia cuprina* are peritrophic matrix proteins with particular molecular characteristics (Chapter 4). In looking for protective antigens from other ectoparasites, it would be reasonable to look first at such molecules. Understanding their biochemical nature and biological role offers a potentially more efficient way to purify and test them than the original procedure of fractionation using vaccination trials as an assay system. It is also possible to conduct studies using known specific inhibitors of bioactive molecules to assess the feasibility prior to any immunization (Billingsley, 1994a). The difficulty with this approach is that antibody reactivity may not reflect inhibitor mode of action, but some evidence at least can be gained of the effects of disrupting identified biological processes.

However, for various reasons, such an approach cannot be universally applied. The function of Bm86, for example, is unknown. It is possible with any potential antigen to search for proteins with conserved amino acid sequences prior to testing. The suitability of this method is untried although, with the Bm91 antigen, the occurrence of 42% sequence identify between humans and ticks would strongly suggest that similar enzymes with conserved sequences would be found in other arthropods (Lamango and Isaac, 1994; Cornell *et al.*, 1995). There is already evidence that this is the case (Wijffels *et al.*, 1996). Conservation has also been observed for the first extracellular domain of Na^{+}/K^{+}ATPase (Emery *et al.*, 1995).

On a more speculative level it is possible to develop a list of potential target molecules which, because of their function or biochemical activity or both, might be appropriate targets for immunological attack in the gut of an insect (Billingsley, 1994a; Elvin and Kemp, 1994). The effort required to test such speculations is significant, and there are few historical successes to offer encouragement. The extra complication that there may be numerous variants of a given type of molecular – receptor, serine protease, etc. – means that picking the appropriate molecule and its gene sequence may be difficult.

The importance of glycosylation in midgut function is intuitively clear but its involvement in protective immunity is a matter for debate. It is reasonable to assume that the question will be solved on a case-by-case basis, whereas advances in glycobiological techniques offer promise that progress in this area will be more rapid.

Obviously, the approach described in this chapter can be envisaged with any blood feeder, but it is attractive to speculate that the potential application may have even broader ramifications. It is now possible to make transgenic plants which produce specific antibodies. As a non-chemical method of dealing with major pest species of plants, why should transgenic plants producing an anti-pest antibody not be a feasible approach? Such a method would be competitive with transgenic plants producing insect-specific toxins, but the fact that antibody is intrinsically non-toxic and readily digestible by mammals could have many attractions. Furthermore, the massive strides made in recombinant antibody techniques (Winter *et al.*, 1994) over the last few years mean that screening for effective antibodies and their subsequent use in transgenic technologies may become a matter of routine.

ACKNOWLEDGEMENTS

PFB was supported by the Royal Society during the preparation of this manuscript. The authors thank Dr A.P.G. Almeida for the use of data presented in Figure 12.1, and Mr S. Wilkins for preparing Figure 12.2.

REFERENCES

Alger, N.E. and Cabrera, E.J. (1972) An increase in death rate of *Anopheles stephensi* fed on rabbits immunized with mosquito antigen. *J. Econ. Entomol.*, **65**, 165–8.

Allingham, P.G., Kerlin, R.L., Tellam, R.L. *et al.* (1992) Passage of host immunoglobulin across the mid-gut epithelium into the haemolymph of blood-fed buffalo flies *Haematobia irritans exigua. J. Insect Physiol.*, **38**, 9–17.

Almeida, A.P.G. (1994) Production and activity of antisera and monoclonal antibodies against the malaria vector *Anopheles stephensi*. PhD Thesis, University of London.

Barillas-Mury, C.V., Noriega, F.G. and Wells, M.A. (1995) Early trypsin activity is part of the signal transduction system that activates transcription of the late trypsin gene in the midgut of the mosquito. *Aedes aegypti. Insect Biochem. Mol. Biol.*, **25**, 241–6.

Barriga, O.O., Andujar, F., Sahibi, H. and Andrzejewski, W.J. (1991) Antigens of *Amblyomma americanum* ticks recognized by repeatedly infested sheep. *J. Parasitol.*, **77**, 710–6.

Ben-Yakir, D., Fox, C.J., Homer, J.T. and Barker, R.W. (1987) Quantification of host immunoglobulin in the hemolymph of ticks. *J. Parasitol.*, **73**, 669–71.

Ben-Yakir, D., Mumcuoglu, K.Y., Manor, O. *et al.* (1994) Immunization of rabbits with a midgut extract of the human body louse *Pediculus humanus*

humanus: the effect of induced resistance on the louse population. *Med. Vet. Entomol.*, **8**, 114–18.

Billingsley, P.F. (1994a) Approaches to vector control: new and trusted. 2. Molecular targets in the insect midgut. *Trans. R. Soc. Trop. Med. Hyg.*, **88**, 136–40.

Billingsley, P.F. (1994b) Vector–parasite interactions for vaccine development. *Int. J. Parasitol.*, **24**, 53–8.

Billingsley, P.F. and Rudin, W. (1992) The role of the mosquito peritrophic membrane in blood meal digestion and infectivity of *Plasmodium* species. *J. Parasitol.*, **78**, 430–40.

Brown, S.J. (1988) Evidence for regurgitation by *Amblyomma americanum*. *Vet. Parasitol.*, **28**, 335–42.

Brummer-Korvenkontio, H., Lappalainen, P., Reunala, T. and Palosuo, T. (1994) Detection of mosquito saliva-specific IgE and IgG_4 by immunoblotting. *J. Allergy Clin. Immunol.*, **93**, 551–5.

Casu, R.E., Jarmey, J.M., Elvin, C.M. and Eisemann, C.H. (1994a) Isolation of a trypsin-like serine protease gene family from the sheep blowfly *Lucilia cuprina*. *Insect Mol. Biol.*, **3**, 159–70.

Cobon, G.S. and Willadsen, P. (1990) Vaccines to prevent cattle tick infestations, in *New Generation Vaccines* (eds C. Woodrow and M.M. Levine), Marcel Dekker, New York and Basel, pp. 901–17.

Cornell, M.J., Williams, T.A., Lamango, N.S. *et al.* (1995) Cloning and expression of an evolutionary conserved single-domain angiotensin converting enzyme from *Drosophila melanogaster*. *J. Biol. Chem.*, **27**, 13613–19.

den Hollander, N. and Allen, J.R. (1986) Cross-reactive antigens between a tick, *Dermacentor variabilis* (Acari: Ixodidae), and a mite, *Psoroptes cuniculi* (Acari: Psoroptidae). *J. Med. Entomol.*, **23**, 44–50.

East, I.J. and Eisemann, C.H. (1993) Vaccination against *Lucilia cuprina*: the causative agent of sheep blowfly strike. *Immunol. Cell Biol.*, **71**, 453–62.

East, I.J., Fitzgerald, C.J., Pearson, R.D. *et al.* (1993) *Lucilia cuprina*: inhibition of larval growth induced by immunization of host sheep with extracts of larval peritrophic matrix. *Int. J. Parasitol.*, **23**, 221–9.

Eisemann, C.H., Johnston, L.A.Y., Broadmeadow, M. *et al.* (1990) Acquired resistance of sheep to larvae of *Lucilia cuprina*, assessed *in vivo* and *in vitro*. *Int. J. Parasitol.*, **20**, 299–305.

Elvin, C.M. and Kemp, D.H. (1994) Generic approaches to obtaining efficacious antigens from vector arthropods. *Int. J. Parasitol.*, **24**, 67–79.

Elvin, C.M., Vuocolo, T., Smith, W.J.M. *et al.* (1994) An estimate of the number of serine protease genes expressed in sheep blowfly larvae (*Lucilia cuprina*) *Insect Mol. Biol.*, **3**, 105–15.

Emery, A.M., Ready, P.D., Billingsley, P.F. and Djamgoz, M.B.A. (1995) A single isoform of the Na^+/K^+-ATPase a-subunit in Diptera: evidence from characterization of the first extracellular domain. *Insect Mol. Biol.*, **4**, 179–92.

Essuman, S., Dipeolu, O.O. and Odhiambo, T.R. (1991) Immunization of cattle with a semi-purified fraction of solubilized membrane-bound antigens extracted from the midgut of the tick *Rhipicephalus appendiculatus*. *Exp. Appl. Acarol.*, **13**, 65–73.

Felix, C.R., Betschart, B., Billingsley, P.F. and Freyvogel, T.A. (1991) Post-feeding induction of trypsin in the midgut of *Aedesaegypti* L. (Diptera: Culicidae) is separable in two cellular phases. *Insect Biochem.*, **21**, 197–203.

Floyd, R.B., Sutherst, R.W. and Hungerford, J. (1995) Modelling the field efficacy of a genetically engineered vaccine against the cattle tick, *Boophilus microplus*. *Int. J. Parasitol.*, **25**, 285–91.

Gough, J.M. and Kemp, D.H. (1993) Localization of a low abundance membrane protein (Bm86) on the gut cells of the cattle tick *Boophilus microplus* by immunogold labeling. *J. Parasitol.*, **79**, 900–7.

Grotendorst, C.A. and Carter, R. (1987) Complement effects on the infectivity of *Plasmodium gallinaceum* to *Aedes aegypti* mosquitoes. II. Changes in sensitivity to complement-like factors during zygote development. *J. Parasitol.*, **73**, 980–4.

Grotendorst, C.A., Carter, R., Rosenberg, R. and Koons, L.B. (1986) Complement effects on the infectivity of *Plasmodium gallinaceum* to *Aedes aegypti* mosquitoes. I. Resistance of zygotes to the alternative pathway of complement. *J. Immunol.*, **136**, 4270–74.

Heath, A.W., Arfsten, A., Yamanaka, M. *et al.* (1994) Vaccination against the cat flea *Ctenocephalides felis felis. Parasite Immunol.*, **16**, 187–91.

Hollingdale, M.R. and do Rosario, V.E. (1987) Malaria-enhancing activity in mosquitoes by mammalian host anti-sporozoite antibodies. *Exp. Parasitol.*, **68**, 365–70.

Jacobs-Lorena, M. and Lemos, F.J.A. (1995) Immunological strategies for control of insect disease vectors: a critical assessment. *Parasitol. Today*, **11**, 144–7.

Jarmey, J., Riding, G.A., Pearson, R.D. *et al.* Carboxydipeptidase from *Boophilus microplus*: a novel antigen with similarity to angiotensin converting enzyme. *Insect Biochem. Mol. Biol.*, **25**, 969–74.

Jarvis, L.M. and Pritchard, D.I. (1992) An evaluation of the role of carbohydrate epitopes in immunity to *Trichinella spiralis. Parasite Immunol.*, **14**, 489–501.

Johnston, L.A.Y., Kemp, D.H. and Pearson, R.D. (1986) Immunization of cattle against *Boophilus microplus* using extracts derived from adult female ticks: effects of induced immunity on tick populations. *Int. J. Parasitol.*, **16**, 27–34.

Kemp, D.H., Agbede, R.I.S., Johnston, L.A.Y. and Gough, J.M. (1986) Immunization of cattle against *Boophilus microplus* using extracts derived from adult female ticks: feeding and survival of the parasite on vaccinated cattle. *Int. J. Parasitol.*, **16**, 115–20.

Kemp, D.H., Pearson, R.D., Gough, J.M. and Willadsen, P. (1989) Vaccination against *Boophilus microplus*: localization of antigens on tick gut cells and their interaction with the host immune system. *Exp. Appl. Acarol.*, **7**, 43–58.

Lamango, N.S. and Isaac, R.E. (1994) Identification and properties of a peptidyl dipeptidase in the housefly, *Musca domestica*, that resembles mammalian angiotensin-converting enzyme. *J. Biochem.*, **299**, 651–7.

Lee, R.P., Jackson, L.A. and Opdebeeck, J.P. (1991) Immune responses of cattle to biochemically modified antigens from the midgut of the cattle tick, *Boophilus microplus. Parasite Immunol.*, **13**, 661–72.

Lee, R.P., and Opdebeeck, J.P. (1991) Isolation of protective antigens from the gut of *Boophilus microplus* using monoclonal antibodies. *Immunology*, **72**, 121–6.

Lehane, M.J. (1994) Digestive enzymes, haemolysins and symbionts in the search for vaccines against blood-sucking insects. *Int. J. Parasitol.*, **24**, 27–32.

Munn, A., Smith, S., Graham, *et al.* (1993) The potential value of integral membrane proteins in the vaccination of lambs against *Haemonchus contortus. Int. J. Parasitol.*, **23**, 261–9.

Müller, H.M., Crampton, J.M., Torre, A. *et al.* (1993) Members of a trypsin gene family in *Anopheles gambiae* are induced in the gut by blood meal. *EMBO J.*, **12**, 2891–900.

Newton, S.E. (1995) Progress on vaccination against *Haemonchus contortus. Int. J. Parasitol.*, **25**, 1281–9.

Nogge, G. and Giannetti, M. (1980) Specific antibodies: a potential insecticide. *Science*, **209**, 1028–9.

Opdebeeck, J.P. (1994) Vaccines against blood-sucking arthropods. *Vet. Parasitol.*, **54**, 205–22.

Opdebeeck, J.P., Wong, J.Y.M., Jackson, L.A. and Dobson, C. (1988) Vaccines to protect Hereford cattle against the cattle tick *Boophilus microplus*. *Immunology*, **63**, 363–7.

Purcell, J.P., Greenplate, J.T. and Sammons, R.D. (1992) Examination of midgut luminal proteinase activities in six economically important insects. *Insect Biochem. Mol. Biol.*, **22**, 41–7.

Ramasamy, M.S., Spikrishnaraj, K.J., Wijekoone, S. *et al.* (1992) Host immunity to mosquitoes: effect of anti-mosquito antibodies on *Anopheles tesselatus* and *Culex quinquefasciatus* (Diptera: Culicidae). *J. Med. Entomol.*, **29**, 934–8.

Ranawaka, G., Alejo-Blanco, R. and Sinden, R.E. (1993) The effect of transmission blocking antibody ingested in primary and secondary bloodfeeds, upon the development of *Plasmodium berghei* in the mosquito vector. *Parasitology*, **107**, 225–31.

Riding, G.A., Jarmey, J., McKenna, R.V. *et al.* (1994) A protective 'concealed' antigen from *Boophilus microplus*: purification, localization and possible function. *J. Immunol.*, **153**, 5158–66.

Schlein, Y. and Lewis, C.T. (1976) Lesions in haematophagous flies after feeding on rabbits immunized with fly tissues. *Physiol. Entomol.*, **1**, 55–9.

Sinden, R.E. and Smalley, M.E. (1976) Gametocytes of *Plasmodium falciparum*: phagocytocis by leucocytes *in vivo* and *in vitro*. *Trans. R. Soc. Trop. Med. Hyg.*, **70**, 344–5.

Smith, T.S., Munn, E.A., Graham, M. *et al.* (1993) Purification and evaluation of the integral membrane protein H11 as a protective antigen against *Haemonchus contortus*. *Int. J. Parasitol.*, **23**, 271.

Spates, G.E. (1979) Fecundity of the stable fly: effects of soya bean trypsin inhibitor in phospholipase inhibitor on the fecundity. *Ann. Entomol. Soc. Amer.*, **72**, 845–52.

Sutherland, G.B. and Ewen, A.B. (1974) Fecundity decrease in mosquitoes ingesting blood from specifically sensitized mammals. *J. Insect Physiol.*, **20**, 655–60.

Tellam, R.L., Eisemann, C.H. and Pearson, R.D. (1994) Vaccination of sheep with purified serine proteases from the secretory and excretory material of *Lucilia cuprina* larvae. *Int. J. Parasitol.*, **24**, 757–64.

Tellam, R.L., Smith, D., Kemp, D.H. and Willadsen, P. (1992) *Vaccination Against Ticks, in Animal Parasite Control Utilizing Biotechnology* (ed. W.K. Yong), CRC Press, Boca Raton, pp. 303–31.

Trager, W. (1939) Acquired immunity to ticks. *J. Parasitol.*, **53**, 1106–7.

Vaughan, J.A., Wirtz, R.A., do Rosario, V. E., and Azad, A.F. (1990) Quantitation of antisporozoite immunoglobulins in the haemolymph of *Anopheles stephensi* after bloodfeeding. *Am. J. Trop. Med. Hyg.*, *42*, 10–12.

Walker, A.R. and Fletcher, J.D. (1987) Histology of digestion in nymphs of *Rhipicephalus appendiculatus* fed on rabbits and cattle naive and resistant to the ticks. *Int. J. Parasitol.*, **17**, 1393–411.

Walker, A.R. and Fletcher, J.D. (1990) *Rhipicephalus appendiculatus* feeding on rabbits and cattle: salivary-gland responses to varying host resistance. *Exp. Appl. Acarol.*, **8**, 285–90.

Webster, K.A., Rankin, M., Goddard, N. *et al.* (1992) Immunological and feeding studies on antigens derived from the biting fly, *Stomoxys calcitrans*. *Vet. Parasitol.*, **44**, 143–50.

Wijfells, G., Fitzgerald, C., Gough, J. *et al.* (1996) The cloning and characterization of angiotensin-converting enzyme from the dipteran species *Haematobia irritans exigua* and its expression in the maturing male reproductive system. *Eur. J. Biochem.*, **273**, 414–23.

Wikel, S.K. (1982) Immune responses to arthropods and their products. *Ann. Rev. Entomol.*, **27**, 21–48.

Wikel, S.K. (1988) Immunological control of hematophagous arthropod vectors: utilization of novel antigens. *Vet. Parasitol.*, **29**, 235–64.

Wilkins, S. and Billingsley, P.F. (1996) Oligosaccharides on midgut microvillar glycoproteins of the mosquito, *Anopheles stephensi* Liston, (submitted).

Willadsen, P. (1994) Vaccination against ectoparasites, in *Vaccines in Agriculture* (eds P.R. Wood, P. Willadsen, J.E., Vercoe, R.M. Hoskinson and D. Demeyer) CSIRO, Australia, pp. 171–6.

Willadsen, P., Bird, P., Cobon, G.S. and Hungerford, J. (1995) Commercialisation of a recombinant vaccine against *Boophilus microplus*. *Parasitology*, **110** (Suppl.), S43–50.

Willadsen, P., Eisemann, C.H. and Tellam, R.L. (1993) 'Concealed' antigens: expanding the range of immunological targets. *Parasitol. Today*, **9**, 132–5.

Willadsen, P. and McKenna, R.V. (1991) Vaccination with 'concealed' antigens: myth or reality? *Parasite Immunol.*, **13**, 605–16.

Winter, G., Griffiths, A.D., Hawkins, R.E. and Hoogenboom, H.R. (1994) Making antibodies by phage display technology. *Annu. Rev. Immunol.*, **12**, 433–55.

Wong, J.Y.M. and Opdebeeck, J.P. (1988) Protective efficacy of antigens solubilized from gut membranes of the cattle tick, *Boophilus microplus*. *Immunology*, **66**, 149–55.

Xu, B. and Powell, M.R. (1991) Carbohydrate epitopes are responsible for antibody cross-reactivity in *Trypanosoma cruzi*-infected mice. *J. Parasitol.*, **77**, 808–10.

13

Bacillus thuringiensis endotoxins: action on the insect midgut

P.V. Pietrantonio and S.S. Gill

13.1 INTRODUCTION

Bacillus thuringiensis is a soil-borne Gram-positive bacterium which produces proteinaceous parasporal crystals during sporulation. These crystals are insecticidal primarily to lepidopterans, dipterans or coleopterans, and some crystals have dual insect order selectivity (e.g. lepidopteran/dipteran) (Höfte and Whiteley, 1989). The scope of this manuscript is to summarize our current knowledge on the interactions of these toxins with the insect midgut epithelium, discussing the various methodologies applied to study these interactions and their limitations. A description of the midgut *B. thuringiensis* toxin-binding proteins or putative receptors currently identified is presented. A discussion of toxin interactions with midgut proteins of resistant insects is also included.

13.2 TOXIN CLASSIFICATION AND TOXIN STRUCTURE

Numerous *B. thuringiensis* toxin genes have been cloned and sequenced, and these are classified according to a taxonomic system proposed by Höfte and Whiteley (1989) which takes into account size, spectrum of activity and primary structure of the genes. More recently a new classification based on the primary amino acid sequence has been proposed (Crickmore *et al.*, 1995) and will be published by the time this review is in print. This new classification places the known toxins into 16 different homology groups.

Biology of the Insect Midgut.
Edited by M.J. Lehane and P.F. Billingsley.
Published in 1996 by Chapman & Hall, London.
ISBN 0 412 61670 X.

Two types of proteins are present in the crystals, the Cry and Cyt toxins. The former vary in size from 65 to 138 kDa, whereas the Cyt toxins are about 28 kDa in size. The Cry and Cyt protoxins are dissolved in the insect midgut where they are activated by insect proteases. After proteolytic processing from the N- and C-terminus, a protease-resistant core constitutes the active toxin. Hence the smaller Cry toxins (65 kDa), which are naturally truncated, require less proteolytic processing, whereas the larger Cry toxins require significant proteolytic processing in the insect midgut for activity (Gill *et al.*, 1992; Pietrantonio *et al.*, 1993 for a detailed review).

The crystal structure of one protein, the coleopteran-specific CryIIIA toxin, has been resolved at 2.5Å (Li *et al.*, 1991). Three domains are present in the toxin: domain I is a 7 α-helix bundle, domain II is a Greek key *β* barrel and domain III is a *β* sandwich (Li *et al.*, 1991). Domain I, consisting of the N-terminus until residue 293, has been proposed as a pore-forming domain. After the toxin binds to an insect midgut receptor, it is thought that there is a change in toxin conformation so that the hydrophobic surfaces of the helices would face the exterior of the bundle, and thus the toxin could then penetrate the cell membrane. This domain I contains blocks I and II of the five conserved blocks among *B. thuringiensis* toxins (Höfte and Whiteley, 1989; Hodgman and Ellar, 1990). Domain II of CryIIIA consists of amino acid residues 294–452, and corresponds to the hypervariable region, which is thought to be involved in receptor binding in most toxins. Sequences identified as responsible for selectivity towards lepidopterans and dipterans correspond to specific *β* strands in this domain (Schnepf *et al.*, 1990; Widner and Whiteley, 1990; Li *et al.*, 1991). Domain III comprising the remainder of the C-terminus includes the conserved blocks III, IV and V. A portion of this domain is in contact with domain I, and thus is thought to be important for the stability and integrity of the toxin (Li *et al.*, 1991). In addition it is possible that the C-terminus sequences beyond amino acid residue 450 (domain III) of CryIA(a) may influence selectivity (Masson *et al.*, 1994b). The last *β* strand of the toxic fragment in CryIA(b) (*β*23) is critical for toxin stability and consequently toxicity (Martens *et al.*, 1995). Denaturation of the CryIIIA toxin with guanidine hydrochloride showed that the N-terminus up to residues 279 is the less stable fragment, and that the C-terminus formed by two structural *β* domains, II and III, may constitute a superdomain (Ort *et al.*, 1995).

On the basis of this crystal structure, which is assumed to reflect the structure of other *B. thuringiensis* Cry toxins, numerous studies have focused on the construction of mutant and chimeric toxins. As a result of these studies and others addressing toxin binding to midgut brush border membrane vesicles (BBMV), a model for the mode of action of these toxins has been proposed and it has been summarized in various papers and reviews (Hodgman and Ellar, 1990; Li *et al.*, 1991; Gill *et al.*, 1992; Knowles, 1994; Gazit and Shai, 1995).

13.3 TOXIN PROCESSING

The proteinaceous *B. thuringiensis* crystals require solubilization in the insect midgut. The low or negligible insecticidal activity of some strains can be explained by their low crystal solubility (Pietrantonio and Gill, 1992). Further, most *B. thuringiensis* endotoxins are activated by proteases (trypsin-, chymotrypsin- and thermolysin-like) in the insect midgut. The 130–140 kDa Cry toxins are cleaved at specific sites yielding 60–70 kDa protease-resistant toxin fragments which are derived from the N-terminal half of the protoxin (see Pietrantonio *et al.*, 1993 for a more detailed review). The size of the smallest active fragment, the protease-resistant core, varies with the *B. thuringiensis* protein. In addition the crystals can also undergo proteolytic processing before solubilization, and these products are different from those obtained when solubilization precedes processing (Dai and Gill, 1993).

Differential processing of the same toxin in different insect species can account for the spectrum of activity of some *B. thuringiensis* proteins. One of the best examples, the CryIA(b) protein of *B. thuringiensis aizawai*, is lepidopteran-active when it is activated by lepidopteran proteases or dipteran-active when the lepidopteran-active toxin is further processed with dipteran proteases. This change in specificity is determined by amino acids 524–595 on the C-terminal end of the toxin. Whereas amino acids 524–558 are important for dipteran toxicity, amino acids 558–595 are critical for lepidopteran toxicity (Haider *et al.*, 1989).

Although there is considerable information on the processing of *B. thuringiensis* toxins, there is insufficient information on the enzymes that are involved. Sequencing of activated toxin N-termini suggests that trypsin-, chymotrypsin- or thermolysin-like enzymes are the principal endoproteases involved (Dai and Gill, 1993). For example, trypsin appears to mediate the proteolytic removal of the 28 N-termini amino acid residues of CryIA(b) since a R28D mutation, which prevents trypsin-mediated activation, decreases the insecticidal activity to *M. sexta* (Martens *et al.*, 1995). Minor truncations could be caused by exoproteases such as aminopeptidases.

13.4 MODE OF ACTION

13.4.1 Selective toxin binding

An early observation (Heimpel and Angus, 1959) indicated that the midgut is disrupted during insect intoxication with *B. thuringiensis*.

More recent research demonstrates that the apical brush border of columnar cells is the site of toxin lesion (Percy and Fast, 1983; Singh *et al.*, 1986), and that toxin binds to the microvillar region (Bravo *et al.*, 1992; Ravoahangimalala and Charles, 1995). Consequently, the *B. thuringiensis* toxin receptor is thought to be localized to the microvilli of the columnar cell.

In lepidopterans, the midgut is composed of a unicellular layer of columnar and goblet cells, although small regenerative and triangular endocrine cells are also present. Hofmann *et al.* (1988a) using BBMV, which contain the apical brush border of columnar cells, first reported that a *B. thuringiensis* toxin specifically bound to a site in these vesicles. In larvae of *Pieris brassicae* the binding site was sensitive to proteases and mixed glycosidases, and the toxin binding was directly correlated with toxicity. This information encouraged the performance of binding assays using different toxins and insect species (Table 13.1). Subsequent studies focused on estimating binding parameters such as dissociation constant K_d) and receptor site concentration (R_t), and many reported a correlation between these binding parameters and toxicity (Hofmann *et al.*, 1988b; Van Rie *et al.*, 1989; Lee *et al.*, 1992). The first reports on toxin binding assays with *B. thuringiensis*-resistant strains supported the idea that binding correlates with susceptibility since resistant insects showed loss of toxin binding sites (Ferré *et al.*, 1991) or decreased binding (Van Rie *et al.*, 1990b). A more confusing picture on the relationship between kinetics of binding and toxicity arose with the gypsy moth, *Lymantria dispar*, in which toxin binding did not correlate with toxicity (Wolfersberger, 1990) (Table 13.1). This observation was subsequently confirmed by Liang *et al.* (1995) for CryIA(b) in *L. dispar* which shows lower binding (higher K_d) than expected for its high toxicity. In the potato tubermoth, *Phthorimaea operculella*, CryIA(b) is five-fold more toxic than CryIC, however only small differences in binding kinetics or binding sites for these two toxins were observed (Escriche *et al.*, 1994). In addition, in one strain of *Heliothis virescens* resistant to CryIA(c) toxin, resistance is not correlated with the loss of CryIA(c) binding sites to this toxin (Gould *et al.*, 1992; Lee *et al.*, 1995).

Binding assays with iodinated toxin present a number of intrinsic methodological difficulties and theoretical limitations. First, a pure toxin must be obtained. Except for the HD-73 strain of *B. thuringiensis* subsp. *kurstaki* that only expresses the CryIA(c) protein, other *B. thuringiensis* strains express a number of toxins so that the cloning and expression of single toxins is necessary to perform binding assays. Second, the protoxin must be activated, and iodinated with high specific activity. The iodination process increases toxin hydrophobicity, which in turn increases toxin non-specific binding to the assay tubes (Dai, personal communication), or through hydrophobic interactions with the

Table 13.1 Binding characteristics of Cry toxins to the BBMV from the midgut of susceptible and resistant lepidopterans

	CryIA(a)			*CryIA(b)*			*CryIA(c)*			*CryIC*			
Species	LC_{50}	K_D	R_t	LC_{50}	K_d	R_t	LC_{50}	K_D	R_t	LC_{50}	K_D	R_t	*Reference*
Bombyx mori	28[a]	3.29											Lu *et al.* (1994)
	2[a]	3.5	7.95				421[a]	50.4	2.85				Lee *et al.* (1992)
								33.7[b]	1.95[b]				
	1.9[a]	0.89	7.68	27.8	1.46	5.22							Ihara *et al.* (1993)
Heliothis virescens							2.26[c]	2.09	38.0				Garczynski *et al.* (1991)
							0.09[d]	0.97	16.3				Cowles *et al.* (1995)
Solubilized BBMV								0.74	19.2				
Susceptible				0.21[e]	5.92	0.37	0.11[d]	1.29	0.37				MacIntosh *et al.* (1991)
Resistant to CryIA(b) and Dipel				15[e]	9.48	2.14	1.8[d]	4.93	1.4				
Susceptible				1.12[d]	2.6	3.3	0.1[d]	0.95	3.4				Gould *et al.* (1992)
Resistant to CryIA(c) strain CP73-3				14.5[d]	1.1	1	5[d]	0.76	2.9				
Susceptible strain YDK		0.64	4.9		0.53	14.1		0.21	37.6				Lee *et al.* (1995)
Resistant to CryIA(c) strain YHD2	RR	NB	NB	R	0.61	10.5	R	0.35	30.1				

Table 13.1 *Continued*

Species	*CryIA(a)*			*CryIA(b)*			*CryIA(c)*			*CryIC*			
	LC_{50}	K_D	R_t	LC_{50}	K_d	R_t	LC_{50}	K_D	R_t	LC_{50}	K_D	R_t	*Reference*
Lymantria dispar		0.44	9.26		3.65	3.33		1.25	5.72				Kwak and Dean (1995)
	77.3[f]	0.3[g]	0.24	157[f]	0.19[g]	0.87	187[f]	0.15[g]	0.36				Liang *et al*. (1995)
			0.24[h]			0.46[h]			0.34[h]				
				1.08[c]	19.8	2.7	425[c]	2.03	3.69				Wolfersberger (1990)
Manduca sexta													
				62°	0.61	2.1							Rajamohan *et al*. (1995)
				154°	5.2	3.3							Chen *et al*. (1995)
					3.6[i]								
							4.5[c]	1.58	20.2				Garczynski *et al*. (1991)
								0.78	2.85				Cowles *et al*. (1995)
Purified APN		28.4			42.8			95.3[j]					Masson *et al*. (1995a)
								299.3[j]					
Purified 210 kDa protein				7.5[c]	0.7								Vadlamudi *et al*. (1993)
Cloned 210 kDa protein					1.0								Vadlamudi *et al*. (1995)
Plutella xylostella													
Susceptible				6.7[c]	4.2	1.6				88.9[c]	6.5	10.8	Ferré *et al*. (1991)
Resistant to Dipel				>1350[c]	NB	NB				46.5[c]	7.6	2.9	

Susceptible	16.9[d]	1.95	1.1	6.8[d]	8.8	3.2	Tabashnik *et al.* (1994)
Resistant to Dipel	>1000[d]	NB	NB	16.2[d]	8.5	3.5	
Revertant	51.3[d]	0.79	1.2	8.8[d]	9.4	3.9	
Susceptible	69%[k]	7.1	121				Masson *et al.* (1995b)
Resistant to Dipel	0%[k]	8.93	44.5				

Values for the dissociation constant, K_d, and the receptor site concentration, R_t are in nM and pmol/mg protein, respectively.
RR, no mortality observed at high concentrations but 100 μg/ml caused small reductions in growth; R > 2000-fold resistance CryIAc and > 2000-fold resistance to CryIAb.
[a]ng toxin/larvae.
[b]heterologous competition.
[c]ng/cm^2.
[d]μ/ml of diet.
[e]CryIA(b) toxin expressed in *Pseudomonas fluorescens* mg dried weight/ml.
[f]ID_{50}, ng toxin per os that causes growth inhibition to 50% of the larvae.
[g]k_2 = toxin insertion rate constant.
[h]irreversible B_{max}.
[i]Heterologous competition with Y153D mutant (K_d 5.4, R_t 6.2).
[j]Two binding sites for CryIAc on APN, high affinity site, K_d 95.3, and low affinity site, K_d 299.3.
[k]% mortality at 100 μg/ml diet.

membrane (Hofmann *et al.*, 1988a). Furthermore, the iodination of tyrosine residues may lead to reduced toxicity (Yan and McCarthy, 1991) although the highest specific activity of toxins routinely used in binding assays is far below the theoretical value of 2200 Ci/mmol, if each molecule is radiolabelled once. For these reasons Cummings and Ellar (1994) developed a sandwich radioimmunoassay using anti-CryIA(c) antibodies and iodinated protein G to detect toxin binding to *Manduca sexta* BBMV. However, this approach does not allow measurement of the kinetic parameters of toxin binding. Third, toxin–toxin interactions can result in the formation of *B. thuringiensis* toxin aggregates (Hofmann *et al.*, 1988b; Chow *et al.*, 1989; Masson *et al.*, 1995b). Finally, heterogeneity among various BBMV preparations may also affect the binding data.

With respect to theoretical limitations, the validity of binding assay results is questionable, considering that data analysis is based on the assumption of reversible binding and equilibrium kinetics (Chen *et al.*, 1995; Liang *et al.*, 1995). *B. thuringiensis* toxins are known to interact with receptors in the midgut epithelium; however, following initial binding the toxin appears to bind the membrane irreversibly (Hofmann *et al.*, 1988a; Ihara *et al.*, 1993), and it is this irreversible step that is routinely ignored when binding data are analysed by Scatchard (Hofmann *et al.*, 1988b; Van Rie *et al.*, 1989, 1990b). The analysis of binding data that takes into account the amount of irreversibly bound toxin, which would not be displaced in competition assays either by homologous or heterologous toxin, seems to be more useful to understand changes in toxin binding (Chen *et al.*, 1995; Liang *et al.*, 1995). Measurements of association and dissociation rates of *B. thuringiensis* toxins can also be achieved by using an optical biosensor detection system that exploits surface plasmon resonance (Masson *et al.*, 1994a). The increase in mass due to protein–protein interactions can be measured by changes in the refractive index of incident polarized light when proteins are monitored on a sensor chip (Masson *et al.*, 1994a). Results using this technique will be further discussed.

Although initial binding of the toxin to the receptor seems to be required for toxicity, it is not sufficient. Post-binding events such as irreversible binding or subsequent insertion seem to be more correlated with toxicity than reversible binding (Liang *et al.*, 1995). Data from Wolfersberger (1990) support the idea that toxicity involves a reversible binding step and a second non-reversible binding step. Toxins from two strains, HD-73 (CryIA(c))) and HD1-9 (CryIA(b)), which share the same binding site in the gypsy moth (*L. dispar*), differ in their toxicity by 400-fold with CryIA(b) being the most potent. Although the number of binding sites is similar for both toxins, the less potent toxin binds with higher affinity to the site (Table 13.1). This suggests that the following steps of irreversible binding, insertion and pore formation may be more important than affinity binding in its correlation with toxicity. CryIA(a) is 17-fold more toxic to *Bombyx mori* that CryIA(b) and shows much higher irreversible binding; however, both toxins exhibit similar binding affinity and binding site concentrations (Ihara *et al.*, 1993).

Competition binding assays only provide an idea of the number of sites on BBMV and the binding affinity or selectivity of binding sites to various toxins, and have more often reflected the initial association between toxins and BBMV (Ferré *et al.*, 1991; MacIntosh *et al.*, 1991). However, if performed differently, they could reflect the contribution of the initial binding and irreversible binding (Liang *et al.*, 1995).

The hypervariable region of *B. thuringiensis* toxins, identified as the selectivity region, is accepted to be the main binding domian or receptor recognition domain (Li *et al.*, 1991), at least for the insects investigated to date. This corresponds to the second domain of *B. thuringiensis* toxins (Li *et al.*, 1991). Within domain II, the selectivity region has been identified for CryIA(a) in relationship to *B. mori* (Ge *et al.*, 1989), in CryIIA for mosquito (Schnepf *et al.*, 1990; Widner and Whiteley, 1990), and in CryIA(c) for *Trichoplusia ni* and *Heliothis virescens* (Ge *et al.*, 1991). The role of domain II in insect selectivity has been most recently studied in two toxins, CryIIA and CryIIB. Both toxins are toxic to lepidopteran larvae (*M. sexta* and *L. dispar*), with CryIIB being three times more toxic to *L. dispar*. In contrast to CryIIB, to which it is 87% homologous, CryIIA is also toxic to mosquito larvae (Widner and Whiteley, 1989). In order to analyse the basis for this selectivity further, domain II was divided in three regions of almost equal length, and chimeric constructs between the two toxins were produced (Liang and Dean, 1994). Bioassays with these constructs products confirmed that mosquito selectivity is provided by region 1 in domain II in CryIIA and that lepidopteran selectivity lies within region 2 (Widner and Whiteley, 1989), corresponding to amino acid residues 238–340 in the 633 amino acid sequence (Liang and Dean, 1994).

With respect to the toxin interaction with insect midgut proteins, studies have shown that toxin binding kinetics are affected by specific amino acids in the selectivity domain. In CryIA(a), deletion or alanine substitution of residues 365–371 determine a 1000-fold decrease in toxicity towards *B. mori* (Lu *et al.*, 1994). This loss of toxicity was not related to instability of the mutant toxin but was correlated with reduced affinity (Lu *et al.*, 1994).

Chemical modification of tyrosine and arginine residues in the CryIA(c) toxin has been shown to decrease toxicity to *M. sexta* by decreasing binding to its receptor (Cummings and Ellar, 1994). Similarly chemical modification of tryptophan rendered a toxin-derivative that retains its binding characteristics but is not toxic, indicating that tryptophan is involved in a post-binding event (Cummings and Ellar, 1994).

Binding assays can also be misleading in estimating the number of toxin binding sites. Different numbers of sites for CryIA(c) toxin in *M. sexta* have been reported. Van Rie *et al.* (1989) reported three CryIA(c) binding sites using BBMV binding assays. In contrast Garczynski *et al.* (1991) found that binding of CryIA(c) to *M. sexta* BBMV was compatible with the 'one site' model. Although not strictly comparable, using a different technology (an optical biosensor), the CryIA(c) toxin was shown to bind to two sites within the purified *M. sexta* 120 kDa aminopeptidase receptor (Masson *et al.*, 1995a). The optical biosensor eliminates the need for toxin radiolabelling and allows real time measurements. BBMV proteins, putative purified receptors or toxin can be immobilized on a chip to which a solution of toxin, or BBMV proteins or solubilized receptor is applied, respectively. The interaction of toxin with midgut proteins can be optically measured. In the same study (Masson *et al.*, 1995a) it was shown that binding of CryIA(a) and CryIA(b) toxins to the 120 kDA purified aminopeptidase type N (APN) from *M. sexta* occurs to a single site.

13.4.2 Toxin overlay assays

Toxin overlay assays (ligand blotting) have been used to identify the midgut binding proteins or putative receptors. Toxin overlay assays rely on the blotting of brush border membrane proteins onto nitrocellulose or nylon (PDVF) membranes which are then probed with radiolabelled toxin, or mixtures of cold and labelled toxin in competition or displacement experiments, in order to identify those midgut proteins that specifically bind the toxin. Antibodies against the toxin have also been used in overlay assays instead of an iodinated toxin to detect the toxin-binding protein complexes (Oddou *et al.*, 1991). The information obtained from these assays, however, may or may not be significant with respect of the role of these toxin-binding proteins for toxicity *in vivo*. In overlay assays, for example, epitopes for toxin binding may become available on BBMV proteins, which may not be available *in vivo* due to the denaturing conditions in which most of these experiments are performed. Calculation of binding parameters from these assays does not always help in identifying putative receptors since toxin binding appears to be necessary but not sufficient for toxicity: in overlay assays several *Heliothis zea* BBMV proteins bound CryIA(c) toxin (Garczynski *et al.*, 1991). Furthermore, binding assays with BBMV of this species indicated the presence of a high affinity binding site (K_d 5.66 nM) existing in high concentration (R_t 44.4 pmol/mg), similar in concentration to the sites present in the susceptible insects *H. virescens* and *M. sexta*.

However, CryIA(c) toxin is about 200 times less toxic to *H. zea* than to the other two species (Garczynski *et al.*, 1991). These results indicate that high affinity binding is not always characteristic of a true functional receptor which upon toxin binding would lead to intoxication. Similarly, in the naturally tolerant *Spodoptera frugiperda*, overlay assays identified a 148 kDa CryIA(c) binding protein (Garczynski *et al.*, 1991). BBMV binding assays indicated that the CryIA(c) binding site had high affinity (K_d, 0.68 nM) but was present in low concentration (R_t, 4.5 pmol/mg) (Garczynski *et al.*, 1991). Although a critical concentration of binding sites could be necessary to achieve toxicity, the 5–10-fold lower concentration of binding sites in *S. frugiperda* was not considered sufficient to explain the high tolerance of this species to CryIA(c) (Garczynski *et al.*, 1991), indicating that other factors that follow binding play a role in biological activity.

In the coleopteran *Tenebrio molitor*, a 144 kDa band that specifically binds CryIIIA was identified in toxin overlay assays (Belfiore *et al.*, 1994). In *P. brassicae*, a 125 kDa glycoprotein is a candidate for CryIA(c) toxin receptor. GalNac did not have any effect on toxin-BBMV interactions in a direct binding assay but in toxin overlay assays GalNac blocked binding (Knowles *et al.*, 1991). Toxin overlay assays with the mosquitocidal CryIVD toxin showed binding to a 148 kDa band in BBMV from the mosquito *Anopheles stephensi*, and to a 78 kDa band in BBMV of the crane fly *Tipula oleracea*, both species being susceptible to CryIVD (Feldmann *et al.*, 1995). In these last two studies kinetic parameters were not measured.

CryIC and CryIA(c) are toxic to the noctuid *Spodoptera littoralis* (Van Rie *et al.*, 1990a). CryIA(c) overlay assay of *S. littoralis* BBMV showed strong labelling of a 40 kDa band and faint labelling of a 120 kDa band. CryIC bound to two bands of 40 and 65 kDa, respectively, and the 40 kDa protein is thought to be a strong candidate for the *in vivo* CryIC toxin receptor (Sanchis and Ellar, 1993). The presence of these two CryIC binding proteins was in accord with a previous report by Van Rie *et al.* (1990a) on the existence of two different binding sites for this toxin (Sanchis and Ellar, 1993).

13.4.3 Receptor cloning and characterization

Toxin overlay assays have, however, helped in the identification of three putative receptors for *B. thuringiensis* toxins that have been recently cloned. Vadlamudi *et al.* (1993) first identified a 210 kDa protein from *M. sexta* as the specific receptor of CryIA(b) toxin, which bound the toxin with high affinity (K_d 0.71 nM). Martínez-Ramírez *et al.* (1994) also reported binding of CryIA(a), CryIA(b) and CryIA(c) to a 210 kDa in toxin-overlay assays of *M. sexta* BBMV. Two bands of 120 kDa and

145 kDa that also bound CryIA(b) were thought to be degradation products of the 210 kDa binding protein (Martínez-Ramírez *et al.*, 1994). Subsequent cloning revealed that this 210 kDa receptor is a novel cadherin-like glycosylated protein (Vadlamudi *et al.*, 1995). The predicted molecular mass deduced from the cloned cDNA is 172 kDa and the difference in molecular mass is due to glycosylation, as proven by treatment of the native protein with N-glycanase F (Vadlamudi *et al.*, 1995). Iodinated CryIA(b) toxin labelled a 210 kDa band in toxin overlay assays of membranes prepared from COS-7 cells that had been transfected with the cDNA clone, indicating specific recognition of the expressed product (Vadlamudi *et al.*, 1995).

Some of the proteins in the cadherins superfamily are involved in calcium-dependent cell adhesion. This CryIA(b) receptor possesses 16 putative glycosylation sites. The deduced amino acid sequence shows a putative peptide signal of 20 residues and one transmembrane region of 22 residues starting at residue 1406 in the 1528 amino acid sequence. There are 11 repeats in the putative extracellular region and the C-terminal cytoplasmic domain (110 residues) is smaller than the corresponding region in vertebrate cadherins (160 residues). This difference from the vertebrate cadherins may account for its unique function (Vadlamudi *et al.*, 1995). It was postulated by Vadlamudi *et al.* (1995) that the receptor could act in a similar fashion to the human intestinal peptide transporter protein, another member of the cadherin superfamily, which transports antibiotic peptides inside the cell. The 210 kDa protein is thought to be the receptor that contains a common binding site for the three CryIA toxins (a, b and c) (Van Rie *et al.*, 1989; Vadlamudi *et al.*, 1993, 1995; Martínez-Ramírez *et al.*, 1994). A 210 kDa protein binds CryIA(a) and CryIA(b) but not CryIA(c) in the gypsy moth (Valaitis *et al.*, 1995).

The *M. sexta* CryIA(c) toxin receptor, a 120 kDa GalNac-bearing glycoprotein, was also identified by these methods (Knowles *et al.*, 1991; Garczynski *et al.*, 1991; Sangadala *et al.*, 1994). Purification of this 120 kDa glycoconjugate by protoxin affinity chromatography and anion exchange chromatography followed by N-terminal (Table 13.2) and internal amino acid sequencing revealed that the putative receptor is aminopeptidase N (APN) (Knight *et al.*, 1994). In an independent study with *M. sexta* BBMV, using CryIA(c) overlay assays, monoclonal antibody affinity purification and N-terminal sequencing, the 120 kDa protein was also identified as APN (Sangadala *et al.*, 1994). DNA probes to this *M. sexta* receptor were developed using polymerase chain reaction primers corresponding to internal amino acid sequences obtained by tryptic digestion (Knight *et al.*, 1994, 1995). These probes were then used to screen a *M. sexta* midgut cDNA library. The predicted molecular mass of the receptor deduced from the partial cDNA clone is

Table 13.2 N-terminal sequences of *H. virescens* toxin-binding proteins and its comparison to other known sequences

Source	*Amino acid sequence*
H. virescens, 120 kDa protein	N V V A S P Y R L P T
H. virescens, 170 kDa protein	D P A Y R L P T L
M. sexta, 120 kDa protein[a]	D P S Y R L P T T
L. dispar, 120 kDa protein[b]	N S E L Y R L P T T
Human aminopeptidase N[c]	S K A W N R Y R L P N T
Consensus	Y R L P T

Sequences from [a]Knight *et al.* (1994); [b]Valaitis *et al.* (1995); [c]Olsen *et al.* (1988).

105 kDa, in contrast to the 120 kDa mass that is apparent in SDS-polyacrylamide gel electrophoresis. Most of this difference is due to glycosylation. Aminopeptidase activity can be removed by treating BBMV with phosphatidylinositol-specific phospholipase C (Knight *et al.*, 1995). The putative receptor, an ectoenzyme, is linked to the cell membrane by a glycosyl-phosphatidylinositol (GPI) anchor (Garczynski and Adang, 1995).

The APN from the gypsy moth was purified and shown to specifically bind CryIA(c) toxin. The purified enzyme with an apparent molecular weight of 100 kDa constitutes 4% of the total BBMV proteins, and the native molecule may be a dimer as indicated by gel-filtration experiments (Valaitis *et al.*, 1995). N-terminal sequencing of this putative receptor (Table 13.2) showed that it lacks the 70 residues present in the N-terminus of mammalian enzymes, suggesting that a GPI anchor linkage may be present in *L. dispar* APN as is the case in *M. sexta* APN (Valaitis *et al.*, 1995).

In *H. virescens*, CryIA(c) toxin binding has been reported to BBMV proteins of various sizes: of 68 and 50 kDa (Knowles *et al.*, 1991); of 140 and 120 kDa (Oddou *et al.*, 1991); of 155, 120, 103, 90, 81 and 63 kDa (Garczynski *et al.*, 1991); and 170, 140, 120, 90, 75, 60 and 50 kDa (Cowles *et al.*, 1995). The N-terminal sequences of the 120- and 170-kDa toxin-binding protein from this later study showed that they both share the sequence YRLPT corresponding to residues 7–11 in the 120 kDa protein and residues 4–8 in the 170 kDa protein (Table 13.2).

The nature of the 170 kDa protein is unknown (Cowles *et al.*, 1995). In *H. virescens*, Oddou *et al.* (1991) reported binding of CryIA(a) and CryIA(b) to a 170 kDa protein band which appears to be a single protein. Cross-linking experiments with the same toxin identified a complex of 230 kDa indicating that the toxin binds to the 170 kDa protein. Using a glycan detection kit a 170 kDa glycoprotein band was identified; however, it was not clear if the 170 kDa glycoprotein observed was the same protein that bound the toxin. On treatment of BBMV with N-glycanase and subsequent toxin overlay assay, two bands of 170 and 120 kDa bound the toxin. The authors speculated that the 120 kDa protein is the result of incomplete deglycosylation of the 170 kDa protein (Oddou *et al.*, 1991). None of the glycosidases tested inhibited the binding of the CryIA(b) toxin to the 170 kDa protein. It was concluded that CryIA(a) and CryIA(b) bind the same 170 kDa protein, and based on the binding data of van Rie *et al.* (1989), possibly at different sites (Oddou *et al.*, 1991). After protease V8 treatment of BBMV, the CryIA(b) toxin binding was to a 50–55 kDa band, demonstrating that only one fragment of the 170 kDa protein is sufficient for toxin binding (Oddou *et al.*, 1991).

Isolation of the cDNA encoding the 120 kDa protein from *H. virescens* showed the protein to be an aminopeptidase-N like protein (Gill *et al.*, 1995). The precursor protein of 1009 amino acids has a calculated molecular weight of 113 kDa and a pI of 5.29. The N-terminal sequence of the 120 kDa toxin binding protein in the midgut corresponds to amino acids 53–63 of the cDNA deduced amino acid sequence. This indicates that in the mature toxin binding protein 52 amino acid residues containing a hydrophobic sequence are cleaved from the N-terminus.

In the cDNA deduced amino acid sequence the presence of a long hydrophobic C-terminus preceded by hydrophilic residues suggests that the 120 kDa protein in *H. virescens* is linked by a GPI anchor through its C-terminus. A putative amino acid sequence signalling the addition of GPI anchors, DSA, is observed at residues 987–989. Proteolytic cleavage at N-(between Ser52 and Asn53) and C-termini (between Asp987 and Ser988) would result in a 105 kDa protein containing 935 amino acid residues. The difference in size between this protein and the mature 120 kDa toxin-binding protein in *H. virescens* midgut is probably due to glycosylation. Two potential N-glycosylation sites are observed at amino acid residues 581 and 906.

The *H. virescens* 120 kDa protein has 42% and 62% identity and similarity, respectively, to the *M. sexta* protein. Similar levels of identity, although slightly lower, are observed with other eukaryote aminopeptidases N such as those identified from human and *Caenorhabditis elegans* (Figure 13.1). The conserved amino acid motif, HEXXH, indicative of all zinc-dependent metalloproteases, is found in both *H. virescens* and *M. sexta* CryIA(c) toxin binding APNs.

Masson *et al.* (1995a), using an optical biosensor, demonstrated the binding of CryIA(a) and CryIA(b) toxins to the 120 kDa purified aminopeptidase type N (APN) from *M. sexta* to a single site. These

Figure 13.1 Comparison of *Heliothis virescens* and *Manduca sexta* CryIA(c) toxin-binding proteins (Knight *et al.*, 1995) with human (Olsen *et al.*, 1988) and *C. elegans* (Wilson *et al.*, 1994) aminopeptidases N. The *M. sexta* sequence used was as in Knight *et al.* (1995).

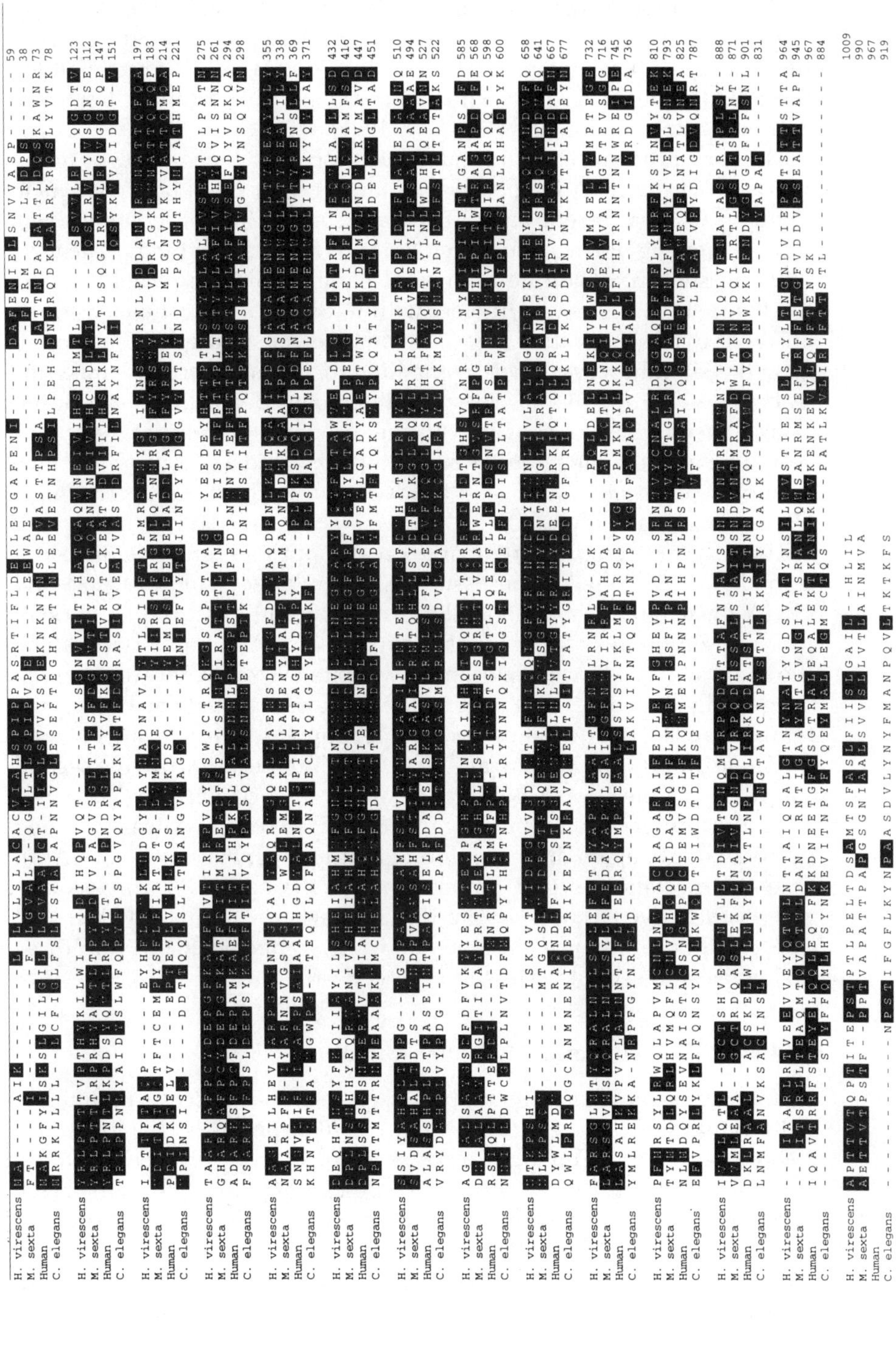

authors also found that binding of CryIA(a) and CryIA(b) to APN was not affected by the presence of sugars. Similarly, sugars did not affect the binding of these toxins to *H. virescens* 170 kDa BBMV-protein and to the 120 kDa BBMV-protein which appears after N-glycanase treatment and binds both toxins (Oddou *et al.*, 1991).

The use of the term 'binding site' with respect to *B. thuringiensis* toxins needs some clarification in order to interpret early binding assay data in the light of more recent data. The term receptor in this discussion will refer to a single molecule to which *B. thuringiensis* toxin binds. The term 'binding site' may indicate a physical place within the receptor where the toxin is more likely to interact with it. Sites can be competed for by different toxins in binding assays (toxins are able to displace each other from the site); however, these competing toxins may not share the same 'subsite' which for this discussion will be considered the smallest and exact physical place of contact with the receptor. For practical reasons it would be useful to imagine the site as a pocket and a subsite as a branch of sugars or certain amino acid residues within the pocket, very much like an enzyme catalytic site.

The CryIA(c) binds to two sites on the purified aminopeptidase N from *M. sexta*. One of these sites is also shared with CryIA(a) and CryIA(b), whereas the second site is not shared (Masson *et al.*, 1995a). By competition analysis it was shown that CryIA(a) and CryI(b) share the same site. N-acetylgalactosamine inhibited 90% of CryIA(c) binding, suggesting that other sugars or other factors are responsible for the remnant 10% CryIA(c) binding. CryIA(c) bound to one site in APN with the same affinity as CryIA(a) and CryIA(b), but the affinity to the second site was 3–4 times lower and due to a faster dissociation rate (Masson *et al.*, 1995a). Interestingly, binding of CryIA(c) to both sites was inhibited by sugars but sugars did not inhibit binding of CryIA(a) and CryIA(b). One interpretation is that the site that is shared by the three toxins possesses a glycosylated subsite that interacts only with CryIA(c), but due to steric hindrance the overall site is competed for by the three toxins, as suggested by Masson *et al.* (1995a).

13.4.4 Toxin aggregation and pore formation

Numerous studies demonstrated that trypsin activated Cry toxins interact with planar lipid bilayers (Slatin *et al.*, 1990), liposomes (English *et al.*, 1991) or reconstituted midgut proteins (English *et al.*, 1991; Sangadala *et al.*, 1994) to form cation-selective channels. The toxin forms specific pores in BBMV which seem to be non-selective with respect to anions, cations or small molecules (Hendrickx *et al.*, 1990; Carroll and Ellar, 1993). The N-terminus of the CryIA(c) toxin, lacking the receptor binding domain, has an ability to form pores in lipid bilayers (Walters *et al.*, 1993).

After receptor binding the toxin inserts into the plasma membrane creating a pore of about 1–2 nm in diameter, as estimated by osmoprotection experiments (Knowles and Ellar, 1987). Hodgman and Ellar (1990) predicted, by computational analysis, the presence of putative amphipathic helices in the N-terminus of *B. thuringiensis* endotoxins. These helices had the potential to form pores of 0.6–6.5 nm in diameter, depending on the set of helices chosen for these predictions. Pores formed by 5–10 molecules would be consistent with the pore size estimated in osmoprotection studies (Hodgman and Ellar, 1990). Resolution of the CryIIIA toxin corroborated the existence of a potential pore-forming region, domain I. Six amphipathic α helices surround an hydrophobic helix, α5, which is the most conserved in sequence among *B. thuringiensis* endotoxins, and mutations in this region affected toxicity in lepidopterans (Li *et al.*, 1991; Wu and Aronson, 1992). After toxin-binding helices α6 and α7 are thought to flatten on the membrane surface, perpendicularly to the hairpin putatively formed by α4 and α5, following an 'umbrella model' (Li *et al.*, 1991; Knowles, 1994; Gazit and Shai, 1995).

To investigate the role of helix α5, a synthetic peptide corresponding to this helix of the CryIIIA coleopteran-specific toxin sequence (residues 193–215) was synthesized, and its ability to form channels in lipid membranes and activity *in vitro* was investigated (Gazit *et al.*, 1994). Labelled peptides bound to membranes of the lepidopteran *S. littoralis* where they were protected from proteolysis indicating that they had penetrated the membrane. The peptide was toxic to lepidopteran Sf-9 cells and formed ion channels in planar lipid membranes. These results suggested that helix α5 may be one of the transmembrane helices that forms the pore (Gazit *et al.*, 1994). Studies with this same peptide, and another peptide corresponding to α helix 7, showed that the α helix 5 peptide binds to phospholipid vesicles in a fast and co-operative manner and that it self-associates, possibly in a parallel organization, within the lipid membrane (Gazit and Shai, 1995). Although the α helix 5 peptide self-associates and the α helix 7 peptide does not, the α helix 5 peptide can co-assemble with α helix 7 peptide but not with unrelated helical peptides such as cecropin B (Gazit and Shai, 1995). These results were interpreted as supporting the umbrella model (Li *et al.*, 1991; Knowles, 1994) in which α helix 5 penetrates the membrane, aggregates with other helices 5 from other monomers to form a pore, whereas alpha helix 7 is the 'sensor' of the membrane surface and binds to phospholipids in the membrane in a random fashion (Gazit and Shai, 1995). The proposed model in which the role of helix 7 is to 'sense' the membrane is supported by the fact that helix 7 is located between the N-terminal pore-forming domain I and the specificity of binding domain (Gazit and Shai, 1995).

B. thuringiensis subsp. *israelensis* CytA toxin is believed to act by binding to phospholipids, followed by toxin aggregation and insertion forming a pore in the cell membrane leading to osmotic lysis (Chow *et al.*, 1989; Maddrell *et al.*, 1989). The pore size has been estimated at 0.6–1 nm in radius (Drobniewski and Ellar, 1988) and is also thought to be cation selective (Knowles *et al.*, 1989). Experiments with synthetic helices corresponding to amino acid residues 55–71 (helix 1) and 110–131 (helix 2) showed that both helices bind strongly to phospholipid small unilamellar vesicles, with helix 2 more active than helix 1. The evidence supports the possibility that these two hydrophobic helices may play a role in toxin aggregation (Gazit and Shai, 1993). However, all these studies have been performed with non-target cell types. In all probability the CytA toxin may show selective binding in target cells through specific receptors.

13.5 RESISTANCE

B. thuringiensis is the most widely used biopesticide and the appearance of resistance and cross-resistance in some insects is of great concern because it jeopardizes its long-term use (for review see Tabashnik, 1994). In 1985 resistance to *B. thuringiensis* toxins was first reported in a population of the Indian meal moth, *Plodia interpunctella*, which had been subjected to treatments with the commercial formulation 'Dipel' (CryIA(a), CryIA(b), CryIA(c), CryIIA and CryIIB protoxins) (McGaughey, 1985). Resistance developed rapidly, about 30-fold resistance was achieved in two generations of laboratory selection, and reached a plateau at 100-fold in 15 generations (McGaughey, 1985). It was later reported that in the resistant strain UE-343R selected with Dipel, resistance to CryIA(b) was due to a 50-fold decrease in binding affinity while the number of receptors remained unchanged (Van Rie *et al.*, 1990b). Resistance was inherited as a recessive trait (McGaughey, 1985). Interestingly, the resistant strain became more susceptible to another toxin, CryIC, showing increase in the number of high affinity binding sites for CryIC, and simultaneously losing the population of low affinity binding sites which are presumably shared by both toxins in the susceptible strain (Van Rie *et al.*, 1990b). In this study the response to CryIA(a) and CryIA(c) also present in Dipel was not investigated (Van Rie *et al.*, 1990b). There were no differences in proteases or proteolytic activities in midguts from both resistant and susceptible strains, or in the manner in which both strains activated *B. thuringiensis* protoxins, supporting the idea that resistance was not due to differential toxin processing (Johnson *et al.*, 1990). However, in a different strain of *P. interpunctella* selected for resistance to *B. t. entomocidus* (HD 198) the activation of CryIA(c) protoxin by midgut proteases was slower than that produced by a susceptible strain or a strain resistant to *B.*

thuringiensis kurstaki HD 1 (Dipel) (Oppert *et al.*, 1994). The expected activated toxin products of 61–63 kDa were only observed after a long period of incubation with the midgut extracts. It was speculated that the reduction in velocity of activation may help reduce toxicity in the resistant strain (Oppert *et al.*, 1994). It was also shown that both resistant strains selected with Dipel and *B. t. entomocidus* (HD 198), respectively, were most resistant to CryIA(c), followed by resistance to CryIA(b). The *B. t. entomocidus* resistant strain also exhibited five-fold less trypsin-like activity and altered proteolytic activation of CryIA(b) and CryIC (Oppert *et al.*, 1994).

In a strain of the mosquito *Culex quinquefasciatus* that exhibited 170-fold resistance to the CryIVD toxin, the proteolytic processing of this endotoxin appears to be slower than in the susceptible strain (Dai and Gill, 1993). Resistance to *B. t. israelensis* whole crystals has not been achieved in the field (Goldman *et al.*, 1986) and only in very low levels (3–16-fold) in the laboratory (Georghiou, 1990). Resistance to CryIVD can be achieved readily in *C. quinquefasciatus* (Dai, 1992). It is commonly accepted that selection of dipterans with whole crystals of *B. thuringiensis* subsp. *israelensis* is slower because it is most likely targeting multiple loci.

The diamondback moth, *Plutella xylostella*, is the first species known to develop resistance to *B. thuringiensis* in open field conditions (Tabashnik *et al.*, 1990). In the resistant diamondback moth, initial studies on CryIA(b) binding revealed that resistant insects had a decreased affinity and number of receptors to CryIA(b), but had increased susceptibility to CryIC (Ferré *et al.*, 1991) (Table 13.1). Similar results were obtained with insects resistant to CryIA(c) in which binding parameters could not be estimated due to reduced toxin binding (Tabashnik *et al.*, 1994) (Table 13.1). The binding kinetics of CryIA(c) in a resistant *P. xylostella* population was calculated using surface plasmon resonance measurements. Binding kinetics of the CryIA(c) resistant population did not differ from those obtained with susceptible insects (Masson *et al.*, 1995b). This was a surprising result considering that the resistant insects were derived from the same population used in previous binding studies (Tabashnik *et al.*, 1994). The number of receptors, however, was reduced three-fold in BBMV from the resistant insects and this decrease was not considered sufficient to account for the level of resistance to CryIA(c), indicating that binding is not sufficient for toxicity (Masson *et al.*, 1995b).

A CryIA(c) selected *H. virescens* strain that exhibited 2000-fold resistance to this toxin also shows similar levels of cross resistance to CryIA(a) and CryIA(b) toxins (Lee *et al.*, 1995) (Table 13.1). The receptor that putatively binds the three toxins in the susceptible strain is altered in the resistant strain, as indicated by the total lack of binding of CryIA(a). The most interesting result, however, is that the resistant strain did not show differences in reversible or irreversible binding with respect to CryIA(c) (Lee *et al.*, 1995). This result may indicate that irreversible binding precedes toxin insertion, which is a distinct process. This differs from the prior assumption that irreversible binding reflected toxin insertion (Hofmann *et al.*, 1988a; Ihara *et al.*, 1993; Liang *et al.*, 1995). Other possibilities may include the improper insertion of the toxin into the membrane, or that toxin insertion *per se* may not be the critical determinant of toxicity, and changes in the interacting receptor(s) are responsible for the lack of toxicity (target site insensitivity). Electrophysiological studies with resistant insects should be conducted to elucidate these aspects.

13.6 PROPOSED MODEL OF *BACILLUS THURINGIENSIS* TOXIN ACTION

The generally accepted model for Cry toxin action is that following toxin binding to a cell membrane receptor, the toxin undergoes a conformational change and inserts into the membrane. Oligomerization of the toxin follows, and this oligomer then forms a pore that leads to osmotic cell lysis (Hodgman and Ellar, 1990; Li *et al.*, 1991; Gill *et al.*, 1992; Knowles and Dow, 1993; Knowles, 1994; Gazit and Shai, 1995). However, *B. thuringiensis* toxin-mediated cell lysis attributed solely to the receptor binding followed by formation of a toxin pore is an oversimplification of the interaction of these toxins with the cell membrane. A number of reports in the literature suggest *B. thuringiensis* toxins can have a variety of effects, including internalization of insect midgut gap junctions (Lane *et al.*, 1989), disruption of midgut cell adhesion (Heimpel and Angus, 1959; Ebersold *et al.*, 1977), inhibition of Na^+/K^+-ATPase (English and Cantley, 1986; Himeno, 1987) and K^+/H^+ exchanger function (English and Cantley, 1985), inhibition of BBMV phosphatase activity (English and Readdy, 1989), alteration of channel function (Schwartz *et al.*, 1991), differential inhibition of amino acid uptake (Reuveni and Dunn, 1991), and activation of adenylate cyclase (Knowles and Farndale, 1988). It is easy to diminish the significance of these observations and consider them as non-specific or secondary effects of intoxication. However, these observations allow us to dissect the intoxication process and enable us to understand that the toxin can affect other molecules, regardless of the pore-forming ability of domian I.

Based on the evidence that is currently available, we propose that the *B. thuringiensis* toxin action is a multistage process.

1. *Receptor binding.* It is evident that toxin action in the insect midgut requires the presence of an extracellular receptor to which the toxin

binds. The identification of two proteins, the 210 kDa cadherin-like CryIA(b) and the 120 kDa aminopeptidase-N CryIA(c) toxin-binding proteins (Sangadala *et al.*, 1994; Gill *et al.*, 1995; Knight *et al.*, 1995; Masson *et al.*, 1995a; Vadlamudi *et al.*, 1995) supports a potential role of these proteins as toxin receptors. Specific amino acid residues primarily in domain II, but also in domains I and III, may be the receptor recognition epitopes. The affinity of a toxin binding protein may, however, not be indicative of the relative importance of that protein as a receptor for toxin intoxication.

2. *Irreversible binding and toxin insertion.* Specific binding involves two steps, one that is reversible and the other irreversible (Liang *et al.*, 1995). Although irreversible binding better correlates with toxicity than reversible binding, it is premature at this stage to consider irreversible binding as a synonym of toxin insertion which leads to toxicity. Irreversible binding can result from electrostatic and hydrophobic interactions, and/or toxin insertion that does not lead to subsequent intoxication. The positive correlation between irreversible binding and toxicity could also indicate that a tighter toxin association with the receptor or binding proteins may facilitate toxin aggregation through toxin–toxin interactions, and favour subsequent insertion.
3. *Toxin oligomerization/aggregation.* Upon receptor binding, a large conformational change is induced in the toxin. Potentially, toxin proteolysis facilitates this change in conformation. It appears that helix 7 'senses' the membrane and facilitates the insertion of $\alpha4$ and $\alpha5$ into the membrane. This insertion mechanism is the 'umbrella' model of Li *et al.* (1991). As other toxin molecules insert into the membrane, they aggregate but the mechanism of aggregation is unknown. Oligomerization probably occurs in the membrane. The toxin domains involved in oligomerization have not been identified.
4. *Function of domain III.* The role of domain III was unknown when the CryIIIA crystal structure was reported, and was thought to be involved in toxin stability and toxin resistance to proteases (Li *et al.*, 1991). More recently it has been shown that a conserved block 4 in domain III, which contains an arginine stretch, is important in toxin-induced channel activity, and it may be the 'voltage sensor' for the toxin-induced channel (Chen *et al.*, 1993; D. Dean, personal communication). Comparison of the CryIIIA toxin (Li *et al.*, 1991) structure with that of diphtheria toxin (Choe *et al.*, 1992) and *Pseudomonas* exotoxin A (Allured *et al.*, 1986) shows that all three toxins have a three domain motif: all have pore-forming domains, receptor-binding domains and a third domain. The third domain in diphtheria toxin and exotoxin A plays a functionally important role, which is ADP ribosylation (Allured *et al.*, 1986; Choe *et al.*, 1992). Consequently, it is possible that domain III in *B. thuringiensis* toxins could functionally

be the most significant domain in the *in vivo* intoxication by Cry toxins. Hence it is essential that additional effort be made to determine the role of this domain in the overall activity of Cry toxins.

5. *Pore formation.* A functional pore is formed by the insertion and aggregation of toxin molecules, specificaly by toxin domains I and III. The entrance to the pore probably contains a charged residue because a Lys165Thr mutant of CryIVD causes a significant change in pore characteristics (Gill *et al.*, unpublished).
6. *Interaction with intramembrane and intracellular proteins.* Although cell lysis occurs rapidly upon toxin ingestion, it is premature at this point to exclude the possibility of toxin internalization into the cell. No rigorous study has been performed that excludes, at this stage, the cellular internalization of the toxin molecule. We propose that part of the toxin molecule, potentially domain III, transverses the cell membrane making it accessible to intracellular proteins. Toxin domains that transverse the membrane can affect signal transduction pathways or lead to the activation of endogenous channels, transporters, voltage-sensitive channels or ion pumps. These interactions could be critical, if not essential, for the full intoxication by *B. thuringiensis* toxins.
7. *Apparent osmotic lysis and cell death.* A few minutes after toxin ingestion by susceptible insects, the midgut columnar cell microvilli swell and break causing leakage of cell contents into the midgut lumen. Cellular damage is so dramatic that in addition to the toxin-induced pore, other proteins in the cell may also contribute to the cell osmotic imbalance that ultimately causes cell death. Although this appears to be the principal mechanism of cell death, other alternative mechanisms cannot be unequivocally excluded.

ACKNOWLEDGEMENTS

Thanks to Dr Donald Dean for helpful discussions on the roles of toxin domains II and III, and for providing unpublished manuscripts.

REFERENCES

Allured, V.S., Collier, R.J., Carroll, S.F. and McKay, D.B. (1986) Structure of exotoxin A of *Pseudomonas aeruginosa* at 3.0-Angstrom resolution. *Proc. Natl Acad. Sci. USA*, **83**, 1320–4.

Belfiore, C.J., Vadlamudi, R.K., Osman, Y.A. and Bulla, L.A., Jr (1994) A specific binding protein from *Tenebrio molitor* for the insecticidal toxin of *Bacillus thuringiensis* subsp. *tenebrionis*. *Biochem. Biophys. Res. Commun.*, **200**, 359–64.

Bravo, A., Jansens, S. and Peferoen, M. (1992) Immunocytochemical localization of *Bacillus thuringiensis* insecticidal crystal proteins in intoxicated insects. *J. Invertebr. Pathol.*, **60**, 237–46.

Carroll, J. and Ellar, D.J. (1993) An analysis of *Bacillus thuringiensis* δ-endotoxin action on insect-midgut-membrane permeability using a light-scattering assay. *Eur. J. Biochem.*, **214**, 771–8.

Chen, X.J., Curtiss, A., Alcantara, E. and Dean, D.H. (1995) Mutations in domain I of *Bacillus thuringiensis* δ-endotoxin CryIAb reduce the irreversible binding of toxin to *Manduca sexta* brush border membrane vesicles. *J. Biol. Chem.*, **270**, 6412–19.

Chen, X.J., Lee, M.K. and Dean, D.H. (1993) Site-directed mutations in a highly conserved region of *Bacillus thuringiensis* δ-endotoxin effect inhibition of short circuit current across *Bombyx mori* midguts. *Proc. Natl Acad. Sci. USA.*, **90**, 9041–5.

Choe, S., Bennett, M.J., Fujii, G. *et al.* (1992) The crystal structure of diphtheria toxin. *Nature*, **357**, 216–22

Chow, E., Singh, G.J.P. and Gill, S.S. (1989) Binding and aggregation of the 25-kilodalton toxin of *Bacillus thuringiensis* subsp. *israelensis* to cell membranes and alteration by monoclonal antibodies and amino acid modifiers. *Appl. Environ. Microbiol.*, **55**, 2779–88.

Cowles, E.A., Yunovitz, H., Charles, J.-F. and Gill, S.S. (1995) Comparison of toxin overlay and solid-phase binding assays to identify diverse CryIA(c) toxin-binding proteins in *Heliothis virescens* midgut. *Appl. Environ. Microbiol.*, **61**, 2738–44.

Crickmore, N., Zeigler, D.R., Feitelson, J. *et al.* (1995) *SIP abstract* (Abstract).

Cummings, C.E. and Ellar, D.J. (1994) Chemical modification of *Bacillus thuringiensis* activated δ-endotoxin and its effect on toxicity and binding to *Manduca sexta* midgut membranes. *Microbiology*, **140**, 2737–47.

Dai, S.-M. (1992) *Bacillus thuringiensis* subsp. *israelensis* CryIVD toxin: resistance development, proteolysis and binding characteristics in *Culex quinquefasciatus*. University of California, Riverside, CA, pp. iv–176.

Dai, S.-M. and Gill, S.S. (1993) *In vitro* and *in vivo* proteolysis of the *Bacillus thuringiensis* subsp. *israelensis* CryIVD protein by *Culex quinquefasciatus* larval midgut proteases. *Insect Biochem. Mol. Biol.*, **23**, 273–83.

Drobniewski, F.A. and Ellar, D.J. (1988) Investigation of the membrane lesion induced *in vitro* by two mosquitocidal delta-endotoxins of *Bacillus thuringiensis*. *Curr. Microbiol.*, **16**, 195–9.

Ebersold, H.R., Luethy, P. and Mueller, M. (1977) Changes in the fine structure of the gut epithelium of *Pieris brassicae* induce by the δ-endotoxin of *Bacillus thuringiensis*. *Bull. Soc. Entomol. Suisse*, **50**, 269–76.

English, L.H. and Cantley, L.C. (1985) Delta endotoxin inhibits Rb^+ uptake, lowers cytoplasmic pH and inhibits a K^+ ATPase in *Manduca sexta* CHE cells. *J. Membr. Biol.*, **85**, 199–204.

English, L.H. and Cantley, L.C. (1986) Delta endotoxin is a potent inhibitor of the (Na,K)-ATPase. *J. Biol. Chem.*, **261**, 1170–3.

English, L.H. and Readdy, T.L. (1989) Delta endotoxin inhibits a phosphatase in midgut epithelial membranes of *Heliothis virescens*. *Insect Biochem.*, **19**, 145–52.

English, L.H., Readdy, T.L. and Bastian, A.E. (1991) Delta-endotoxin-induced leakage of $^{86}Rb^+K^+$ and H_2O from phospholipid vesicles is catalyzed by reconstituted midgut membrane. *Insect Biochem.*, **21**, 177–84.

Escriche, B., Martínez-Ramírez, A.C., Real, M.D. *et al.* (1994) Occurrence of three different binding sites for *Bacillus thuringiensis* δ-endotoxins in the midgut brush border membrane of the potato tuber moth, *Phthorimaea operculella* (Zeller). *Arch. Insect Biochem. Physiol.*, **26**, 315–27.

Feldmann, F., Dullemans, A. and Waalwijk, C. (1995) Binding of the CryIVD toxin of *Bacillus thuringiensis* subsp. *israelensis* to larval dipteran midgut proteins. *Appl. Environ. Microbiol.*, **61**, 2601–5.

Ferré, J., Real, M.D., Van Rie, J. *et al.* (1991) Resistance to the *Bacillus thuringiensis* bioinsecticide in a field population of *Plutella xylostella* is due to a change in a midgut membrane receptor. *Proc. Natl Acad. Sci. USA*, **88**, 5119–23.

Garczynski, S.F. and Adang, M.J. (1995) *Bacillus thuringiensis* CryIA(c) δ-endotoxin binding aminopeptidase in the *Manduca sexta* midgut has a glycosyl-phosphatidylinositol anchor. *Insect Biochem. Mol. Biol.*, **25**, 409–15.

Garczynski, S.F., Crim, J.W. and Adang, M.J. (1991) Identification of putative insect brush border membrane-binding molecules specific to *Bacillus thuringiensis* δ-endotoxin by protein blot analysis. *Appl. Environ. Microbiol.*, **57**, 2816–20.

Gazit, E., Bach, D., Kerr, I.D. *et al.* (1994) The α-5 segment of *Bacillus thuringiensis* δ-endotoxin: *in vitro* activity, ion channel formation and molecular modelling. *Biochem, J.*, **304**, 895–902.

Gazit, E. and Shai, Y. (1993) Structural characterization, membrane interaction, and specific assembly within phospholipid membranes of hydrophobic segments from *Bacillus thuringiensis* var. *israelensis* cytolytic toxin. *Biochemistry*, **32**, 12363–71.

Gazit, E. and Shai, Y. (1995) The assembly and organization of the α5 and α7 helices from the pore-forming domain of *Bacillus thuringiensis* α-endotoxin. *J. Biol. Chem.*, **270**, 2571–8.

Ge, A.Z., Rivers, D., Milne, R. and Dean, D.H. (1991) Functional domains of *Bacillus thuringiensis* insecticidal crystal proteins. Refinement of *Heliothis virescens* and *Trichoplusia ni* specificity domains on CryIA(c). *J. Biol. Chem.*, **266**, 17954–8.

Ge, A.Z., Shivarova, N.I. and Dean, D.H. (1989) Location of the *Bombyx mori* specificity domain on a *Bacillus thuringiensis* δ-endotoxin protein. *Proc. Natl Acad. Sci. USA*, **86**, 4037–41.

Georghiou, G.P. (1990) Resistance potential to biopesticides and consideration of countermeasures, in *Pesticides and Alternatives* (ed. J.E. Casida), Elsevier Science, Amsterdam, pp. 409–20.

Gill, S.S., Cowles, E.A. and Francis, V. (1995) Identification, isolation and cloning of a *Bacillus thuringiensis* CryIAc toxin binding protein from the midgut of the lepidopteran insect *Heliothis virescens*. *J. Biol. Chem.*, **270**, 27277–82.

Gill, S.S., Cowles, E.A. and Pietrantonio, P.V. (1992) The mode of action of *Bacillus thuringiensis* endotoxins. *Annu. Rev. Entomol.*, **37**, 615–36.

Goldman, I.F., Arnold, J. and Carlton, B.C. (1986) Selection for resistance to *Bacillus thuringiensis* subspecies *israelensis* in field and laboratory populations of the mosquito *Aedes aegypti*. *J. Invertebr. Pathol.*, **47**, 317–24.

Gould, F., Martínez-Ramírez, A., Anderson, A. *et al.* (1992) Broad-spectrum resistance to *Bacillus thuringiensis* toxins in *Heliothis virescens*. *Proc. Natl Acad. Sci. USA*, **89**, 7986–90.

Haider, M.Z., Smith, G.P. and Ellar, D.J. (1989) Delineation of the toxin coding fragments and an insect-specificity region of a dual toxicity *Bacillus thuringiensis* crystal protein gene. *FEMS Microbiol. Lett.*, **58**, 157–64.

Heimpel, A.M. and Angus, T.A. (1959) The site of action of crystalliferous bacteria in lepidopteran larvae. *J. Insect Pathol.*, **1**, 152–70.

Hendrickx, K., De Loof, A. and Van Mellaert, H. (1990) Effects of *Bacillus thuringiensis* delta-endotoxin on the permeability of brush border membrane vesicles from tobacco hornworm (*Manduca sexta*) midgut. *Comp. Biochem. Physiol.*, **95C**, 241–5.

Himeno, M. (1987) Mechanism of *Bacillus thuringiensis* insecticidal delta-endotoxin action on insect cells *in vitro*, in *Biotechnology in Invertebrate Pathology and Cell Culture* (ed. K. Maramorosch), Academic Press, San Diego, pp. 29–43.

Hodgman, T.C. and Ellar, D.J. (1990) Models for the structure and function of the *Bacillus thuringiensis* δ-endotoxins determined by computational analysis. *J. DNA Sequence Mapping*, **1**, 97–106.

Hofmann, C., Lüthy, P., Hütter, R. and Pliska, V. (1988a) Binding of the delta endotoxin from *Bacillus thuringiensis* to brush-border membrane vesicles of the cabbage butterfly (*Pieris brassicae*). *Eur. J. Biochem.*, **173**, 85–91.

Hofmann, C., Vanderbruggen, H., Höfte, H. *et al.* (1988b) Specificity of *Bacillus thuringiensis* δ-endotoxins is correlated with the presence of high-affinity binding sites in the brush border membrane of target insect midguts. *Proc. Natl Acad. Sci. USA*, **85**, 7844–8.

Höfte, H. and Whiteley, H.R. (1989) Insecticidal crystal proteins of *Bacillus thuringiensis*. *Microbiol. Rev.*, **53**, 242–55.

Ihara, H., Kuroda, E., Wadano, A. and Himeno, M. (1993) Specific toxicity of delta-endotoxins from *Bacillus thuringiensis* to *Bombyx mori*. *Biosci. Biotech. Biochem.*, **57**, 200–4.

Johnson, D.E., Brookhart, G.L., Kramer, K.J. *et al.* (1990) Resistance to *Bacillus thuringiensis* by the Indian meal moth, *Plodia interpunctella:* comparison of midgut proteinases from susceptible and resistant larvae. *J. Invertebr. Pathol.*, **55**, 235–44.

Knight, P.J.K., Crickmore, N. and Ellar, D.J. (1994) The receptor for *Bacillus thuringiensis* CryIA(c) delta-endotoxin in the brush border membrane of the lepidopteran *Manduca sexta* is aminopeptidase N. *Mol. Microbiol.*, **11**, 429–36.

Knight, P.J.K., Knowles, B.H. and Ellar, D.J. (1995) Molecular cloning of an insect aminopeptidase N that serves as a receptor for *Bacillus thuringiensis* CryIA(c) toxin. *J. Biol. Chem.*, **270**, 17765–70.

Knowles, B.H. (1994) Mechanism of action of *Bacillus thuringiensis* insecticidal δ-endotoxins. *Adv. Insect Physiol.*, **24**, 275–308.

Knowles, B.H., Blatt, M.R., Tester, M. *et al.* (1989) A cytolytic delta-endotoxin from *Bacillus thuringiensis* var. *israelensis* forms cation-selective channels in planar lipid bilayers. *FEBS Lett.*, **244**, 259–62.

Knowles, B.H. and Dow, J.A.T. (1993) The crystal δ-endotoxins of *Bacillus thuringiensis*: models for their mechanism of action on the insect gut. *BioEssays*, **15**, 469–76.

Knowles, B.H. and Ellar, D.J. (1987) Colloid-osmotic lysis is a general feature of the mechanism of action of *Bacillus thuringiensis* δ-endotoxins with different insect specificity. *Biochim. Biophys. Acta*, **924**, 509–18.

Knowles, B.H. and Farndale, R.W. (1988) Activation of insect cell adenylate cyclase by *Bacillus thuringiensis* δ-endotoxins and melittin. *Biochem. J.*, **253**, 235–41.

Knowles, B.H., Knight, P.J.K. and Ellar, D.J. (1991) *N*-acetyl galactosamine is part of the receptor in insect gut epithelia that recognizes an insecticidal protein from *Bacillus thuringiensis*. *Proc. R. Soc. London. B*, **245**, 31–5.

Kwak, I.-S and Dean, D.H. (1995) Exploration of receptor binding of *Bacillus thuringiensis* toxins. *Mem. Inst. Oswaldo Cruz.*, **90**, 75–9.

Lane, N.J., Harrison, J.B. and Lee, W.M. (1989) Changes in microvilli and Golgi-associated membranes of lepidopteran cells induced by an insecticidally active bacterial δ-endotoxin. *J. Cell Sci.*, **93**, 337–47.

Lee, M.K., Milne, R.E., Ge, A.Z. and Dean, D.H. (1992) Location of a *Bombyx mori* receptor binding region on a *Bacillus thuringiensis* δ-endotoxin. *J. Biol. Chem.*, **267**, 3115–21.

Lee, M.K., Rajamohan, F., Gould, F. and Dean, D.H. (1995) Resistance to *Bacillus thuringiensis* CryIA δ-endotoxins in a laboratory-selected *Heliothis virescens* strain is related to an alteration in one of the receptors. *Appl. Environ. Microbiol.*, **61**, 3826–42.

Li, J., Carroll, J. and Ellar, D.J. (1991) Crystal structure of insecticidal δ-endotoxin from *Bacillus thuringiensis* at 2.5 Å resolution. *Nature*, **353**, 815–21.

Liang, Y. and Dean, D.H. (1994) Location of a lepidopteran specificity region in insecticidal crystal protein CryIIA from *Bacillus thuringiensis. Mol. Microbiol.*, **13**, 569–75.

Liang, Y., Patel, S.S. and Dean, D.H. (1995) Irreversible binding kinetics of *Bacillus thuringiensis* CryIA δ-endotoxins to gypsy moth brush border membrane vesicles is directly correlated to toxicity. *J. Biol. Chem.*, **270**, 24719–24.

Lu, H., Rajamohan, F. and Dean, D.H. (1994) Identification of amino acid residues of *Bacillus thuringiensis* δ-endotoxin CryIAa associated with membrane binding and toxicity to *Bombyx mori. J. Bacteriol.*, **176**, 5554–9.

MacIntosh, S.C., Stone, T.B., Jokerst, R.S. and Fuchs, R.L. (1991) Binding of *Bacillus thuringiensis* proteins to a laboratory-selected line of *Heliothis virescens. Proc. Natl Acad. Sci. USA*, **88**, 8930–3.

Maddrell, S.H.P., Overton, J.A., Ellar, D.J. and Knowles, B.H. (1989) Action of activated 27 000 Mr toxin from *Bacillus thuringiensis* var. *israelensis* on Malpighian tubules of the insect, *Rhodnius prolixus. J. Cell Sci.*, **94**, 601–8.

Martens, J.W.M., Visser, B., Vlak, J.M. and Bosch, D. (1995) Mapping and characterization of the endomocidal domain of the *Bacillus thuringiensis* CryIA(b) protoxin. *Mol. Gen. Genet.*, **247**, 482–7.

Martínez-Ramírez, A.C., González-Nebauer, S., Escriche, B. and Real, M.D. (1994) Ligand blot identification of a *Manduca sexta* midgut binding protein specific to three *Bacillus thuringiensis* CryIA type ICPs. *Biochem. Biophys. Res. Commun.*, **201**, 782–7.

Masson, L., Lu, Y.-J., Mazza, A. *et al.* (1995a) The CryIA(c) receptor purified from *Manduca sexta* displays multiple specificities. *J. Biol. Chem.*, **270**, 20309–15.

Masson, L., Mazza, A. and Brousseau, R. (1994a) Stable immobilization of lipid vesicles for kinetic studies using surface plasmon resonance. *Anal. Biochem.*, **218**, 405–12.

Masson, L., Mazza, A., Brousseau, R. and Tabashnik, B.E. (1995b) Kinetics of *Bacillus thuringiensis* toxin binding with brush border membrane vesicles from susceptible and resistant larvae of *Plutella xylostella. J. Biol. Chem.*, **270**, 11887–96.

Masson, L., Mazza, A., Gringorten, L. *et al.* (1994b) Specificity domain localization of *Bacillus thuringiensis* insecticidal toxins is highly dependent on the bioassay system. *Mol. Microbiol.*, **14**, 851–60.

McGaughey, W.H. (1985) Insect resistance to the biological insecticide *Bacillus thuringiensis. Science*, **229**, 193–5.

Oddou, P., Hartman, H. and Geiser, M. (1991) Identification and characterization of *Heliothis virescens* midgut membrane proteins binding *Bacillus thuringiensis* δ-endotoxins. *Eur. J. Biochem.*, **202**, 673–80.

Olsen, J., Cowell, G.M., Konigshofer, E. *et al.* (1988) Complete amino acid sequence of human intestinal aminopeptidase N as deduced from cloned cDNA. *FEBS Lett.*, **238**, 307–14.

Oppert, B. Kramer, K.J., Johnson, D.E. *et al.* (1994) Altered protoxin activation by midgut enzymes from a *Bacillus thuringiensis* resistant strain of *Plodia interpunctella*. *Biochem. Biophys. Res. Commun.*, **198**, 940–7.

Ort, P., Zalunin, I.A., Gasparov, V.S. *et al.* (1995) Domain organization of *Bacillus thuringiensis* CryIIIA δ-endotoxin studied by denaturation in guanidine hydrochloride solutions and limited proteolysis. *J. Protein Chem.*, **14**, 241–9.

Percy, J. and Fast, P.G. (1983) *Bacillus thuringiensis* crystal toxin: ultrastructural studies of its effect on silkworm midgut cells. *J. Invertebr. Pathol.*, **41**, 86–98.

Pietrantonio, P.V., Federici, B.A. and Gill S.S. (1993) Interaction of *Bacillus thuringiensis* endotoxins with the insect midgut epithelium, in *Parasites and Pathogens of Insects* Vol. 2 (eds N.E. Beckage, S.N. Thompson and B.A. Federici), Academic Press, San Diego, pp. 55–79.

Pietrantonio, P.V. and Gill, S.S. (1992) The parasporal inclusion of *Bacillus thuringiensis* subsp. *shandongiensis*: Characterization and screening for insecticidal activity. *J. Invertebr. Pathol.*, **59**, 295–302.

Rajamohan, F., Alcantara, E., Lee, M.K., Chen, X.J. *et al.* (1995) Single amino acid changes in domain II of *Bacillus thuringiensis* CryIA(b) δ-endotoxin affect irreversible binding to *Manduca sexta* midgut membrane vesicles. *J. Bacteriol.*, **177**, 2276–82.

Ravoahangimalala, O. and Charles, J.-F. (1995) *In vitro* binding of *Bacillus thuringiensis* var. *israelensis* individual toxins to midgut cells of *Anopheles gambiae* larvae (Diptera: Culicidae). *FEBS Lett.*, **362**, 111–15.

Reuveni, M. and Dunn, P.E. (1991) Differential inhibition by *Bacillus thuringiensis* δ-endotoxin of leucine and aspartic acid uptake into BBMV from midgut of *Manduca sexta*. *Biochem Biophys. Res. Commun.*, **181**, 1089–93.

Sanchis, V. and Ellar, D.J. (1993) Identification and partial purification of a *Bacillus thuringiensis* CryIC δ-endotoxin binding protein from *Spodoptera littoralis* gut membranes. *FEBS Lett.*, **316**, 264–8.

Sangadala, S., Walters, F.S., English, L.H. and Adang, M.J. (1994) A mixture of *Manduca sexta* aminopeptidase and phosphatase enhances *Bacillus thuringiensis* insecticidal CryIA(c) toxin binding and $^{86}Rb^{+}$-K^{+} efflux *in vitro*. *J. Biol. Chem.*, **269**, 10088–92.

Schnepf, H.E., Tomczak, K., Ortega, J.P. and Whiteley, H.R. (1990) Specificity-determining regions of a lepidoteran-specific insecticidal protein produced by *Bacillus thuringiensis*. *J. Biol. Chem.*, **265**, 20923–30.

Schwartz, J.-L. Garneau, L., Masson, L. and Brousseau, R. (1991) Early response of cultured lepidopteran cells to exposure to δ-endotoxin from *Bacillus thuringiensis*: involvement of calcium and anionic channels. *Biochim. Biophys. Acta*, **1065**, 250–60.

Singh, G.J.P., Schouest, L.P., Jr and Gill, S.S. (1986) The toxic action of *Bacillus thuringiensis* var. *israelensis* in *Aedes aegypti in vivo*. 1. The relevance of midgut lesions to its poisoning syndrome. *Pestic. Biochem. Physiol.*, **26**, 36–46.

Slatin, S.L. Abrams, C.K. and English, L. (1990) Delta-endotoxins form cation-selective channels in planar lipid bilayers. *Biochem. Biophys. Res. Commun.*, **169**, 765–72.

Tabashnik, B.E. (1994) Evolution of resistance to *Bacillus thuringiensis*. *Annu. Rev. Entomol.*, **39**, 47–79.

Tabashnik, B.E., Cushing, N.L., Finson, N. and Johnson, M.W. (1990) Field development of resistance to *Bacillus thuringiensis* in diamondback moth (Lepidoptera: Plutellidae). *J. Econ. Entomol.*, **83**, 1671–6.

Tabashnik, B.E., Finson, N., Groeters, F.R. *et al.* (1994) Reversal of resistance to *Bacillus thuringiensis* in *Plutella xylostella*. *Proc. Natl Acad. Sci. USA*, **91**, 4120–4.

Vadlamudi, R.K., Ji, T.H. and Bulla, L.A. Jr (1993) A specific binding protein from *Manduca sexta* for the insecticidal toxin of *Bacillus thuringiensis* subsp. *berliner*. *J. Biol. Chem.*, **268**, 12334–40.

Vadlamudi, R.K., Weber, E., Ji, I. *et al.* (1995) Cloning and expression of a receptor for an insecticidal toxin of *Bacillus thuringiensis*. *J. Biol. Chem.*, **270**, 5490–4.

Valaitis, A.P., Lee, M.K., Rajamohan, F. and Dean, D.H. (1995) Brush border membrane aminopeptidase-N in the midgut of the gypsy moth serves as the receptor for the CryIA(c) δ-edotoxin of *Bacillus thuringiensis*. *Insect Biochem. Mol. Biol.*, **25**, 1101–114.

Van Rie, J., Jansens, S., Höfte, H. *et al.* (1989) Specificity of *Bacillus thuringiensis* δ-endotoxins – importance of specific receptors on the brush border membrane of the mid-gut of target insects. *Eur. J. Biochem.*, **186**, 239–47.

Van Rie, J., Jansens, S., Höfte, H. *et al.* (1990a) Receptors on the brush border membrane of the insect midgut as determinants of the specificity of *Bacillus thuringiensis* delta-endotoxins. *Appl. Environ. Microbiol.*, **56**, 1378–85.

Van Rie, J., McGaughey, W.H. Johnson, D.E. *et al.* (1990b) Mechanism of insect resistance to the microbial insecticide *Bacillus thuringiensis*. *Science*, **247**, 72–4.

Walters, F.S., Slatin, S.L., Kulesza, C.A. and English, L.H. (1993) Ion channel activity of N-terminal fragments from CryIA(c) delta-endotoxin. *Biochem. Biophys. Res. Commun.*, **196**, 921–6.

Widner, W.R. and Whiteley, H.R. (1989) Two highly related insecticidal crystal proteins of *Bacillus thuringiensis* subsp. *kurstaki* possess different host range specificities. *J. Bacteriol.*, **171**, 965–74.

Widner, W.R. and Whiteley, H.R. (1990) Location of the dipteran specificity region in a lepidopteran–dipteran crystal protein from *Bacillus thuringiensis*. *J. Bacteriol.*, **172**, 2826–32.

Wilson, R., Ainscough, R., Anderson, K. *et al.* (1994). 2.2 Mb of contiguous nucleotide sequence from chromosome III of *C. elegans*. *Nature*, **368**, 32–8.

Wolfersberger, M.G. (1990) The toxicity of two *Bacillus thuringiensis* δ-endotoxins to gypsy moth larvae is inversely related to the affinity of binding sites on midgut brush border membranes for the toxins. *Experientia*, **46**, 475–7.

Wu, D. and Aronson, A.I. (1992) Localized mutagenesis defines regions of the *Bacillus thuringiensis* δ-endotoxin involved in toxicity and specificity. *J. Biol. Chem.*, **267**, 2311–17.

Yan, X. and McCarthy, W.J. (1991) Chemical modification of *Bacillus thuringiensis* subsp. *thuringiensis* (HD-524) tryspin-activated endotoxin: implication of tyrosine residues in lepidopteran cell lysis. *J. Invertebr. Pathol.*, **57**, 101–8.

14

Antinutritive plant defence mechanisms

G.W. Felton and J.A. Gatehouse

14.1 INTRODUCTION

Plants deploy an arsenal of chemical defences against invading herbivores (e.g. Rosenthal and Berenbaum, 1991). Although hundreds of natural products have been characterized, the mode of actions is known for only a few. Our focus is on 'antinutrients', a term that has been used in the entomological literature without sufficient definition. Antinutrients diminish nutrient bioavailability by chemically modifying the nutrient, forming less active complexes with the nutrient, or by hindering digestion, absorption or utilization of the nutrient. Defining criteria for an antinutrient are (1) the symptoms produced by the antinutrient resemble those caused by insufficiency of a given nutrient and (2) nutrient treatment alleviates the effect of the antinutrient.

Previous coverage of the subject is limited (Ishaaya, 1986; Duffey and Felton, 1989, 1991), although the antinutritive plant defence may be more widespread than generally recognized. The intention of this chapter is not to imply that the sole mode of action of these compounds is antinutritive nor that it is independent of the insect species or of the chemical context of the host plant. Various antinutrients are discussed together with the properties of the insect digestive system that may mitigate or accelerate their adverse effects. The chapter discusses the use of antinutrients in the development of insect-resistant host plants.

Biology of the Insect Midgut.
Edited by M.J. Lehane and P.F. Billingsley.
Published in 1996 by Chapman & Hall, London.
ISBN 0 412 61670 X.

14.2 INSECT HERBIVORE NUTRITION

The nutritional requirements of insects are comparable to many mammals in that they require water, amino acids (10 essential), carbohydrates, lipids, sterols, vitamins and minerals. Insects are unique among animals in requiring ascorbic acid and sterols (McFarlane, 1985). However, as insect nutritional requirements are known for only a small fraction of the extant species, assumptions are made for the remaining species.

The inter/intraplant variation in amino acids, vitamins, minerals, water and lipids may be of sufficient magnitude to contribute to poor performance by herbivores (Haukioja, 1991; Berenbaum, 1995; Felton, 1996). For virtually every nutrient class, there is a corresponding antinutrient found in plants (Table 14.1) that may further exacerbate the effects of suboptimal nutrient levels.

14.3 ANTINUTRITIVE PLANT DEFENCE MECHANISMS

14.3.1 Low-molecular-weight antinutrients

(a) Amino acid availability

Many structurally diverse and taxonomically widespread phytochemicals can interfere with amino acid availability by affecting digestive proteinases or their substrates. Common plant alkylating agents include sesquiterpene lactones, epoxides, quinones (e.g. benzoquinones, naphthoquinones, anthraquinones), pyrrolizidine alkaloids, α,β-unsaturated aldehydes, hydroxamic acids (e.g. DIMBOA or 2,4-dihydroxy-7-methoxy-2H-1,4-benzoxazin-3(4)-one) and sesquiterpene aldehydes (e.g. gossypol) (Figure 14.1) (Nelson and Pearson, 1990). Nucleophilic side chains of proteins (i.e. -SH of Cys, sulphur of Met, ε-NH_2 of Lys, and -NH of His) contain highly polarized electronegative donor atoms that react readily with the electrophilic, alkylating compounds containing acceptor atoms of low positive charge with excitable outer electrons (Hinson and Roberts, 1992). The alkylated side chains of amino acids are thus rendered unavailable for nutritional use (Felton, 1996). Moreover, the amino acids most susceptible to alkylation are among the most nutritionally limiting in plant tissues. Alkylation may be just one of several possible modes of action of these compounds.

In some instances compounds such as cycasin, methylenedioxybenzenes (e.g. sesamol) and pyrrolizidine alkaloids (e.g. seneciocine) undergo metabolic activation to yield electrophiles that bind covalently to electron-rich macromolecules, but their impact on nutrition is unexplored. In other cases, the specific mechanisms of interaction with

Table 14.1 Nutritional requirements and potential antinutrients

Nutrient class	*Potential antinutrient*
Amino acids	Alkylating agents (quinones, epoxides, aldehydes, sesquiterpene lactones, etc.); protease inhibitors; thiaminase; transglutaminase; diamine oxidase, myrosinase; peroxidase, polyphenol oxidase
Carbohydrates	Amylase inhibitors, glycosidase inhibitors (e.g. polyhydroxylated alkaloids)
Fatty acids (linoleic/linolenic acid)	Lipoxygenase, peroxidase
Minerals	Oxalates, phytate, phenolics, polyphenols
Sterols	Saponins
Vitamins	Ascorbate oxidase, thiaminase, lipoxygenase, phenolics, quinones
Water	Diuretics(?)

protein are unknown. Saponins are widely distributed triterpenoid glycosides that complex with sterols but can also complex with protein and decrease the free amino content of protein (Potter *et al.*, 1993). Soybean saponins inhibit the midgut proteinases of *Tribolium castaneum, Tenebrio molitor* and *Dermestes maculatus*, whereas alfalfa saponins inhibit chymotryptic activity in *T. castaneum* (Ishaaya, 1965, 1986; Ishaaya and Birk, 1965).

DIMBOA normally occurs in graminaceous plants as a glycoside which is degraded to the aglycone by β-glycosidases following tissue disruption. DIMBOA and its decomposition product, 6-MBOA, reduce the digestibility of food in insects (Campos *et al.*, 1989; Houseman *et al.*, 1992), and in larval *Ostrinia nubilalis*, the reduction is due to inhibition of chymotrypsin and trypsin. In addition to this covalent interaction of DIMBOA with digestive enzymes, DIMBOA may also interact directly with dietary protein (Figure 14.2).

Gossypol and related terpene aldehydes are toxic for several insect species. Nevertheless their toxicity can be overcome by supplemental protein (Moore, 1983), suggesting that protein utilization is compromised by gossypol–protein binding (Figure 14.3). Significant amounts of gossypol are excreted in a bound form (Montandon *et al.*, 1987; Rojas *et*

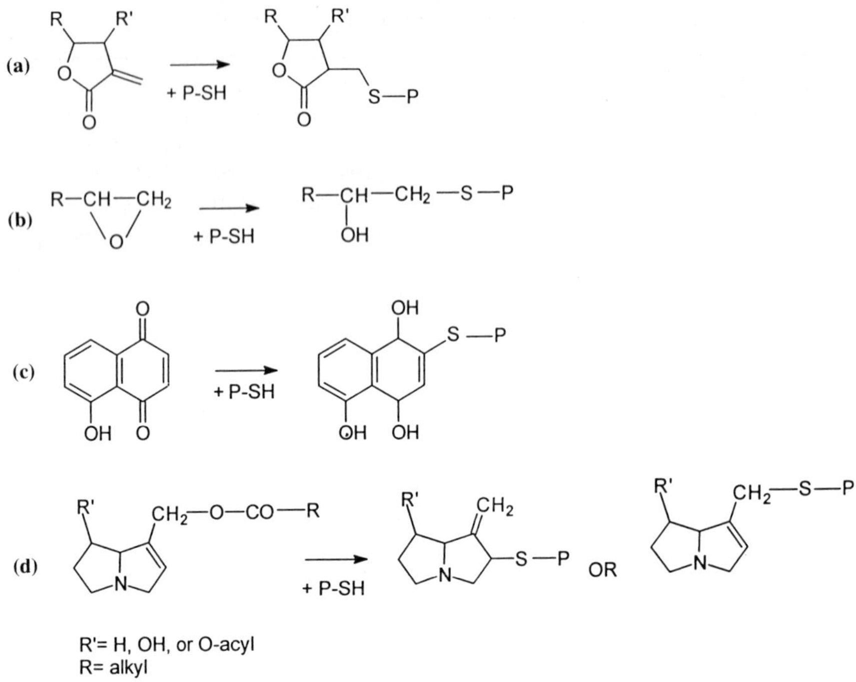

Figure 14.1 Alkylation of protein by plant natural products. (a) Reaction of a sesquiterpene lactone with a protein thiol (P-SH); (b) reaction of an epoxide with a protein thiol; (c) reaction of juglone with a protein thiol; (d) reaction of a pyrrolizidine alkaloid with a protein thiol. (Data from Gershenzon and Croteau, 1991; Gardner, 1979; O'Brien, 1991; Culvenor *et al.*, 1962.)

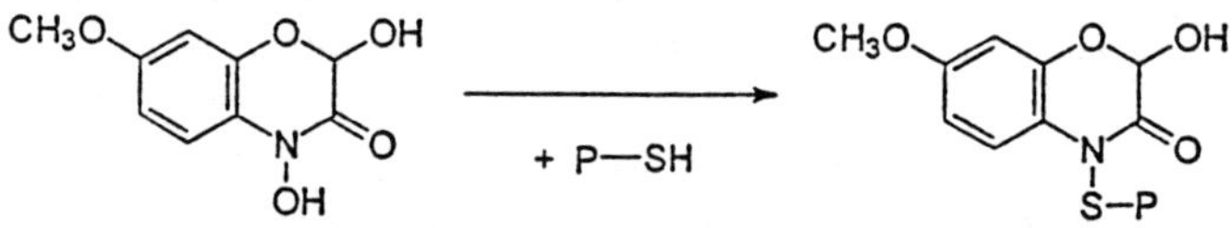

Figure 14.2 Reaction of DIMBOA with protein. P-SH = protein bound thiol. (Data from Perez and Niemeyer, 1985.)

al., 1992) but the presence of specific gossypol–protein conjugates and the impact of gossypol on protein utilization were not reported. Gossypol in artificial diet or high gossypol-producing cotton cultivars inhibited proteinase activity in *Spodoptera littoralis* (Meisner *et al.*, 1978); however, it is uncertain if gossypol inhibits the enzyme directly or reacts with the protein substrate.

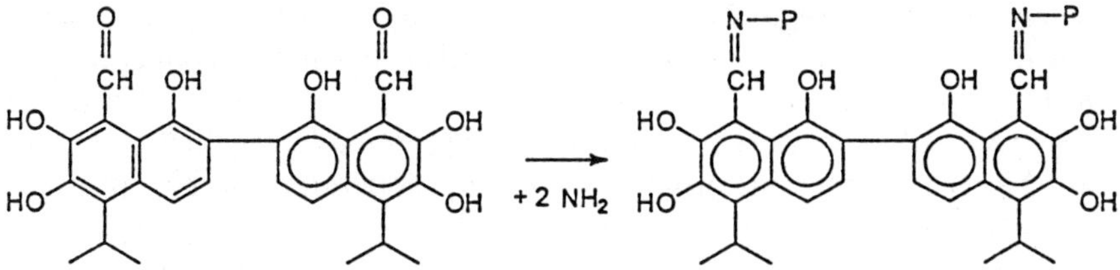

Figure 14.3 Reaction of gossypol with protein. (Data from Gershenzon and Croteau, 1991.)

Tannins are not covered in this discussion. Once considered the archetypal antinutrients, more recent evidence suggests that the primary mode of action of tannins may not be antinutritive (Martin *et al.*, 1987; Karowe, 1989; for excellent reviews see Bernays *et al.*, 1989; Clausen *et al.*, 1992).

(b) Sterol availability

Saponins occur in more than 80 plant families (Ishaaya, 1986) and may be of broad antinutritional importance due to their complex formation with sterols. Saponins are toxic to insects (Potter and Kimmerer, 1989) and their antinutritive action is counteracted by sterol supplements. Toxicity of alfalfa saponins to larvae *T. castaneum* was attenuated by increasing dietary cholesterol (Ishaaya *et al.*, 1969), whereas the toxicity of medicagenic acid (from lucerne or alfalfa) was alleviated by cholesterol, stigmasterol, *β*-sitosterol, or campesterol (Shany *et al.*, 1970). These sterols also alleviate the toxicity of tomatine, a steroidal glycoalkaloid-type saponin, to *Helicoverpa zea* and *Spodoptera exigua* (Bloem *et al.*, 1989; Duffey, unpublished data). Similarly, the toxicity of spirostane-type saponin, aginosid, from leek is reduced in the leek-moth (*Acrolepiopsis assectella*) by cholesterol supplement (Harmatha *et al.*, 1987). In the same insect, digitonin inhibited ecdysis, but the effect was blocked by cholesterol (Arnault and Mauchamp, 1985). We are unaware of investigations that assess directly the effect of saponins on sterol uptake.

(c) Vitamin and mineral availability

Many plant products may interfere with vitamin and mineral availability but few investigations have examined their role in insect nutrition. Several compounds merit further attention as insect antinutrients due to their common distribution and accumulation in certain tissues.

Phytates and oxalates are mammalian antinutrients, although little evidence exists regarding their effects on insects. Phytic acid is distributed widely among plant species and occurs at high concentrations in seeds, particularly of legumes. Phytate is a strong acid that forms salts with divalent metal ions potentially limiting mineral availability (Harland and Morris, 1995). Salt formation is highly dependent on pH, presence of secondary cations, and concentrations of phytate and metal ions.

Certain plant species accumulate relatively large amounts of oxalic acid (up to 30% dry weight), mainly as soluble potasium or sodium salts or as insoluble calcium salts. Oxalic acid may be responsible for chronic calcium deficiency when dietary levels are high (Libert and Franceschi, 1987).

Other compounds such as phenolics (*ortho*-dihydroxyphenolics) may interfere with ascorbate, thiamine and iron availability (Summers and Felton, 1994). Owing to the relatively large amounts of these compounds in plant tissues, a concerted effort to examine their antinutritive effects is merited.

14.3.2 Antinutrient proteins: digestive enzyme inhibitors

Proteins are normally considered as part of the primary metabolism, and thus were not considered to be involved in host selection (Ehrlich and Raven, 1964). However, use of proteins as defensive compounds rather than the means of synthesizing defensive compounds is now well established, although there remains controversy over how important these molecules are in the defensive arsenal of plants. The majority of proteins involved in defence of plants against pests and pathogens have an assayable biochemical activity, so classification of these proteins is based on function rather than structure. Many of these proteins also form families (or superfamilies) which can be related by structure (Garcia-Olmedo *et al.*, 1987). It is noteworthy that the classifications by structure and function do not entirely correspond.

In order for a defensive protein to affect an insect pest, it must interact with a target molecule; for a phytophagous insect, this target is likely to be a dietary component or a component of the insect feeding system. The protein could interact directly with a nutrient, deter feeding by interacting with taste receptors, interfere with digestion by interaction with digestive enzymes, or disrupt gut tissues by interaction with the midgut cell surfaces. It is also possible that the target molecule is 'inside' the insect, but to reach the target, the defensive protein would have to cross the gut wall without being inactivated. This requirement is not easily met, and thus the most common protein-based strategy for plant defence against insects seems to be direct interference with normal digestive processes in the gut.

(a) Serine proteinase inhibitors

The major digestive endoproteinase enzymes in insect gut are the serine proteinases trypsin and chymotrypsin (Chapter 6). Most plant proteinase inhibitors affect one or more of these enzymes, trypsin inhibitors being most common. At least seven families of serine protease inhibitor have been described in plants. The inhibitors bind to the active site on the enzyme to form a complex with a very low dissociation constant (10^{-7}–10^{-14} M at neutral pH values), effectively blocking the active site. A binding loop on the inhibitor, usually 'locked' into conformation by a disulphide bond, projects from the surface of the molecule, and contains a peptide bond ('reactive site') cleavable by the enzyme. This peptide bond may be cleaved in the enzyme–inhibitor complex, but cleavage does not affect the interaction, so that a hydrolysed inhibitor molecule is bound just as well as an unhydrolysed one. The inhibitor thus directly mimics a normal substrate for the enzyme, but does not allow the normal enzyme mechanism of peptide bond cleavage to proceed to completion (dissociation of the product). Specificity of the inhibitor–enzyme interaction is determined primarily by the specificity of the enzyme; for example, trypsin cleaves C-terminally to a basic amino acid residue, so the reactive site in a trypsin inhibitor will comprise an Arg-X or Lys-X dipeptide. However, the strength of interaction, and thus the effectiveness of the inhibitor, is not solely determined by the reactive site, since other residues are also important in stabilizing the enzyme–inhibitor complex. Extensive structure–function studies have been carried out on certain inhibitors (for review see Garcia-Olmedo *et al.*, 1987).

(b) Carboxypeptidase inhibitors

Potatoes and other Solanaceae contain carboxypeptidase inhibitors which rely on a metal ion for catalytic activity. These inhibitors are small polypeptides of approximately 35 amino acid residues. The mechanism of action is similar to that of serine proteinase inhibitors, but the enzyme–inhibitor complex is formed by interaction of a protruding C-terminal 'tail' on the inhibitor with the enzyme active site; this complex is then stabilized by interactions between other residues on the inhibitor and enzyme, so that dissociation is prevented.

(c) Thiol proteinase inhibitors

Thiol proteinases are the main type of degradative proteinase used by plants. The absence of serine endoproteinases allows large amounts of serine proteinase inhibitors to be tolerated in plant tissues. Conversely, accumulation of thiol proteinase inhibitors in plant tissues is not a 'risk-free' strategy for the plant. Nevertheless, thiol proteinase inhibitors have been isolated from a number of plant sources, including pineapple stems, rice seeds, legume seeds and potato tubers, although they are present at low concentrations. Thiol proteinase inhibitors may play important roles in regulating endogenous proteolysis, although this has yet to be demonstrated directly. A role for these inhibitors in defence against pests is less likely on the basis of their low abundance. Most thiol protease inhibitors characterized in plants belong to the cystatin family, which is also present in animals and micro-organisms.

(d) Induction of proteinase inhibitor synthesis by insect attack

Plant proteinase inhibitors are usually accumulated in such storage tissues as seeds or tubers and are absent from, or present at low concentrations in, non-storage tissues. This pattern of accumulation reinforces a role in defence against pests, since plant survival will depend on the ability to protect storage tissues. In potato and other Solanaceae, and in certain legumes, proteinase inhibitors are produced in the leaves in response to mechanical damage. The synthesis of proteinase inhibitors occurs initially at the wound but expression of the proteinase inhibitor genes spreads systemically, and causes synthesis of inhibitors throughout the plant. This phenomenon, termed the 'wound response', provides strong evidence for the role of proteinase inhibitors in defence against pests; feeding by phytophagous insects such as lepidopteran and coleopteran larvae provides the mechanical damage necessary to induce protease inhibitor synthesis. The molecular basis for gene regulation in the wound response has been studied in great detail; the signalling system is complex, and the reader is referred to relevant reviews (Ryan, 1984, 1989; Farmer and Ryan, 1992). Whatever the signalling process, the end result is significant; proteinase inhibitors accumulate at the site of attack, reaching maximum concentrations within 12–24 h, and throughout the plant, over several days. The inhibitors are then able to protect the plant against insect feeding. This induction mechanism means proteinase inhibitors can be considered as one class of genes encoding pathogenesis-related (PR) proteins. Other PR-proteins are induced by fungal or bacterial attack, and include chitinases, glucanases and various proteins of unknown function, some of which have sequence homology to certain proteinase inhibitors.

(e) Effects of proteinase inhibitors on insects

Since the early observation that soybean trypsin inhibitors were toxic to larvae of the flour beetle *Tribolium confusum* (Lipke *et al.*, 1954), many *in vivo* and *in vitro* studies have examined the effects of protease inhibitors on insect survival and development. *In vivo* assays have used artificial

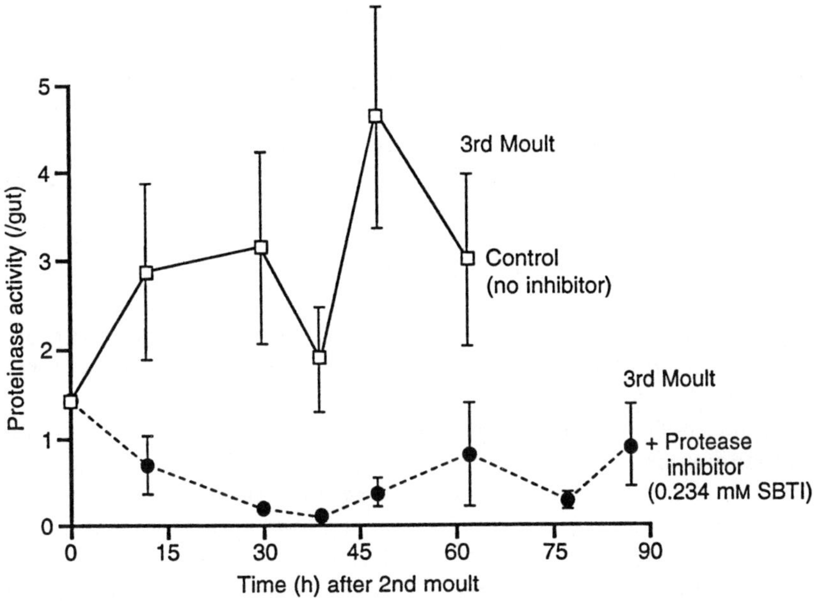

Figure 14.4 Suppression of midgut proteinase activity in larvae of *Helicoverpa armigera* during the third instar by proteinase inhibitor (SBTI, soybean Kunitz trypsin inhibitor) administered in the diet. Note the extended developmental period in the presence of the inhibitor. (Adapted from Johnston *et al.*, 1993.)

diets, incorporating inhibitors at known concentrations, as the basis for bioassays. Effective proteinase inhibitors decrease insect survival, biomass (both of individual insects and the total insect biomass), and the rate of development in terms of time taken to pass through defined stages of the life cycle (Figure 14.4). Proteinase inhibitors are not toxic enough to produce high levels of mortality in a short assay period; their effects are longer term, and are thus termed antinutritional, antinutritive or antimetabolic, rather than toxic. In enzyme assay *in vitro* plant proteinase inhibitors are as effective against insect digestive proteinases as against those of higher animals; however, the effectiveness of inhibition for a given inhibitor against similar insect or mammalian enzymes may differ. Further, an inhibitor may be more effective at inhibiting proteolysis in some insect species than others.

(f) Mechanism of action of proteinase inhibitors

The antimetabolic action of proteinase inhibitors against insects is far from clear. Inhibition of digestive enzymes is unlikely to be the only effect, and in insects as in mammals, the major deleterious effect may be loss of nutrients through overproduction of digestive enzymes (Liener, 1980). However, whereas synthesis and secretion of digestive proteinases and their control are well understood in higher animals, there is little knowledge about analogous processes in insects (Chapter 7). Direct measurements of the response of proteinase synthesis to the presence of inhibitors have rarely been made.

Some proteinase inhibitors in certain insect species result in an increase of endogenous proteolytic activity, suggesting that synthesis of proteinases increases to compensate for the inhibition, whereas other combinations of proteinase inhibitor and insect species result in a depression of proteolytic activity *in vivo*. In larvae of the lepidopterans *H. zea* and *S. exigua*, the presence in the diet of soybean Kunitz trypsin inhibitor (SKTI) or potato proteinase inhibitor II (PI-II) stimulated endogenous tryptic activity (Broadway and Duffey, 1986, 1988). In contrast, SKTI incorporated in the diet significantly reduces the levels of endogenous tryptic activity in larvae of *H. armigera* (Johnston *et al.*, 1993). In an extended study, Dymock *et al.* (1992) found that whereas SKTI initially depressed serine proteinase activity in the coleopteran *Costelytra zealandica*, there was a marked increase in trypsin, elastase and leucine aminopeptidase activity after 14 days. A consensus would suggest that proteinase inhibitors cause short-term depression of digestive proteinase activity in insects, which is reversed after longer exposure, possibly due to increased proteinase synthesis. Proteinase inhibitors clearly do affect overall digestive biochemistry in insects; the antinutritional effects of cowpea trypsin inhibitor on larvae of the coleopteran *Callosobruchus maculatus* could be overcome by supplementation of the diet with methionine, the limiting amino acid in the cowpea seed (Gatehouse and Boulter, 1983). Similar results (Broadway and Duffey, 1986, 1988) suggested that protease inhibitors are part of the complex interaction between plant nutritional value and the insects' digestive physiology. In this model, insect growth is determined by the nutrients that can be obtained from the plant host; proteinase inhibitors limit nutrient availability, by interfering with protein digestion. However, their contribution to plant defence must be considered alongside factors such as protein and free amino acid content, the inherent digestibility of different plant proteins and other compounds that inhibit proteolysis, such as condensed tannins. This may explain why there are some varieties of cowpea with high levels of trypsin inhibitors which bruchids are able to infest (Xavier-Filho *et al.*, 1989).

(g) Insect adaptation to proteinase inhibitors

Insects may adapt to dietary protease inhibitors through changes in their digestive proteinases. Such adaptations can be either inherent in the adapted species, or induced by the presence of inhibitors. Many phytophagous coleopteran species avoid the toxicity of serine proteinase inhibitors in their plant hosts by relying on thiol proteinases for protein digestion. Thiol proteinases were first demonstrated in the seed weevil *Callosobruchus maculatus* (Gatehouse *et al.*, 1985), where the host is rich in serine proteinase inhibitors. Subsequently, other agronomically important coleopteran pests such as corn rootworm (*Diabrotica* spp.) and Colorado potato beetle (*Leptinotarsa decemlineata*) have been shown to rely on thiol proteases as major digestive enzymes. This limits the capacity of the plant to produce suitable inhibitors since the plant itself relies on thiol proteinases. However, there is presumably a cost to the insect in the efficiency of these enzymes as digestive proteinases, or possibly in the gut conditions required for them to function effectively.

Adaptations in proteinases can also be induced. Although cabbage proteinase inhibitors caused substantial inhibition of midgut serine proteinase activity in larval *H. zea* and *Plutella xylostella,* chronic ingestion of these inhibitors failed to affect subsequent growth and development (Broadway, 1995). These larvae were able to overcome the antimetabolic effects of the protease inhibitors by synthesizing proteases insensitive to cabbage proteinase inhibitors. The adaptation hypothesis has been supported by subsequent results demonstrating the synthesis of PI-II insensitive proteinases in *S. exigua* larvae exposed to transgenic tobacco plants expressing the gene for potato PI-II (Jongsma *et al.*, 1995). Similarly larvae of the Colorado potato beetle, *L. decemlineata,* were exposed to potato plants treated with gaseous methyl jasmonate used to induce cysteine and aspartic proteinase inhibitors. Cysteine proteinases insensitive to the cysteine proteinase inhibitor papain were induced in larvae ingesting treated leaves (Bolter and Jongsma, 1995).

(h) Amylase inhibitors

Protein inhibitors of mammalian α-amylases are abundant in cereal grains and present in other seeds (Richardson, 1991). The activity of α-amylase inhibitors is based on their non-covalent interaction with the active site on an amylase molecule, forming a complex with a low dissociation constant. This type of inhibition is inherently less efficient than that of proteinases, since the amylase substrate (starch) cannot be mimicked accurately by a polypeptide chain. Consequently, such inhibitors fit poorly into the enzyme active site, and some amylase inhibitors do not completely inhibit their target enzymes. Specificity is exercised through the residues responsible for the contact between inhibitor and target enzyme. Several distinct sequence families of inhibitors have been described (for review see Garcia-Olmedo *et al.*, 1987).

Induced synthesis of amylase inhibitors by insect attack has not been observed, and therefore the physiological role of these proteins remains open to speculation. However, α-amylase inhibitors purified from wheat and *Phaseolus vulgaris* are insecticidal to non-pest coleopteran species when tested in artificial diet (Gatehouse *et al.*, 1986; Ishimoto and Kitamura, 1988).

A specific and potent α-amylase inhibitor has been isolated from a wild line of *P. vulgaris* (G12953), resistant to attack by a major coleopteran pest, *Zabrotes subfasciatus* (Minney *et al.*, 1990). Inhibitor presence correlated with seed resistance to this particular pest species. This inhibitor is toxic to *Z. subfasciatus* when tested in artificial seeds. The normal α-amylase inhibitor of *P. vulgaris* is toxic to a range of coleopteran pests which do not normally infest this species, but is not effective against *Z. subfasciatus*; thus this example shows adaptation by both the pest and the host.

Different types of α-amylase inhibitor present in wheat endosperm are differentially active against different lepidopterans. Wheat amylase inhibitor has been tested in artificial diet against a range of phytophagous Lepidoptera and Coleoptera; effects *in vivo* range from little or no effect, to significant effects on mortality and development.

Amylase inhibitors are relatively resistant to digestion, and thus can inhibit their target enzymes throughout the digestive process. However, whether any effect other than direct enzyme inhibition is involved remains a matter for speculation.

14.3.3 Antinutrient proteins: enzymes

(a) Polyphenol oxidase (PPO)

Polyphenol oxidases (including monophenolase EC 1.14.18.1, diphenolase EC 1.10.3.1, and laccase EC 1.10.3.2 activities) are typically plastid enzymes that catalyse the oxidation of mono-, di- and polyhydric phenols to quinones (Figure 14.5). The nomenclature is confusing because the same PPO enzyme frequently possesses both mono- and diphenolase action. The substrates are normally compartmentally separated from the enzyme and do not mix unless tissues are disrupted. Quinones formed by these reactions are potent electrophiles that bind irreversibly to the nucleophilic side chains (-SH, $-NH_2$, -NH) of proteins (Figure 14.6). Consequently, dietary proteins treated with PPO and the diphenol chlorogenic acid show significant losses in lysine, histidine and thiol amino acids, with a directly correlated loss in nutritive value of proteins (Felton *et al.*, 1992b). Furthermore, reactive oxygen species (e.g. superoxide radical, hydrogen peroxide, hydroxyl radical) may be formed, and can contribute to loss of amino acids via carbonyl formation

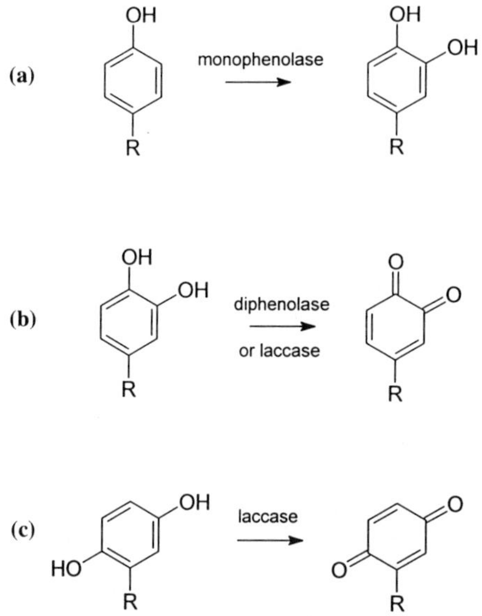

Figure 14.5 Reaction of polyphenol oxidases and laccases with phenols. (a) Monophenolase action; (b) diphenolase action by polyphenol oxidase or laccase; (c) action of laccase on *p*-diphenol.

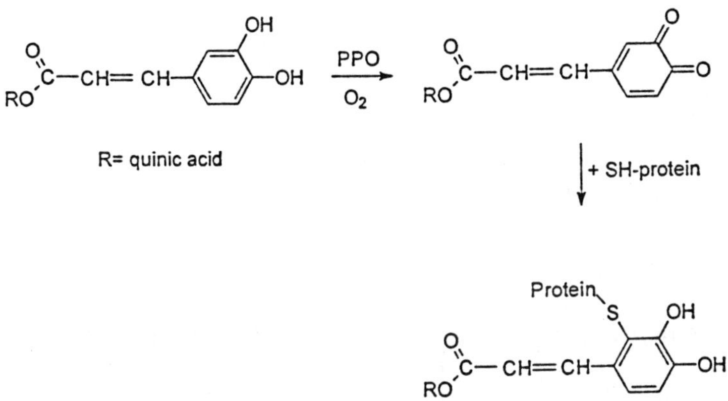

Figure 14.6 Oxidation of chlorogenic acid and subsequent alkylation of protein. PPO = polyphenol oxidase; S = thiol.

and to protein damage by polymerization and/or fragmentation (O'Brien, 1991; Stadtman, 1993).

Catechols that are attached to proteins by alkylation or are formed by radical attack on the proteins may accumulate and then form reactive oxygen species which in turn attack other biomolecules (Figure 14.7).

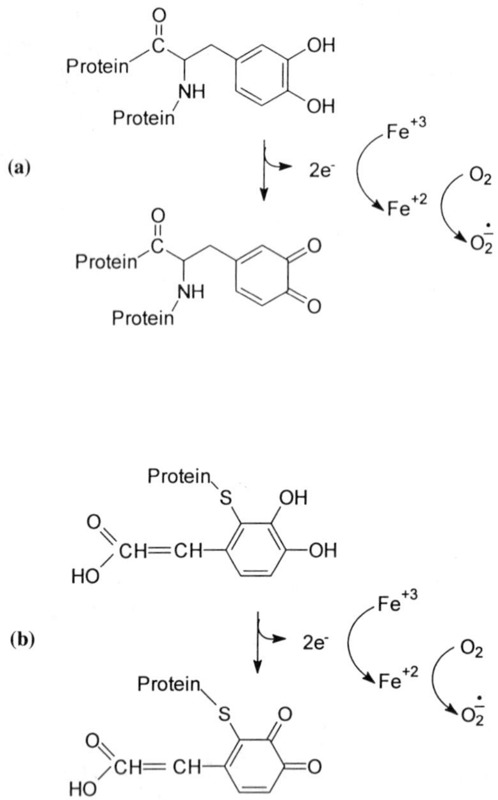

Figure 14.7 Formation of reactive oxygen species by iron-catalysed oxidations of protein catechols. (a) Formation by protein catechols formed by radical attack on tyrosine; (b) formation by catechols (e.g. caffeic acid) bound to protein. (Data from Dean *et al.*, 1993.)

Protein hydroperoxides may also be formed as products of radical attack on proteins and contribute to further protein damage (Dean *et al.*, 1993).

Oxidative deamination of free amino acids (i.e. Strecker degradation) to form corresponding aldehydes is also catalysed by PPO (Saijo and Takeo, 1970) (Figure 14.8). *Ortho*-dihydroxyphenols (e.g. chlorogenic acid) are more effective than triphenols in forming aldehydes (Motoda, 1979). Not only are essential amino acids lost in this process, but some of the aldehyde products may undergo Schiff base formation with proteins. The significance of these reactions in plant defence merits future study considering the double-edged effect of losing essential nutrients and simultaneously forming toxins.

The modifications of proteins produced by phenolic oxidation are

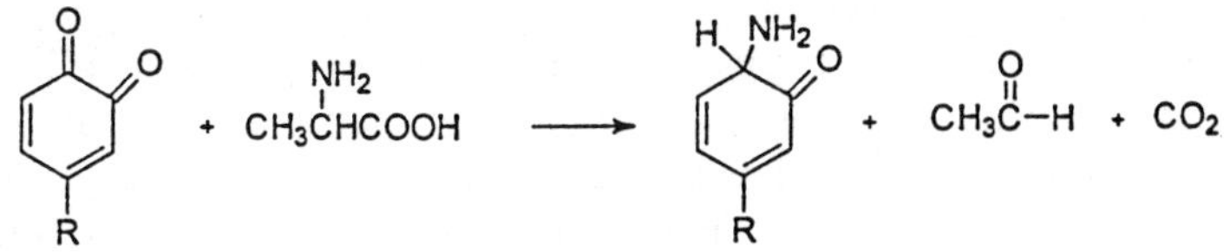

Figure 14.8 Oxidative deamination of amino acids by quinones. (Data from Motoda, 1979.)

numerous; these include phenolic–amino acid conjugates, protein carbonyl formation, protein hydroperoxides, thiol oxidations, hydroxylated aromatic amino acids (e.g. DOPA), and aldehydes bound as Schiff bases (Motoda, 1979; Barbeau and Kinsella, 1983; Richard *et al.*, 1991; Dean *et al.*, 1993). The contribution of these individual modifications to the loss in nutritive quality of dietary protein is unknown. Moreover, owing to co-oxidation processes, ascorbic acid is rapidly oxidized by *o*-quinones (Rouet-Mayer *et al.*, 1990; Nicolas *et al.*, 1994) and the dehydroascorbic acid product may exert still further antinutritive effects on proteins (section 14.3.5d).

Evidence for the role of PPO in plant defence is mounting. PPO and the lignan catechol, nordihydroguaiaretic acid, from creosote repelled leaf-feeding insects and was negatively correlated with growth of the grasshopper *Astroma quadrilobatum* (Rhoades, 1977). The reducing action of the catechol on *in vitro* protein digestion was enhanced by PPO (Rhoades, 1977). Growth rates of *H. zea* were also negatively correlated with PPO activity in tomato foliage and fruit (Felton *et al.*, 1989).

The antinutritive effects of PPO depend on the quantity and quality of dietary protein; high quality or increased quantity of protein spares the antinutritive effects (Felton *et al.*, 1992b). The antinutritive properties of PPO in noctuid herbivores result from reduced uptake of amino acids conjugated with phenolics (Felton, 1996). PPO and phenolic substrates are also found in the trichomes of solanaceous plants. Upon disruption of the trichomes, the enzyme and substrates mix to form a strong adhesive gum that entraps small insects on the leaf surface (Steffens *et al.*, 1994).

Foliar- or phloem-feeding arthropods cause systemic increases in PPO activity in numerous plant species including tomato, soybean, cotton, lemon, sugar beet, Chinese cabbage and red clover (Hori, 1973; Hori and Atalay, 1980; Raman *et al.*, 1984; Felton *et al.*, 1992a; Stout *et al.*, 1994; Bi and Felton, unpublished data). In tomato, the peptide systemin acts as signal compound for the induction of both PPO and a protease inhibitor by stimulating jasmonic acid biosynthesis. Transgenic plants constitutively expressing the prosystemin gene showed nearly 10-fold increases in PPO activity. PPO activity in these plants was several times higher than PPO activity induced by wounding in wild-type plants (Constabel *et al.*, 1995). This illustrates a valuable strategy for simultaneously enhancing the induction of at least two classes of plant defences by a single genetic modification. Expression of antisense prosystemin in tomato plants markedly reduced resistance towards *Manduca sexta* larvae, but it is impossible to determine how much of the larval growth increase was due to reduced PPO or protease inhibitor (Orozco-Cadenas *et al.*, 1993). The development of transgenic plants that specifically overexpress PPO would provide more definitive proof for a defensive role of PPO against herbivores (Hunt *et al.*, 1993). Transgenic tomato plants with three-fold overexpression of PPO, PPO-null plants, and non-transformed plants have been tested for suitability to *L. decemlineata* larvae. Mortality was lowest and weight gain highest on the PPO-null whereas mortality was highest and weight gain lowest on the PPO-overexpressing plants (Steffens *et al.*, 1994). Interestingly, fitness of the PPO-null plants was similar to control plants when tested in the absence of herbivory.

Surprisingly, PPO is extremely resistant to proteolytic enzymes (King and Flurkey, 1987; Marques *et al.*, 1994). Resistance to purified proteinases and to gut homogenates of *H. zea* has been tested using a partially purified tomato PPO preparation (ca. 2500 chlorogenic acid oxidase units/mg protein (Felton *et al.*, 1992a)). At least 90% of the PPO activity was retained when the PPO was treated for 60 min with purified proteinases (cathepsin B, pH 6; trypsin, pH 8.0; chymotrypsin, pH 8.0; pronase, pH 6.0; and pepsin, pH 5.8) or larval midgut proteinases at a proteinase concentration of 1 mg proteinase/mg tomato foliar protein (unpublished data). PPO retains substantial activity in the alkaline midgut lumen of *H. zea, Trichoplusia ni, M. sexta, S. exigua* and in the acidic lumen of the beetle *L. decemlineata* (Felton *et al.*, 1989, 1992a; unpublished data). Ongoing oxidase activity in the midgut lumen would allow for the alkylation of proteins by quinones in the midgut (Felton *et al.*, 1989).

PPO is active across a broad pH range (5–10 in many species), but the alkylating properties of the products are strongly influenced by pH. Under acidic conditions, most protein nucleophilic side chains are protonated and not susceptible to attack by electrophilic quinones (Barbeau and Kinsella, 1983). The fate of the phenolic substrate chlorogenic acid in tomato foliage can be quite different in herbivores with varying midgut pHs. In *L. decemlineata*, which possesses a mildly acidic midgut (pH 6.0–6.5), very little protein is alkylated by chlorogenic acid. PPO is very active in the midgut of *L. decemlineata*, but most of the oxidized chlorogenic acid appears to form phenolic polymers rather than phenolic–protein conjugates. However, in lepidopterans with alkaline midguts (*H. zea, S. exigua, M. sexta, T. ni*; pH 8.5 to ~10.0), a significant amount of ingested chlorogenic acid is covalently bound to protein (Felton *et al.*, 1989, 1992a; Felton and Duffey, 1991b; unpublished data).

The midgut redox environment of insect herbivores is quite variable among and within species (Appel and Maines, 1995; Johnson and Felton, 1996). Redox conditions are influenced by diet and lumen pH (Appel and Maines, 1995; Johnson and Felton, 1996). However, specific plant factors affecting the redox potential of the midgut are unknown. The quinones formed from PPO are easily reduced to the hydroquinone by compounds of lower redox potential (ascorbate, thiols, certain flavonoids). Thus foliage with high levels of these reductants may inhibit formation of quinone–protein adducts. However, when quinones are reduced by ascorbate, dehydroascorbic acid, an undesirable product, is formed (see below). The co-oxidation processes with dihydroxyphenols and flavonoids (Figure 14.9) may yield significant amounts of cross-linked proteins and/or mixed phenolic dimers (Cheynier *et al.*, 1988; 1989; Rouet-Mayer *et al.*, 1990; Macheix *et al.*, 1991; Nicolas *et al.*, 1994). The coupled reactions allow the indirect oxidation of phenolic compounds that are not normally substrates for PPO (Cheynier *et al.*, 1988). Considering that plant tissues contain a complex mixture of phenolics

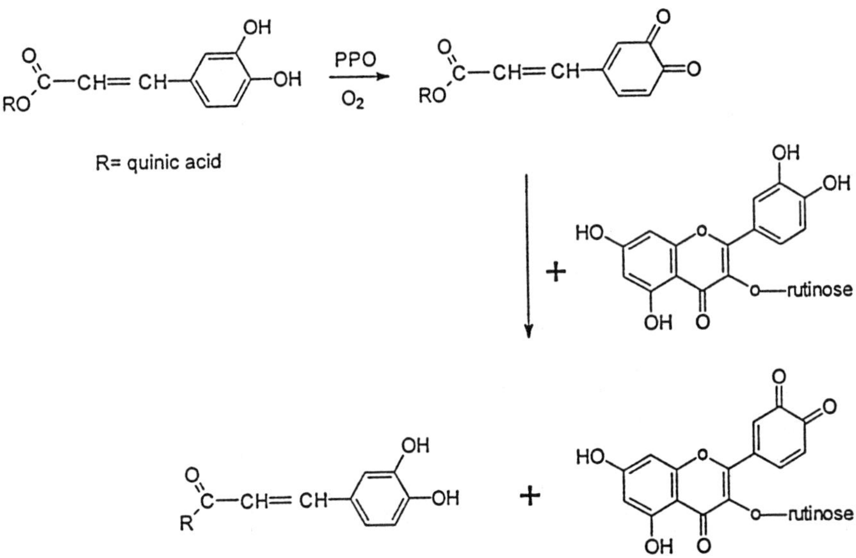

Figure 14.9 Coupled oxidation of chlorogenic acid (cinnamic acid derivative) and rutin (flavonoid). PPO = polyphenol oxidase. (Data from Cheynier *et al.*, 1989.)

and flavonoids, it is most difficult to predict the final oxidative products formed. We currently have very limited knowledge of the influence of midgut redox potential on the toxic and antinutritive properties of PPO.

Finally, gut surfactants such as lysolecithin may inhibit PPO activity in the midgut (Felton and Duffey, 1991b). Certain surfactants can alter the normal tertiary structure of enzymes, affecting the molecular interactions of the enzyme–substrate complex.

(b) Peroxidases (POD)

POD (guaiacol-type, EC 1.11.1.7) oxidizes an array of substrates (phenolics, alcohols, aromatic amino acids, thiols, ascorbic acid) using H_2O_2 as the oxidizing agent. POD in plants includes multiple isoenzymes coded by separate genes on different chromosomes (Robinson, 1991; Marañon and van Huystee, 1994). The isoenzymes vary by tissue and cellular location and may be differentially induced by wounding and other environmental stresses.

The POD-mediated oxidation products of dihydroxyphenolics are similar to those produced by PPO, but many additional oxidized products are possible. The formation of quinone methides from

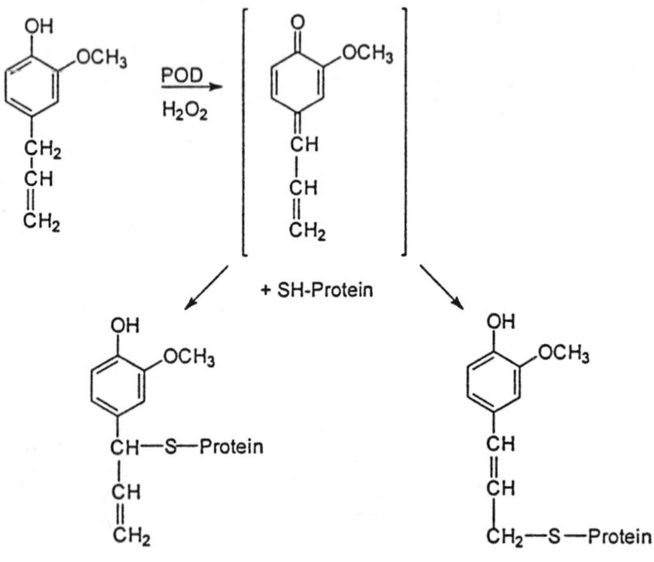

Figure 14.10 Formation of a quinone methide from eugenol by peroxidase (POD) and subsequent alkylation of a protein thiol. (Data from Thompson *et al.*, 1990.)

compounds such as eugenol (Figure 14.10) are potent alkylating agents of protein. Another important reaction contributing to nutritional deficiency is the deamination of protein lysyl ε-amino groups by POD, catechols and H_2O_2 (Stahmann and Spencer, 1977; Figure 14.11). The *o*-quinones products are also available to alkylate protein or to oxidize ascorbic acid. Furthermore, linoleic and arachidonic acids are oxidized to lipid hydroperoxides by POD (Garner, 1984), but the relative significance of POD-mediated lipid peroxidation versus the more widely recognized lipoxygenase pathway is unknown. Although POD may catalyse a multitude of substrates *in vitro*, very few substrates have been shown to be important *in situ*.

POD is inducible by herbivory and its oxidation products are toxic to several insects (Felton and Duffey, 1991a; Dowd, 1994; Dowd and Norton, 1995). More conclusive evidence for its defensive role has been obtained using transgenic tobacco and tomato plants with a chimeric tobacco POD gene, which had increased resistance to *H. zea* (Lagrimini, 1991; Dowd and Lagrimini, 1995).

A key constraint on POD activity in the midgut is the quantity of H_2O_2. Insect H_2O_2-degrading enzymes, such as catalase, in the midgut can limit the POD activity (Felton and Duffey, 1991a), whereas ascorbate peroxidase reduces peroxides at the expense of ascorbate oxidation (Felton, 1995; Felton and Summers, 1995; Mathews and Felton, unpublished data). These enzymes which depend on their relative affinities for H_2O_2, may mitigate the activity of POD by competing for H_2O_2. Again, the midgut pH and redox conditions probably influence these reactions.

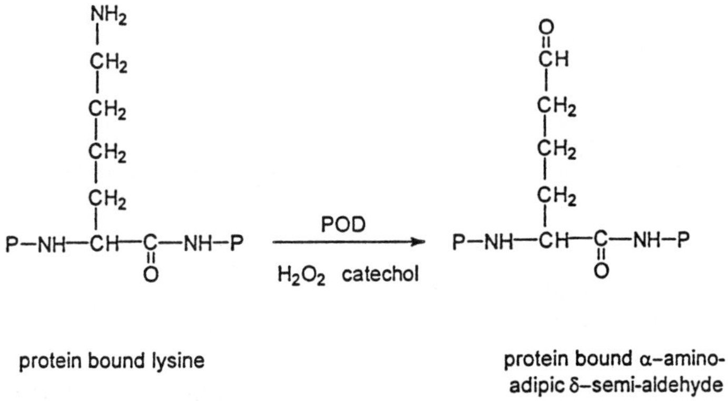

Figure 14.11 Oxidative deamination of protein bound lysine by peroxidase (POD) and catechol. (Data from Stahmann and Spencer, 1977.)

(c) Lipoxygenase (LOX)

LOX (EC 1.13.11.12) occurs widely and catalyses the oxidation of polyunsaturated fatty acids containing a *cis,cis*-1,4-pentadiene moiety to conjugated dienoic hydroperoxides (Gardner, 1979, 1991; Figure 14.12). Multiple LOX isoenzymes exist that vary in pH optima, substrate specificity, product formation and tissue or cellular location. The hydroperoxide products are metabolized by lyases, dehydrases, isomerases, peroxygenases and epoxygenases resulting in epoxide, alcohol, aldehyde and acidic products (Gardner, 1991), one of which is jasmonic acid, an important signal eliciting multiple defence genes.

LOX exerts multiple antinutritive effects. Direct removal of fatty acid substrates such as linoleic and linolenic acid contributes to nutritional deficiencies, as their hydroperoxide products are nutritionally inert. The primary and secondary lipid oxidation products – (C_6 aldehydes, hydroxynonenal, malondialdehyde – may then complex with proteins (Gardner, 1979; Uchida and Stadtman, 1992; Figure 14.13). Linoleic oxidation products adversely affect protein quality and harm growth in *M. sexta* and *H. zea* (Shukle and Murdoch, 1983; Bi *et al.*, 1994; Felton *et al.*, 1994b). Soybean LOX presented in artificial diets to rice brown planthopper, *Nilaparvata lugens*, causes approximately 80% mortality at 0.1% w/v (Powell *et al.*, 1993, 1995).

Type-2 LOX isoenzymes (pH optima 6–7) are very effective at catalysing the co-oxidation of carotenoids and α-tocopherol (Gordon and Barimalaa, 1989). In wounded tissues, loss of total carotenoids is correlated with increased LOX activity (Hildebrand *et al.*, 1988; Bi and Felton, 1995); this loss may also exacerbate the effects of other oxidatively mediated plant defences (PPO, POD). The antioxidants are

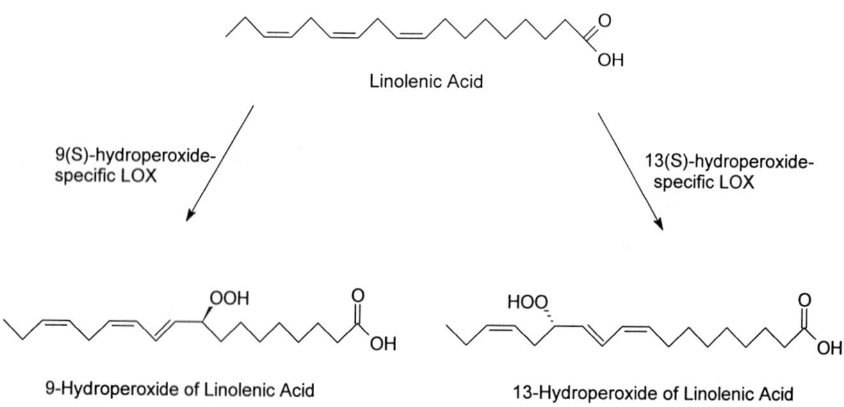

Figure 14.12 Linolenic acid hydroperoxide formation by lipoxygenase (LOX). (Data from Gardner, 1991.)

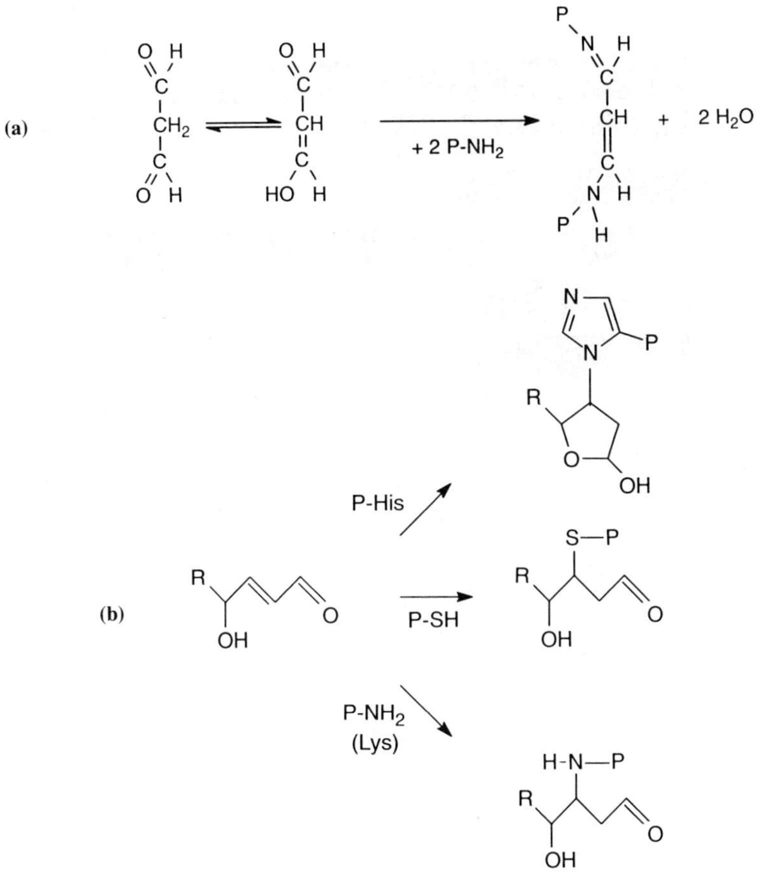

Figure 14.13 Reaction of lipid peroxidation products with protein. (a) Reaction of malondialdehyde with protein amino groups; (b) reaction of α,ß-unsaturated aldehyde (e.g. hydroxynonenal) with histidine, cysteine and lysine. (Data from Gardner, 1979 and Uchida and Stadtman, 1992.)

also essential vitamins, but the oxidative products of carotenoids or vitamin E are of unknown nutritional value.

LOX is inducible in many plant species by arthropod feeding, but inducibility varies among cultivars and among specific isoenzymes (Hildebrand *et al.*, 1986; Felton *et al.*, 1994a,b). Regardless of the herbivore species – foliar feeding insects of varying feeding patterns (older vs. younger leaf, leaf margins vs. near midrib) or a phloem-feeding insect – type-2 LOX (pH optimum 7.0) is induced in soybeans within 48 h following feeding (Figure 14.14; Thompson and Felton, unpublished data). By contrast, feeding by *L. decemlineata* or *H. zea* on

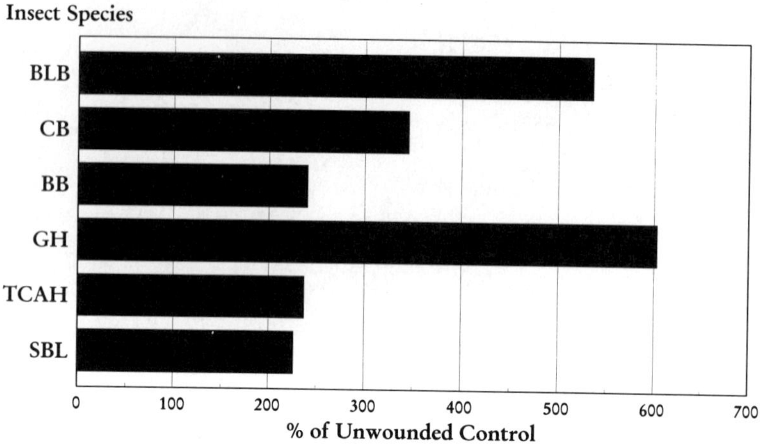

Figure 14.14 Induction of type-2 lipoxygenases by insect herbivores. BLB, bean leaf beetle (*Ceratoma trifurcata*); CB, cucumber beetle (*Diabrotica undecimpunctata howardi*); BB, blister beetle (*Epicauta vittata*); GH, grasshopper (*Melanoplus differentialis*); TCAH, threecornered alfalfa hopper (*Spissistilus festinus*); SBL, soybean looper (*Pseudoplusia includens*). Lipoxygenase assayed in foliage from V-6 stage soybean plants (cv. Hutcheson) following 72 h of herbivory.

tomato foliage has minimal effect on type-2 LOX, but tomato russet mites strongly induce activity within 72 h of wounding (Stout *et al.*, Felton and Duffey, unpublished data). Simultaneously damage by tomato russet mites and *L. decemlineata* or mites and *H. zea* on tomato foliage results in suppression of LOX induction by mites (Figure 14.15). These results show that multiple herbivore species may interfere with the induction of resistance factors.

LOX also mediates resistance indirectly. Jasmonic acid and other LOX products are potent inducers of multiple plant defences against insects (Enyedi *et al.*, 1992; Hildebrand *et al.*, 1993; Constabel *et al.*, 1995). Fatty acid substrates of LOX applied to *Phaseolus vulgaris* foliage reduced the fecundity of spider mites relative to the untreated controls (Kasu *et al.*, 1994), whereas foliar applications of 100 μM jasmonic acid to cotton and soybean plants elicits resistance to *H. zea* (Felton *et al.*, unpublished observations).

Tobacco plants have been transformed to express soybean type-2 LOX at high levels, but leaves from regenerated transformed plants showed no increase in LOX activity, presumably due to a high background of LOX activity in the control leaves (Deng *et al.*, 1992; Hildebrand, 1992). None the less, C_6-aldehyde formation in the transgenic line was substantially greater, but relative insect resistance of the transgenic versus the control plants was not reported (Deng *et al.*, 1992). A much greater

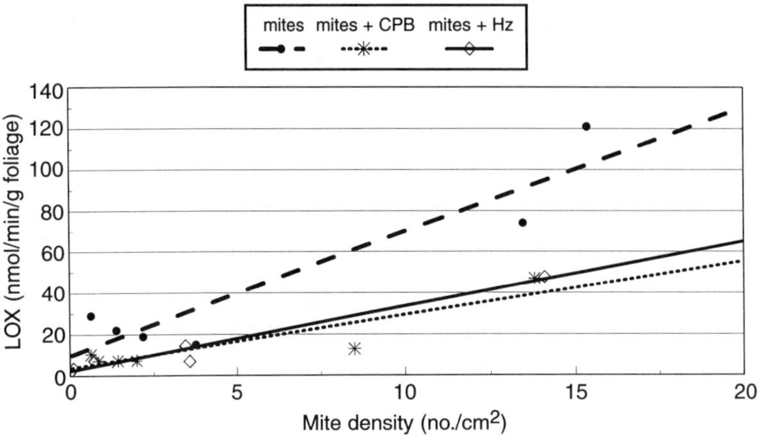

Figure 14.15 Induction of foliar lipoxygenase (LOX) by tomato russet mites, colorado potato beetle and corn earworm herbivory in tomato. CPB, Colorado potato beetle; Hz, corn earworm *Helicoverpa zea*. LOX measurements made at 72 h after herbivory by the beetle or earworm. Mite population was endemic on all plants tested. Preflowering stage tomato plants (cv. Castlemart) were used.

increase in LOX activity was obtained by treating tobacco plants with methyl jasmonate (Hildebrand, 1992). This enhancement of signal pathways for defensive genes may be a parsimonious approach for enhancing expression of multiple resistant genes (Constabel *et al.*, 1995).

Unlike many of the proteins discussed in this chapter, LOX is rapidly degraded and inactivated by larval midgut proteinases (Felton *et al.*, 1994b; unpublished data). Consequently, antinutritive effects are exerted prior to the ingestive process. Lipid hydroperoxides and C_6-aldehydes are increased in plant tissues damaged by herbivore feeding (Kasu *et al.*, 1995), so that herbivores probably ingest increased amounts of lipid peroxidation products. Furthermore, foliar protein extracted from previously damaged soybean plants is of poor quality and shows evidence of hydroperoxide damage (Bi *et al.*, 1994).

(d) Ascorbate oxidase (AOX)

AOX (EC 1.10.3.3) is ubiquitous among plant species and catalyses the reaction:

$$2\ \text{L-ascorbic acid} + O_2 + 2H^+ \rightarrow 2\ \text{L-dehydroascorbic acid} + 2\ H_2O$$

Ascorbate oxidation may impose nutritional stress on an insect due to the loss of the vitamin, although insects possess dehydroascorbate reductase activity that regenerates ascorbate at the expense of glutathione

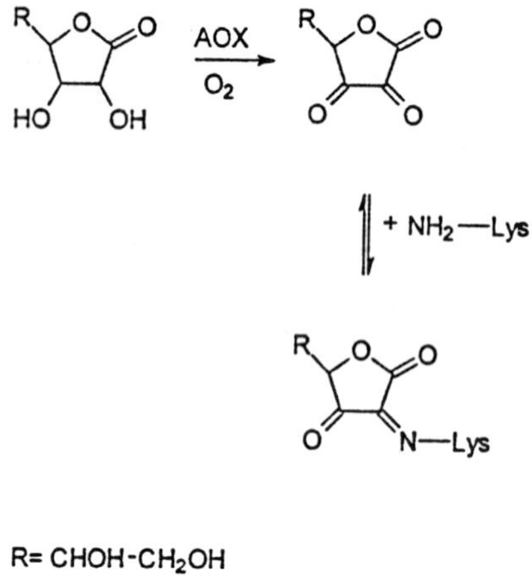

Figure 14.16 Oxidation of ascorbic acid by ascorbate oxidase (AOX) and subsequent binding of dehydroascorbate with lysine.

(Summers and Felton, 1993). Dehydroascorbic acid is unstable at the alkaline pH found in the midguts of lepidopterans and may undergo rapid, irreversible degradation to diketogulonic acid and other products that lack nutritive value. Furthermore, dehydroascorbic acid may bind to proteins (ε-NH_2 of lysine; Figure 14.16) and reduce protein digestibility (Felton and Summers, 1993). Oxygen radicals may be formed during these reactions and contribute to protein polymerization and/or fragmentation (Felton and Summers, 1993). AOX in cotton, soybean and tomato is systemically induced by insect feeding and results in the decline of ascorbate levels in wounded tissues (Felton *et al.*, 1994a; Bi and Felton, 1995; unpublished data). In summary, the dual properties of AOX, loss of ascorbate and formation of a protein-binding electrophile, indicate a possible role in antinutritive-based resistance.

(e) Copper-amine oxidase, polyamine oxidase or diamine oxidase (DAO)

DAO (ED 1.4.3.6) catalyses the oxidation of assorted diamines, polyamines and histamine that are converted into the corresponding amino aldehydes, ammonia and H_2O_2 (Figure 14.17). The aminoaldehydes formed from common plant amines such as putrescine, cadaverine and spermidine may spontaneously cyclize to Δ^1-pyrroline, Δ^1-piperidine and 1,5-diazabicyclononane, respectively (Medda *et al.*, 1995) or form Schiff bases with free amino groups of proteins (Figure 14.17). The physiological function of DAO is unclear, although H_2O_2 production for POD reactions in the cell wall is indicated. DAO is induced by *H. zea* feeding on foliar and reproductive structures in cotton and soybean (Bi

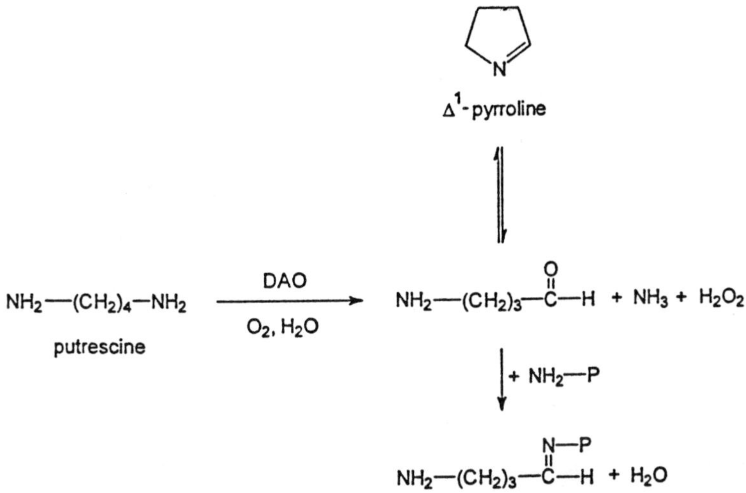

Figure 14.17 Oxidation of putrescine by diamine oxidase (DAO) and subsequent Schiff base formation of product with protein amino group. (Data from Medda *et al.*, 1995.)

and Felton, 1995; unpublished data), but the impact of DAO on herbivores is unknown.

(f) Transglutaminase (TGA)

TGA (EC 2.3.2.13) catalyses post-translational modifications of glutamine residues of proteins producing cross-links either by covalent incorporation of small molecular weight amines (e.g. putrescine, spermidine, cadaverine, etc.) or by cross-linking of γ-glutamyl ε-lysine peptide bridges (Figure 14.18). TGA is present in most plant tissues where it polymerizes many proteins including chlorophyll-binding proteins and RUBISCO (Icekson and Apelbaum, 1987; Serafini-Fracassini *et al.*, 1988, 1995; Margosiak *et al.*, 1990; Del Duca *et al.*, 1994). Plant TGA cross-links globular proteins more efficiently than animal TGA (Siepaio and Meunier, 1995a,b), and may function in enzyme regulation, cell extensibility, defence, progammed cell death and wound healing. TGA may block the access of trypsin to lysyl NH_2 groups and, thus, may reduce protein digestibility and lysine availability. Its role in plant defence awaits evaluation.

(g) Myrosinase (MYR)

Glucosinolates and MYR (= thioglycosidase, EC 3.2.3.1) are restricted to the order Capparales including crucifers (Louda and Mole, 1991). The

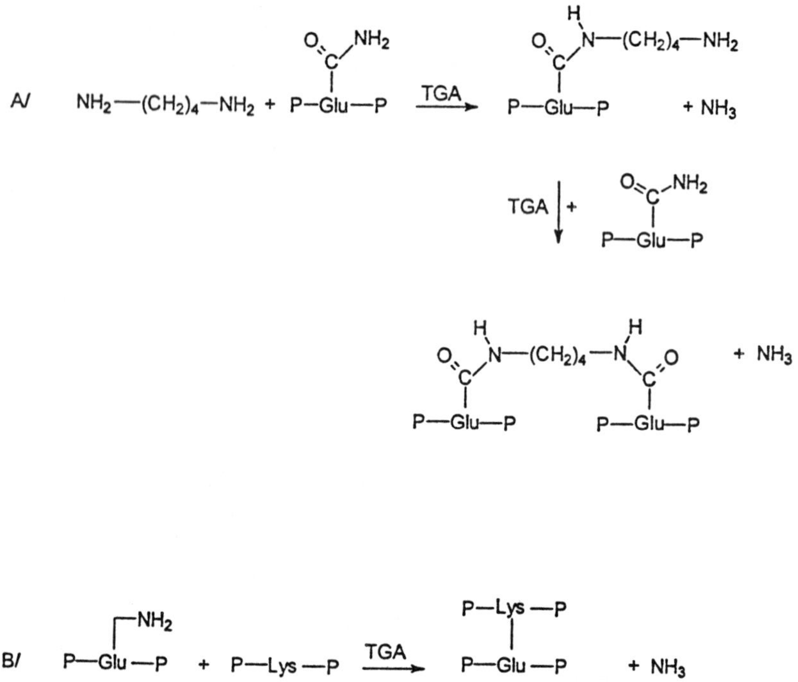

Figure 14.18 Modifications of glutamine residues of proteins by transglutaminase (TGA). (a) Covalent incorporation of putrescine with protein; (b) cross-linking of glutamyl-lysyl peptide bridges. (From Serafini-Fracassini *et al.*, 1995.)

glucosinolates are stored as glucosides (probably in vacuoles) and separated from MYR in the cytosol (Luthy and Matile, 1984). Upon tissue disruption, MYR cleaves the glucosinolate to glucose and an unstable aglycone which decomposes to form sulphate and several products including isothiocyanates, thiocyanates, nitriles, cyanoepithioalkanes and oxazolidine-2-thiones (Chew, 1988). The isothiocyanates may bind to protein.

Glucosinolates may serve as ovipositional cues and/or feeding stimulants in crucifer-adapted specialists (Chew, 1988), but field data on glucosinolates are consistent with a defensive function (Louda and Mole, 1991). Gut pH was suggested as an important adaptation for feeding on glucosinolate-containing plants (Louda and Mole, 1991) as several adapted species such as *Plutella xylostella* and *Pieris rapae* have a comparatively neutral midgut pH (pH ~7). However, other insects such as *T. ni* successfully utilize crucifers despite having an alkaline gut pH (~10.0).

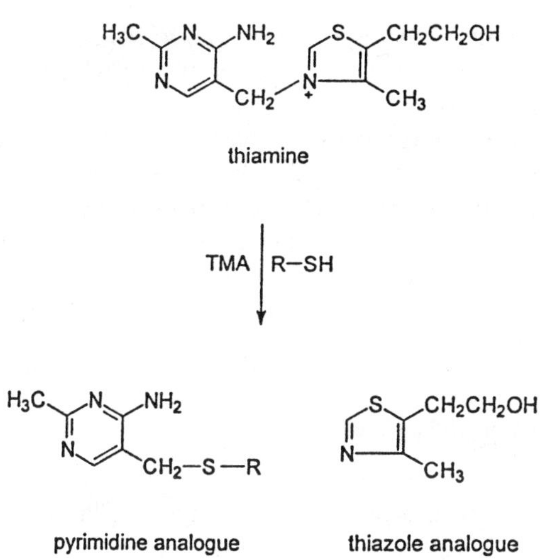

Figure 14.19 Decomposition of thiamine by thiaminase (TMA) and formation of pyrimidine and thiazole analogues with thiols (R-SH). (Data from Evans, 1976.)

(h) Thiaminase (TMA)

TMA (EC 2.5.1.2) catalyses the decomposition of thiamin by producing pyrimidine and thiazole analogues (Figure 14.19), but requires a co-substrate amine or a sulphydryl (e.g. cysteine). Consequently, an additional nutritional amino acid loss is exerted (Evans, 1976). TMA is probably restricted to pteridophytes where it can exert severe thiamin deficiency in mammals consuming bracken ferns (Evans, 1976). TMA may induce resistance to *Spodoptera eridania* and *T. ni* larvae (Soo-Hoo and Fraenkel, 1964; Hendrix, 1975; Jones, 1983), but more conclusive evidence for its antinutritional role in insects is needed.

14.3.4 Antinutrient proteins: lectins

Lectins form a large, diverse group of proteins, identified by a common property of binding to carbohydrate residues, either as free sugars or as part of oligo- or polysaccharides. They are distinguished from enzymes by having no action on the carbohydrate other than binding to it. Most lectin molecules contain multiple binding sites, and thus can cross-link oligo- or polysaccharides.

Plants were the first known source of lectins, and accumulate lectins in many storage tissues; seeds are an abundant source, but other storage tissues such as bulbs, or bark, also contain lectins. They can accumulate to 1% or more of total protein, but their role in plants has been a source of much speculation. In legumes, they are involved in the interaction between the plant and the symbiotic nitrogen-fixing bacterium, *Rhizogenes* spp. (Diaz *et al.*, 1989). Roles as storage proteins, or cross-linking agents for storage proteins have also been proposed, as well as roles in defence against pests and pathogens. The binding specificities of lectins vary greatly (Pusztai, 1991), and although lectins from related plant species often show similar binding specificities, similarity of carbohydrate binding does not necessarily imply that lectins are related in sequence. A number of families of lectins, related in sequence, have been identified in plants.

(a) Lectins active against Coleoptera

The first experiment showing insecticidal activity in a lectin was made with the lectin from common bean, *P. vulgaris* (PHA), which was toxic to the cowpea bruchid, *C. maculatus* (Janzen *et al.*, 1976). Although the PHA preparation used was probably contaminated with α-amylase inhibitor, which is highly active as an insecticidal protein, bioassays using purified PHA lectins also have an insecticidal effect (Gatehouse *et al.*, 1984). Certain lines of *P. vulgaris* contain a related insecticidal protein, termed arcelin, which largely replaces the normal seed storage protein. Arcelin has a high sequence homology with PHA, but does not possess agglutination activity. Arcelin is only insecticidal at high (>5%) dietary concentrations and prevents insect development by being resistant to digestion (Minney *et al.*, 1990). It is worthy of comment that *P. vulgaris* contains three different types of insecticidal protein (α-amylase inhibitor, lectin and storage protein resistant to digestion), which are homologous in sequence, and are products of the same gene family; this illustrates the plasticity of the defence response in plants.

In a screening experiment five lectins, at dietary levels of 0.2% and 1.0% (w/w), caused significant delays in larval *C. maculatus* development (Murdoch *et al.*, 1990). The lectins were specific for either *N*-acetylgalactosamine/galactose (GalNAc) or *N*-acetylglucosamine (GlcNAc). The winged bean (*Psophocarpus tetragonolobus*) lectin (GlcNAc-specific) was also shown to be toxic to *C. maculatus* (Gatehouse *et al.*, 1989), as were rice and stinging nettle lectins (UDA), also specific for GlcNAc (Huesing *et al.*, 1991). All the lectins effective against the southern corn rootworm (*Diabrotica undecimpunctata*) were specific for either GalNAc or GlcNAc (Czapla and Lang, 1990).

The specificity of a lectin is a poor indicator of its potential insecticidal properties, and it is thus still necessary to test each lectin against a target pest on a case by case basis. For example, the toxic effects of mannose/glucose specific lectins from different sources towards *C. maculatus* differ considerably; the lectin from garden pea has little or no toxic effect, whereas lectins from *Dioclea* spp. are significantly toxic, as is the mannose-specific lectin from snowdrop (GNA) (Gatehouse, unpublished results).

(b) Lectins active against Lepidoptera

Comparatively few lectins have been tested by bioassay in artificial diet and found to be toxic to lepidopterans. The lectins from castor bean (*Ricinus communis*), Camel's foot tree (*Bauhinia purpurea*) and wheatgerm (WGA), specific for GalNAc, GalNAc and GlcNAc respectively, produced 100% mortality after 7 days when administered to European cornborer, *O. nubilalis*, larvae as a 2% topical application (Czapla and Lang, 1990). WGA and *R. communis* agglutin also inhibited larval weight gain by >50% at 0.1% topical applications. Soybean lectin increased larval weights of *O. nubilalis* by >25% compared with control larvae (Czapla and Lang, 1990), whereas its addition at 1% protein in the diet was detrimental to larval growth of *M. sexta* (Shukle and Murdock, 1983).

(c) Lectins active against Homoptera

The insecticidal activity of lectins against homopteran pests is now receiving much attention. The two most effective lectins tested against the rice brown planthopper (*Nilaparvata lugens*) were GNA and WGA, each of which gave approximately 80% corrected mortality at a concentration of 0.1% (w/v) in the diet (Powell *et al.*, 1993). The LC_{50} value for GNA against brown planthopper was 0.02%, or approximately 6 mM (Powell *et al.*, 1994). GNA was also toxic to another sucking pest of rice, the rice green leafhopper, *Nephotettix cinciteps*.

Lectins specific for glucose/mannose, GlcNAc or GalNAc were tested against the potato leafhopper (*Empoasca fabae* at dietary levels of 0.2–1.5% (w/w). Lectins from jackfruit, pea, lentil and horse gram and also PHA and WGA caused significant reductions in insect survival. *Canavalia ensiformis* (Con A) lectin is a potent toxin to the pea aphid *Acyrthosiphon pisum*, having a significant effect on both survival and growth, whereas WGA was relatively non-toxic (Rabhé and Febvay, 1993). GNA is also inhibitory to aphid development, causing a significant reduction in growth and female fecundity of aphid *Myzus persicae*.

(d) Lectins active against Diptera

Recently, the effects of plant lectins on larvae of the blowfly, *Lucilia cuprina*, have been assayed, in an attempt to identify possible control strategies for this pest (Eisemann *et al.*, 1994, Chapter 12). The larvae feed on tissue and tissue fluids of susceptible sheep. Both WGA and Con A caused concentration-dependent inhibition of larval growth and significant mortality. WGA was the most potent, resulting in 50% inhibition of larval growth at 2 mM and 100% mortality at 25 mM concentrations. Deleterious effects caused by the lectins could be prevented by the appropriate sugars.

(e) Mechanism of lectin toxicity

In higher animals, lectin toxicity involves:

1. resistance of ingested lectin protein to proteolysis;
2. binding of the lectin to glycoproteins on the surfaces of cells in the gut wall;
3. lectin-mediated interactions between the gut microflora and cells in the gut wall;
4. mitogenic effects of lectins on gut cells;
5. passage of lectins across the gut wall, possibly by endocytosis, and migration of intact lectin to the circulatory system, causing systemic effects on other organs.

Of these 2 and 3 are mediated by the carbohydrate-binding capacity of the lectin, but the others are not; there is thus no simple correlation between the specificity of a lectin and its toxicity.

Binding of lectins to cells in the gut wall is a feature of their effects on higher animals, and similarly in insects. PHA binds to midgut epithelial cells of *C. maculatus*, and is toxic to this species (Gatehouse *et al.*, 1984), whereas PHA did not bind to midgut epithelial cells of *Acanthoscelides obtectus* which are able to tolerate moderately high levels of PHA (Gatehouse *et al.*, 1989). Lectins may also bind to the peritrophic matrix which is composed largely of GlcNAc in the form of chitin (Chapter 4). Many of the insecticidal lectins are GlcNAc specific (Czapla and Land, 1990; Huesing *et al.*, 1991), but apart from the work of Eisemann and co-workers (summarized in Chapter 2), there is little direct evidence to correlate lectin binding to the peritrophic matrix with toxicity.

A feeding deterrent effect caused by lectins has been observed. When adults of the rice brown planthopper were fed a diet containing GNA (0.1% w/v) for 24 h, the volume of honeydew produced was reduced to less than 10% of that collected from control insects (Powell *et al.*, 1994, 1995). Since the volume of honeydew excreted is roughly proportional to the volume of fluid ingested, it appears that the lectin is a feeding deterrent. GNA almost completely abolished the normal feeding behaviour of this insect in a 4 h exposure period. Similarly, blowfly larvae, when offered a free choice between pads containing bovine serum albumin (5 mg/ml) in the presence or absence of 50 mM WGA were nine times more likely to choose the non-lectin-treated pad (Eisemann, personal communication).

The data reported suggest that there may be several mechanisms involved in the toxicity of lectins to insects. Any causal relationship between these mechanisms, and their relative importance, has not yet been determined.

14.4 INSECT RESISTANT PLANTS

Production of insect-resistant transgenic plants became possible almost as soon as the technology became available. Attention has focused on genes encoding the insecticidal toxins from *Bacillus thuringiensis*, which have had an established track record of use in agriculture as sprays and dusts. However, an alternative strategy is to exploit defensive proteins found in plants. High expression of foreign proteins are relatively easy to attain, since the gene transfer is from one plant species to another, and the transcription and translation systems are similar. The drawback is that insect predators of plants have already been exposed to the defensive proteins being introduced and may already have mechanisms to counter their effects. Fortunately, the diversity of defensive proteins found in plants, and the differing specificities shown by proteins of the same type from different plant sources, means that transfer of defensive proteins from one plant species to another can give effective levels of resistance, particularly against pests that are essentially monophagous.

14.4.1 Insect resistant transgenic plants expressing serine proteinase inhibitors

The first example of a plant gene conferring resistance to insects in a different plant species was the transfer of a chimaeric cowpea trypsin inhibitor (CpTI) gene to tobacco (Hilder *et al.*, 1987). Expression of the gene was driven by the constitutive CaMV 35S promoter, and transfer was carried out using an *Agrobacterium tumefaciens* vector system. The transformed plants expressed CpTI in the leaves up to nearly 1% of total soluble protein, and expression was confirmed by a direct *in vitro* assay of leaf extracts for bovine trypsin inhibition. Bioassay of clones of selected transformants was carried out using first instar larvae of the tobacco budworm (*H. virescens*); this insect was chosen as it is a serious pest of tobacco, cotton and other plants. CpTI expressing plants showed only minor damage compared to the control plants, which in some instances were very severely damaged. Although the larvae begin to feed on the CpTI-expressing plants, causing some limited damage, some die and most fail to develop as they would on control plants. The protection afforded by CpTI has subsequently been demonstrated for other lepidopteran pests including *H. zea*, *Spodoptera littoralis* and *M. sexta*. Statistical analysis of the bioassay in terms of plant damage by leaf area, and insect survival and biomass confirmed the highly significant protection afforded by CpTI. Expression of CpTI in tobacco afforded significant protection in the field against *H. zea* (Hoffman *et al.*, 1991), results from these trials closely resembling those obtained in growth chambers.

The CpTI gene has been engineered into many different crops, including potato, oil seed rape, rice and soft fruits such as strawberry. Transgenic strawberry plants expressing CpTI are highly resistant to the vine weevil (Graham *et al.*, 1995), and CpTI-expressing transgenic rice show enhanced levels of resistance to rice stem borers (D. Xu, personal communication). Despite CpTI being an effective antimetabolite against a wide spectrum of insect pests, mammalian feeding trials incorporating the purified protein at levels of 10% of the total protein have failed to demonstrate acute toxicity (Pusztai *et al.*, 1992); this is an important consideration for the use of this gene in crop plants, and the result reflects differences in the organization of the insect and mammalian digestive systems.

Other proteinase inhibitor-encoding genes have also been used to produce plants with enhanced insect resistance. The tomato inhibitor II (TI-II) gene (similar to the potato PI-II inhibitor), when constitutively expressed in tobacco, provides increased levels of protection against *M. sexta* (Johnson *et al.*, 1989). The decrease in larval weight was roughly proportional to the amount of inhibitor expressed, with significant effects on larval development observed at 100 μg TI-II/g leaf tissue. However, tobacco plants expressing tomato inhibitor I at levels of 130 μg/g had no deleterious effects on larval development, illustrating the specificity of the inhibitors–insect interaction. Growth of the noctuid lepidopteran *Chrysodeixis eriosoma* (green looper) was slowed significantly when fed leaf tissue from transgenic tobacco plants expressing the PI-II gene, an inhibitor most active against chymotrypsin-like proteases (McManus *et al.*, 1994).

Expression of other plant serine protease inhibitors in transgenic plants has not resulted in enhanced resistance to insects, and this failure has greatly hindered the wider acceptance of this method of plant protection. Concern has also been expressed about the possible effects of protease inhibitors on beneficial insects. Soybean Kunitz trypsin inhibitor causes mortality when fed to adult bees in sugar syrup (Malone *et al.*, 1995). These problems could be avoided by the use of tissue-specific promoters in transgenic plants.

14.4.2 Insect resistant transgenic plants expressing cysteine proteinase inhibitors

Genes encoding cysteine proteinase inhibitors have been suggested for use in transgenic plants for control of coleopteran insects. These inhibitors are effective *in vitro* and when incorporated into artificial diets. The gene encoding oryzacystatin has been expressed constitutively in transgenic poplar trees, conferring resistance towards the coleopteran *Chrysomela tremulae* (Leplé *et al.*, 1995).

14.4.3 Insect resistant transgenic plants expressing α-amylase inhibitors

The α-amylase inhibitor of *P. vulgaris* is encoded by a gene designated LLP (Moreno and Chrispeels, 1989). A chimaeric gene, consisting of the coding sequence of LLP and the 5′ and 3′ flanking sequences of the gene that encodes a lectin subunit, PHA-2, has been constructed and expressed in tobacco (Altabella and Chrispeels, 1990). The promoter in this construct is seed-specific. Seeds from these transgenic plants expressed the bean α-amylase inhibitor, and contained inhibitory activity against both porcine pancreatic α-amylase and the α-amylase present in the midgut of meal worm, *T. molitor*. Although suitable insect bioassays could not be carried out with tobacco seeds, the inhibitory activity of the transgene product against insect α-amylase led to the suggestion that the introduction of the bean amylase inhibitor gene into other leguminous plants may be a strategy to protect the seeds from seed-eating larvae of Coleoptera. This suggestion was verified in a series of experiments where transgenic garden peas and other legumes were produced, using a construct similar to that described above. Transformation was by an improved *Agrobacterium tumefaciens* vector system. Seeds of these plants contained about 0.3% total protein as bean amylase inhibitor, and were highly resistant to attack by *Bruchus pisorum* and *C. maculatus*. Unfortunately, this inhibitor is unlikely to be useful against lepidopteran pests as it is inactive at the alkaline pH of the lepidopteran gut.

14.4.4 Insect resistant transgenic plants expressing lectins

A gene encoding the pea lectin (P-Lec) has been expressed in transgenic tobacco plants using the constitutive CaMV 35S promoter. Plants expressing the pea lectin at up to 1.0% of total protein were then tested in bioassay for resistance to *H. virescens*; larval biomass and leaf damage were significantly reduced (Boulter *et al.*, 1990). Transgenic tobacco plants expressing both the cowpea trypsin inhibitor (CpTI) and P-Lec were obtained by crossing plants derived from the two primary transformed lines, and screened for resistance to *H. virescens*. The insecticidal effects of the two genes were additive, with insect biomass on plants expressing both transgenes reduced by nearly 90% compared to control plants, whereas plants expressing either CpTI or P-Lec alone only reduced biomass by about 50%. Leaf damage was also most reduced on the double-expressing plants. Thus the products of lectin genes in transgenic plants can enhance resistance to insect attack, and additive protective effects are obtained from different plant-derived insect-resistance genes.

A gene encoding the snowdrop lectin (GNA) has also been engineered into transgenic plants (van Damme *et al.*, 1987). Initial experiments placed the GNA coding sequence under control of the CaMV 35S promoter. Transgenic potato plants expressing GNA at 0.5–1.5% of total protein were significantly protected against attack by larvae of the tomato moth, *Lacanobia oleracea*, larval survival was reduced by less than 25% and there were significant reductions in larval biomass (>50%) and leaf damage (>70%). Similar results were obtained both in the laboratory (growth cabinet bioassay) and in a large-scale bioassay in the glasshouse (Gatehouse *et al.*, unpublished results). GNA-expressing transgenic tobacco and potato plants also reduced the development and fecundity of two aphid species, the green peach *Myzus persicae* (Hilder *et al.*, 1995; Gatehouse *et al.*, 1996) and the glasshouse potato aphid, *Aulacorthum solani*. GNA-expressing transgenic potatoes significantly reduced the population build-up of *A. solani* in a glasshouse experiment (Gatehouse, personal communication). Phloem-specific GNA expression in transgenic tobacco and rice plants has been achieved, using a gene construct containing the GNA coding sequence driven by the promoter from the rice sucrose synthase gene, Rss1 (Shi *et al.*, 1994). Further work with plants expressing GNA is in progress.

14.5 SUMMARY

Several opportunities for using antinutritive defences in host resistance exist. The overexpression of antinutritive proteins has been used for PPO, POD, LOX, protease inhibitors and lectins. Alternatively, expression of novel genes for expressing antinutritional proteins is feasible (Morgan *et al.*, 1993), as the biotin-binding animal proteins, avidin and streptavidin, were toxic to several lepidopteran and coleopteran insects. The antinutritive properties of these proteins were demonstrated by the fact that inclusion of elevated dietary levels of biotin effectively eliminated toxicity. Cholesterol oxidase, a product of *Streptomyces* spp., exerts potent toxicity to several insects including the boll weevil *Anthonomus grandis*. This enzyme oxidizes ingested cholesterol and interferes with sterol nutrition, but the mode of action is direct lysis of the midgut epithelium induced by the oxidation of membrane cholesterol (Purcell *et al.*, 1993). In adult beetles, cholesterol oxidase significantly reduced ovarian development and disrupted normal fecundity and larval survival (Greenplate *et al.*, 1995), presumably by the reduced ability to assimilate nutrients. There are numerous other potential antinutritional proteins (e.g. glucose oxidase, sulphydryl oxidase, lysyl oxidase) found in animal and microbial species that may prove beneficial for engineering insect resistance.

A parsimonious approach to engineering resistance is illustrated by work in C.A. Ryan's laboratory where overexpression of a single signal peptide results in the induction of at least two proteinase inhibitors and PPO (Constabel *et al.*, 1995). How widespread peptide signal compounds occur in the plant kingdom is unknown, but further research on signal compounds used by plants to recognize herbivore damage would be valuable.

Adaptations to antinutritive plant defences are known (Broadway, 1995), consequently strategies such as pyramiding multiple resistance traits may be necessary to avoid or delay pest adaptations. Many of the antinutrients discussed in this chapter interfere with amino acid nutrition, but we have little knowledge about their effectiveness in combination. The rational development of resistant host plants based on antinutrients depends on a greater fundamental knowledge of their mechanism of action against target pests.

ACKNOWLEDGEMENTS

The support of the United States Department of Agriculture was greatly appreciated.

REFERENCES

Appel, H.M. and Maines, L.W. (1995) The influence of host plant on gut conditions of gypsy moth caterpillars. *J. Insect Physiol.*, **41**, 241–6.

Arnault, C. and Mauchamp, B. (1985) Ecdysis inhibition in *Acrolepiopsis assectella* larvae by digitonin: antagonistic effects of cholesterol. *Experientia*, **41**, 1074–7.

Atabella, T. and Chrispeels, M.J. (1990) Tobacco plants transformed with the bean ai gene express an inhibitor of insect α-amylase in their seeds. *Plant Physiol.*, **93**, 805–10.

Barbeau, W.E. and Kinsella, J.E. (1983) Factors affecting the binding of chlorogenic acid to fraction 1 leaf protein. *J. Agric. Food Chem.*, **31**, 993–8.

Berenbaum, M.R. (1995) Turnabout is fair play: secondary roles for primary compounds. *J. Chem. Ecol.*, **21**, 925–40.

Bernays, E.A., Cooper-Driver, G. and Bilgener, M. (1989) Herbivores and plant tannins. *Adv. Ecol. Res.*, **19**, 263–302.

Bi, J.L. and Felton, G.W. (1995) Foliar oxidative stress and insect herbivory: primary compounds, secondary metabolites and reactive oxygen species as components of induced resistance. *J. Chem. Ecol.*, **21**, 1511–30.

Bi, J.L., Felton, G.W. and Mueller, A.J. (1994) Induced resistance in soybean to *Helicoverpa zea*: role of plant protein quality. *J. Chem. Ecol.*, **20**, 183–98.

Bloem, K.A., Kelley, K.C. and Duffey, S.S. (1989) Differential effect of tomatine and its alleviation by cholesterol on larval growth and efficiency of food utilization between *Heliothis zea* and *Spodoptera exigua*. *J. Chem. Ecol.*, **15**, 387–98.

Boulter, D., Edwards, G.A., Gatehouse, A.M.R. *et al.* (1990) Additive protective effects of incorporating two different higher plant derived insect resistance genes in transgenic tobacco plants. *Crop Protect.*, **9**, 351–4.

Bolter, C.J. and Jongsma, M.A. (1995) Colorado potato beetles (*Leptinotarsa decemlineata*) adapt to proteinase inhibitors induced in potato leaves by methyl jasmonate. *J. Insect Physiol.*, **41**, 1071–8.

Broadway, R.M. (1995) Are insects resistant to plant proteinase inhibitors? *J. Insect Physiol.*, **41**, 107–16.

Broadway, R.M. and Duffey, S.S. (1986) Plant proteinase inhibitors: mechanism of action and effect on the growth and digestive physiology of larval *Heliothis zea* and *Spodoptera exigua*. *J. Insect Physiol.*, **32**, 827–33.

Broadway, R.M. and Duffey, S.S. (1988) The effect of plant protein quality on insect digestive physiology and the toxicity of plant proteinase inhibitors. *J. Insect Physiol.*, **34**, 1111–17.

Campos, F., Atkinson, J., Arnason, J.T. *et al.* (1989) Toxicokinetics of 2,4-dihydroxy-7-methoxy-1,4-benzoxazin-3-one (DIMBOA) in the European corn borer, *Ostrinia nubilalis* (Hubner). *J. Chem. Ecol.*, **15**, 1989–2001.

Chew, F.S. (1988) Searching for defensive chemistry in the Cruciferae, or, do glucosinolates always control interactions of Cruciferae with their potential herbivores and symbionts? No!, in *Chemical Mediation of Coevolution* (ed. K.C. Spencer), Academic Press, San Diego, pp. 81–112.

Cheynier, V., Basire, N. and Rigaud, J. (1989) Mechanism of *trans*-caffeoyltartaric acid and catechin oxidation in model solutions containing grape polyphenoloxidase. *J. Agric. Food Chem.*, **37**, 1069–71.

Cheynier, V., Osse, C. and Rigaud, J. (1988) Oxidation of grape juice phenolic compounds in model solutions. *J. Food Sci.*, **53**, 1729–32.

Clausen, T.P., Reichardt, P.B., Bryant, J.P. and Provenza, F. (1992) Condensed tannins in plant defense: a perspective on classical theories, in *Plant Polyphenols* (eds R.W. Hemingway and P.E. Laks), Plenum Press, New York, pp. 639–51.

Constabel, C.P., Bergey, D.R. and Ryan, C.A. (1995) Systemin activates synthesis of wound-inducible tomato leaf polyphenol oxidase via the octadecanoid defense signaling pathway. *Proc. Natl Acad. Sci. USA*, **92**, 407–11.

Culvenor, C.C.J., Dann, A.T. and Dick, A.T. (1962) Alkylation as the mechanism by which the hepatotoxic pyrrolizidine alkaloids act on cell nuclei. *Nature*, **195**, 570–3.

Czapla, T.H. and Lang, B.A. (1990) Effect of plant lectins on the larval development of European corn borer (Lepidoptera: Pyralidae) and Southern corn rootworm (Coleoptera: Chrysomelidae). *J. Econ. Entomol.*, **83**, 2480–5.

Dean, R.T., Gieseg, S. and Davies, M.J. (1993) Reactive oxygen species and their accumulation on radical-damaged proteins. *Trends Biol. Sci.*, **18**, 437–41.

Del Duca, S., Tidu, V., Bassi, R. *et al.* (1994) Identification of chlorophyll-a/b proteins as substrates of transglutaminase activity in isolated chloroplasts of *Helianthus tuberosus* L. *Planta*, **193**, 283–9.

Deng, W., Grayburn, W.S., Hamilton-Kemp, T.R., Collins, G.B. and Hildebrand, D.F. (1992) Expression of soybean-embryo lipoxygenase 2 in transgenic tobacco tissue. *Planta*, **187**, 203–8.

Diaz, L.C., Melchers, L.S., Hooykaas, P.J.J. *et al.* (1989) Root lectin as a determinant of host–plant specificity in the Rhizobium–legume symbiosis. *Nature*, **338**, 579–81.

Dowd, P.F. and Lagrimini, L.M. (1996) Examination of transgenic tobacco and tomato expressing high levels of tobacco anionic peroxidase for resistance to insects. *Environ. Entomol.* (in press).

Dowd, P.F. and Norton, R.A. (1995) Browning-associated mechanisms of resistance to insects in corn callus tissue. *J. Chem. Ecol.*, **21**, 583–600.

Duffey, S.S. and Felton, G.W. (1989) Plant enzymes in resistance to insects, in *Biocatalysis in Agricultural Biotechnology* (eds J.R. Whitaker and P.E. Sonnet), American Chemical Society, Washington, DC, pp. 288–313.

Duffey, S.S. and Felton, G.W. (1991) Enzymatic antinutritive defenses of the tomato plant against insects, in *Naturally Occurring Pest Bioregulators* (ed. P.A. Hedin), American Chemical Society, Washington, DC, pp. 166–97.

Dymock, J.J., Laing, W.A., Shaw, B.D. *et al.* (1992) Behavioural and physiological responses of grass grub larvae (*Costelytra zealandica*) feeding on protease inhibitors. *NZ J. Zool.*, **19**, 123–31.

Ehrlich, P.R. and Raven, P.H. (1964) Butterflies and plants: a study in coevolution. *Evolution*, **18**, 586–608.

Eisemann, C.H., Donaldson, R.A., Pearson, R.D. *et al.* (1994) Larvicidal activity of lectins on *Lucilia cuprina*: mechanism of action. *Entomol. Exp. Appl.*, **72**, 1–10.

Enyedi, A.J., Yalpani, N., Silverman, P. and Raskin, I. (1992) Signal molecules in systemic plant resistance to pathogens and pests. *Cell*, **70**, 879–86.

Evans, C.W. (1976) Bracken thiaminase-mediated neurotoxic syndromes. *Bot. J. Linn. Soc.*, **73**, 113–31.

Farmer, E.E. and Ryan, C.A. (1992) Octadecanoid precursors of jasmonic acid activate the synthesis of wound-inducable proteinase inhibitors. *Plant Cell*, **4**, 129–34.

Felton, G.W. (1995) Oxidative stress of vertebrates and invertebrates, in *Oxidative Stress and Antioxidant Defenses in Biology* (ed. S. Ahmad), Chapman & Hall, New York, pp. 356–434.

Felton, G.W. (1996) Nutritive quality of plant protein: sources of variation and insect herbivore responses. *Arch. Insect Biochem. Physiol.*, **32**, 107–30.

Felton, G.W., Bi, J.L., Summers, C.B. *et al.* (1994b) Potential role of lipoxygenases in defense against insect herbivory. *J. Chem. Ecol.*, **20**, 651–66.

Felton, G.W., Donato, K.K., Broadway, R.M. and Duffey, S.S. (1992b) Impact of oxidized plant phenolics on the nutritional quality of dietary protein to a noctuid herbivore, *Spodoptera exigua*. *J. Insect Physiol.*, **38**, 277–85.

Felton, G.W., Donato, K., Del Vecchio, R.J. and Duffey, S.S. (1989) Activation of plant foliar oxidases by insect feeding reduces the nutritional quality of foliage for herbivores. *J. Chem. Ecol.*, **15**, 2667–94.

Felton, G.W. and Duffey, S.S. (1991a) Protective action of midgut catalase in lepidopteran larvae against oxidative plant defenses. *J. Chem. Ecol.*, **17**, 1715–32.

Felton, G.W. and Duffey, S.S. (1991b) Reassessment of the role of gut alkalinity and detergency in insect herbivory. *J. Chem. Ecol.*, **17**, 1821–36.

Felton, G.W. and Summers, C.B. (1993) Potential role of ascorbate oxidase as a plant defense protein against insect herbivory. *J. Chem. Ecol.*, **19**, 1553–68.

Felton, G.W. and Summers, C.B. (1995) Antioxidant systems in insects. *Arch. Insect Biochem. Physiol.*, **29**, 187–97.

Felton, G.W., Summers, C.B. and Mueller, A.J. (1994a) Oxidative responses in soybean foliage to herbivory by bean leaf beetle and three-cornered alfalfa hopper. *J. Chem. Ecol.*, **20**, 639–50.

Felton, G.W., Workman, J. and Felton, G.W. (1992a) Avoidance of antinutritive plant defense: role of midgut pH in Colorado potato beetle. *J. Chem. Ecol.*, **18**, 571–83.

Garcia-Olmedo, F., Salcedo, G., Sanchez-Monge, R. *et al.* (1987) Plant proteinaceous inhibitors of proteinases and α-amylases. *Oxford Surv. Plant Mol. Cell Biol.*, **4**, 275–334.

Gardner, H.W. (1979) Lipid hydroperoxide reactivity with proteins and amino acids: a review. *J. Agric. Food Chem.*, **27**, 220–8.

Gardner, H.W. (1991) Recent investigation into the lipoxygenase pathway of plants. *Biochim. Biophys. Acta*, **1084**, 221–39.

Garner, C.W. (1984) Peroxidation of free and esterified fatty acids by horseradish peroxidase. *Lipids*, **19**, 863–8.

Gatehouse, A.M.R. and Boulter, D. (1983) Assessment of the anti-metabolite effects of trypsin inhibitors from cowpea (*Vigna unguiculata*) and other legumes on development of the bruchid beetle *Callosobruchus maculatus*. *J. Sci. Food Agric.*, **34**, 345–50.

Gatehouse, A.M.R., Butler, K.J., Fenton, K.A. and Gatehouse, J.A. (1985) Presence and partial characterisation of a major proteolytic enzyme in the larval gut of *Callosobruchus maculatus*. *Entomol. Exp. Appl.*, **39**, 279–86.

Gatehouse, A.M.R., Dewey, F.M., Dove, J. *et al.* (1984) Effect of seed lectin from *Phaseolus vulgaris* on the development of larvae of *Callosobruchus maculatus*; mechanism of toxicity. *J. Sci. Food Agric.*, **35**, 373–80.

Gatehouse, A.M.R., Down, R.E., Powell, K.S. *et al.* (1996) Transgenic potato plants with enhanced resistance to the peach-potato aphid, *Myzus persicae*. *Entomol. Exp. Appl.* (in press).

Gatehouse, A.M.R., Fenton, K.A., Jepson, I. and Pavey, D.J. (1986) The effects of α-amylase inhibitors on insect storage pests; inhibition of α-amylase *in vitro* and effects on development *in vitro*. *J. Sci. Food Agric.*, **37**, 727–34.

Gatehouse, A.M.R, Howe, D.S., Flemming, J.E. *et al.* (1991) Biochemical basis of insect resistance in winged bean (*Psophocarpus tetragnobolus*) seeds. *J. Sci. Food Agric.*, **55**, 63–74.

Gatehouse, A.M.R., Shackley, S.J., Fenton, K.A. *et al.* (1989) Mechanism of seed lectin tolerance by a major insect storage pest of *Phaseolus vulgaris*, *Acanthoscelides obtectus*. *J. Sci. Food Agric.*, **47**, 269–80.

Gershenzon, J. and Croteau, R. (1991) Terpenoids, Vol. 1, in *Herbivores – Their Interactions with Secondary Metabolites*, Academic Press, San Diego, pp. 165–220.

Gordon, M.H. and Barimalaa, I.S. (1989) Co-oxidation of fat-soluble vitamins by soybean lipoxygenase. *Food Chem.*, **32**, 31–7.

Graham, J., McNicol, R.J. and Greig, K. (1995) Towards genetic based insect resistance in strawberry using cowpea trypsin inhibitor. *Ann. Appl. Biol.*, **127**, (in press).

Greenplate, J.T., Duck, N.B., Pershing, J.C. and Purcell, J.P. (1995) Cholesterol oxidase: an oöstatic and larvicidal agent active against the cotton boll weevil, *Anthonomus grandis*. *Entomol. Exp. Appl.*, **74**, 253–8.

Harding, J.J. (1985) Nonenzymatic covalent posttranslational modification of proteins *in vivo*. *Adv. Protein Chem.*, **37**, 247–334.

Harland, B.F. and Morris, E.R. (1995) Phytate: a good or bad food component. *Nutr. Res.*, **15**, 733–54.

Harmatha, J., Mauchamp, B., Arnault, C. and Slama, K. (1987) Identification of a spirostane-type saponin in the flowers of leek with inhibitory effects on growth of leek-moth larvae. *Biochem. Syst. Ecol.*, **15**, 113–16.

Haukioja, E., Ruohomaki, K., Suomela, J. and Vuorisalo, T. (1991) Nutritional quality as a defense against herbivores. *Forest Ecol. Manag.*, **39**, 237–45.

Hendrix, S.D. (1975) The resistance of *Pteridium aquilinum* (L.) Kuhn (bracken fern) to insect attack. PhD Thesis, University of California, Berkeley.

Hildebrand, D.F. (1992) Altering fatty acid metabolism in plants. *Food Technol.*, **46**, 71–4.

Hildebrand, D.F., Brown, G.C., Jackson, D.M. and Hamilton-Kemp, T.R. (1993) Effects of some leaf-emitted volatile compounds on aphid population increase. *J. Chem. Ecol.*, **19**, 1875–87.

Hildebrand, D.F., Hamilton-Kemp, T.R, Legg, C.S. and Bookjans, G. (1988) Plant lipoxygenases: occurrence, properties and possible function. *Curr. Topics Plant Biochem. Physiol.*, **7**, 201–19.

Hildebrand, D.F., Rodriguez, J.G., Brown, G.C. *et al.* (1986) Peroxidative responses of leaves in two soybean genotypes injured by two-spotted spider mites (Acari: Tetranychidae). *J. Econ. Entomol.*, **79**, 1459–65.

Hilder, V.A., Gatehouse, A.M.R., Sherman, S.E. *et al.* (1987) A novel mechanism for insect resistance engineered into tobacco. *Nature*, **330**, 160–3.

Hilder, V.A, Powell, K.S., Gatehouse, A.M.R. *et al.* (1995) Expression of snowdrop lectin in transgenic tobacco plants results in added protection against aphids. *Transgenic Res.*, **4**, 18–25.

Hinson, J.A. and Roberts, D.W. (1992) Role of covalent and noncovalent interactions in cell toxicity: effects on proteins. *Annu. Rev. Pharmacol. Toxicol.*, **32**, 471–510.

Hoffman, M.P., Zalom, F.G., Smilanick, J.M. *et al.* (1991) Field evaluation of transgenic tobacco containing genes encoding *Bacillus thuringiensis* δ-endotoxin or cowpea trypsin inhibitor: efficacy against *Helicoverpa zea* (Lepidoptera: Noctuidae). *J. Econ. Entomol.*, **85**, 2516–22.

Hori, K. (1973) Studies on the feeding habits of *Lygus disponsi* Linnavuori (Hemiptera: Miridae) and the injury to its host plant. III. Phenolic compounds, acid phosphatase and oxidative enzymes in the injured tissue of sugar beet leaf. *Appl. Entomol. Zool.*, **8**, 103–13.

Hori, K. and Atalay, R. (1980) Biochemical changes in the tissue of Chinese cabbage injured by the bug *Lygus disponsi. Appl. Entomol. Zool.*, **15**, 234–41.

Houseman, J.G., Campos, F., Thie, N.M.R. *et al.* (1992) Effect of maize-derived compounds DIMBOA and MBOA on growth and digestive processes of European corn borer (Lepidoptera: Pyralidae). *J. Econ. Entomol.*, **85**, 669–74.

Huesing, J.E., Murdock, L.L. and Shade, R.E. (1991) Rice and stinging nettle lectins: insecticidal activity similar to wheat germ agglutinin. *Phytochemistry*, **30**, 3565–8.

Hunt, M.D., Eanetta, N.T., Yu, H. *et al.* (1993) cDNA cloning and expression of potato polyphenol oxidase. *Plant Mol. Biol.*, **21**, 59–68.

Icekson, I. and Apelbaum, A. (1987) Evidence for transglutaminase activity in plant tissue. *Plant Physiol.*, **84**, 972–4.

Ishaaya, I. (1965) Significance of soybean saponins in animal nutrition and their effects on digestive enzymes. PhD Thesis, The Hebrew University of Jerusalem.

Ishaaya, I. (1986) Nutritional and alellochemic insect–plant interactions relating to digestion and food intake: some examples, in *Insect–Plant Interactions* (eds J.R. Miller and T.A. Miller), Springer-Verlag, New York, pp. 191–223.

Ishaaya, I. and Birk, Y. (1965) Soybean saponins IV. The effects of proteins on the inhibitory activity of soybean saponins on certain enzymes. *J. Food Sci.*, **30**, 118–20.

Ishaaya, I., Birk, Y., Bondi, A. and Tencer, Y. (1969) Soybean saponins. IX. Studies of their effects on birds, mammals and cold-blooded organisms. *J. Sci. Food Agric.*, **20**, 433–6.

Ishimoto, M. and Kitamura, K. (1988) Identification of the growth inhibitor on azuki bean weevil in kidney bean (*Phaseolus vulgaris* L.). *Jpn J. Breeding*, **38**, 367–70.

Janzen, D.H., Juster, H.B. and Liener, I.E. (1976) Insecticidal action of the phytohaemagglutinin in black bean on a bruchid beetle. *Science*, **192**, 795–6.

Johnson, K.S. and Felton, G.W. (1996) Physiological and dietary influences on midgut redox conditions in generalist lepidopteran larvae. *J. Insect Physiol.*, **42**, 191–8.

Johnson, R., Narvaez, J., An, G. and Ryan, C. (1989) Expression of proteinase inhibitors I and II in transgenic tobacco plants: effects on natural defense against *Manduca sexta* larvae. *Proc. Natl Acad. Sci. USA*, **86**, 9871–5.

Johnston, K.A., Gatehouse, J.A. and Anstee, J.H. (1993) Effects of soybean protease inhibitors on the growth and development of larval *Helicoverpa armigera*. *J. Insect Physiol.*, **39**, 657–64.

Jones, C.G. (1983) Phytochemical variation, colonization, and insect communities: the case of bracken fern, in *Variable Plants and Herbivores in Natural and Managed Systems* (eds R.F. Denno and M.S. McClure), Academic Press, New York, pp. 513–58.

Jongsma, M.A., Bakker, P.L., Peters, J. *et al.* (1995) Adaptations of *Spodoptera exigua* larvae to plant proteinase inhibitors by induction of gut proteinase activity insensitive to inhibition. *Proc. Natl Acad. Sci. USA*, **92**, 8041–5.

Karowe, D.N. (1989) Differential effect of tannic acid on two tree-feeding Lepidoptera: implications for theories of plant anti-herbivore chemistry. *Oecologia*, **80**, 507–12.

Kasu, T., Brown, G.C. and Hildebrand, D.F. (1994) Application of fatty acids to elicit lipoxygenase-mediated host-plant resistance to twospotted spider mites (Acari: Tetranychidae) in *Phaseolus vulgaris* L. *Environ. Entomol.*, **23**, 437–41.

Kasu, T., Brown, G.C. and Hildebrand, D.F. (1995) Formation of lipoxygenase products in *Phaseolus vulgaris* L. leaves as a response to twospotted spider mite (Acari: Tetranychidae) feeding and their effect on spider mite populations. *J. Kansas Entomol. Soc.*, **68**, 27–34.

King, R.S. and Flurkey, W.H. (1987) Effects of limited proteolysis on broad bean polyphenol oxidase. *J. Sci. Food Agric.*, **41**, 231–40.

Lagrimini, L.M. (1991) Wound-induced deposition of polyphenols in transgenic plants overexpressing peroxidase. *Plant Physiol.*, **96**, 577–83.

Lane, B.G. (1994) Oxalate, germin and the extracellular matrix of higher plants. *FASEB J.*, **8**, 294–301.

Leplé, J.C., Bonadé-Bottino, M., Augustin, S. *et al.* (1995) Toxicity to *Chrysomela tremulae* (Coleoptera: Chrysomelidae) of transgenic poplars expressing a cysteine proteinase inhibitor. *Mol. Breeding*, (in press).

Libert, B. and Franceschi, V.R. (1987) Oxalate in crop plants. *J. Agric. Food Chem.*, **35**, 926–38.

Liener, I.E. (1980) *Toxic Constituents of Plant Foodstuffs*, 2nd edn, Academic Press, New York.

Lipke, H., Fraenkel, G.S. and Liener, I.E. (1954) Effect of soybean inhibitors on growth of *Tribolium confusum*. *J. Agric. Food Chem.*, **2**, 410–15.

Louda, S. and Mole, S. (1991) Glucosinolates: chemistry and ecology, in *Herbivores: Their Interactions With Secondary Plant Metabolites*, 2nd edn, Vol. I *The Chemical Participants* (eds G.A. Rosenthal and M.R. Berenbaum), Academic Press, San Diego, pp. 123–64.

Luthy, B. and Matile, P. (1984) The mustard oil bomb: rectified analysis of the subcellular organization of the myrosinase system. *Biochem. Physiol. Planz*, **179**, 5–12.

Macheix, J.-J., Spais, J.-C. and Fleuriet, A. (1991) Phenolic compounds and polyphenoloxidase in relation to browning in grapes and wines. *Crit. Rev. Food Sci. Nutr.*, **30**, 441–86.

Malone, L.A., Giacon, H.A., Burgess, E.P.J. *et al.* (1995) Toxicity of trypsin endopeptidase inhibitors to honey bees (Hymenoptera: Apidae). *J. Econ. Entomol.*, **88**, 46–50.

Marañon, M.J.R. and van Huystee, R.B. (1994) Plant peroxidases: interactions between their prosthetic groups. *Phytochemistry*, **37**, 1217–25.

McFarlane, J.E. (1985) Nutrition and digestive organs, in *Fundamentals of Insect Physiology* (ed. M.S. Blum), John Wiley, New York, pp. 59–90.

McManus, M.T., White, D.W.R. and McGregor, P.G. (1994) Accumulation of a chymotrypsin inhibitor in transgenic tobacco can affect the growth of insect pests. *Transgenic Res.*, **3**, 50–8.

Margosiak, S.A., Dharma, A., Bruce-Carver, M.R. *et al.* (1990) Identification of the large subunit of ribulose 1,5-bisphosphate carboxylase/oxygenase as a substrate for transglutaminase in *Medicago sativa* L. (alfalfa). *Plant Physiol.*, **92**, 88–96.

Marques, L., Fleuriet, A., Cleyet-Marcel, J. and Macheix, J. (1994) Purification of an apple polyphenoloxidase isoform resistant to SDS-proteinase K digestion. *Phytochemistry*, **36**, 1117–21.

Martin, J.S., Martin, M.M. and Bernays, E.A. (1987) Failure of tannic acid to inhibit digestion or reduce digestibility of plant protein in gut fluids of insect herbivores: implications for theories of plant defense. *J. Chem. Ecol.*, **13**, 605–21.

Medda, R., Padiglia, A. and Floris, G. (1995) Plant copper-amine oxidases. *Phytochemistry*, **39**, 1–9.

Meisner, J., Ishaaya, I., Ascher, K.R.S. and Zur, M. (1978) Gossypol inhibits protease and amylase activity of *Spodoptera littoralis* Boisduval larvae. *Ann. Entomol. Soc. Am.*, **71**, 5–8.

Minney, B.H.P., Gatehouse, A.M.R., Dobie, P. *et al.* (1990) Biochemical base of seed resistance to *Zabrotes subfasciatus* (bean weevil) in *Phaseolus vulgaris* (common bean); a mechanism for arcelin toxicity. *J. Insect Physiol.*, **36**, 757–67.

Montandon, R., Stipanovich, R.D., Williams, H.J. *et al.* (1987) Nutritional indices and excretion of gossypol by *Alabama argillacea* (Hubner) and *Heliothis virescens* (F.) (Lepidoptera: Noctuidae) fed glanded and glandless cotyledonary cotton leaves. *J. Econ. Entomol.*, **80**, 32–6.

Moore, R.F. (1983) Effect of dietary gossypol on the boll weevil (Coleoptera: Curculionidae). *J. Econ. Entomol.*, **76**, 696–9.

Moreno, J. and Chrispeels, M.J. (1989) A lectin gene encodes the α-amylase inhibitor of the common bean. *Proc. Natl Acad. Sci. USA*, **86**, 7885–9.

Morgan, T.D., Oppert, B., Czapla, T.H. and Kramer, K.J. (1993) Avidin and streptavidin as insecticidal and growth inhibiting proteins. *Entomol. Exp. Appl.*, **69**, 97–108.

Motoda, S. (1979) Formation of aldehydes from amino acids by polyphenol oxidase. *J. Ferment. Technol.*, **57**, 395–9.

Murdoch, L.L., Huesing, J.E., Nielsen, S.S. *et al.* (1990) Biological effects of plant lectins on the cowpea weevil. *Phytochemistry*, **29**, 85–9.

Nelson, S.D. and Pearson, P.G. (1990) Covalent and noncovalent interactions in acute lethal cell injury caused by chemicals. *Annu. Rev. Pharmacol. Toxicol.*, **30**, 169–95.

Nicolas, J.J., Richard-Forget, F.C., Goupy, P.M. *et al.* (1994) Enzymatic browning reactions in apple and apple products. *Crit. Rev. Food Sci. Nutr.*, **34**, 109–57.

O'Brien, P.J. (1991) Molecular mechanisms of quinone cytotoxicity. *Chem-Biol Interactions*, **80**, 1–41.

Orozco-Cardenas, M., McGurl, B. and Ryan, C.A. (1993) Expression of an antisense prosystemin gene in tomato plants reduces resistance toward *Manduca sexta* larvae. *Proc. Natl Acad. Sci. USA*, **90**, 8273–6.

Perez, F.J. and Niemeyer, H.M. (1985) The reduction of 2,4-dihydroxy-7-methoxy-1,4-benzoxazin-3-one by thiols. *Phytochemistry*, **24**, 2963–6.

Potter, D.A. and Kimmerer, T.W. (1989) Inhibition of herbivory on young holly leaves: evidence for the defensive role of saponins. *Oecologia*, **78**, 322–9.

Potter, S.M., Jimenez-Flores, R., Pollack, S.M. *et al.* (1993) Protein–saponin interaction and its influence on blood lipids. *J. Agric. Food Chem.*, **41**, 1287–91.

Powell, K.S., Gatehouse, A.M.R., Hilder, V.A. and Gatehouse, J.A. (1993) Antimetabolic effects of plant lectins and plant and fungal enzymes on the nymphal stages of two important rice pests, *Nilaparvata lugens* and *Nephotettix cincticeps*. *Entomol. Exp. Appl.*, **66**, 119–26.

Powell, K.S., Gatehouse, A.M.R., Hilder, V.A. and Gatehouse, J.A. (1995) Antifeedant effects of plant lectins and an enzyme on the adult stage of the rice brown planthopper, *Nilaparvata lugens*. *Entomol. Exp. Appl.*, **75**, 51–9.

Purcell, J.P., Greenplate, J.T., Jennings, M.G. *et al.* (1993) Cholesterol oxidase: a potent insecticidal protein active against boll weevil larvae. *Biochem. Biophys. Res. Commun.*, **196**, 1406–13.

Pusztai, A. (1991) *Plant Lectins*, Cambridge University Press, Cambridge.

Pusztai, A., Grant, G., Stewart, J.C. *et al.* (1992) Nutritional evaluation of the trypsin inhibitor from cowpea. *Br. J. Nutr.*, **68**, 783–91.

Rahbe, Y. and Febvay, G. (1993) Protein toxicity to aphids: an *in vitro* test on *Acyrthosiphon pisum*. *Entomol. Exp. Appl.*, **67**, 149–60.

Raman, K., Sanjayan, K.P. and Suresh, G. (1984) Impact of feeding injury of *Cyrtopeltis tenuis* Reut. (Hemiptera: Miridae) on some biochemical changes in *Lycopersicon esculentum* Mill. (Solanaceae). *Curr. Sci.*, **20**, 1092–3.

Rhoades, D.F. (1977) The antiherbivore chemistry of *Larrea*, in *Creosote Bush* (eds T.J. Mabry, J.H. Hunziker and D.R. DiFeo), Dowden, Hutchinson and Ross, New York, pp. 135–75.

Richard, F.C, Goupy, P.M., Nicolas, J.J. *et al.* (1991) Cysteine as an inhibitor of enzymatic browning. 1. Isolation and characterization of addition compounds formed during oxidation of phenolics by apple polyphenol oxidase. *J. Agric. Food Chem.*, **39**, 841–7.

Richardson, M.J. (1991) Seed storage proteins: the enzyme inhibitors, in *Methods in Plant Biochemistry* Vol. 5 (ed. L.J. Rogers), Academic Press, New York, pp. 259–305.

Robinson, D.S. (1991) Peroxidases and catalases in food, in *Oxidative Enzymes in Foods* (eds D.S. Robinson and N.A.M. Eskin), Elsevier Applied Science, London, pp. 1–47.

Rojas, M.G., Stipanovich, R.D., Williams, H.J. and Vinson, S.B. (1992) Metabolism of gossypol by *Heliothis virescens* (F.) (Lepidoptera: Noctuidae). *Environ. Entomol.*, **21**, 518–26.

Rosenthal, G.A. and Berenbaum, M.R. (eds) (1991) *Herbivores – Their Interactions with Secondary Metabolites*, 2nd edn, Academic Press, San Diego.

Rouet-Mayer, M.A., Ralambosoa, J. and Phillipon, J. (1990) Roles of *o*-quinones and their polymers in the enzymic browning of apples. *Phytochemistry*, *29*, 435–40.

Ryan, C.A. (1984) Defense responses of plants, in *Plant Gene Research: Genes*

Involved in Microbe–Plant Interactions (eds D.P.S. Verma and T. Hohn), Springer-Verlag, Wien, New York, pp. 375–86.

Ryan, C.A. (1989) Proteinase inhibitor gene families: strategies for transformation to improve plant defenses against herbivores. *BioEssays*, **10**, 20–4.

Saijo, R. and Takeo, T. (1970) The production of phenylacetaldehyde from L-phenylalanine in tea fermentation. *Agric. Biol. Chem.*, **34**, 222–6.

Serafini-Fracassini, D., Del Duca, S. and Beninati, S. (1995) Plant transglutaminases. *Phytochemistry*, **40**, 355–65.

Serafini-Fracassini, D., Del Duca, S. and D'Orazi, D. (1988) First evidence for polyamine conjugation mediated by an enzymic activity in plants. *Plant Physiol.*, **87**, 757–61.

Shany, S., Gestetner, B., Birk, Y. and Bondi, A. (1970) Lucerne saponins III. Effect of lucerne saponins on larval growth and their detoxification by various sterols. *J. Sci. Food Agric.*, **21**, 508–10.

Shi, Y., Wang, M.B., Hilder, V.A. *et al.* (1994) Use of the rice sucrose synthase-1 promoter to direct phloem-specific expression of ß-glucuronidase and snowdrop lectin genes in transgenic tobacco plants. *J. Exp. Bot.*, **45**, 623–31.

Shukle, R.H. and Murdock, L.L. (1983) Lipoxygenase, trypsin inhibitor, and lectin from soybeans: effects on larval growth of *Manduca sexta* (Lepidoptera: Sphingidae). *Environ. Entomol.*, **12**, 787–91.

Siepaio, M.P. and Meunier, J.-C.F.(1995a) Polymerization of soybean proteins and spinach RuBisCO by FXIIIa with respect to the labelling or reactive glutaminyl residues. *J. Agric. Food Chem.*, **43**, 568–73.

Siepaio, M.P. and Meunier, J.-C.F. (1995b) Diamine oxidase and transglutaminase activities in white lupine seedlings with respect to cross-linking of proteins. *J. Agric. Food Chem.*, **43**, 1151–6.

Soo-Hoo, C.F. and Fraenkel, G.S. (1964) Resistance of ferns to the feeding of *Prodenia eridania* larvae. *Ann. Entomol. Soc. Am.*, **25**, 790–1.

Stadtman, E.R. (1993) Oxidation of free amino acids and amino acid residues in proteins by radiolysis and by metal-catalysed reactions. *Annu. Rev. Biochem.*, **62**, 797–821.

Stahmann, M.A. and Spencer, A.K. (1977) Deamination of protein lysyl ε-amino groups by peroxidase *in vitro*. *Biopolymers*, **16**, 1299–306.

Steffens, J.C., Harel, E., and Hunt, M.D. (1994) Polyphenol oxidase, in *Genetic Engineering of Plant Secondary Metabolism, Recent Advances in Phytochemistry* Vol. 28 (eds B.E. Ellis, G.W. Kuroki, and H.A. Stafford), Plenum Press, New York, pp. 275–312.

Stout, M.J., Workman, J. and Duffey, S.S. (1994) Differential induction of tomato foliar proteins by arthropod herbivores. *J. Chem. Ecol.*, **20**, 2575–94.

Summers, C.B. and Felton, G.W. (1993) Antioxidant role of dehydroascorbic acid reductase in insects. *Biochim. Biophys. Acta*, **1156**, 235–8.

Summers, C.B. and Felton, G.W. (1994) Prooxidant effects of phenolic acids on the generalist herbivore *Helicoverpa zea* (Lepidoptera: Noctuidae): potential mode of action for phenolic compounds in plant anti-herbivore chemistry. *Insect Biochem. Mol. Biol.*, **24**, 943–53.

Thompson, D., Constantin-Teodosiu, D., Egestad, B. *et al.* (1990) Formation of glutathione conjugates during oxidation of eugenol by microsomal fractions of rat liver and lung. *Biochem. Pharmacol.*, **39**, 1587–95.

Uchida, K. and Stadtman, E.R. (1992) Modification of histidine residues in proteins by reaction with 4-hydroxynonenal. *Proc. Natl Acad. Sci. USA*, **89**, 4544–8.

van Damme, E.J.M., Allen, A.K. and Peumans, W.J. (1987) Isolation and characterization of a lectin with exclusive specificity towards mannose from snowdrop (*Galanthus nivalis*) bulbs. *FEBS Lett.*, **215**, 140–4.

Xavier-Filho, J., Campos, F.A.P., Ary, M.B. *et al.* (1989) Poor correlation between levels of proteinase inhibitors found in seeds of different cultivars of cowpea (*Vigna unguiculata*) and the resistance/susceptibility to predation by *Callosobruchus maculatus*. *J. Agric. Food Chem.*, **37**, 1139–43.

Part Four

The Midgut as an Environment for Other Organisms

15

Microbial symbioses in the midgut of insects

A.E. Douglas and C.B. Beard

15.1 INTRODUCTION

15.1.1 Concepts and definitions

Micro-organisms are ubiquitous, and the guts of animals are particularly favoured sites for microbial colonization. Gut symbionts are those micro-organisms which persist for extended periods in the intestinal tract. They may be located in the lumen of the gut, associated with the gut wall or in the animal cells lining the gut. Animal guts also bear 'transients', i.e. micro-organisms that gain access with food but are killed (e.g. by digestive enzymes) or pass out with the faeces. Although the distinction between transients and the resident gut symbionts may be blurred in some systems, the difference can be illustrated by the study on *Escherichia coli* in the gut of, not an insect, but a single human (research scientist) (Caugant *et al.*, 1981). Of the 54 electrophoretic variants of *E. coli* identified, only two persisted for the full 11 months of the study and the remainder had a maximal residence time of 5 days. These are the resident gut symbionts and the transients, respectively.

The definition of gut symbionts as residents of animal guts makes no assumptions about the advantage or harm of the relationship to either animal or micro-organisms. This is consistent with the original definition of symbiosis 'the living together of differently named animals', as popularized by Anton de Bary in the last century (Sapp, 1994) (but not

Biology of the Insect Midgut.
Edited by M.J. Lehane and P.F. Billingsley.
Published in 1996 by Chapman & Hall, London.
ISBN 0 412 61670 X.

with the view, widespread in the biological literature, of symbioses as mutually beneficial associations).

15.1.2 Context and scope of this chapter

The microbiology of insect guts has been studied for many decades (Steinhaus, 1946; Buchner, 1965). Until recently progress has been slow, largely because many gut symbionts are unculturable and are consequently intractable to the traditional methods of microbiology based on axenic cultivation. The study of unculturable micro-organisms, including gut microbiota, has been revitalized by the advent of molecular biology, which provides the techniques to classify and study micro-organisms in their natural habitats (O'Neill *et al.*, 1992; Olsen *et al.*, 1994). Molecular advances have also suggested novel ways in which micro-organisms might be utilized to control insect pests and vectors of disease (Beard *et al.*, 1993b; Richards, 1993).

In this chapter three aspects of midgut symbioses in insects will be reviewed: (1) the incidence of gut symbionts in various regions of the insect midgut (section 15.2), (2) the impact of micro-organisms on the nutrition and fitness of the insect (section 15.3), and (3) the modes of acquisition of micro-organisms by insects (section 15.4). The systems considered will be those in which the insect requires its microbial partner (usually because the micro-organisms provide nutrients) and those in which the micro-organisms are of no detectable significance to the insect (although it is impossible to establish that micro-organisms are of no significance). Certain insect–microbial relationships are not considered here. These include overt pathogens of insects (reviewed in Beckage *et al.* (1993a,b)) and the various micro-organisms which specifically distort the sex ratio or reduce the fecundity of their insect hosts (e.g. agents of cytoplasmic incompatibility, male killers) (reviewed in Hurst, 1993). A further group of microbial associates of insects are those which have little or no effect on the insect but are transmitted from the insect (the vector) to other organisms, in which they are pathogenic. These agents include various trypanosomes: for example, those borne by tsetse flies *Glossina* and by kissing bugs *Rhodnius*, which cause sleeping sickness and Chagas' disease, respectively, in mammals; and the interactions between trypanosomes and the resident midgut microbiota of the insect vectors are reviewed in section 15.3.2 (also Chapter 16).

15.2 LOCATION OF MICRO-ORGANISMS IN THE MIDGUT

In most insects there are two main regions of the midgut. The tubular ventriculus along which ingesta is passed and blind-ended evaginations called caeca (see Chapter 1). The microbial symbioses in insect midguts can be classified by their location into three broad groups; in the lumen of the ventriculus, in epithelial cells lining the ventriculus and in the caeca. The insects with these associations are surveyed in Table 15.1.

15.2.1 Lumen of the ventriculus

Substantial microbial populations have been described in the lumen of the midgut ventriculus of a few insect groups. For example, the soil-feeding termite *Cubitermes* has actinomycetes in the ectoperitrophic space (i.e. between the gut epithelium and peritrophic matrix) (Bignell *et al.*, 1980); the blood-feeding heteropteran bugs of the subfamily Triatominae have a diversity of Gram-positive bacteria, mostly in the proximal segment of the midgut (Cavanagh and Marsden, 1969; Nyirady, 1973); and the pyrrhocorid bug *Pyrrhocoris apterus* bears an obligate anaerobe allied to the Gram-positive *Bifidobacterium* in the third (of four) midgut segment (Haas and Konig, 1987).

The midgut microbiota in the heteropteran bug *Rhodnius prolixus* has been investigated in detail, largely due to its importance as a vector of Chagas' disease. Much emphasis has been given to an actinomycete *Rhodococcus* (= *Nocardia*) *rhodnii*, on which the insect is apparently dependent (Brecher and Wigglesworth, 1944; Baines, 1956). However, (and contrary to some indications in the literature), this actinomycete is not the sole symbiont; a variety of bacteria, including *Streptococcus*, *Staphylococcus*, *Corynebacterium*, *Mycobacterium*, *Pseudomonas* and *E. coli*, have been isolated from the midguts of *Rhodnius* in some studies (Gampert and Schwartz, 1962; Cavanagh and Marsden, 1969).

15.2.2 Epithelium of the ventriculus

Several groups of insects have bacteria within cells lining the ventriculus of the midgut. Those insect cells containing bacteria that are considerably larger than adjacent uninfected cells are known as mycetocytes. In some groups, including the anopluran *Haematopinus*, the heteropteran *Chilocoris* and ants of the genus *Camponotus*, the mycetocytes are irregularly distributed among the secreting and absorbing cells of the midgut. In other insects they are aggregated together and restricted to specific zones of the gut. For example, the mycetocytes in *Glossina* (tsetse flies) form two conspicuous longitudinal bands along the central portion of the midgut, whereas the midgut of the ant *Formica fusca* has a single linear array of mycetotcytes (Buchner, 1965).

The best-studied association in midgut epithelium is in *Glossina*, which contains at least two bacterial forms. The first are 5–9 μm-long Gram-negative, rod-shaped bacteria, known as P-endosymbionts, which

Table 15.1 Survey of midgut symbiosis in insects

Insect	*Lumen*	*Epithelial cells lining the ventriculus*	*Caeca (evaginations of the midgut)*	*Incidence symbiosis*	*Mode of transmission*
Heteroptera					
Coreidae	√			Irregular	Egg smearing
Lygaeidae	√			Irregular	Egg smearing
Pentatomidae	√			Widespread	Egg smearing
Pyrrhocoridae	√			Widespread	Egg smearing
Triatominae	√			Universal	Faecal
Diptera					
Tephritidae		√		Universal	Egg smearing
Glossinidae and Hippoboscidae		√		Universal	Milk gland secretions
Coleoptera					
Anobiidae			√	Universal	Egg smearing
Cerambycidae			√	Widespread	Egg smearing
Chrysomelidae			√	Irregular	Egg smearing
Curculionidae			√	Cleoninae only	Egg smearing
Hymenoptera					
Formicidae		√		Universal in Camponoti, irregular in Formicinae	Via egg in maternal ovary

are restricted to the insect's mycetocytes (Rheinhardt *et al.*, 1972). These bacteria have not been cultured. The second form of bacteria, the S-endosymbionts, are 2–3 μm rods, and are located in many cells of the midgut epithelium, other tissues and haemolymph of the insect (Huebner and Davey, 1974; Pinnock and Hess, 1974; Pell and Southern, 1975a,b). These bacteria are culturable (Figure 15.1) (Welburn *et al.*, 1987; Beard *et al.*, 1993a). From 16S rRNA sequence analysis, P- and S-endosymbionts have been assigned to distinct lineages in the γ-3 subgroup of the Proteobacteria (Beard *et al.*, 1993b; Aksoy, 1995; Askoy *et al.*, 1995). The frequent description in the literature of S-endosymbionts as 'rickettsia-like organisms' is now known to be erroneous (rickettsias are in the α-subgroup of Proteobacteria).

15.2.3 Associations in midgut caeca

Micro-organisms in the midgut caeca are well-documented in two groups of insects, the Coleoptera and tephritid flies, and those are described here. Other insects almost certainly bear micro-organisms in these locations; these remain to be studied. Of particular potential interest are the many phytophagous Heteroptera, especially lygaeids, coreids and pentatomids, which have substantial caeca, reportedly bearing dense bacterial populations (Buchner, 1965).

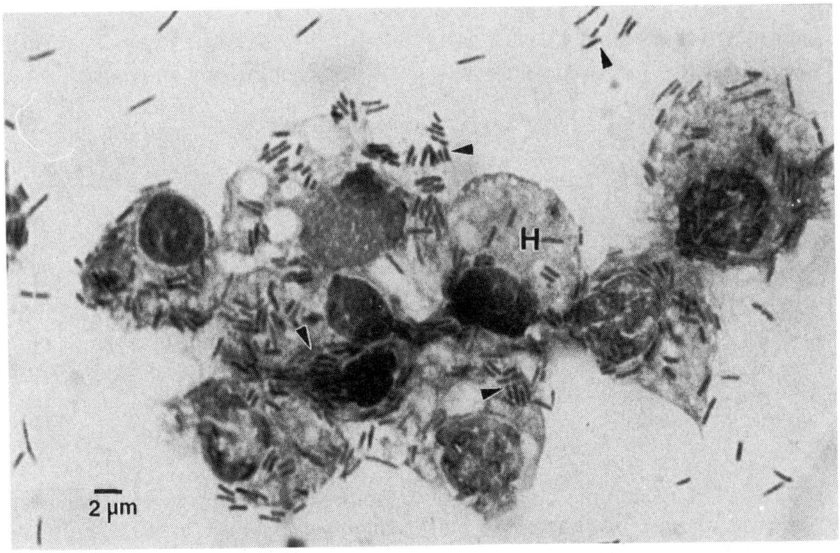

Figure 15.1 S-endosymbionts of tsetse growing intra- and extracellularly in an *Aedes albopictus* (C6/36) cell line. Host cell (H), symbionts (arrows). (Unpublished micrograph of C.B. Beard.)

Larvae of at least four families of Coleoptera bear micro-organisms in midgut caeca. Most (and possibly all) Anobiidae, many species of Cerambycidae (in association with yeasts), the cleonine curculionids and a few Chrysomelidae (including the Donaciinae and some *Cassida* species) contain bacteria. When the larval gut is destroyed at metamorphosis, the symbiosis breaks down. However, many of the micro-organisms survive in female insects and become incorporated into structures associated with the ovipositor and, in anobiids and chrysomelids, in caeca of the adult midgut. The associations are entirely absent from adult males.

There has been virtually no recent research on the caecal associations in Coleoptera, but many descriptive studies were conducted earlier in this century. The caeca are located at the anterior end of the midgut, and their general form is illustrated in Figure 15.2. The single cellular layer bounding each caecum comprises two morphologically distinct cell types: epithelial cells that lack the micro-organisms and bear microvilli along the apical border and larger mycetocytes which contain the micro-organisms and have few or no microvilli. In most species, the epithelial cells are irregularly distributed among the more abundant mycetocytes (for example, the mycetocytes account for 79–90% of all cells in the larval caeca of the anobiid *Stegobium paniceum* (Kiefer, 1932)), but in some anobiids the mycetocytes are restricted to the terminus of the caeca and are linked to the gut lumen by a narrow canal bounded by epithelial cells (Koch, 1960).

Jurzitza (1979) provides an ultrastructural study of the association in larvae of the anobiid *Lasioderma serricorne*, including a description of

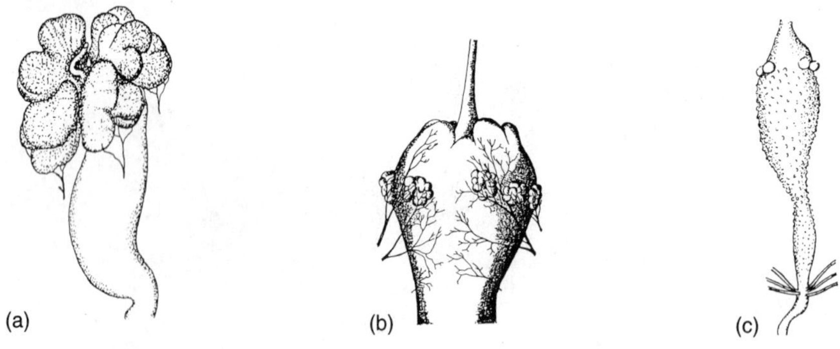

Figure 15.2 Morphology of the midgut in Coleoptera bearing caecal symbioses. (a) Larval midgut of the anobiid *Stegobium paniceum* with four large and convoluted caeca; (b) larval midgut of the cerambycid *Leptura rubra* bearing eight 'rosette'-shaped caeca; (c) midgut of adult female of the chrysomelid *Cassida viridis* with two sets of paired sac-like caeca. (Reproduced from Buchner, 1965, figs 17a, 27a, 41, with permission.)

spatial differences in the association within each mycetocyte. At the base of the mycetocytes the yeasts are structurally intact and include many budding forms, indicative of a proliferating population. However, the yeasts in the apical region of the cells are in a disintegrating condition. Lysed yeast cells, presumably derived from the mycetocytes, are also evident in the lumen of the caeca.

The yeasts in the anobiid *Stegobium paniceum* have been isolated into culture and they have been assigned to the genus *Torulopsis* (Bismanis, 1976). Laboratory cultures of *S. paniceum* can be infected with a variety of yeasts other than *Torulopsis*, but these 'foreign' forms infect both mycetocytes and epithelial cells of the larval caeca and are lost from the insect at metamorphosis (i.e. adult females are yeast-free). These experiments, incidentally, indicate that the epithelial and mycetocytes are developmentally distinct, and that differentiation of the mycetocytes is not dependent on infection by yeasts. The microbiology of the caecal symbioses has not been addressed in any other species.

In larvae of tephritid flies, bacteria are borne in the caeca arising from the proximal region of the midgut, but (in contrast to the Coleoptera above) the micro-organisms are housed in the lumen of the caeca and are not intracellular. This midgut association in tephritids is lost at metamorphosis and adult insects may contain the bacteria in an evagination of the foregut (known as the cephalic bulb).

The bacteria in tephritids are culturable. Early researchers reported that the associations were highly specific and that the bacteria in some species are important pathogens of the fruit on which the flies feed. Recent research cannot substantiate these claims (review by Drew and Lloyd, 1991). In particular, the bacterial associate of the olive fruit fly *Dacus oleae* has been reported as *Pseudomonas savastanoi*, the agent of olive-knot; but the bacteria isolated by Luthy *et al.* (1983) from the cephalic bulb of *D. oleae* did not include this species. In an extensive study of the microbiota in seven *Rhagoletis* species, Howard and Bush (1989) isolated at least 12 species of bacteria, including *Klebsiella oxytoca*, *K. pneumoniae*, *Enterobacter agglomerans* and *E. cloacae*, from the cephalic bulbs of adults. However, only *K. oxytoca* was isolated from the larval guts, possibly suggesting that the association may be more specific in the larvae than in the more extensively studied adults.

15.3 THE SIGNIFICANCE OF MIDGUT MICRO-ORGANISMS TO INSECTS

15.3.1 Microbial provision of nutrients to insects

The nutritional significance of micro-organisms to insects can very profitably be explored by 'performance experiments'. In these experiments, the survival, larvel growth, adult fecundity, etc. of insects with and without the complement of micro-organisms are determined on diets that contain and lack the nutrients of interest. If the performance of the microbial-free insects, but not the insects with the micro-organisms, is depressed on the nutrient-deficient diet then it is very likely that the insect derives those nutrients from the micro-organisms. Further experiments can then be conducted to establish directly whether the nutrients of interest are synthesized by the micro-organisms and translocated to the insect.

The value of performance experiments can be illustrated by the classic experiments of Fraenkel and Blewitt (1943) on the anobiid beetle *Stegobium paniceum*. Larvae with midgut caeca devoid of yeasts were generated by surface-sterilization of the eggs (section 15.4). The growth of these larvae, but not those containing yeasts, was impaired on diets lacking riboflavin, nicotinic acid, pyridoxine and pantothenic acid, indicating that the yeasts provide these four vitamins. Performance experiments of a similar design have demonstrated that the blood-feeding insects *Glossina* (Nogge, 1976) and *Rhodnius* derive B-vitamins from their midgut micro-organisms.

Performance experiments have demonstrated that not all insects derive nutritional advantage from their associated microbiota. For example, Howard and Bush (1989) found that the performance of the tephritid *Rhagletis pomonella* feeding on its natural diet of apples was not affected by elimination of the bacteria.

A major limitation of performance experiments, however, is that they can be conducted only on insects that are readily cultured in the laboratory, ideally on chemically defined diets. Many insects with microbial symbioses are intractable to these experimental approaches and for these species the significance of the micro-organisms can be deduced only from their nutritional ecology. Thus, micro-organisms associated with haematophagous insects are likely to be of nutritional significance to those species which depend on blood through their life cycle (e.g. *Glossina*, *Rhodnius*), but not to insects (e.g. fleas, mosquitoes) which use blood only as adults; and micro-organisms are probably of greater nutritional significance to Coleoptera feeding on nutrient-poor wood than on living plant tissues. However, nutritional factors may not account for the distribution of micro-organisms in all insects. For example, there is, at present, no convincing ecological explanation for the distribution of caecal symbioses in cerambycid beetles and there is no apparent relationship between diet and the possession of midgut symbionts in ants.

In summary, both performance experiments and study of the nutritional ecology of insects indicate that some, but not all, insects with a midgut microbiota derive nutrients from the association. The demonstration of a substantial microbial population in a particular site in an insect does not necessarily indicate that these micro-organisms are advantageous to the insect. For certain micro-organisms, the guts and associated tissues of insects may merely provide a favourable habitat (Douglas, 1995).

15.3.2 Interactions of midgut symbionts with vector-borne pathogens of other animals.

The transient micro-organisms in certain insects include important pathogens of other organisms that are transmitted by the insects. Recent studies suggest that the resident midgut microbiota may interact with these transients, influencing their probability of transmission. As indicated in section 15.1.2, this possibility has been developed for the transmission of trypanosomes by *Glossina* and *Rhodnius* (both of which have midgut symbioses, Table 15.1).

In *Glossina,* the susceptibility of teneral females to infection by trypanosomes is substantially reduced by elimination of the midgut endosymbionts (reviewed by Maudlin, 1991). Maudlin and Welburn (1988) and Welburn *et al.* (1994) suggest that the midgut symbionts degrade the chitin of the peritrophic matrix producing D-glucosamine, which inhibits the binding capacity of midgut lectins and that this results in increased susceptibility of the insects to trypanosomes. In support of this hypothesis, cultures of the S-endosymbionts have been shown to produce chitinase (Welburn *et al.,* 1993). Such interactions between the trypanosomes and the midgut symbionts, however, may not be universal. Studies on the potential interaction between the midgut actinomycete *Rhodococcus rhodnii* of *Rhodnius* and the trypanosome *Trypanosoma cruzi* have been inconclusive (Muhlpfordt, 1959; Jadin, 1967; Fennie, Beard and Tesh, unpublished data).

Genetic modification of micro-organisms in the midguts of insect vectors may prove to be a very valuable approach to reduce trypanosome transmission. *Rhodococcus nocardii* in *Rhodnius* has been used as a vehicle for expressing a selectable marker gene in the gut of *R. prolixus* (Beard *et al.,* 1992) and it may be possible very soon to stably express and secrete an antitrypanosomal agent, thus establishing lines of *Rhodnius* are incapable of transmitting *T. cruzi.*

15.4 ACQUISITION OF SYMBIONTS

The various ways in which insects acquire their complement of midgut symbionts is summarized in Table 15.1. Most insects with midgut symbionts acquire the micro-organisms via feeding. The insects vary, primarily, in the location of the micro-organisms that they ingest (in faeces, associated with the eggshell, etc.). Exceptionally, the midgut symbionts of a few insects, notably the ants, are not acquired by feeding but by the direct insertion of the micro-organisms into the cytoplasm of the unfertilized egg in the ovary. An excellent description of the mode of transmission of the ant symbionts is provided by Buchner (1965) and is not considered further here.

15.4.1 From the external environment

In general, symbiotic micro-organisms are neither widely distributed nor abundant apart from their animal hosts (Douglas, 1995) and therefore insects are unlikely to acquire midgut symbionts by the 'chance' association of the micro-organisms with their food. Faeces, however, are a potentially rich source of gut symbionts and infection by ingestion of the faeces of conspecifics occurs in many insects. Gut symbionts can spread very rapidly between insects in this way. In one study (Beard, unpublished data), two separate cartons were prepared, each containing 65 aposymbiotic first instar *R. prolixus* nymphs. To each carton 10 third instar nymphs were added which contained a population of genetically distinguishable *R. rhodnii* symbionts. After 5 months with successive monthly feedings, 25 of the younger nymphs were assayed from each group. The guts of all the insects assayed were positive exclusively for the genetically distinguishable symbiont introduced to their particular carton.

15.4.2 Egg smearing

The deposited eggs of those Coleoptera which have caecal symbionts usually bear symbionts on their surface. As each larval insect hatches, it feeds on the eggshell and becomes infected. The symbionts are derived directly from the larva's mother. This is because, during the metamorphosis of females, the symbionts become transferred to sacs associated with the ovipositor; and, as each egg is deposited, a small volume of micro-organisms is smeared onto the egg surface.

15.4.3 Acquisition by *Glossina* – a special case

The mode of symbiont acquisition by *Glossina* is unusual because the insects are viviparous and most of larval development occurs *in utero*. The developing larvae are nourished by secretions of the milk gland that are rich in the bacterial symbionts and the larvae acquire P-endosymbionts as they feed on these secretions. The S-endosymbionts may be acquired with the P-endosymbionts by larval feeding but this has not been demonstrated directly. Indeed, S-endosymbionts could not be confirmed by PCR amplification of ovary tissues, suggesting that they are not transferred to the eggs in the ovary (O'Neill *et al.*, 1992).

15.5 CONCLUSIONS

Two issues arise from this review of midgut symbioses in insects. First, there is a considerable body of information on the distribution of these symbioses among insects and the morphology of the associations and this classical literature is readily available to the modern entomologist, especially through the excellent book of Buchner (1965). Second, midgut symbioses are of immediate importance to applied entomologists. The midgut microbiota are crucial to the growth and reproduction of many insect pests, especially many timber-infesting beetles and, as indicated in section 15.3.2, they are of potential value in the development of novel methods to reduce the transmission of pathogenic trypanosomes by insect vectors.

Despite the availability of much basic information and the economic importance of many insects bearing midgut symbionts, the level of research into these systems is currently very low. With the advent of advanced molecular techniques to study both culturable and non-culturable micro-organisms, there is no scientific reason for the neglect of this topic – the time is perhaps right for a resurgence of research activity into the microbiology of insects.

REFERENCES

Aksoy, S. (1995) Molecular analysis of the endosymbionts of tsetse flies: 16S rDNA locus and over-expression of a chaperonin. *Insect Mol. Biol.*, **4**, 23–9.

Aksoy, S., Pourhosseini, A.A. and Chow, A. (1995) Mycetome endosymbionts of tsetse flies constitute a distinct lineage related to Enterobacteriaceae. *Insect Mol. Biol.*, **4**, 15–22.

Baines, S. (1956) The role of the symbiotic bacteria in the nutrition of *Rhodnius prolixus* (Hemiptera). *J. Exp. Med.*, **33**, 533–41.

Beard, C.B., Mason, P.W., Aksoy, S. *et al.* (1992) Transformation of an insect symbiont and expression of a foreign gene in the Chagas' disease vector *Rhodnius prolixus*. *Am. J. Trop. Med. Hyg.*, **46**, 195–200.

Beard, C.B., O'Neill, S.L., Mason, P. *et al.* (1993a) Genetic transformation and phylogeny of bacterial symbionts from tsetse. *Insect Mol. Biol.*, **1**, 123–31.

Beard, C.B., O'Neill, S.L., Tesh, R.B. *et al.* (1993b) Modification of arthropod vector competence via symbiotic bacteria. *Parasitol. Today*, **9**, 179–83.

Beckage, N.E., Thompson, S.N. and Federici, B.A. (eds) (1993a) *Parasites and Pathogens of Insects* Vol. 1, *Parasites*, Academic Press, San Diego, CA.

Beckage, N.E., Thompson, S.N. and Federici, B.A. (eds) (1993b) *Parasites and Pathogens of Insects* Vol. 2, *Pathogens*, Academic Press, San Diego, CA.

Bignell, D.E., Oskarsson, H. and Anderson, J.M. (1980) Colonisations of the epithelial face of the peritrophic membrane and the ectoperitrophic space by actinomycetes in a soil-feeding termite. *J. Invertebr. Pathol.*, **36**, 426–8.

Bismanis, J.E. (1976) Endosymbionts of *Sitodrepa panicea*. *Can. J. Microbiol.*, **22**, 1415–24.

Brecher, G. and Wigglesworth, V.B. (1944) The transmission of *Actinomyces rhodnii* Erikson in *Rhodnius prolixus* Stal (Hemiptera) and its influence on the growth of the host. *Parasitology*, **35**, 220–4.

Buchner, P. (1965) *Endosymbiosis of Animals with Palant Microorganisms*, Wiley, London.

Caugant, D.A., Levin, B.R. and Selander, R.K. (1981) Genetic diversity and temporal variation in the *Escherichia coli* population of a human host. *Genetics*, **98**, 467–90.

Cavanagh, P. and Marsden, P.D. (1969) Bacteria isolated from the gut of some reduviid bugs. *Trans. R. Soc. Trop. Med. Hyg.*, **63**, 415–16.

Douglas, A.E. (1996) The ecology of symbiotic micro-organisms. *Adv. Ecol. Res.*, (in press).

Drew, R.A.I. and Lloyd, A.C. (1991) Bacteria in the life cycle of tephritid fruit flies, in *Microbial Mediation of Plant–Herbivore Interactions* (eds P. Barbosa, V.A. Krischik and C.G. Jones), Wiley, London, pp. 441–65.

Fraenkel, G. and Blewitt, M. (1943) Intracellular symbionts of insects as a source of vitamins. *Nature*, **152**, 506–7.

Gumpert, J. and Schwartz, W. (1962) Untersuchungen uber die symbiose von tieren mit pilzen und bakterien, X. Die symbiose der triatominen 1. Aufzucht symbiontenhaltiger und symbiontenfreier triatominen und eigenschaften der bei triatominen vorkommenden mikroorganismen. *Z. Allg. Mikrobiol.*, **2**, 209–302.

Haas, F. and Konig, H. (1987) Characterisation of an anaerobic symbiont and the associated aerobic bacterial flora of *Pyrrhocoris apterus* (Heteroptera: Pyrrhocoridae). *FEMS Microbiol. Ecol.*, **45**, 99–106.

Howard, D.J. and Bush, G.L. (1989) Influence of bacteria on larval survival and development in *Rhagoletis* (Diptera: Tephritidae). *Ann. Entomol. Soc. Am.*, **82**, 633–40.

Huebner, E. and Davey, K.G. (1974) Bacteroids in the ovaries of the tsetse fly. *Nature*, **249**, 260–1.

Hurst, L. (1993) The incidences, mechanisms and evolution of cytoplasmic sex ratio distorters in animals. *Biol. Rev.*, **68**, 121–93.

Jadin, J. (1967) Du role des bacteries dans le tube digestif des insectes vecterus des plasmodidae et des trypanosomidae. *Ann. Soc. Belge Med. Trop.*, **47**, 331–42.

Jurzitza, G. (1979) The fungi symbiotic with anobiid beetles, in *Insect Fungal Symbiosis* (ed. L.R. Batra), Wiley, New York.

Kiefer, H. (1932) Der Einfluss von Kalte und Hunger aud die Symbionten der Anobiiden- und Cerambycidenlarven. *Zentralhl Bakteriol.*, 86.

Koch, A. (1960) Intracellular symbiosis in insects. *Annu. Rev. Microbiol.*, **14**, 121–40.

Luthy, P.D., Studer, D., Jaquet, F. and Yamvrias, C. (1983) Morphology and *in vitro* cultivation of the bacterial symbiote of *Dacus oleae*. *Schweiz Entomol. Geschalt*, **56**, 67–72.

Maudlin, I. (1991) Transmission of African trypanosomiasis: interactions among tsetse immune systems, symbionts and parasites, in *Advances in Disease Vector Research* (ed. K.F. Harris), Springer, New York, pp. 117–48.

Maudlin, I. and Welburn, S.C. (1988) Tsetse immunity and the transmission of trypanosomiasis. *Parasitol. Today*, **4**, 109–11.

Muhlpfordt, V.H. (1959) Der einfluss der darmsymbionten von *Rhoxnius prolixus* auf *Trypanosoma cruzi*. *Z. Tropenmed. Parasitol.*, **10**, 314–27.

Nogge, G. (1976) Sterility in tsetse fly (*Glossina morsitans* Westwood) caused by loss of symbionts. *Experientia*, **32**, 995.
Nyirady, S.A. (1973) The germfree culture of three species of Triatominae: *Triatoma protracta* (Uhler), *Triatoma rubida* (Uhler), and *Rhodnius prolixus* Stal. *J. Med. Entomol.*, **10**, 417–48.
O'Neill, S.L., Giordana, R., Colbert, A.M.E. *et al.* (1992) 16S rRNA phylogenetic analysis of the bacterial endosymbionts associated with cytoplasmic incompatibility in insects. *Proc. Natl Acad. Sci. USA*, **89**, 2699–702.
O'Neill, S.L., Gooding, R.H. and Aksoy, S. (1993) Phylogenetically distant symbiotic micro-organisms reside in *Glossina* midgut and ovary tissues. *Med. Vet. Entomol.*, **7**, 377–83.
Olsen, G.J., Woese, C.R. and Overbeck, R. (1994) The winds of (evolutionary) change: breathing new life into microbiology. *J. Bacteriol.*, **176**, 1–6.
Pell, P.E. and Southern, D.I. (1975a) Symbionts in the female tsetse fly *Glossina morsitans morsitans* (Dipt. Glossinidae). *Experientia*, **31**, 650–1.
Pell, P.E. and Southern, D.I. (1975b) Maternal transmission of symbiotic bacteroids in *Glossina morsitans morsitans*. *Trans. R. Soc. Trop. Med. Hyg.*, **169**, 283.
Pinnock, D.E. and Hess, R.T. (1974) The occurrence of intracellular rickettsia-like organisms in the tsetse flies, *Glossina morsitans, G. fuscipes, G. brevipalpis* and *G. pallidipes*. *Acta Trop.*, **31**, 70–9.
Reinhardt, C., Steiger, R. and Hecker, H. (1972) Ultrastructural study of the midgut mycetome-bacteroids of the tsetse flies *Glossina morsitans, G. fuscipes* and *G. brevipalpis* (Diptera, Brachycera). *Acta Trop.*, **29**, 280–8.
Richards, F.F. (1993) An approach to reducing arthropod vector competence. *Am. Soc. Microbiol. News*, **59**, 509–14.
Sapp, J. (1994) *Evolution by Association*, Oxford University Press, New York.
Steinhaus, E.A. (1946) *Insect Microbiology*, Comstock Publishing, Ithaca.
Welburn, S.C., Arnold, K., Maudlin, I. and Gooday, G.W. (1993) Rickettsia-like organisms and chitinase production in relation to transmission of trypanosomes by tsetse flies. *Parasitology*, **107**, 141–5.
Welburn, S.C., Maudlin, I. and Ellis, D.S. (1987) *In vitro* cultivation of rickettsia-like organisms from *Glossina* spp. *Ann. Trop. Med. Parasitol.*, **81**, 331–5.
Welburn, S.C., Maudlin, I. and Molyneux, D.H. (1994) Midgut lectin activity and sugar specificity in teneral and fed tsetse. *Med. Vet. Entomol.*, **8**, 81–7.

16

Insect-transmitted pathogens in the insect midgut

D.C. Kaslow and S. Welburn

16.1 INTRODUCTION

Pathogens are transmitted from one host to another by several routes: airborne, faecal–oral, sexual, direct contact, water-borne and by insects. Other than perhaps airborne transmission, insect-transmitted pathogens may have the greatest capability to spread rapidly through a susceptible host population, particularly when an animal reservoir is involved (perhaps the most dramatic example being the insect and then airborne transmission of Black Death in the Middle Ages).

For successful transmission, most blood-borne insect-transmitted pathogens must face the inhospitable environment of the insect gut which can be critically important to pathogen transmission. The transition from vertebrate bloodstream to the insect midgut might appear to be the most drastic that a pathogen faces in its life cycle requiring major biochemical, physiological and/or morphological changes. However, for many pathogens the difficult transition is from the insect back to the vertebrate host. As the disease in the vertebrate host is the major focus of concern, it is easy to forget that the insect is most likely the primary host and the vertebrate the incidental, accidental or secondary host.

This chapter will focus on the transitions that occur when the pathogen establishes itself in the invertebrate vector. We also briefly

Biology of the Insect Midgut.
Edited by M.J. Lehane and P.F. Billingsley.
Published in 1996 by Chapman & Hall, London.
ISBN 0 412 61670 X.

review some vector genetic and/or susceptibility factors that contribute to the establishment of midgut infections.

16.2 PROTOZOANS

16.2.1 Trypanosomes

We will concentrate on the two genera of the trypanosomatidae, *Trypanosoma* and *Leishmania*, which are of the greatest economic importance. The trypanosome cell surface changes as it moves between vertebrate and invertebrate hosts; glycoconjugates on the surface often playing a crucial role in determining parasite survival and infectivity both to its mammalian and insect hosts (McConville and Ferguson, 1993). These changes are not simply constitutive but may be induced by interactions between the parasite and the insect midgut environment.

(a) African trypanosomes

We shall concentrate here on the Salivaria which are transmitted during feeding via the mouthparts (subgenus *Duttonella* and *Nannomonas*) or salivary glands (subgenus *Trypanozoon*). Of particular interest here are *Trypanosoma brucei rhodesiense* and *T. b. gambiense* (*Trypanozoon*), the major pathogens of humans, and *T. b. brucei* (*Trypanozoon*) and *Trypanosoma congolense* (*Nannomonas*), pathogens of domestic livestock. These trypanosomes are transmitted exclusively by tsetse, both sexes of which are haematophagous, and the parasites spend a significant part of their lifetime in the tsetse midgut.

(i) Establishment of trypanosome infections in the midgut of tsetse
In the mammalian host the entire surface of the African trypanosome is covered in variant surface glycoprotein (VSG) which varies antigenically to protect against the immune system. On entering the tsetse midgut the parasite faces radically different environmental conditions: a drop in temperature from 37°C to about 25°C, a battery of digesive enzymes and a much more primitive, non-specific, 'immune' system. Glucose rapidly disappears from the blood meal so trypanosomes must develop the ability to oxidize alternative substrates. In response, bloodstream form trypanosomes differentiate into procyclic forms; this is accompanied by ultrastructural changes in the mitochondrion and activation of oxidase systems and a cytochrome electron transport system (Vickerman, 1985). The trypanosome apparently anticipates these impending changes in the bloodstream by differentiation from dividing long slender forms to non-dividing short stumpy forms pre-adapted to the rapid loss of glucose they will experience in the tsetse midgut (Matthews and Gull,

1994). The trypanosome sheds its VSG coat in about 4 h and replaces it with an invariant coat made from procyclins (or PARPs) (Roditi *et al.*, 1989) attached to the plasma membrane by a glycosylphosphatidylinositol (GPI) anchor (Ferguson *et al.*, 1994). There are about 5×10^6 copies of PARP per cell and the PARP-GPI anchor is thought to provide a protective glycocalyx for the midgut form of the parasite (Ferguson *et al.*, 1994).

Procyclic transformation takes place much more rapidly in the tsetse midgut than *in vitro*; Turner *et al.* (1988) found that half the bloodstream-form trypanosomes in the posterior midgut (PMG) lose their VSG coat within 2–3 h of ingestion compared with 7–8 h in the anterior midgut (AMG). The AMG of tsetse is involved solely in water absorption, no digestion takes place in AMG and a proteinase inhibitor is also active (Houseman, 1980). Proteinases, largely trypsin, are secreted in the PMG; it has been suggested that a midgut proteinase acts as the trigger for differentiation from bloodstream to procyclic forms (Imbuga *et al.*, 1992) and it may be that proteinases speed up the process. However, as differentiation takes place quite rapidly in the absence of such stimuli *in vitro* simply by dropping the temperature of cultures, it cannot be said that proteinases are essential triggers for transformation from bloodstream forms to procyclics (Brun and Schonenburger, 1981).

Wild tsetse populations are intrinsically refractory to the establishment of trypanozoon and *Nannomonas* midgut infections and much effort has gone into determining the basis of this refractoriness (for review see Maudlin, 1991). Many factors have been shown to influence establishment, the most obvious being fly age; teneral flies (which have not taken a blood meal) are far more susceptible to infection than flies which have previously been fed (Welburn and Maudlin, 1992). Most bloodstream form trypanosomes entering the fly midgut fail to transform and simply die (Turner *et al.*, 1988) but it takes flies about 5 days to completely clear trypanosomes from their midguts (Welburn *et al.*, 1989). As the surviving trypanosomes (and there need only be a single survivor to establish a midgut infection; Maudlin and Welburn, 1989) have already transformed to procyclics within a few hours of entering the midgut, it is to this form that our attention must turn in searching for refractory mechanisms.

The central role of tsetse midgut lectins in refractoriness was revealed through lectin inhibition studies *in vitro* and *in vivo*. *In vitro* studies showed the presence of a lectin in tsetse midguts with specificity for glucosamine and *N*-acetyl-D-glucosamine (GlcNAc), the agglutinating activity of which could be inhibited more strongly by glucosamine than GlcNAc (Ibrahim *et al.*, 1984). Similarly *in vivo* feeding of glucosamine (Maudlin and Welburn, 1987) and GlcNAc (Welburn *et al.*, 1993, 1994) to flies along with the infective feed dramatically increased infection rates in tsetse, suggesting that the midgut lectin was responsible for the death of incoming trypanosomes. A glucosyl lectin, secreted in response to the blood meal, is clearly involved in the death of procyclics in *Glossina m. morsitans*. Several molecules with glucosamine specificity have been isolated from tsetse midguts; a blood meal-induced lectin from *G. tachinoides* (Grubhoffer *et al.*, 1994) and a further molecule from *G. longipennis* (Osir *et al.*, 1995). Feeding purified procyclin with the infective feed also significantly increases midgut infection rates in tsetse (Welburn and Pearson, unpublished observation) suggesting that trypanosome death normally results from midgut lectin binding to PARP on the procyclic surface.

It has been suggested that glucosamine simply inhibits midgut trypsin activity, implying that this protease was normally responsible for procyclic death (Osir *et al.*, 1993). However, doses of glucosamine which reduce trypsin activity by less than 5% produce massive increases in midgut infection rates (7.5 mM, Maudlin and Welburn, 1987); similarly GlcNAc which does not inhibit midgut protease activity (Osir *et al.*, 1993), promotes midgut infections in tsetse in the same manner as D-glucosamine (Welburn *et al.*, 1993). Moreover, procyclic trypanosomes show remarkable tolerance to trypsin *in vitro* and feeding soybean trypsin inhibitor or anti-trypsin antibody to infected flies had no effect on infection rates (Welburn, unpublished observations).

(ii) Genetics and susceptibility to midgut infection in tsetse

Selection experiments showed that susceptibility to midgut infection in tsetse was a maternally inherited character (Maudlin, 1982) related to the inheritance of symbiotic rickettsia-like organisms (RLO). The relationship between RLO and susceptibility to infection was shown to be quantitative (Welburn and Maudlin, 1991). It was suggested that GlcNAc, which would specifically inhibit the potentially lethal midgut lectin, would be produced by the chitinolytic action of the fly's symbiotic bacteria on chitin (Welburn *et al.*, 1993). RLO numbers increase during larval/pupal development in tsetse (Welburn *et al.*, 1993) and the greater susceptibility of the teneral fly may be due to the action of RLO on chitin during pupation.

(iii) Peritrophic matrix and trypanosome infections

Trypanosomes entering the midgut of the tsetse find themselves enclosed in a chitinous tube – the peritrophic matrix – which is open only at the distal end in the hindgut. Trypanosomes have to by-pass this membrane to establish and divide in the ectoperitrophic space. The route of establishment – whether the trypanosomes enter the ecto-peritrophic space at the anterior end of the midgut or simply pass around the open end of the peritrophic matrix – is still, after all this time, questionable. The enhanced susceptibility to infection of teneral tsetse has been linked to the semifluid structure of the peritrophic matrix in unfed flies providing easier access to the ectoperitrophic space (Willett, 1966). Lehane and Msangi (1991) also suggested that the peritrophic matrix acted as a barrier to infection in non-teneral flies. As flies can be infected at any age (i.e. with fully formed peritrophic matrices), given that a suitable lectin inhibitor is added to the blood meal, it seems that the peritrophic matrix is not a mechanical barrier to establishment of infections in tsetse (Welburn and Maudlin, 1992).

(iv) Maturation of trypanosome infections in the midgut

Maturation of established midgut infections is not automatic as only a proportion of flies with midgut infections develop into mature infections. Several factors influence this transmissibility including fly species, fly sex and trypanosome genotype (for review see Maudlin and Welburn, 1994).

Maturation has been shown to be dependent on a signal from the tsetse midgut lectin. Although the midgut lectin promotes cell death in the midgut lumen it is also essential for trypanosome maturation. Maturation can be blocked by feeding lectin inhibitor for as little as 5 days post-infection, suggesting that trypanosomes receive the signal to mature soon after establishment (Welburn and Maudlin, 1989).

(b) New World trypanosomes

Trypanosoma cruzi is the causal organism of Chagas' disease, one of the most significant causes of mortality and morbidity in South and Central America.

The life cycle of *T. cruzi* in the insect vector is confined to the gut of triatomine bugs and parasites are transmitted in the faeces; both sexes of the bug vector are exclusively haematophagous. Hemipteran bugs do not produce a peritrophic matrix but instead produce an extracellular midgut coating in response to the blood meal (Billingsley and Downe, 1983, 1986). Unlike the holometabolous Diptera which blood feed only as adults, triatomine bugs are hemimetabolous and pass through five larval instars before moulting to the adult form. Each of these larval stages requires a blood meal to promote the next moult and larvae may pick up a trypanosome infection at any of these immature stages. Infections established during the first instar can persist through the moulting process; a bug such as *Panstrongylus megistus* may become infected after emerging from the egg and carry the same infection 9 months later as a mature adult. Adult bugs feed far less frequently than for example tsetse flies, which feed about every 3 days; *Rhodnius prolixus* females will feed in the laboratory about once a fortnight. The opportunity to acquire infections is less in triatomine trypanosome vectors than in tsetse but this is balanced by the ability of infected bugs to vector these infections for long periods.

(i) Establishment of infections in the gut of triatomines

Trypanosoma cruzi, unlike its African relatives, does not undergo antigenic variation via a variant surface glycoprotein (VSG) as it escapes from the humoral response of its mammalian hosts by invading the hosts cells and multiplying as amastigotes. Amastigotes subsequently give rise to trypomastigotes which can invade other cells or be ingested by bugs. On entering the bug gut with the blood meal, trypomastigotes transform within a few hours into epimastigotes which then multiply within the crop and midgut. In contrast to tsetse, infection rates in wild bugs can be very high (>70%, Pless *et al.*, 1992) and in the laboratory infection rates of over 90% are readily achieved in some species (Phillips and Bertram, 1967). Given these data it seems that there are few barriers to establishment of *T. cruzi* infections in midguts of triatomines. It has, however, long been known that some triatomine species are better vectors than others when infected with the same trypanosome strain (Zeledon and Vieto, 1957), but the basis for these interspecific differences remains unclear.

Like the African trypanosomes, the midgut form of *T. cruzi* is covered in glycoproteins anchored to the plasma membrane. The major molecule on the *T. cruzi* epimastigote cell surface is lipopeptidophosphoglycan (LPPG), with around 1.5×10^7 molecules/epimastigote. LPPG is found only in the epimastigote form and acts as a GPI anchor (Previato *et al.*, 1990; De Lederkremer *et al.*, 1991) for the 35–50 kDa mucin-like glycoproteins (Schenkman *et al.*, 1993). Whether lectins identified in the bug gut (Pereira *et al.*, 1981) specifically bind to these glycoproteins is unknown and lectin inhibition experiments, of the sort carried out in tsetse, have not been done.

Most interesting is the observation that azadirachtin (isolated from seeds of the neem tree) reduces the numbers of trypanosomes in an infected bug gut (Garcia *et al.*, 1989). Azadirachtin is an insect growth regulator and it has been suggested that interference with neuroendocrine function somehow disturbs trypanosome development in azadirachtin-treated bugs (Garcia and Azambuja, 1991).

(ii) Maturation of midgut infections

In the bug's rectum, epimastigotes mature into non-dividing metacyclic trypomastigotes which attach to the cuticular lining of the rectum (Boker and Schaub, 1984; Zeledon *et al.*, 1984). However, this attachment is not necessary for metacyclogenesis. De Isola *et al.* (1981) added midguts from fed bugs to *in vitro* cultures and found this increased the proportion of epimastigotes transforming to metacyclics suggesting, like Pereira *et al.* (1981), that some midgut factor was involved in metacyclogenesis. It has been suggested that bug species differ in their ability to promote maturation (Perlowagoraszumlewicz and Moreira, 1994) but it seems that it is the nature of the parasite which most affects metacyclogenesis in triatomines (Garcia and Dvorak, 1982). As with the tsetse fly, proteases secreted in the midgut of bugs have apparently no effect on the development of parasites in the bug (Garcia and Gilliam, 1980).

(iii) Genetics and susceptibility to infection in bugs

Given the long generation time of triatomine bugs it is hardly surprising that little published information is available on the genetics of susceptibility to *T. cruzi* infection in relation to Mendelian control. Selection experiments have indicated that the number of trypanosomes maturing and excreted by *Rhodnius prolixus* was heritable (Maudlin, 1976) which is interesting in view of the effects of a growth regulator on trypanosome numbers (Garcia and Azambuja, 1991).

16.2.2 *Leishmania*

Leishmaniasis affects the health of people throughout the tropics and subtropics in both Old and New Worlds. All *Leishmania* species are transmitted by sandflies (Phlebotominae) but considerable specificity has evolved between parasite and vector species so that 'good' and 'poor' vector species are found. Unlike tsetse and triatomine bugs, both sandfly sexes feed on sugar solutions but only the female feeds on blood, most species requiring a blood meal for egg production.

The genus *Leishmania* has been divided into three sections, Hypopolaria, Peripylaria and Suprapylaria, depending on where they develop in the insect gut. The Hypopolaria develop in the midgut and hindgut, the Peripylaria spend some time in the hindgut before migrating forwards, whereas the Suprapylaria do not enter the hindgut (Molyneux and Ashford, 1983). Most information is available on the life cycle of the Suprapylaria which do not have a midgut stage in sandflies (see review by Schlein, 1993).

(a) Establishment and maturation of midgut infections in sandflies

Leishmania are almost entirely intracellular in the vertebrate host, living as amastigotes in macrophages. Amastigotes enter the sandfly midgut in infected macrophages and are surrounded by a Type I peritrophic matrix which is secreted by cells along the entire length of the midgut only when a blood meal is taken; this contrasts with the continuously produced, open-ended, Type II peritrophic matrix of tsetse (Lehane, 1991). Parasites are released from the macrophages and after several divisions as amastigotes transform to free-swimming, long slender, nectomonad promastigotes within 2–3 days, confined within the peritrophic matrix. The peritrophic matrix breaks down as digestion of the blood meal is completed, releasing promastigotes which migrate to the thoracic midgut and cardiac valve. Here they attach, by their flagella, to the microvilli of the gut lining and in the region of the cardiac valve transform to short haptomonad promastigotes. Colonization of the oesophagus and pharynx by compact paramastigotes is followed by metacyclogenesis to produce metacyclics (Sacks, 1989).

As with the African trypanosomes, changes in the structure of the parasite membrane accompany the progression of the parasite through its vertebrate–invertebrate cycle. *Leishmania* amastigotes, unlike the mammalian form of African trypanosomes, have almost no surface protein or lipophosphoglycan (LPG) but are covered in a glycocalyx of sugar containing glycoinositolphospholipids (GIPLs) (McConville and Blackwell, 1991; Kelleher *et al.*, 1994). The *Leishmania* promastigote surface contains several GPI-anchored proteins, lipophosphoglycan (LPG) and GIPLs (McConville and Ferguson, 1993). The major surface glycoprotein is a protease (gp63), and there are $3–5 \times 10^6$ LPG molecules anchored by GPI and with terminal mannose on each promastigote. Only procyclic stages can bind to the insect midgut epithelium; maturation to metacyclic forms is accompanied by detachment from the midgut. The change from non-infective procyclic stages to infective metacyclics is accompanied by critical changes in surface LPG in *Leishmania donovani*. During metacyclogenesis the LPG molecule gets much bigger due to a doubling in the number of disaccharide repeats and the number of sugars terminally exposed decreases. The secret of attachment to the insect midgut is thought to lie in the capping sugars (galactose) which are exposed on procyclic LPG – these sugars are then masked on the metacyclic LPG surface (McConville *et al.*, 1992; Sacks *et al.*, 1995). Binding of *Leishmania major* promastigotes to *P. papatasi* midguts can be inhibited *in vitro* by phosphoglycan derived from procyclic LPG, suggesting that binding of promastigotes is mediated through LPG terminally exposed galactose residues (Pimenta *et al.*, 1994). This model implies that there are galactose-binding lectins on the insect midgut epithelium to trap procyclic LPG. Lectin-like receptors have been demonstrated in the midgut of sandflies which can be specifically inhibited by galactosamine, glucosamine and mannosamine (Volf *et al.*, 1994).

Studies of infected sandflies present great technical problems (Schlein, 1993). It is well established that many Old World *Leishmania* parasites are only transmissible by certain vector species (for review see Killick-Kendrick, 1979). Again, LPG is thought to mediate this specificity as there are variations between *Leishmania* species in exposed LPG terminal sugars which would affect midgut binding (Sacks *et al.*, 1994).

The proteinase on the surface of procyclics (gp63) is thought to be involved in blood meal digestion before the peritrophic matrix breaks down (Davies *et al.*, 1990). Escape from the peritrophic matrix is in turn promoted by chitinolytic enzymes secreted by procyclics (Schlein *et al.*, 1991).

Sugar feeds are important in transmission of *Leishmania*-infected flies fed on sugars are better vectors than flies fed exclusively on blood. Parasites are pumped out of the gut and into the mammalian host in the opposite direction to the incoming host blood – an apparent anomaly which is dependent on damage to the cardiac valve which holds the parasites in the insect gut. Chitinolytic enzymes secreted by procyclics are again involved in digesting the chitinous lining of the valve exposing the underlying tissues to proteolytic action. The result is a leaky valve which allows parasites to escape during sugar feeds (Schlein *et al.*, 1992). Haemoglobin has been shown to inhibit chitinolytic activity which accounts for the poor transmission rates of flies fed exclusively on blood (Schlein and Jacobson, 1994a). Certain plant foods have been shown to either promote or inhibit midgut infections in sandflies again suggesting the importance of parasite–carbohydrate midgut interactions in determining infections (Schlein and Jacobson, 1994b).

(b) Genetics of susceptibility of sandflies to Leishmania *infection*

Lines of *Phlebotomus papatasi* have been selected for refractoriness and susceptibility to infection with *Leishmania major*. The genetic basis of the relationship is not yet known but is not thought to be a single major gene (Wu and Tesh, 1990a,b).

16.2.3 *Plasmodium* and Haemosporina

Of the three family members of Haemosporina (Plasmodiidae, Haemoproteidae and Leukocytozoidae), the four major species of *Plasmodium* (*P. falciparum*, *P. vivax*, *P. malariae* and *P. ovale*) that infect humans are the most important pathogens in humans.

Almost all human *Plasmodium* infections are transmitted by *Anopheles* mosquitoes, whereas avian malaria parasites are transmitted primarily by *Culex* and *Aedes* (Garnham, 1980). Most mosquito species do not transmit malaria parasites; of those species that do transmit, most wild caught mosquitoes are uninfected (Wernsdorfer, 1980). The restricted mosquito host range and seemingly tenuous transmission of malaria parasites in susceptible vectors are undoubtedly due to the hostile environment the parasite faces in the mosquito midgut.

(a) Establishment of infection in the midgut of mosquitoes

For successful transmission malaria parasites must penetrate three potential barriers in the midgut: (1) the peritrophic matrix – a chitinous sac-like structure that forms around the blood meal; (2) the midgut epithelium; and (3) the basal lamina that surrounds the midgut. The parasite must also evade the defence mechanism(s) (e.g. melanization and lytic peptides) mosquitoes have evolved to protect themselves against infection by micro-organisms (bacterial, viral and parasitic) ingested in the blood meal.

To infect a mosquito successfully the malaria parasite must undergo a series of developmental changes that are undoubtedly designed to protect the parasite against the changing environment of the mosquito midgut during a blood meal. The male and female *Plasmodium* gametes emerge from erythrocytes in the midgut in response to a drop in temperature and an increase in pH (Nijhout and Carter, 1978); the now extracellular parasites must be resistant to complement fixation and an array of antibody responses that might have developed during the course of infection in the vertebrate host. Despite the rapid diaphoresis of the food bolus that occurs almost immediately upon ingestion of a blood meal, the gametes must come together and fertilize. The parasite must then avoid the coming onslaught of digestive enzymes secreted by the mosquito midgut. The parasite must move to the peritrophic matrix that has started to form around the periphery of the food bolus. The motile parasite must then breach the peritrophic matrix (PM) (Sieber *et al.*, 1991). Once it passes through the chitinous sac-like PM, the ookinete must recognize and bind to the surface of the midgut epithelium and then penetrate it, protecting itself against whatever lytic factors and digestive enzymes are present within the midgut epithelial cells. The parasite having reached the basal lamina must rapidly switch to a non-invasive form rounding up to form a cyst that will accumulate sporozoites. For the most part, the molecular mechanisms involved in these processes have not yet been identified.

(b) Parasite changes during transition from vertebrate to vector host

Immediately upon ingestion by the mosquito, the parasite emerges from within the erythrocyte to form gametes. Several prominent gamete surface proteins have been characterized (e.g. Pfs230, Pfs48/45, Pfs40/10) but their function is unknown (Kaslow, 1993). The gamete surface must be involved in fertilization and in protecting the parasite from complement-mediated lysis. Antibodies to Pfs230 may render the parasite susceptible to complement fixation (Quakyi *et al.*, 1987) and antibodies to Pfs48/45 may interfere with fertilization (Vermeulen *et al.*, 1985). The presence of the gamete surface proteins is short-lived as they are replaced on the zygote surface by a series of new proteins, the most prominent of which appears to be a family of GPI-anchored proteins comprising four tandem epidermal growth factor (EGF)-like domains (Duffy *et al.*, 1993). The EGF-like domains may promote binding to the midgut epithelium. Alternatively these cysteine-rich proteins, which are highly resistant to proteolysis (Kaslow *et al.*, 1988), may protect against insect proteolytic enzymes. In the avian-type malaria parasites (*P. falciparum* and *P. gallinaceum*) the P25-type EGF-like proteins found on zygotes, ookinetes and early ookinetes appear to be replaced by the P28-type EGF-like proteins found on ookinetes and oocysts (Kaslow *et al.*, 1989; Duffy *et al.*, 1993).

(c) Changes in the cytoplasmic-RNA and microtubules

The most intriguing changes within the parasite occur in the transcription of ribosomal RNA (Waters *et al.*, 1989; Zhu *et al.*, 1990; Li *et al.*, 1994). Two genes encode the small subunit rRNAs of *Plasmodium*: an A gene transcribed predominantly in the asexual blood-stage parasites; and a C gene transcribed mainly in the sporogonic forms that develop in the mosquito. The switch from A to C gene expression involves rRNA processing (Waters *et al.*, 1989). The precursor C-gene transcripts accumulate in gametocytes and these precursors are processed to mature size in the zygote and the early ookinete. As the C-gene precursor rRNA appears, breakdown of the dominant A-gene rRNA occurs. By late in the oocyst stage, the A-gene transcripts are virtually replaced by mature C-gene transcripts. Other than the drop in temperature and the rise in pH in the midgut, it is unclear what effect the signals the parasite receives from the midgut have on inducing these changes in transcription and processing.

In the blood meal, the parasite develops from a round sedentary zygote into a highly invasive oblong ookinete. The morphological changes and the ability to locomote are undoubtedly mediated by rearrangement of the cytoskeleton of the parasite. The signal(s) that causes these changes in which filamentous structures are involved have not yet been identified.

(d) Interaction of parasite with the midgut

The malaria parasite interfaces with the mosquito midgut at several levels: biochemically (resistance to digestion by the proteolytic, lipolytic and glycolytic enzymes that may be secreted by the epithelium), structurally (penetration of peritrophic matrix) and cellularly (with the epithelial cells and perhaps endocrine cells).

The ingestion of a blood meal by a mosquito induces a cascade of insect biochemical activity, including secretion of a series of enzymes. Some of the enzymatic activity is anabolic, e.g. chitin synthase, but most is catabolic, to digest the cellular and particulate blood meal into a form that can be readily absorbed for the nutritional needs required by the gravid female. The most abundant proteolytic enzymes are the trypsins, chymotrypsins and carboxypeptidase. To survive in this hostile environment the parasite must have developed resistance factors although the identity of these putative parasite surface and/or secreted molecules is unknown.

In response to midgut distension, the mosquito forms a chitinous sac-like PM around the blood meal. Unlike some pathogens that escape the blood meal before the PM fully forms or after the PM disintegrates, the malaria parasite must cross this potential barrier after the PM has fully formed and solidified. Stohler (1957) and Sieber *et al.* (1991) observed *P. gallinaceum* ookinetes traversing the PM of *Ae. egypti* and observed that many parasites appeared trapped within the PM, suggesting that the PM presents a significant barrier to the parasites to further develop in the mosquito midgut. Sieber *et al.* (1991) also observed parasites focally disrupting the chitin microfibrils and noted that the disruption was associated with electron-dense material near the anterior end of the parasite, suggestive of a parasite-secreted substance mediating digestion of the peritrophic matrix. Chitinolytic activity had been measured in the culture supernatant of mature ookinetes (Huber *et al.*, 1991) and inhibition of chitinase activity using the inhibitor allosamidin was found to block the parasite's ability to infect mosquitoes. Shahabuddin *et al.* (1993) found that the blocking activity of allosamidin could be reversed by artificially disrupting the PM *in vivo*. In an intricate cross-talk between parasite and vector, the parasite chitinase appears to be secreted as a zymogen, requiring activation by a late trypsin produced by the vector. Inhibition of the mosquito trypsin using chemical inhibitors or antitrypsin antibodies, rendered the parasite avirulent (Shahabuddin *et al.*, 1994). The production by the parasite of an inactive chitinase which requires activation by an insect enzyme may be of use as a timing device to co-ordinate invasion of the PM and midgut epithelium with the mosquito's nutritional, hormonal or physiological state.

(e) Interaction with the midgut epithelium

Few of the ingested infectious gametocytes penetrate the midgut epithelium to form oocysts and Vaughan *et al.* (1992) observed a 69-fold reduction in the number of *P. falciparum* parasites making the transition. The peritrophic matrix does not appear to be a major barrier to mosquito infection as studies in which the PM is either artificially removed (through enzymatic digestion by exogeneous chitinase) or is prevented from forming (using the chitin synthase inhibitor, polyoxin D) show no significant increase in oocyst development (Shahabuddin *et al.*, 1994).

The midgut epithelium, however, may present a barrier to infection due to the presence of a lytic factor responsible for the refractoriness of Old World anopheline species to avian malaria parasites. Vernick *et al.* (1995) observed that refractoriness of *P. gallinaceum* to *An. gambiae*, appears due to ookinete lysis with non-comitant vacuolation of the infected midgut epithelial cell. The ability of the midgut epithelium to lyse ookinetes (i.e. the refractory phenotype) is controlled by a single dominant genetic locus that is linked to the esterase allele, *EstA* (Vernick *et al.*, 1988). The exact biochemical reaction that mediates lysis is not yet known but it is clear that this process is different from the previously described refractory mechanism by which malaria parasites are encapsulated by a melanization reaction (Collins *et al.*, 1986).

Parasites may also be encapsulated by melanization outside the midgut epithelium in the space between the basolateral epithelial cell membrane and basal lamina. The parasite gets through the midgut epithelium but is neutralized just as it begins to form an oocyst. In contrast to ookinete lysis, encapsulation is a recessive or codominant phenotype which maps to at least two genetic loci, one of which is linked to *EstA* (Vernick and Collins, 1989; Vernick *et al.*, 1989). The *EstA* linkage common to encapsulation and ookinete lysis may suggest a common biochemical pathway; however, the 2La inversion that carries this allele contains up to 10% of the mosquitoes genome so the linkage may be simply fortuitous.

To evade lysis and encapsulation, it is likely that the successful parasite has developed resistance mechanisms based on surface or secreted molecules, but their identity and function has not yet been elucidated. Disrupting the normal function of this process using specific antibodies may provide novel candidates for transmission-blocking vaccines.

(f) Midgut basal lamina

The basal lamina separates the midgut epithelial cells from the haemolymph and haemocoel, and comprises a complex mixture of extracellular matrix proteins secreted by the midgut epithelium. The basal lamina that ensheathes the midgut epithelium may provide a signal to the highly invasive, oblong ookinete to switch back to the sedentary, round form that characterizes both zygote and oocyst.

Neither the parasite receptor nor the basal lamina ligand have been identified and again these molecules may present potent targets for malaria transmission-blocking vaccines.

16.3 FILARIA

Filarial worms are transmitted by a variety of insects – mosquitoes, blackflies, ceratopogonid midges, tabanids, fleas – but we shall concern ourselves with mosquito and blackfly vectors about which most is known.

16.3.1 Lymphatic filaria

Mosquitoes are vectors of several species of filarial nematodes the life cycles of which are similar. Young nematodes, microfilariae (mf), develop in the female uterus, are released into the lymph and migrate to the blood where they are taken up by the mosquito. On entering the mosquito gut the young mf (L_1) shed the eggshell relic or sheath (exsheathment is essential for the subsequent development of the worm in the mosquito). Worms then pass through the midgut wall using piercing stylets for penetration. *Wuchereria bancrofti* and *Brugia malayi*, two of the most important human filarial parasites, mf migrate through the haemocoel to the thoracic muscles where they moult twice (to L_3), re-enter the haemocoel and move towards the mouthparts.

Development in the flight muscle takes up to 12 days and migration to the head region begins 13 days after the initial blood meal (Paily *et al.*, 1995). The L_3 forms enter the mammalian host by breaking through the labial membrane at the next feed. With *Dirofilaria immitis*, a parasite of domestic animals, mf enter the mosquito midgut then migrate to the Malpighian tubules where they penetrate the cells and emerge into the haemocoel as second stage larvae. The relationships between filarial worms and their mosquito vectors have been reviewed by Townson and Chaithong (1991).

(a) Establishment of filarial infections in mosquitoes

Only 50% of *Brugia pahangi* mf entering an *Aedes aegypti* midgut reach the haemocoel (Ewert, 1965). The first obstacle for incoming worms is the cibarial and buccopharyngeal armature of some *Culex* and *Anopheles* mosquitoes which act as a physical and sometimes lethal barrier (Coluzzi and Trabucchi, 1968; Bryan *et al.*, 1974). *Brugia* mf retain a remnant of the eggshell, the extracuticular sheath, in the vertebrate host and exsheathment takes place soon after entering the mosquito gut. Exsheathment, which is essential for further development of the worm, is thought to be promoted by a larval chitinase acting on *N*-acetyl-D-glucosamine oligomers of the larval sheath (Fuhrman *et al.*, 1992). Ivermectin has been shown to inhibit mf exsheathment, blocking further development in mosquitoes (Rao *et al.*, 1992). The next midgut hurdle for worms might be thought to be the peritrophic matrix – a Type I membrane produced along the length of the midgut in response to the blood meal (Lehane, 1991) – but filarial worms have usually left the midgut before the peritrophic matix has hardened (Ewert, 1965; Sutherland *et al.*, 1986).

Ham *et al.* (1991) have shown that the penetration of *B. pahangi* mf from *Aedes aegypti* midgut to haemocoel can be increased by the addition of GlcNAc to the infective feed. These results suggest that a midgut lectin normally blocks midgut migration, perhaps by binding GlcNAc moieties on the mf sheath (Fuhrman *et al.*, 1992). The midgut wall clearly presents a barrier to mf entering the haemocoel but little is known of this process (Townson and Chaithong, 1991).

(b) Genetics of mosquito filarial susceptibility

Aedes aegypti is used as the favoured laboratory model for the investigation of filarial susceptibility in mosquitoes but is of no importance as a vector of human filariae in the wild (Townson and Chaitong, 1991). Numerous lines of susceptible and refractory *A. aegypti* have been selected, refractoriness being controlled by a sex-linked dominant gene. Attempts to select for refractoriness to filarial infection in the human vector, *Culex quinquefasciatus*, have been largely unsuccessful (for a detailed review of the genetics of susceptibility to filarial infection see Curtis and Graves, 1983).

The most important barrier to infection in mosquitoes is the response provoked by mf in the haemocoel, particularly the encapsulation and melanization of mf. This is clearly beyond the scope of the present review but those interested should see the detailed review by Christensen and Severson (1993). There is no direct evidence linking a particular gene product to mosquito refractoriness/susceptibility to mf infection and it appears that vector competence may be a quantitative trait involving many genes (Severson, 1994).

16.3.2 *Onchocerca*

Onchocerca spp. are transmitted by blackflies (Simuliidae), the most notorious being *O. volvulus* which causes onchocerchiasis or river blindness in the tropics. Blackflies scarify the skin of the mammalian host allowing microfilariae (mf) to migrate into the pool of blood for uptake by the insect – mosquitoes with piercing mouthparts cannot, for this simple reason, act as vectors of onchocerchiasis. The life cycle of the worms within blackflies takes about 12 days and is very similar to that of filariae in mosquitoes: mf rapidly leave the midgut and enter the thoracic musculature via the haemocoel; following two moults the L_3 larvae migrate to the mouthparts.

As with *Leishmania* and its sandfly vectors, the relationship between different *Onchocerca* species and their blackfly vectors is highly specific but whether this involves the midgut barrier is unclear.

(a) Establishment of Onchocerca *infections in blackflies*

As in the mosquito, the first obstacle for blood meal-borne mf is the cibarial armature of the blackfly; these teeth can be lethal to *Onchocerca* mf. It appears that the majority of ingested mf are trapped in the coagulating blood meal and go no further. Those mf escaping coagulation are trapped within the peritrophic matrix, considered to be a potent barrier in blackflies to mf migration (Orihel, 1975). Mf are thought to use a cephalic hook and proteinases to escape these barriers. It is also suggested that a midgut-secreted lectin is involved in mf-midgut binding in Simuliidae (Ham *et al.*, 1991).

The major barrier to *Simulium* infection lies in the haemolymph where a battery of defence molecules awaits mf entering the haemocoel. This is a subject beyond the scope of the present work but readers are referred to the comprehensive review of Ham *et al.* (1995).

16.4 ARBOVIRUSES

16.4.2 Overviews of biology and ecology of arboviruses

Insect-borne viral-mediated diseases, such as yellow fever, have had a substantial impact on man's health and mobility during history. For example, the first attempts to construct the Panama Canal ended in failure in large part because of infection of canal workers with yellow fever. After control of the mosquito vector in Panama in 1903 the canal was finally completed. Numerous other examples exist of the spread of yellow fever and/or its insect host, *Aedes aegypti*, by ship or other carrier and underlie the requirement of many tropical countries for an international certificate of vaccination against yellow fever for incomers. With the possible exception of airborne transmission of emerging viruses, transmission of viruses by the ubiquitous insects of tropical and subtropical regions of the world poses a constant health threat to the human populations living in these regions (for review see Manson-Bahr and Apted, 1982). Recent examples of such outbreaks include a variety of encephalitides (Murray Valley encephalitis in Australia, Japanese encephalitis in Asia, Western equine encephalitis in the Americas).

Insect transmission of pathogenic viruses can be either mechanical (non-replicative in the insect vector) or biological (replicative in both the vertebrate and invertebrate hosts). Viruses that replicate in the insect host are referred to as arboviruses. Despite unsubstantiated reports in the popular press that the human immunodeficiency virus (HIV) can be transmitted mechanically by mosquitoes, none of the important human pathogenic viruses is known to be transmitted mechanically by insects. Thus, all vector-borne viruses pathogenic in humans appear to be arboviruses and most are assumed to be transmitted to humans and other vertebrates by the adult insect. Even viruses that are passed in insects from one generation to the next by transovarian transmission are likely to encounter the insect midgut at some period in their life cycle.

A wide variety of insects transmit arboviruses. The four major families of important arboviruses are the Togaviridae, Bunyaviridae, Rhabdoviridae and Reoviridae. The Togaviridae includes such members as yellow fever virus, Dengue virus and the Alphaviruses and Flaviviruses that cause the encephalitides and the tick-borne (e.g. Kyasanur Forest disease) and mosquito-borne (e.g. Ross River, Chikungunya, Sindbis, O'Nyong-nyong and west Nile) diseases. The family to which a virus belongs is not predictive of the family of invertebrate hosts in which the virus is competent to replicate.

16.4.3 Examples of molecular interactions of arboviruses with insect midgut structures

Unlike many eukaryotic pathogens that (a) have some means of motility; (b) can survive extracellularly; or (c) have the ability to change surface molecules in real time in the vector gut, arboviruses are non-motile, obligate intracellular organisms that do not appear to be able to rapidly change the virion envelope in response to transition from vertebrate to invertebrate. Viruses in the blood meal must rely on the chance interaction with a gut epithelial cell and must be able to replicate within the vertebrate and vector hosts with impunity. Paradoxically, arboviruses, which must be able to invade and replicate in a variety of vertebrate and vector host cells, have an amazingly restrictive host range. Despite the widespread accessibility to anopheline mosquitoes, most pathogenically important arboviruses are transmitted by culicine mosquitoes, predominately *Culex* and *Aedes* (for review, see Leake, 1992).

It is not clear what underlies *in vivo* host specificity. A number of vertebrate host, ecological and environmental factors have been suggested that might determine host range specificity (see Hardy, 1983, 1988 for reviews). It is well established that the oral route is the principal means of infection of mosquitoes by arboviruses and a number of clear barriers might determine the observed host range specificity of arbovirus transmission. The first barrier would be the ability of the virus to survive the change in the environment from vertebrate host to the vector midgut. Second would be the ability of the virus to specifically invade the vector alimentary tract, replicate therein and perhaps secondarily infect surrounding alimentary cells by cell-to-cell infection. Third would be the ability of the virus to escape the alimentary tract and be released in the haemolymph/haemocoel.

With regard to alimentary tract invasion, the midgut seems to be the operationally important barrier. Although anterior areas of the alimentary tract do become infected with viruses most probably due to infected blood collected in the diverticula (Kuberski, 1979; Romoser *et al.*, 1987) and, in some cases, the virus may pass directly into the haemolymph without replicating in midgut cells (Weaver, 1986), it is usually the posterior midgut epithelial cells that are the first to become infected. Thus, it is likely that one determinant of host range specificity of arboviruses is the presence or absence of specific viral receptors on the midgut cells for one or more of the virion envelope proteins that encapsulate most arboviruses. Again, for reasons that would seem paradoxical, this specificity may be tightly regulated. Woodward (1991) recently found that a single amino acid change in the glycoprotein E2h epitope (MARV 1A3B-7) of Venezuelan equine encephalitis (VEE) virus decreased virus growth when compared with the wild-type using an *Aedes albopictus* cell line, C6/36. The MARVs replicated as efficiently as the parent virus when inoculated into *Aedes aegypti* mosquitoes, but MARV 1A3B-7 was restricted in its ability to infect and disseminate from the midgut following oral infection, demonstrating this single amino acid change in the E2 glycoprotein can affect the ability of VEE virus to replicate and disseminate in *Ae. aegypti* mosquitoes.

The specific receptors on the midgut cell surface recognized by virions have not yet been identified; however, the brush border is a likely site of initial attachment of some if not most arboviruses. Houk *et al.* (1990) recently compared viral binding of brush border fragments (BBF) isolated from epithelial cells of mosquitoes that are either susceptible (*Culex tarsalis*) or refractory (*Culex pipiens*) to oral infection by western equine encephalomyelitis (WEE) virus. BBF isolated from susceptible mosquitoes specifically bound significantly greater amounts of virus, compared to BBF isolated from refractory mosquitoes. Scatchard analysis of the binding data estimated that more than a million binding sites are present on each epithelial cell and the binding affinity is quite high ($K_a = 2.2 \times 10^{11}$). Whether the specific receptor is a protein, lipid, carbohydrate or a combination thereof, the broad host range of many arboviruses (e.g. Semiliki Forest virus and Sindbis) *in vitro* but not necessarily *in vivo* suggests that other factors come into play in the midgut (see below) and also determines host range specificity.

Once the virus binds the first-point-of-contact cell, the virus must invade and replicate. If the virus enters the vector cell then it most likely involves endocytosis. The acidic environment of the endosome appears to induce conformational changes in viral proteins that eventually lead to release of the viral nucleic acids into the cytoplasm. Such a phenomenon has been observed *in vitro* for a variety of arboviruses during invasion of mosquito cells (for review see Taddieu *et al.*, 1982). The next step is where the similarities between the biology of arboviruses in mammalian and mosquito cells diverge. The classic cytopathological effects (CPE) of viruses in mammalian cells does not appear to occur in mosquito cells, at least *in vitro* (however, evidence is accumulating that cytopathology occurs in gut and salivary gland cells *in vivo*, affecting blood feeding, fecundity and mosquito longevity). Rather than rapid replication within and lysis of the host cell, mosquito cells have persistent arbovirus infections that are not usually associated with CPE which suggests a close evolutionary relationship between virus and vector. The mosquito cell inherently regulates viral replication, possibly by regulating nucleic acid replication.

The next barrier for the arbovirus to breach after initial binding and replication in a midgut cell, is either to invade surrounding cells, spreading the infection throughout the midgut and/or to be released from the basolateral aspect of the midgut cells to infect the haemolymph. For the latter to occur, the arbovirus would have to contend with penetrating the basal lamina. The mechanism the arboviruses use to breach this barrier is basically unknown; however, cytopathic effects of some arboviruses (e.g. alphavirus (Weaver *et al.*, 1988)) on midgut structures may contribute. That the escape from the midgut cells into the haemolymph represents a real barrier comes from studies such as those by Miller and Mitchell (1991). Using two inbred *Aedes aegypti* mosquito lines, one that manifested a resistant and another a susceptible phenotype following ingestion of flaviviruses (e.g. yellow fever and Dengue), resistance to infection of the haemolymph was pinpointed to lack of virus movement from the midgut. Genetic crosses between the two lines indicated: (a) the phenotypes were codominant; and (b) there is a single, major controlling genetic locus and a second locus capable of modulating the phenotype of the major gene. The identity of the products of the controlling and the modulating genetic loci has not yet been reported.

16.4.3 Exploitation of arboviruses in research: transfecting midgut structures

Arboviruses can infect insect cells and seize complete control of the molecular and biochemical machinery of the cell. Often this leads to destruction and lysis of the infected mammalian cells. Some arboviruses are, however, highly promiscuous, infecting a broad range of vertebrate and invertebrate host cells, and causing few cytopathogenic changes in infected insect cells or can be attenuated to limit their cytopathogenic effects. These viruses can be exploited for transferring genetic or biochemical information into insect cells, enabling one to study a variety of cellular events. Viruses that look particularly promising are Sindbis and Semliki Forest virus.

16.5 BACTERIA

Insect-transmitted pathogens include the genera *Rickettsia, Coxiella, Ehrlichia, Anaplasma, Borrelia, Yersinia* and *Francisella*. A wide range of human pathogens are also mechanically transmitted by flies. An intricate relationship has developed between the insect-borne pathogens and their arthropod vectors, particularly with regard to the vector gut. Vector guts differ markedly and it is not surprising that establishment of infection by bacterial pathogens in mosquitoes is fundamentally different from ticks. For each of the major pathogens, the disease process is outlined and what is known about the mechanisms that the bacterial pathogens use to successfully evade the barriers presented by the gut.

16.5.1 Rickettsiales

The order Rickettsiales consists of Gram-negative intracellular organisms all of which are associated with arthropods (Moulder, 1974). There are three recognized families: the Anaplasmatacae, the Bartonellacea and the Rickettsiaceae.

Anaplasma are important pathogens for ruminants, especially cattle (see Ristic, 1977 for review). *Anaplasma* is biologically transmitted by the ixodid ticks and mechanically by blood-sucking insects including mosquitoes (*Psorophora*) and horse flies. *Bartonella bacilliformis* is transmitted by the phlebotomine sandfly (*Lutzomyia verrucarum*) and is highly pathogenic to humans (see Schultz, 1968 for review). Humans appear to be a reservoir for this organism which causes two successive stages of Carrion's disease: local verruga nodules and Oroya. The Rickettsiaceae, which includes Wolbachieae, Ehrlichieae and Rickettsieae, are the most important family in terms of human pathogens and are transmitted by ticks. The Wolbachieae (which do not occur in vertebrates) are being actively studied as potential vehicles for introducing recombinant genes into arthropods.

16.5.2 *Borrelia*

Epidemics of louse-borne relapsing fever, caused by *Borrelia recurrentis*, have incapacitated millions of people in the early part of this century

(Burgdorfer, 1976). The Gram-negative spirochaete, transmitted by *Pediculus humanus* and related species and ticks, circulates in the vertebrate host's bloodstream until ingested in a blood meal. *Borrelia burgdorferi* causes Lyme disease, first described in the 1970s (Steere *et al.*, 1977) when a cluster of arthritis and skin eruptions occurred in Lyme, Connecticut, USA. Transmission of *Borrelia* has been extensively studied in the tick.

(a) Establishment of infection in the tick gut

Most spirochaetes of *B. recurrentis* die in the midgut after ingestion, but those that survive to invade and traverse the epithelium, multiply in the haemolymph. As the spirochaete does not invade the salivary gland and for the most part is dead when excreted in the faeces, transmission appears to occur when an infected louse is crushed or dies on the skin of the vertebrate host (Buxton, 1950).

Spirochaetes of *B. duttoni* readily invade and traverse the midgut (probably prior to formation of the peritrophic matrix) and enter the haemocoel where they multiply and invade several organs including the salivary and coxal glands where they continue to multiply. When the tick feeds, borrelias are introduced into the vertebrate blood via the saliva and through the host's skin from the excreted infected coxal fluid (Geigy, 1968).

Changes in *B. burgdorferi* surface proteins are associated with invasion of the tick gut. The spirochaetes remain along the microvilli in the gut lumen until structural changes in the gut occur which allow invasion (Burgdorfer *et al.*, 1982, 1989; Zung *et al.*, 1989). The extensive epithelial cell and spirochaete changes appear to be triggered by the blood meal; an increase in temperature and blood feeding trigger a major alteration to the spirochaetal outer membrane. In unfed ticks, spirochaetes produce outer surface protein A (OspA), but not OspC (Fikrig *et al.*, 1992). In the mammal, spirochaetes express OspC but little or no OspA (Roehrig *et al.*, 1992). Schwan *et al.* (1995) found *in vitro* that the change in surface proteins is in part temperature regulated (OspC is produced by spirochaetes at 32–37°C but not at 24°C) and by blood-feeding (spirochaetes in the midgut of fully engorged ticks no longer have OspA but present OspC on their surface) suggesting two environmental cues. This change is associated with the migration of the spirochaete out of the gut into the haemolymph and finally into the tick's salivary gland. Antibodies to the OspC form of the spirochaete (infected via ticks) do not protect against pathogenicity in mammals whereas antibodies to OspA/B forms (intradermally infected or immunized) may prevent symptoms (Kurtenbach *et al.*, 1994). An OspA and B vaccine offered protection against *B. burgdorferi* infection in a beagle model (Coughlin *et al.*, 1995). It appears that as the tick takes in a blood meal, it also ingests anti-OspA antibodies and these prevent the spirochaete from switching to the OspC; in the absence of OspC surface antigen further development of the spirochaete in the tick is blocked.

16.5.3 *Yersinia*

Yersinia pestis ssp., the aetiological agent of plague, is a Gram-negative facultative anaerobe which is transmitted by fleas or may be air-borne. Plague can be bubonic, septicaemic and pneumonic. Virulence is associated with three antigens (Fraction 1, V and W antigens) and exo- and endotoxins (Chen and Meyer, 1954; Burrows and Bacon, 1956). In bubonic plague the pathogen causes buboes (swollen, suppurative lymph glands in the axillary, inguinal and femoral regions); once released from the lymph system, the disease disseminates into the bloodstream to cause septicaemic plague. Circulating organisms are taken up by blood-sucking fleas and transmitted by the insect vector. Once the pathogen invades the respiratory system, it becomes highly infectious and can then be transmitted in air-borne droplets.

(a) *Establishment of infection in the flea gut*

The transmission of *Y. pestis* spp. by fleas (e.g. *Xenopsylla cheopis*, the rat flea; *Diamanus montanus*, the squirrel flea; *Pulex irritans*, the human flea) is curious (Bacot and Martin, 1914). When the flea feeds, a blood bolus is transported to the midgut via the oesophagus by the pharyngeal pump. The proventriculus acts as a valve to prevent regurgitation. Plague organisms multiply in the midgut and form an expanding gelatinous mass of coagulated blood and bacteria that fills the midgut and may occlude the lumen of the proventriculus and protrude into the oesophagus. When the proventricular lumen is occluded, the flea is said to be 'blocked'. When a blocked flea attempts to feed, the pharyngeal pump dilates the oesophagus with the blood meal that cannot pass into the 'blocked' proventriculus and midgut, the pump is unable to function and blood that has accumulated in the oesophagus, along with pieces of the gelatinous mass containing 25 000 to 1 000 000 infectious *Yersinia*, is forcefully regurgitated back into the vertebrate host. As a 'blocked' flea cannot successfully engorge it passes from one host to another in an attempt to feed, transmitting plague. Completely 'blocked' fleas are not considered the best vectors of plague because if the flea remains blocked it will die of dehydration. Partially blocked fleas are considered better vectors of plague (Burroughs, 1947) as in this case, the midgut recanalizes, but because the proventriculus is damaged, these fleas will regurgitate organisms during feeding. These fleas are less susceptible to desiccation and can survive and transmit for prolonged periods.

Coagulase and fibrinolytic associated activity are considered to contribute to the induction of 'blockage' in plague bacillus-infected fleas. Using insertional mutagenesis to disrupt the gene encoding (*pla*) in *Y. pestis*, fleas infected with a *pla*- strain of *Y. pestis* had a significantly lower mortality than *pla*+ strains, suggesting that expression of the coagulase- and fibrinolytic-associated phenotype contributes to the deleterious effect of *Y. pestis* on fleas (McDonough and Falkow, 1989; McDonough *et al.*, 1993).

16.5.4 *Francisella* (tularaemia)

Tularaemia is a sporadic disease limited usually to lagomorphs (rabbits) and rodents but humans can act as accidental hosts. *Francisella tularensis* is transmitted from host to host by several routes including contaminated food, water, inhalation and blood-sucking arthropods. Lice, mites, ticks and deerflies are vectors. Direct contact with infected animals or with the discharges or bites of arthropods is the major route of pathogenic infections to humans. The mortality rate in untreated humans is approximately 7.5% (Boyce, 1975).

16.5.5 Other pathogens

The housefly is a potential vector of a wide range of human pathogens, particularly viral (e.g. hepatitis and polio viruses) and bacterial (e.g. *Vibrio, E. coli, Chlamydia, Diphtheria, Mycobacteria, Salmonella* and *Shigella*) but also protozoan (e.g. *Entamoeba histolytica* which causes amoebic dysentery), nematode and cestode pathogens (e.g. hookworm) (for review see Greenberg, 1965, 1973). The continual movement of houseflies between faeces/garbage and food makes them superb carriers of a variety of human diseases, and more than 100 different pathogens have been recovered from this ubiquitous pest. Whether the fly actually transmits all of the pathogens it harbours is not clear. Of the three potential means (carriage on the body and legs, regurgitation and defecation) by which houseflies can transmit pathogens, one involves passage through the gut. During passage through the housefly, pathogens may replicate within the gut. Many bacterial pathogens are not very successful as they rely on the pathogen's ability to compete with the normal flora of the housefly gut. In the case of *Salmonella*, the initial inoculum of bacteria is important in order to out-compete the normal flora of the housefly gut. Ingestion of more than a thousand *Salmonella* organisms is necessary for the pathogen to appear in the housefly's faeces.

It is unlikely the feeding larva carry over bacterial pathogens to the adult stage. Although the larvae thrive in environments rich in micro-organisms, several factors reduce the vertical transmission of pathogens: (a) prepupae stop feeding and thus starve the bacterial pathogen; (b) the midgut is highly acidic (pH 3–3.5); and (c) during pupation the gut lining and its contents are shed.

16.6 CONCLUSION

We have attempted to present a summary of diseases caused by insect-transmitted human and, in some cases, animal pathogens. In the few examples in which anything is known about their interactions, we have briefly highlighted some of the important mechanisms (molecular and/or physical) used by those pathogens when interacting with the insect midgut. As evidenced by the paucity of published reports cited in this chapter, much of the molecular, biochemical and physical interactions of pathogens with the insect midgut remains unexplored. A more thorough understanding of these interactions will undoubtedly reveal intricate mechanisms that are perhaps vulnerable to relatively easily induced immune or chemical interventions. The power of modern biotechnology is now being harnessed and directed towards elucidating these interactions.

REFERENCES

Bacot, A.W. and Martin, C.J. (1914) Observations on the mechanism of the transmission of plague by fleas. *J. Hyg.*, **13**, Plague Supplement III, 423–39.

Billingsley, P.F. and Downe, A.E.R. (1983) Ultrastructural-changes in posterior midgut cells associated with blood-feeding in adult female *Rhodnius prolixus* Stål (Hemiptera, Reduviidae). *Can. J. Zool.*, **61**, 2574–86.

Billingsley, P.F. and Downe, A.E.R. (1986) Nondigestive cell-types in the midgut epithelium of *Rhodnius prolixus* (Hemiptera, Reduviidae). *J. Med. Entomol.*, **23**, 212–16.

Boker, C.A. and Schaub, G.A. (1984) Scanning electron-microscopic studies of *Trypanosoma cruzi* in the rectum of its vector *Triatoma infestans*. *Z. Parasit.*, **70**, 459–69.

Boyce, J.M. (1975) Recent trends in the epidemiology of tularemia in the United States. *J. Infect. Dis.*, **131**, 197–9.

Brun, R. and Schonenberger, M. (1981) Stimulating effect of citrate and *cis*-aconitate on the transformation of *Trypanosoma brucei* bloodstream forms to procyclic forms *in vitro*. *Z. Parasit.*, **66**, 17–24.

Bryan, J.H., Oothman, P., Andrews, B.J. and McGreevy, P.B. (1974) Effects of pharyngeal armatures of mosquitoes on microfilariae of *Brugia pahangi*. *Trans. R. Soc. Trop. Med. Hyg.*, **68**, 14–15.

Burgdorfer, W. (1976) The epidemiology of the relapsing fevers, in *The Biology of Parasitic Spirochetes* (ed. R.C. Johnson), Academic Press, New York, pp. 191–200.

Burgdorfer, W., Barbour, A.G., Hayes, S.F. *et al.* (1982) Lyme disease – a tick-borne spirochetosis? *Science*, **216**, 1317–19.

Burgdorfer, W., Hayes, S.F. and Corwin, D. (1989) Pathophysiology of the Lyme disease spirochete, *Borrelia burgdorferi*, in ixodid ticks. *Rev. Infect. Dis.*, **11**, S1442–50.

Burroughs, A.L. (1947) Sylvatic plague studies. The vector efficiency of nine species of fleas compared with *Xenopsylla cheopis*. *J. Hyg.*, **45**, 371–96.

Burrows, T.W. and Bacon, G.A. (1956) The basis of virulence in *Pasteurella pestis*. An antigen determining virulence. *Br. J. Exp. Pathol.*, **37**, 481–93.

Buxton, P.A. (1950) *The Louse*, Edward Arnold, London.

Chen, T.H. and Meyer, K.F. (1954) Studies on immunization against plague. VII. A haemagglutinin test with the protein fraction of *Pasteurella pestis*: a serologic comparison of virulent and avirulent strains with observations on the structure of the bacterial cells and its relationship to infection and immunity. *J. Immunol.*, **72**, 282–98.

Christensen, B.M. and Severson, D.W. (1993) Biochemical and molecular basis of mosquito susceptibility to *Plasmodium* and Filarioid nematodes, in *Parasites and Pathogens of Insects* (eds N.E. Beckage, S.N. Thompson and B.A. Federici), Academic Press, San Diego, California, pp. 245–66.

Collins, F.S, Sakai, R.K., Vernick, K.D. *et al.* (1986) Genetic selection of a *Plasmodium*-refractory strain of the malaria vector *Anopheles gambiae*. *Science*, **234**, 607–10.

Coluzzi, M. and Trabucchi, S. (1968) Importanza dell'armatura bucco-faringea in *Anophelese Culex* in relazione alle infezioni con *Dirofilaria*. *Parassitologia*, **10**, 47–59.

Coughlin, R.T., Fish, D., Mather, T.N. *et al.* (1995) Protection of dogs from Lyme disease with a vaccine containing outer surface protein (Osp) A, OspB, and the saponin adjuvant QS21. *J. Infect. Dis.*, **171**, 1049–52.

Curtis, C.F. and Graves, P.M. (1983) Genetic variation in the ability of insects to transmit filariae, trypanosomes and malarial parasites. *Curr. Top. Vector Res.*, **1**, 31–62.

Davies, C.R., Cooper, A.M., Peacock, C. *et al.* (1990) Expression of LPG and gp63 by different developmental stages of *Leishmania major* in the sandfly *Phlebotomus papatasi*. *Parasitology*, **101**, 337.

De Isola, E.L.D., Lammel, E.M., Katzin, V.J. and Gonzalez-Cappa, S.M. (1981) Influence of organ extracts of *Triatoma infestans* on differentiation of *Trypanosoma cruzi*. *J. Parasitol.*, **67**, 53–8.

De Lederkremer, R.M., Lima, C., Ramirez, M.I. *et al.* (1991) Complete structure of the glycan of lipopeptidophosphoglycan from *Trypanosoma cruzi* epimastigotes. *J. Biol. Chem.*, **266**, 23670–5.

Duffy, P.E., Pimenta, P. and Kaslow, D.C. (1993) Pgs28 belongs to a family of epidermal growth factor-like antigens that are targets of malaria transmission-blocking antibodies. *J. Exp. Med.*, **177**, 505–10.

Ewert, A. (1965) Exsheathment of the microfilariae of *Brugia pahangi* in susceptible and refractory mosquitoes. *Am. J. Trop. Med. Hyg.*, **14**, 260–2.

Ferguson, M.A.J., Brimacombe, J.S., Cottaz, S. *et al.* (1994) Glycosyl-phosphatidylinositol molecules of the parasite and the host. *Parasitology*, **108**, S45–S54.

Fikrig, E., Telford, S.R. III, Barthold, S.W. *et al.* (1992) Elimination of *Borrelia burgdorferi* from vector ticks feeding on OspA-immunized mice. *Proc. Natl Acad. Sci. USA*, **89**, 5418–21.

Fuhrman, J.A., Lane, W.S., Smith, R.F. *et al.* (1992) Transmission-blocking

antibodies recognize microfilarial chitinase in brugian lymphatic filariasis. *Proc. Natl Acad. Sci. USA*, **89**, 1548–52.
Garcia, E.S. and Azambuja, P. (1991) Development and interactions of *Trypanosoma cruzi* within the insect vector. *Parasitol. Today*, **7**, 240–4.
Garcia, E.S. and Dvorak, J.A. (1982) Growth and development of 2 *Trypanosoma cruzi* clones in the arthropod *Dipetalogaster maximus*. *Am. J. Trop. Med. Hyg.*, **31**, 259–62.
Garcia, E.S. and Gilliam, F.C. (1980) *Trypanosoma cruzi* development is independent of protein digestion in the gut of *Rhodnius prolixus*. *J. Parasitol.*, **66**, 1052–3.
Garcia, E.S., Gonzalez, M.S., Azambuja, P. and Rembold. H. (1989) Chagas-disease and its insect vector-effect of azadirachtin-A on the interaction of a triatomine host (*Rhodnius prolixus*) and its parasite (*Trypanosoma cruzi*). *Z. Naturforsch.* [C], **44**, 317–22.
Garnham, P.P.C. (1980) Malaria in its various vertebrate hosts, in *Malaria* Vol. I (ed. J.P. Kreier), Academic Press, New York, pp. 95–144.
Geigy, R. (1968) Relapsing fevers, *Infectious Blood Diseases of Man and Animals* Vol. II (eds D. Weinman and M. Ristic), Academic Press, New York, pp. 175–216.
Greenberg, B. (1965) Flies and disease. *Sci. Am.*, **213**, 92–9.
Greenberg, B. (1973) *Flies and Disease*. II. *Biology and Disease Transmission*, Princeton University Press, Princeton, New Jersey.
Grubhoffer, L., Muska, M. and Volf, P. (1994) Midgut haemagglutinins in five species of tsetse flies (*Glossina* spp.): two different lectin systems in the midgut of *Glossina tachinoides*. *Folia Parasitol.*, **41**, 229–32.
Ham, P.J., Hagen, H.E., Baxter, A.J. and Grunewald, J. (1995) Mechanisms of resistance to onchocerca infection in blackflies. *Parasitol. Today*, **11**, 63–7.
Ham, P.J., Phiri, J.S. and Nolan, G.P. (1991) Effect of *N*-acetyl-D-glucosamine on the migration of *Brugia pahangi* microfilariae into the haemocoel of *Aedes aegypti*. *Med. Vet. Entomol.*, **5**, 485–93.
Hardy, J.L. (1988) in *The Arboviruses: Ecology and Epidemiology* Vol. I (ed. T.P. Monath), CRC Press, Boca Raton, pp. 87–126.
Hardy, J.L., Houk, E.J., Kramer, L.D. and Reeves, W.C. (1983) Intrinsic factors affecting vector competence of mosquitoes for arboviruses. *Annu. Rev. Entomol.*, **28**, 229–62.
Houk, E.J., Arcus, Y.M., Hardy, J.L. and Kramer, L.D. (1990) Binding of western equine encephalomyelitis virus to brush border fragments isolated from mesenteronal epithelial cells of mosquitoes. *Virus Res.*, **17**, 105–17.
Houseman, J.G. (1980) Anterior midgut proteinase inhibitor from *Glossina morsitans morsitans* Westwood (Diptera: Glossinidae) and its effects upon tsetse digestive enzymes. *Can. J. Zool.*, **58**, 79–87.
Huber, H., Cabib, E. and Miller, L.H. (1991) Malaria chitinase and penetration of the mosquito peritrophic membrane. *Proc. Natl Acad. Sci. USA*, **88**, 2807–10.
Ibrahim, E.A.R., Ingram, G.A. and Molyneux, D.H. (1984) Haemagglutinins and parasite agglutinins in haemolymph and gut of *Glossina*. *Tropenmed. Parasitol.*, **35**, 151–6.
Imbuga, M.O., Osir, E.O., Labongo, V.L. *et al.* (1992) Studies on tsetse midgut factors that induce differentiation of blood-stream *Trypanosoma brucei brucei in vitro*. *Parasitol. Res.*, **78**, 10–15.
Kaslow, D.C. (1993) Transmission-blocking immunity against malaria and other vector-borne diseases. *Curr. Opin. Immunol.*, **5**, 557–65.
Kaslow, D.C., Quakyi, I.A., Syin, C. *et al.* (1988) A vaccine candidate from the

sexual stage of human malaria that contains EGF-like domains. *Nature*, **333**, 74–6.

Kaslow, D.C., Syin, C., McCutchan, T.F. and Miller, L.H. (1989) Comparison of the primary structure of the 25 kDa ookinete surface antigens of *Plasmodium falciparum* and *Plasmodium gallinaceum* reveal six conserved regions. *Mol. Biochem. Parasitol.*, **33**, 283–8.

Kelleher, M., Curtis, J.M., Sacks, D.L. *et al.* (1994) Epitope mapping of monoclonal-antibodies directed against lipophosphoglycan of *Leishmania major* promastigotes. *Mol. Biochem. Parasitol.*, **66**, 187–200.

Killick-Kendrick, R. (1979) in *Biology of the Kinetoplastida* Vol. 2 (eds W.H.R. Lumsden and D.A. Evans), Academic Press, London, pp. 395–499.

Kuberski, T. (1979) Fluorescent antibody studies on the development of dengue-2 virus in *Aedes albopictus* (Diptera: Culicidae). *J. Med. Entomol.*, **16**, 343–9.

Kurtenbach, K., Dizij, A., Seitz, H.M. *et al.* (1994) Differential immune responses to *Borrelia burgdorferi* in European wild rodent species influence spirochete transmission to *Ixodes ricinus* L. (Acari: Ixodidae). *Infect. Immun.*, **62**, 5344–52.

Leake, C.J. (1992) Arbovirus – mosquito interactions and vector specificity. *Parasitol. Today*, **8**, 123–8.

Lehane, M.J. (1991) *Biology of Blood-sucking Insects*, HarperCollins Academic, London.

Lehane, M.J. and Msangi, A.R. (1991) Lectin and peritrophic membrane development in the gut of *Glossina morsitans morsitans* and a discussion of their role in protecting the fly against trypanosome infection. *Med. Vet. Entomol.*, **5**, 495–501.

Li, J., McConkey, G.A., Rogers, M.J. *et al.* (1994) *Plasmodium*: the developmentally regulated ribosome. *Exp. Parasitol.*, **78**, 437–41.

Manson-Bahr, P.E.C. and Apted, F.I.C. (1982) *Manson's Tropical Diseases*, Baillière Tindall, London.

Matthews, K.R. and Gull, K. (1994) Cycles within cycles – the interplay between differentiation and cell-division in *Trypanosoma brucei*. *Parasitol. Today*, **10**, 473–6.

Maudlin, I. (1976) Inheritance of susceptibility to *Trypanosoma cruzi* infection in *Rhodnius prolixus*. *Nature*, **262**, 214–15.

Maudlin, I. (1982) Inheritance of susceptibility to *Trypanosoma congolense* infection in *Glossina morsitans*. *Ann. Trop. Med. Parasitol.*, **76**, 255–27.

Maudlin, I. (1991) Transmission of African trypanosomiasis: interactions among tsetse immune system, symbionts, and parasites. *Adv. Dis. Vector Res.*, **7**, 117–48.

Maudlin, I. and Welburn, S.C. (1987) Lectin mediated establishment of midgut infections of *Trypanosoma congolense* and *Trypanosoma brucei* in *Glossina morsitans*. *Trop. Med. Parasitol.*, **38**, 167–70.

Maudlin, I. and Welburn, S.C. (1989) A single trypansome is sufficient to infect a tsetse-fly. *Ann. Trop. Med. Parasitol.*, **83**, 431–3.

Maudlin, I. and Welburn, S.C. (1994) Maturation of trypanosome infections in tsetse. *Exp. Parasitol.*, **79**, 202–5.

McConville, M.J. and Blackwell, J.M. (1991) Developmental-changes in the glycosylated phosphatidylinositols of *Leishmania donovani* – characterization of the promastigote and amastigote glycolipids. *J. Biol. Chem.*, **266**, 15170–9.

McConville, M.J. and Ferguson, M.A.J. (1993) The structure, biosynthesis and function of glycosylated phosphatidylinositols in the parasitic protozoa and higher eukaryotes. *Biochem. J.*, **294**, 305–24.

McConville, M.J., Turco, S.J., Ferguson, M.A.J. and Sacks, D.L. (1992)

Developmental modification of lipophosphoglycan during the differentiation of *Leishmania major* promastigotes to an infectious stage. *EMBO J.*, **11**, 3593–600.

McDonough, K.A., Barnes, A.M., Quan, T.J. *et al.* (1993) Mutation in the *pla* gene of *Yersinia pestis* alters the course of the Plague Bacillus-Flea (Siphonaptera: Ceratophyllidae) interaction. *J. Med. Entomol.*, **30**, 772–80.

McDonough, K.A. and Falkow, S. (1989) A *Yersinia pestis*-specific DNA fragment encodes temperature-dependent coagulase and fibrinolytic-associated phenotypes. *Med. Microbiol.*, **3**, 776–5.

Miller, B.R. and Mitchell, C.J. (1991) Genetic selection of a flavivirus strain of the yellow fever mosquito *Aedes aegypti. Am. J. Trop. Med. Hyg.*, **45**, 399–407.

Molyneux, D.H. and Ashford, R.W. (1983) *The Biology of* Trypanosoma *and* Leishmania, *Parasites of Man and Domestic Animals*, Taylor and Francis, London.

Moulder, J.W. (1974) Order I. Rickettsiales Gieszczkiewicz 1939, 25, in *Bergey's Manual of Determinative Bacteriology* (eds R.E. Buchanan and N.E. Gibbons), Williams and Wilkins, Baltimore, pp. 882–914.

Nijhout, M.M. and Carter, R. (1978) Gamete development in malaria parasite: bicarbonate dependent stimulation by pH *in vitro. Parasitology*, **76**, 39–53.

Orihel, T.C. (1975) The peritrophic membrane: its role as a barrier to infection of the arthropod host, in *Invertebrate Immunity. Mechanisms of Invertebrate Vector–Parasite Relations* (eds K. Maramarosch and R.E. Shope), Academic Press, New York.

Osir, E.O., Abubakar, L. and Imbuga, M.O. (1995) Purification and characterization of a midgut lectin–trypsin complex from the tsetse-fly *Glossina longipennis. Parasitol. Res.*, **81**, 276–81.

Osir, E.O., Imbuga, M.O. and Onyango, P. (1993) Inhibition of *Glossina morsitans* midgut trypsin activity by D-glucosamine. *Parasitol. Res.*, **79**, 93–7.

Paily, K.P., Hoti, S.L., Manonmani, A.M. and Balaraman, K. (1995) Longevity and migration of *Wuchereria bancrofti* infective larvae and their distribution pattern in relation to the resting and feeding-behaviour of the vector mosquito, *Culex quinquefasciatus. Ann. Trop. Med. Parasitol.*, **89**, 39–47.

Pereira, M.E.A., Andrade, A.F.B. and Ribeiro, J.M.C. (1981) Lectins of distinct specificity in *Rhodnius prolixus* interact selectively with *Trypanosoma cruzi. Science*, **211**, 597–600.

Perlowagoraszumlewicz, A. and Moreira, C.J.D. (1994) *In vivo* differentiation of *Trypanosoma cruzi*. 1. Experimental evidence of the influence of vector species on metacyclogenesis. *Mem. Inst. Oswaldo Cruz*, **89**, 603–18.

Phillips, N.R. and Bertram, D.S. (1967) Laboratory studies of *Trypanosoma cruzi* infections in *Rhodnius prolixus* – larvae and adults, in *Triatoma infestans, T. protracta* and *T. maculata* – adults. *J. Med. Entomol.*, **4**, 168–74.

Pimenta, P.F.P., Saraiva, E.M.B., Rowton, E. *et al.* (1994) Evidence that the vectorial competence of phlebotomine sand flies for different species of *Leishmania* is controlled by structural polymorphisms in the surface lipophosphoglycan. *Proc. Natl Acad. Sci. USA*, **91**, 9155–9.

Pless, M., Juranek, D., Kozarsky, P. *et al.* (1992) The epidemiology of Chagas' disease in a hyperendemic area of Cochabamba, Bolivia – a clinical study including electrocardiography, seroreactivity to *Trypanosoma cruzi*, xenodiagnosis, and domiciliary triatomine distribution. *Am. J. Trop. Med. Hyg.*, **47**, 539–46.

Previato, J.O., Gorin, P.A.J., Mazurek, M. *et al.* (1990) Primary structure of the oligosaccharide chain of lipopeptidophosphoglycan of epimastigote forms of *Trypanosoma cruzi. J. Biol. Chem.*, **265**, 2518–26.

Quakyi, I.A., Carter, R., Rener, J. *et al.* (1987) The 230 kDa gamete surface protein of *Plasmodium falciparum* is also a target for transmission-blocking antibodies. *J. Immunol.*, **139**, 4213–17.

Rao, U.R., Vickery, A.C., Kwa, B.H. and Nayar, J.K. (1992) *Brugia malayi* – ivermectin inhibits the exsheathment of microfilariae. *Am. J. Trop. Med. Hyg.*, **46**, 183–8.

Ristic, M. (1977) Bovine anaplasmosis, in *Parasitic Protozoa* Vol. IV (ed. J.P. Kreier), Academic Press, New York, pp. 235–49.

Roditi, I., Schwarz, H., Pearson, T.W. *et al.* (1989) Procyclin gene-expression and loss of the variant surface glycoprotein during differentiation of *Trypanosoma brucei*. *J. Cell Biol.*, **108**, 737–46.

Roehrig, J.T., Piesman, J., Hunt, A.R. *et al.* (1992) The hamster immune response to tick-transmitted *Borrelia burgdorferi* differs from the response to needle-inoculated, cultured organisms. *J. Immunol.*, **149**, 3648–53.

Romoser, W.S., Faran, M.E. and Bailey, C.L. (1987) Newly recognized route of arbovirus dissemination from the mosquito (Diptera: Culicidae) midgut. *J. Med. Entomol.*, **24**, 431–2.

Sacks, D.L. (1989) Metacyclogenesis in *Leishmania* promastigotes. *Exp. Parasitol.*, **69**, 100–3.

Sacks, D.L., Pimenta, P.F.P., McConville, M.J. *et al.* (1995) Stage-specific binding of *Leishmania donovani* to the sand fly vector midgut is regulated by conformational-changes in the abundant surface lipophosphoglycan. *J. Exp. Med.*, **181**, 685–97.

Sacks, D.L., Saraiva, E.M., Rowton, E. *et al.* (1994) The role of the lipophosphoglycan of *Leishmania* in vector competence. *Parasitology*, **108**, 55–62.

Schenkman, S., Ferguson, M.A.J., Heise, N. *et al.* (1993) Mucin-like glycoproteins linked to the membrane by glycosylphosphatidylinositol anchor are the major acceptors of sialic-acid in a reaction catalyzed by *trans*-sialidase in metacyclic forms of *Trypanosoma cruzi*. *Mol. Biochem. Parasitol.*, **59**, 293–303.

Schlein, Y. (1993) *Leishmania* and sandflies – interactions in the life-cycle and transmission. *Parasitol. Today*, **9**, 255–8.

Schlein, Y. and Jacobson, R.L. (1994a) Hemoglobin inhibits the development of infective promastigotes and chitinase secretion in *Leishmania major* cultures. *Parasitology*, **109**, 23–8.

Schlein, Y. and Jacobson, R.L. (1994b) Mortality of *Leishmania major* in *Phlebotamus papatasi*. *Am. J. Trop. Med. Hyg.*, **50**, 20–7.

Schlein, Y., Jacobson, R.L. and Messer, G. (1992) *Leishmania* infections damage the feeding mechanism of the sandfly vector and implement parasite transmission by bite. *Proc. Natl Acad. Sci. USA*, **89**, 9944–8.

Schlein, Y., Jacobson, R.L. and Shlomai, J. (1991) Chitinase secreted by *Leishmania* functions in the sandfly vector. *Proc. R. Soc. London, B*, **245**, 121–6.

Schultz, M.G. (1968) A history of bartonellosis (Carrion's disease). *Am. J. Trop. Med. Hyg.*, **17**, 503–15.

Schwan, T.G., Piesman, J., Golde, W.T. *et al.* (1995) Induction of an outer surface protein on *Borrelia burgdorferi* during tick feeding. *Proc. Natl Acad. Sci. USA*, **92**, 2909–13.

Severson, D.W. (1994) Applications of molecular marker analysis to mosquito vector competence. *Parasitol. Today*, **10**, 336–40.

Shahabuddin, M., Criscio, M. and Kaslow, D.C. (1994) Unique specificity of *in vitro* inhibition of mosquito midgut trypsin-like activity correlates with *in vivo* inhibition of malaria parasite infectivity. *Exp. Parasitol.*, **80**, 212–19.

Shahabuddin, M., Toyoshima, T., Aikawa, M. and Kaslow, D.C. (1993) Transmission-blocking activity of a chitinase inhibitor and activation of malarial parasite chitinase by mosquito protease. *Proc. Natl Acad. Sci. USA*, **90**, 4266–70.

Sieber, K.P., Huber, M., Kaslow, D. *et al.* (1991) The peritrophic membrane as a barrier: its penetration by *Plasmodium gallinaceum* and the effect of a monoclonal antibody to ookinetes. *Exp. Parasitol.*, **72**, 145–56.

Steere, A.C., Hardin, J.A. and Malawista, S.E. (1977) Erythema chronicum migrans and Lyme arthritis: cryoimmunoglobulins and clinical activity of skin and joints. *Science*, **196**, 1121–2.

Stöhler, H. (1957) Analyse des Infectionsverlanfes von *Plasmodium gallinaceum* in Darme von *Aedes egypti. Acta Trop.*, **14**, 302–52.

Sutherland, D.R., Christensen, B.M. and Lasee, B.A. (1986) Midgut barrier as a possible factor in filarial vector competency in *Aedes trivittatus. J. Invert. Pathol.*, **47**, 1–7.

Tardieu, M., Epstein, R.L. and Weiner, H.L. (1982) Interaction of viruses with cell surface receptors. *Int. Rev. Cytol.*, **80**, 27–61.

Townson, H. and Chaithong, U. (1991) Mosquito host influences on development of filariae. *Ann. Trop. Med. Parasitol.*, **85**, 149–63.

Turner, C.M.R., Barry, J.D. and Vickerman, K. (1988) Loss of variable antigen during transformation of *Trypanosoma brucei rhodesiense* from bloodstream to procyclic forms in the tsetse fly. *Parasitol. Res.*, **74**, 507–11.

Vaughan, J.A., Noden, B.H. and Beier, J.C. (1992) Population dynamics of *Plasmodium falciparum* sporogony in laboratory infected *Anopheles gambiae. J. Parasitol.*, **78**, 716–24.

Vermeulen, A.N., Ponnudurai, T., Beckers, P.J.A. *et al.* (1985) Sequential expression of antigens on sexual stages of *Plasmodium falciparum* accessible to transmission-blocking antibodies in the mosquito. *J. Exp. Med.*, **162**, 1460–76.

Vernick, K.D. and Collins, F.H. (1989) Association of a *Plasmodium*-refractory phenotype with an esterase locus in *Anopheles gambiae. Am. J. Trop. Med. Hyg.*, **40**, 593–7.

Vernick, K.D., Collins, F.H. and Gwadz, R.W. (1989) A general system of resistance to malaria infection in *Anopheles gambiae* controlled by two main genetic loci. *Am. J. Trop. Med. Hyg.*, **40**, 585–92.

Vernick, K.D., Collins, F.H., Seeley, D.C. *et al.* (1988) The genetics and expression of an esterase locus in *Anopheles gambiae. Biochem. Genet.*, **26**, 367–79.

Vernick, K.D., Fujioka, H., Seeley, D.C. *et al.* (1995) *Plasmodium gallinaceum*: a refractory mechanism of ookinete killing in the mosquito, *Anopheles gambiae. Exp. Parasitol.*, **80**, 583–95.

Vickerman, K. (1985) Developmental cycles and biology of pathogenic trypanosomes. *Br. Med. Bull.*, **41**, 105–14.

Volf, P., Killick-Kendrick, R. and Bates, P.A. (1994) Comparison of the haemagglutination activities in gut and head extracts of various species and geographical populations of phlebotomine sandflies. *Ann. Trop. Med. Parasitol.*, **88**, 337–40.

Waters, A.P., Syin, C. and McCutchan, T.F. (1989) Developmental regulation of stage-specific ribosome populations in *Plasmodium. Nature*, **342**, 438–40.

Weaver, S.C. (1986) Electron microscopic analysis of infection pattern for Venezuelan equine encephalomyelitis virus in the vector mosquito, *Culex* (Melanoconion) *taeniopus. Am. J. Trop. Med. Hyg.*, **35**, 624–31.

Weaver, S.C. *et al.* (1988) Togavirus-associated pathologic changes in the midgut of a natural mosquito vector. *J. Virol.*, **62**, 2083–90.

Welburn, S.C., Arnold, K., Maudlin, I. and Gooday, G.W. (1993) Rickettsia-like organisms and chitinase production in relation to transmission of trypanosomes by tsetse flies. *Parasitology*, **107**, 141–5.

Welburn, S.C. and Maudlin, I. (1989) Lectin signalling of maturation of *Trypanosoma congolense* infections in tsetse. *Med. Vet. Entomol.*, **3**, 141–5.

Welburn, S.C. and Maudlin, I. (1991) Rickettsia-like organisms, puparial temperature and susceptibility to trypanosome infection in *Glossina morsitans*. *Parasitology*, **102**, 210–6.

Welburn, S.C. and Maudlin, I. (1992) The nature of the teneral state in *Glossina* and its role in the acquisition of trypanosome infection in tsetse. *Ann. Trop. Med. Parasitol.*, **86**, 529–36.

Welburn, S.C., Maudlin, I. and Ellis, D.S. (1989) Rate of trypanosome killing by lectins in midguts of different species and strains of *Glossina*. *Med. Vet. Entomol.*, **3**, 77–82.

Welburn, S.C., Maudlin, I. and Molyneux, D.H. (1994) Midgut lectin activity and sugar specificity in teneral and fed tsetse. *Med. Vet. Entomol.*, **8**, 81–7.

Wernsdorfer, W.H. (1980) The importance of malaria in the world, in *Malaria* Vol. I (ed. J.P. Kreier), Academic Press, New York, pp. 1–93.

Willett, K.C. (1966) Development of the peritrophic membrane in *Glossina* (tsetse flies) and its relation to infection with trypanosomes. *Exp. Parasitol.*, **18**, 290–5.

Woodward, T.M. (1991) A single amino acid change in the E2 glycoprotein of Venezuelan equine encephalitis virus affects replication and dissemination in *Aedes aegypti* mosquitoes. *J. Gen. Virol.*, **72**, 2431–5.

Wu, W.K. and Tesh, R.B. (1990a) Selection of *Phlebotomus papatasi* (Diptera, Psychodidae) lines susceptible and refractory to *Leishmania major* infection. *Am. J. Trop. Med. Hyg.*, **42**, 320–8.

Wu, W.K. and Tesh, R.B. (1990b) Genetic-factors controlling susceptibility to *Leishmania major* infection in the sandfly *Phlebotomus papatasi* (Diptera, Psychodidae). *Am. J. Trop. Med. Hyg.*, **42**, 329–34.

Zeledon, R., Bolanos, R. and Rojas, M. (1984) Scanning electron-microscopy of the final phase of the life cycle of *Trypanosoma cruzi* in the insect vector. *Acta Trop.*, **41**, 39–43.

Zeledon, R. and Vieto, P.L. (1957) Susceptibilidad de varias especies de triatominos a un cepa costarricense de *Schizotrypanum cruzi* Chagas, 1909. *Rev. Biol. Trop.*, **5**, 195–9.

Zhu, J.D., Waters, A.P., Appiah, A. *et al.* (1990) Stage-specific ribosomal RNA expression switches during sporozoite invasion of hepatocytes. *J. Biol. Chem.*, **265**, 12740–4.

Zung, J.L., Lewengrub, S., Rudzinska, M.A. *et al.* (1989) Fine structural evidence for the penetration of the Lyme disease spirochete *Borrelia burgdorferi* through the gut and salivary tissues of *Ixodes dammini*. *Can. J. Zool.*, **67**, 1737–48.

Index

Page numbers appearing in *italics* refer to tables and page numbers appearing in **bold** refer to figures.